Advanced Techniques for Characterizing Microstructures

Advanced Techniques for Characterizing Microstructures

Edited by

F.W. WIFFEN

Oak Ridge National Laboratory
Oak Ridge, Tennessee

and

J.A. SPITZNAGEL

Westinghouse Research Center
Pittsburgh, Pennsylvania

A Publication of The Metallurgical Society of AIME

D
669.95
A DV

A Publication of The Metallurgical Society of AIME
P.O. Box 430
420 Commonwealth Drive
Warrendale, Pa. 15086
(412) 776-9000

Printed in the United States of America.
Library of Congress Card Catalogue Number 82-81288
ISBN Number   0-89520-390-1

Foreword

While many of the papers in this book deal with microstructures altered by irradiation, the same techniques can be used in the investigation of all microstructural changes. The techniques discussed will thus be of use across a broad range of materials science applications.

Irradiation of metals by fast neutrons or charged particles and modification of metal surfaces by ion implantation often produce microstructural and microchemical changes on a near-atomic scale. Such effects can profoundly alter the properties of the solids. Characterization of the nature and extent of the atomic rearrangements and defect structures responsible for the property changes poses one of the most exciting challenges in material science.

Historically, understanding the physical processes responsible for "irradiation damage" has paralleled the resolution of our instruments. There is every reason to believe that this will continue to be the case. The rapid advances in fabrication and application of microelectronic circuits in the 1970s have resulted in an incredible array of diagnostic tools for the characterization of materials. Along with the appearance of the new techniques and instruments has come a flood of acronyms threatening to instill great insecurity in the materials scientist still grappling with applying two-beam dynamical diffraction theory to interpretation of 100keV transmission electron microscopy images! With this in mind, each author was requested to discuss the limitations as well as the advantages of each microanalysis technique relative to conventional transmission electron microscopy. What has emerged is a very readable and practical discussion of advanced microstructural and microchemical methods that should make this a valuable source book for a number of years. The reader will quickly realize that the diversity and statistical nature of atomic rearrangements and defect production requires the judicious use of combinations of techniques in the hands of specialists. To the extent that no single laboratory or individual is likely to possess all the requisite instrumentation and expertise, this book indirectly (but intentionally) shows the need for more collaborative research to advance our understanding of irradiation effects.

This book evolved from a Department of Energy Workshop initiated by Dr. P. Wilkes and held for a limited audience in March 1979. The symposium "Advanced Techniques for the Characterization of Microstructures" sponsored by the Nuclear Metallurgy Committee of The Metallurgical Society of AIME at the 1980 AIME Annual Meeting in Las Vegas, Nevada, allowed a second interaction of the Workshop participants, added information on techniques overlooked in the Workshop, and served as a forum for a much broader participation of experts interested in the evaluation of the defect state of metals. This volume contains most of the papers presented at this symposium, with revisions through the Fall of 1981.

We wish to express our thanks to the reviewers; to P. Wilkes, T.C. Reuther, and F.V. Nolfi for their encouragement and efforts in helping to organize the conference; and to Frances Scarboro for assistance with the manuscript. On behalf of the irradiation effects community, we thank The Metallurgical Society of AIME for making possible the Symposium and for publication of this volume.

F.W. Wiffen
Oak Ridge National Laboratory
Oak Ridge, Tennessee

J.A. Spitznagel
Westinghouse Research and
Developement Center
Pittsburgh, Pennsylvania

February, 1982

Table of Contents

Data Needs

A SUMMARY OF PROBLEMS ASSOCIATED WITH PRIMARY DAMAGE AND SECONDARY

DEFECT AGGREGATION UNDER CTR IRRADIATION CONDITIONS

M. R. Hayns[*]

UKAEA Harwell

The structural materials forming the first wall of a Controlled Thermo-nuclear Reactor (CTR) will be exposed to a harsher irradiation environment than for similar materials in the core of a fast breeder reactor. At present it is very difficult to simulate the expected high energy (14 MeV) neutron damage and no major experimental facility for materials testing will be available in the near future. Consequently the available facilities — fission reactors, ion beams and electrons — must provide the necessary information upon which to base materials choices for the first large-scale CTRs. Proper use of the available data requires an intimate understanding of physical processes involved to permit extrapolation to a CTR environment. Here we outline some of the mechanisms which play an important role and hence isolate those quantities needing detailed experimental investigations. We concentrate on the primary damage processes, covering two areas in particular — the nature of displacement damage under 14 MeV neutrons and the nucleation and stability of secondary defect aggregates. In the latter we particularly highlight interstitial loop nucleation as the precursor to an irradiation-produced dislocation network and void nucleation. The role of helium, produced via n,α reactions, is discussed as a dominant feature of the void nucleation process through the concept of a critical size for the transition from bubble to void growth.

---

[*]Present address: UKAEA Safety and Reliability Directorate, Wigshaw Lane, Culcheth, Warrington, Cheshire, UK.

Introduction

The characterization of the irradiation damage in structural materials
exposed to neutron irradiation in the first wall of Controlled Thermonuclear
Reactors (CTRs) is an immensely complex and daunting problem. In addition
to the known, but not completely solved problems associated with lower
energy neutron damage in Fast Breeder Reactors (FBRs), no major experimental
facility exists, or will exist in the near term, for the investigation of
the effects of the harder neutron spectrum and its associated helium generation on the irradiation microstructure of structural materials. Consequently, an experimental program of irradiation in fission reactors and simulation environments (heavy or light ions or 1 MeV electrons) is the only means
of obtaining pertinent data. Thus, if these data are to be at all useful,
we must have a sufficiently detailed physical understanding of the processes
to allow quantitative extrapolation to the CTR environment. It is therefore
necessary to isolate those mechanisms which require a detailed knowledge of
materials parameters and to establish procedures with the available equipment which can provide this information. This brief review, and the one
following it, attempts to outline some of the mechanisms which play an important role in the damage process and hence isolate these quantities which
require detailed experimental investigations. Here we are concerned with
the primary stages of the damage process and we cover two area in particular.
The first is the nature of the displacement damage under 14 MeV neutrons,
compared with either fission reactor or simulation conditions. The second
is the broad area of the nucleation and stability of the secondary defects
(aggregates of point defects) which evolve into the irradiation-induced
microstructural features which affect the macroscopic properties of the
material.

Primary Damage and the Importance of Displacement Cascades

The problem of characterizing the lattice atom displacements due to
impinging energetic particles has plagued the study of irradiation damage.
Here we concentrate upon those aspects which specifically relate to CTR
irradiation environments and the correlation with lower energy neutrons,
heavy ions or electrons. Two questions are highlighted: the ramifications
of the harder neutron spectrum under CTR conditions and the pulsed nature of
the irradiation.

In the fission reactor, particularly fast breeder reactor, the principal damage component arises from neutrons with ~1 MeV energy. Collisions
between these particles and lattice atoms dislodge the lattice atoms with
sufficient energy to displace further lattice atoms. This process continues
until the displaced atoms do not have sufficient energy to dislodge further
atoms. The area of crystal undergoing such a process is termed a displacement cascade. In this region both interstitials and vacancies are produced
and, whilst considerable numbers may be annihilated by recombination, a
vacancy-rich core can remain as the self-interstitials rapidly diffuse away.
The vacancy supersaturation in the core is such that vacancy loops can form.
Thus, the number of "free" point defects created by a single collision event
depends upon the energetics of the displaced atoms, the diffusivity of the
defects, and the number of vacancies trapped temporarily in vacancy loops.
For fission-related conditions and for heavy ions (typically 4–40 MeV nickel
ions) the production of vacancy loops is well established (1,2), and some
idea of their size and shrinkage kinetics has been established (3). For
electron irradiation, no displacement cascades occur because the energy
available is only sufficient for Frenkel pair production. In this case, the
threshold energy for displacements and the effects of correlated recombination are the principal problems. Thus, in order to correlate the three
types of irradiation, the displacement events must be understood in some

4

detail. Under 14 MeV neutron irradiation the displacement events, especially because of large contributions from inelastic processes, are expected
to be significantly larger, with the possibility of subcascade formation.
This would lead to more than one vacancy loop per collision event; such a
situation is shown in Figure 1, taken from ref. (1). The principal quantities to be characterized are the number of defects surviving the cascades
and the size, type and concentration of the secondary defects. A range of
measuring techniques could be used. Up to now TEM and electrical resistivity have dominated but, as detailed in a recent review (4), diffuse x-ray
scattering, mechanical property changes, superconductivity, internal friction, and neutron scattering have all been used to some extent. Computer
simulation of displacement cascades also provides a significant source of
information (5). Clearly such processes are most important for projected
CTR materials applications and, without generally available sources of 14
MeV neutrons, their influence can only be understood by extrapolating physical models based upon information gleaned from lower energy neutron or heavy
ion experiments. This places a great burden upon both the basic data and
the theoretical models for some years to come. Some helpful results may be
forthcoming from the limited experimental facilities available for 14 MeV
neutron irradiation. For example, results have already been obtained (6)
from the rotaing target neutron source facility at the Lawrence Livermore
Laboratory for 14 MeV neutron damage in nickel and niobium, and these
results have been compared with data obtained from 16 MeV proton irradiations. This will enable a more quantitative estimation of the usefulness of
the proton irradiations as a "simulation" of the 14 MeV neutrons. Results
obtained so far for both microstructural features and the yield stress of
the irradiated material indicate many similarities between the two irradiation environments. This is very encouraging, but such measurements are only
just beginning and more detailed microstructural analysis as well as more
data at higher fluences and temperatures are needed for a quantitative
analysis.

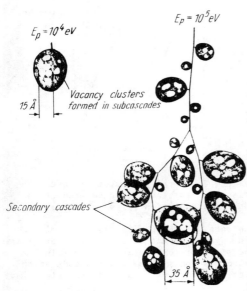

Figure 1. The production of subcascades
in a 14 MeV displacement cascade in
gold [from ref. (1)].

Another possibly important
distinction between CTR environments and those typically existing in FBRs and simulation
experiments is that the latter
produce a very uniform damage as
a function of time (unless specifically altered) whilst most
of the CTR designs, whether they
are magnetic or inertial confine-
ment fusion devices, operate on
a pulsed basis. Plasma burns in
experimental reactors are
expected to occur with cycle
times in the region of a few per
second to a few minutes per
pulse. As the damage will be
imposed as large displacement
cascades, as discussed above, it
is important to understand how
cycling such damage can influence the irradiation microstructure. Experimental information
is very sparse; to date we know
of only one attempt to simulate
pulsing, and this was done using
electron irradiation and consequently does not involve any

cascade effects (7). Theoretical studies have been performed, however, and we shall refer to these in the context of defect *nucleation* in the next section. One study is particularly relevant to the present discussion since it involves the inhomogeneous nature of the damage studies both spatially and temporally (8). The argument can be summarized as follows. At a particular point in space, cascades will occur at random distances and times around it. Large local fluctuations in point defect concentrations occur as the virtually instantaneous "wind" of interstitials from new cascades cross the reference point. The local vacancy concentration varies much more slowly, reflecting the much lower mobility of these defects. Figures 2 and 3 are typical of the computed local point defect concentrations arising from random cascade production for vacancies and interstitials, respectively. Even though the local defect concentrations vary by almost an order of magnitude, comparisons with the spatially and temporally averaged rate theory continuum model traditionally used shows that for most instances of *continuous* irradiation the averaging processes lead to an adequate representation of void growth except at elevated temperatures, and this is illustrated in Figure 4 where the fractional changes in void radius with and without spatial averaging are shown. However, cyclic operation, with a cycle time greater than approximately the relaxation time implied by Figure 2, could lead to an acute sensitivity to this effect and have a significant effect on secondary defect nucleation, as discussed in the following section.

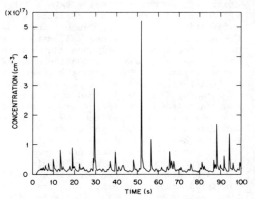

Figure 2. Local vacancy concentrations due to random cascade production during irradiation [from ref. (8)].

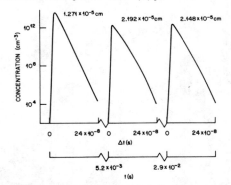

Figure 3. Local interstitial concentrations due to random cascade production during irradiation [from ref. (8)].

Nucleation of Secondary Defect Aggregates

In order that irradiation produced defects should influence the physical properties of the material[*] they must first form clusters of defects and these clusters are called secondary defect aggregates. These aggregates act to separate interstitials and vacancies, thereby reducing the simple recombination of the point defects. However, a stable separation can only occur under steady-state conditions when sinks for both types of defect exist and interstitials have

---

[*]With the exception of irradiation-enhanced diffusion which does not introduce new mechanisms of deformation.

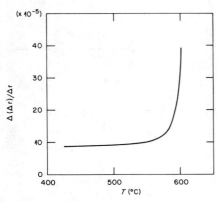

Figure 4. Difference in void radius computed with and without spatially inhomogeneous defect flux, for a fixed growth increment [from ref. (8)].

a bias for one sink type. The initial aggregation under the transient conditions occurring at early times is therefore a prerequisite for irradiation induced deformation. It is a difficult area for experimental observation since, in general, all of the necessary information concerning the properties of small clusters has to be inferred from observations of the macroscopic defects after some growth has been achieved. Here we concentrate upon the nucleation of point defect aggregates; the precipitation of solute and impurity atoms will be treated in the following presentation (9). We shall treat the nucleation of two types of secondary defects — interstitial dislocation loops and cavities.

Nucleation of Dislocation Loops

A great deal of information is available on the nucleation of interstitial loops under electron irradiation in pure metals and alloys. In recent reviews Kiritani (10) and Yoshida (11) cover a wide range of observations. Several features of loop nucleation have now been clarified. Even for ostensibly "pure" metals the nucleation density is very sensitive to impurity atom concentrations, with levels in the ppm range being important. The rate-controlling process is the diffusion of vacancies. At higher temperatures, therefore, the vacancy migration energy controls the rate of growth of loops and the local increase in vacancy concentration adjacent to interstitial sinks can lead to the formation of vacancy clusters by thermal and/or irradiation-induced diffusion of vacancies (10). The observed nucleation density as a function of inverse temperature for gold and molybdenum are shown in Figure 5. The separation into four significantly different temperature regimes is more clearly shown in Figure 6 where the loop densities were calculated (10). Note that there are only three regimes covered in Figure 5 because the densities are very low at the higher temperatures.

The gradient of the Arrhenius curve in Region II is interpreted as arising from trapping (of interstitials) at relatively shallow impurity traps. The apparent activation energies are 0.19 eV for gold and 0.18 eV for molybdenum. Activation energies for other elemental metals are given in ref. (10). These quantities are interpreted as a combination of the interstitial activation energy and the interstitial impurity binding energy. The plateau in Region III is interpreted as heterogeneous nucleation on impurity atoms. This nucleation mechanism is temperature independent over a range of temperatures where the traps are present in large enough numbers to dominate homogeneous nucleation processes. The trap density here was assumed to be 10 ppm. Region IV is normally associated with homogeneous nucleation — that is, the free diffusion of interstitials. However, there is the possibility that it is a combination of free-interstitial diffusion and shallow impurity traps, similar to the interpretation of Region II, but with much lower binding energies for the traps. These results are included to demonstrate the level of complication found for simple electron irradiation under "ideal" conditions for very pure materials. Our concern is for technologically important materials under heavy ion or neutron irradiation

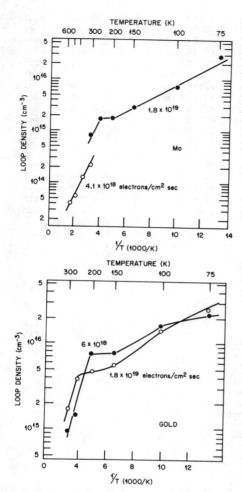

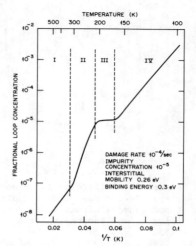

Figure 6. Calculated interstitial loop density as a function of temperature [from ref.
(10)].

and several new features have
to be considered. The important role of impurities as
heterogeneous nucleation sites
has been established for pure
metals; in technological alloys
it can be expected to be dominant. Consequently, careful
measurements are required to
separate the free diffusion of
point defects and point defect
trapping processes in interpreting the loop nucleation
density. The very important
part played by the segregation of impurities to the
nucleating clusters is treated
in the next presentation (9).
Another feature which is especially important under CTR

Figure 5. Interstitial loop densities
in electron irradiated molybdenum and gold
[from ref. (10)].

conditions is the influence of helium gas on the nucleation process. We shall
discuss this further below with reference to cavity nucleation, but there is
clear evidence that the presence of helium can increase the observed loop
number density (12,13). When the damage is caused by 14 MeV neutrons, with
the consequent copious helium generation, the loop nucleation is likely to
be influenced to an extent which at present is not well understood. Further
questions relating to the damage-rate dependence (14), the presence of free
surfaces (in electron and heavy ion simulation experiments) (15), and the presence of large diffusion gradients away from the damage zone for defects (in
heavy ion experiments) (16) must also be considered.

Even though there is general agreement that interstitial loop nucleation
is completed very early in an irradiation, the final loop density and hence
the dislocation density is sensitive to many influences which are not well

characterized, particularly for technological alloys and CTR irradiation
conditions.

Nucleation and Stability of Cavities[*]

The nucleation of voids is a more complicated process than that for
interstitial loops, as discussed above, because it arises from the mutual
aggregation of more than one diffusing species. For interstitial loops the
nucleation process is completed before steady-state conditions are reached;
it is their further growth which requires separation of defects. In the
simplest case, one must account for the flux of two species, interstitials,
and vacancies, whilst in general the simultaneous arrival of helium, or
intrinsic gases, is also necessary for the creation of stable embryos.
Cavities are special in that a three-dimensional collection of vacancies is
not a stable unit. Hence the appearance of two-dimensional vacancy loops is
usually observed from cascade events and from quenching experiments
(although in pure metals vacancy-type stacking fault tetrahedra are also
observed) (11). For stable growth the net vacancy flux to an embryo must be
large enough to overcome the thermodynamic driving force for shrinkage by
vacancy thermal emission. This physical process is simply conceptualized as
a critical size or void nucleation barrier, below which a cavity is unstable
and will shrink, and this size is strongly dependent upon damage rate, local
microstructure, and temperature. At elevated temperatures the critical size
can be large enough to inhibit nucleation. Figure 7 gives examples of the
computed critical size for 316 stainless steel as a function of temperature
for different dose rates (17). There are three ways in which a cavity can
appear in the system with sufficient size to grow as a void. First, by
statistical fluctuations — that is the out-of-equilibrium appearance of
clusters of vacancies by random thermodynamic accretion. Secondly, from
high vacancy supersaturations in damage cascades. Thirdly, by the growth of
gas bubbles up to the required size. At higher temperatures (the peak
swelling temperature and above), it is clear that the latter is the only
viable process. We therefore prefer to separate the void nucleation process
into two stages. The first concerns the aggregation of gas atoms to form
small gas bubbles, the void embryos, the second the growth of bubbles up to
the critical size or their existence in the material prior to irradiation.
The first stage has been treated using classical thermodynamical arguments
(18) and by sophisticated numerical treatments which involve all three diffusing species (19—21). A simpler approach in which only the motion of the
helium was taken into account has also been given (17,22). All of these
methods are complicated and depend upon poorly known quantities. For
example, even the simplest picture of gas bubble nucleation requires a
knowledge of an equation of state for statistically small ensembles of gas
atoms and vacancies. Further, the role of impurity trapping, leading to
heterogeneous nucleation, is thought to be just as important as discussed
previously for interstitial loops (22). Typical calculated bubble densities
as a function of inverse temperature for different assumed gas atom trap
energies are shown in Figure 8 (22). The high apparent Arrhenius energy can
be explained (for interstitial helium motion) by means of a combination of
free helium diffusion and impurity trapping as for interstitial loop
nucleation in Region II of Figure 6. In the calculations leading to Figure
8, no assumptions were made as to the nature of the trap site other than the
binding energy. Thus, the trap could be an interstitial loop, perhaps
leading to the commonly observed feature of loops and voids apparently produced together, single solute atoms, precipitates, dislocation nodes or
other lattice defects. We have insufficient evidence to clearly define the

---

[*]The term cavity is used to cover both gas bubbles and voids; only when it
is clear which type is dominant will the terms "gas bubble" or "void" be
used.

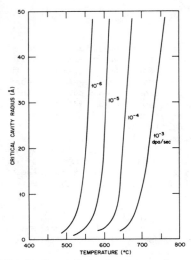

Figure 7. Critical cavity size calculated for 316 SS as a function of temperature [from ref. (17)].

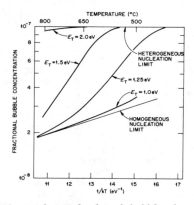

Figure 8. Calculated bubble density in 316 SS as a function of temperature for various gas atom trap-binding energies [from ref. (22)].

nature of the trapping site and all of the above possibilities have been observed. The direct observation of void embryos is not possible and therefore techniques other than TEM need to be brought to bear.

The second part of the void nucleation process, which depends upon the critical size, is rather more amenable to experimental confirmation since, from Figure 7, we see that for an appropriate choice of damage rate and temperature the critical size predicted theoretically is well within observational range in the TEM, and indeed a considerable body of evidence now exists in which the void nucleation can be interpreted in this way. Mazey and Nelson (23) have presented most convincing evidence from heavy ion irradiations, but here we utilize an example from the work of Packan and Farrell (24). The material investigated was a "pure" 316 type austenitic steel (17% Ni, 16.7% Cr, 2.5% Mo) irradiated with 4 MeV Ni ions after being preinjected with 1400 ppm He. Preinjection has long been used to simulate the helium production during neutron irradiation, and to give consistent values of the void density. Notice that preinjection is significantly different from the continuous production of gas under neutrons. The consequences have been discussed in some detail recently (17,25). Upon preinjection we visualize the gas nucleating a large number of small bubbles; the following ion irradiation will allow only those bubbles with a radius in excess of the critical size to grow. Thus, with the large amounts of injected gas in Packan and Farrell's experiments we would predict a bimodal cavity population consisting of growing voids and a stable background population of gas bubbles. Figure 9 shows this very clearly for two different experimental conditions — preinjection in case (b) was at room temperature, and case (a) at the temperature of the irradiation (625°C). The cavity distribution following the low-temperature preinjection is clearly bimodal with a large (dominant) population of small cavities. The case where helium was introduced at 625°C resulted in a background cavity population of less numbers, but larger size. The histograms of cavity size distributions for these two cases are shown in Figure 10 and clearly highlight this effect. The differences between these two experiments are readily interpreted by means of the

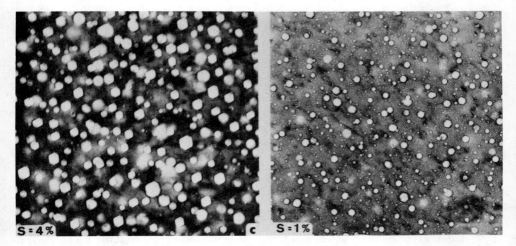

Figure 9. Electron micrograph of bimodal void populations in "pure" 316 SS irradiated at 625°C to 70 dpa. Case (a) is for hot injected helium; case b for cold injected helium, both to 1400 ppm [from ref. (24)].

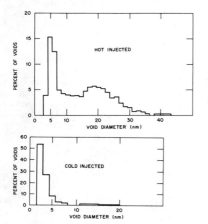

Figure 10. Histograms of the void size distributions obtained from the micrographs shown in Figure 9.

temperature at which the gas bubbles nucleated, the high temperature resulting in a smaller number of larger bubbles — hence in turn leading to a more evenly distributed bimodal population. Calculations using a distribution of cavity sizes show the expected evolution of a bimodal population and the results are summarized in Figure 11, essentially validating the physical model.

The existence of a critical cavity size has also been useful in explaining several other features of the observed void swelling behavior. For example, it was initially introduced as a concept to explain the observed sharp reduction in void numbers at elevated temperatures, which seem to be a feature of all irradiation conditions. For ion beam or electron irradiations, which use the technique of preimplanting the material with helium, the very rapid increase in critical size with temperature (see Figure 7) ensures that the number of pre-existing bubbles of greater than the critical size falls rapidly with increasing temperature. The calculated and observed (17) void concentrations in 316 steel under nickel ion irradiation conditions are shown in Figure 12. The very rapid fall of void numbers is clearly seen above ~625°C. The existence of an incubation dose and a second high-temperature swelling peak in 316 steel under neutron irradiation conditions are also explicable in terms of the requirement that cavities grow to the critical size as helium gas bubbles before void growth is possible. In this situation it is the continuous production of helium via n,α reactions which dominates the evolution of the void population (26). Similar arguments have also proved useful in interpreting some features of the possible effects of

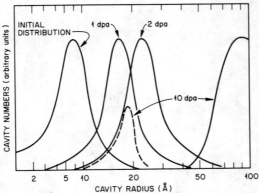

Figure 11. The calculated evolution of the void size distribution in CW 316 steel under neutron irradiation conditions at 550°C. The dotted curve represents that part of the bimodal cavity distribution which is shrinking at 10 dpa [from ref. (27)].

temperature changes on void swelling (27). Useful though these simple arguments have been, the complications arising from dose dependent effects, such as the evolving microstructure and perhaps solute trapping and segregation, requires further detailed measurements like those mentioned above (23,24) to utilize fully the physical model and therefore allow its extrapolation to CTR conditions.

In Figures 10 and 11 we introduced the void size distribution as an observable feature of irradiation damage. The examples cited show that measurements of void size distributions are a source of much more detailed information. However these principally concern questions of growth kinetics and are outside the range of present discussion. Details are available in the literature (28,29).

We have emphasized the role of helium in the nucleation kinetics of voids and shown how physical models based upon it can give a very successful interpretation of the available experimental results. In particular, we would re-emphasize the vital distinction between preinjected (or intrinsic) gas in simulation experiments and the continuous gas production under neutron irradiation conditions. Experimental observations under dual ion gas injection irradiation conditions are vital to simulate CTR, and even FBR environments and there is already a significant body of evidence which emphasizes this (25,32). We therefore highlight the whole area of inert gases in metals as a crucial one for understanding the response of materials to CTR irradiation environments.

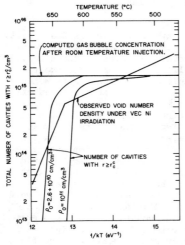

Figure 12. Calculated observable void concentrations for heavy ion irradiation conditions, shown with the observed void densities [from ref. (17)].

In addition to the vital role of continuously produced helium on the cavity nucleation kinetics [see especially ref. (21)], the pulsed nature of the CTR irradiation environment is also expected to cause significant changes in the nucleation microstructure. Several studies have been made of the influence of pulsed irradiations on cavity or void growth (8,30). However, we are concerned here with nucleation processes, and to our knowledge only one theoretical study is available (31). This shows that if the pulse rate is fast compared to the characteristic time for nucleation (typically 1—10 dpa) then the continuous interference in the transient nucleation phase can seriously alter the final predicted cavity density.

Thus, it seems that a full simulation of CTR conditions will require not only the simulation of continuous helium production, but also the pulsed nature of the irradiation.

Summary

In this very brief note we have tried to highlight some of the important new quantities which need experimental verification and characterization to further our understanding of the irradiation damage processes in materials in CTR environments. We have particularly emphasized the need to thoroughly understand the available simulation experiment conditions so that a proper correlation with the as-yet-unobtainable CTR conditions can be made. We have only focused upon the primary stages of the damage process and the nucleation of secondary defects. This should in no way be taken as implying that our understanding of the more basic quantities, such as diffusion in complex alloys, surface energies, stacking-fault energies, dislocation bias values, trap nature and binding energies, phase stability and solute segregation to irradiation-produced sinks is anywhere near adequate and further work is needed here as well as in the other fields discussed in the body of the paper.

References

1. K. L. Merkle, Radiation Damage in Metals, N. L. Peterson and S. D. Harkness, eds., American Society for Metals, Metals Park, Ohio (1976), p. 58.

2. B. L. Eyre and D. M. Maher, Phil. Mag. 24 (1971) 767.

3. R. Bullough, B. L. Eyre, and K. Krishan, Proc. Roy. Soc. London A346 (1975) 81; A. D. Brailsford, J. Nucl. Mater. 84 (1979) 245, 269.

4. A. Goland, J. Nucl. Mater. 85&86 (1979) 453.

5. See, for example, H. L. Heinish, J. O. Schiffgens, and D. M. Schwartz, J. Nucl. Mater. 85&86 (1979) 607.

6. R. H. Jones, D. L. Styris, and E. R. Bradley, Effects of Radiation on Structural Materials, J. A. Sprague and D. Kramer, eds., ASTM-STP-683, (1978), p. 346.

7. R. W. Powell and G. R. Odette, J. Nucl. Mater. 85&86 (1979) 695.

8. L. K. Mansur, W. A. Coghlan, and A. D. Brailsford, J. Nucl. Mater. 85&86 (1979) 591.

9. H. Wiedersich, this symposium.

10. M. Kiritani, Progress in the Study of Point Defects, University of Tokyo Press, Doyama and Yoshida, eds., Tokyo (1977), p. 247.

11. S. Yoshida, Progress in the Study of Point Defects, University of Tokyo Press, Doyama and Yoshida, eds., Tokyo (1977), p. 195.

12. D. S. Gelles and F. A. Garner, J. Nucl. Mater. 85&86 (1979) 689.

13. D. J. Mazey, private communication.

14. M. R. Hayns, J. Nucl. Mater. 56 (1975) 267.

15. M. R. Hayns and R. C. Perrin, AERE Harwell Research Report AERE R7934 (1975), p. 188 (Proceedings of the Consultants Symposium on Irradiation Produced Voids, September 9—11, 1974, eds., R. S. Nelson and M. R. Hayns, USERDA-CONF-751006-P1 (1975), p. 764.

16. L. K. Mansur, Nucl. Technol. 40 (1978) 1 and 5.

17. M. R. Hayns, AERE Harwell Research Report AERE R8806 (1977).

18. K. C. Russell and D. H. Hall, J. Nucl. Mater. 18 (1973) 545.

19. H. Wiedersich and J. L. Katz, Proceedings of the Workshop on Correlation of Neutron and Charged Particle Damage, USERDA-CONF-760673 (1976) p. 21.

20. K. C. Russell, Acta Met. 26 (1978) 1615.

21. B. O. Hall, J. Nucl. Mater. 85&86 (1979) 565.

22. M. H. Hayns and M. H. Wood, Proc. Roy. Soc. London A368 (1979) 331.

23. D. J. Mazey and R. S. Nelson, J. Nucl. Mater. 85&86 (1979) 671.

24. N. H. Packan and K. Farrell, J. Nucl. Mater. 85&86 (1979) 677.

25. R. Bullough, M. R. Hayns, and M. H. Wood, J. Nucl. Mater. 85&86 (1979) 559.

26. M. R. Hayns, J. Gallagher, and R. Bullough, J. Nucl. Mater. 78 (1978) 236.

27. M. R. Hayns, J. Nucl. Mater. 82 (1979) 102.

28. M. H. Wood and M. R. Hayns, AERE Harwell Research Report AERE R9178 (1978).

29. L. K. Mansur, P. R. Okamoto, A. Taylor, and Che-Yu Li, Nucl. Met. 18 (1973) 509.

30. N. Ghoniem and G. L. Kulcinski, J. Nucl. Mater. 69&70 (1978) 816.

31. Y. H. Choi, A. L. Bement, and K. C. Russell, "Radiation Effects and Tritium Technology for Fusion Reactors," USERDA-CONF-750989, Vol. II (1976).

32. See, for example, K. Farrell and N. H. Packan, J. Nucl. Mater. 85&86 (1979) 683.

THE DEVELOPMENT OF MICROSTRUCTURAL FEATURES DURING IRRADIATION*

H. Wiedersich
Materials Science Division
Argonne National Laboratory
Argonne, IL 60439

During elevated temperature irradiation with energetic neutrons, ions or electrons, the microstructure of alloys undergoes significant changes accompanied by corresponding changes in physical properties. The microstructural development is caused by the production of defects throughout the material and the concurrent annihilation of the mobile point defects and small defect clusters on spatially discrete sinks such as dislocations, grain boundaries and voids. We review here the present understanding of the development of the radiation microstructure beyond the nucleation stage of stable defect clusters. After a short characterization of the dynamic state of a crystalline material during elevated temperature irradiation we discuss (a) the void swelling phenomenon; (b) radiation-enhanced, diffusion controlled processes such as precipitate coarsening; and (c) radiation-induced processes such as radiation-induced precipitation and spatial redistribution of phases in multiphase alloys.

*Work supported by the U. S. Department of Energy.

Introduction

During elevated temperature irradiations with energetic neutrons, ions or electrons, the microstructure of alloys undergoes significant changes accompanied by corresponding changes in physical properties. Detailed information on various aspects of radiation-induced microstructural effects and property changes can be found in several recent proceedings (1-3). The preceeding paper by M. R. Hayns summarizes the production of defects and nucleation of small defect aggregates during irradiation. Here, we outline the present knowledge and understanding of the development of the microstructure on a somewhat grosser scale. We intend to give the reader a short overview of the area with the aid of a few characteristic examples rather than an in-depth review of existing knowledge.

After a short characterization of the dynamic state of a crystalline material during irradiation we address, (a) the void swelling phenomenon; (b) radiation-enhanced, diffusion-controlled processes such as precipitate coarsening; and (c) radiation-induced processes such as radiation-induced second phases and the spatial redistribution of phases in multiphase alloys.

The Dynamic State During Irradiation

For a conceptual understanding of the microstructural developments during irradiation a treatment of the production and annihilation of defects and defect clusters by chemical rate theory is most useful (4-7). This theory assumes that defects are produced randomly throughout the material. Defects are classified as "immobile", e.g., collapsed vacancy loops or void embryos in the cores of displacement cascades, and "mobile", e.g., interstitials, vacancies, small clusters of these point defects, point defect-solute complexes, helium and other transmutation products. The interstitial- and vacancy-type mobile defects diffuse, except at low temperatures, and are eliminated by mutual recombination throughout the material, by formation of immobile defect clusters or by annihilation at stationary or slowly moving sinks such as dislocations, voids, free surfaces and grain boundaries. Defect annihilation at sinks induces defect concentration gradients and, hence, defect fluxes from the interiors of grains to spatially discrete sinks. These fluxes are the predominant cause of microstructural development during irradiation. The rate theory describes the state of the system by coupled reaction rate equations for the defect species that are considered. The change of the spatially averaged concentration of a particular species is expressed in terms of the rates of all reactions among defect species that produce, convert to a different species, or annihilate that particular species. The defect losses by fluxes to discrete sinks are taken into account by products of loss-rate constants called "sink annihilation probabilities" (4) or "sink strengths" (5-7) and the average defect concentrations. The growth rate of microstructural features such as voids and dislocation loops is then obtained from the defect fluxes to these particular sinks.

For an appreciation of the types and scale of microstructural changes that can be expected in different temperature regimes, it is useful to consider the quasi-steady-state, average concentrations of mobile defects that are achieved, after an initial transient, during irradiation at a constant displacement rate. At high temperatures, all point defects diffuse fast and, hence, their concentration does not rise significantly above the thermal equilibrium value; the rise is insufficient for nucleation of interstitial loops or voids; moreover, thermal annealing eliminates incipient microstructural changes.

16

At low temperatures, at least one of the two principal types of defects, in a metal usually the vacancy type, is essentially immobile. Its concentration builds to a level limited by diffusive or spontaneous recombination, thereby eliminating any significant long range diffusion processes. The microstructural changes in this temperature range are thus limited to a high density of point defects and small defect clusters, up to sizes on the order of those produced directly by cascades.

Between these two temperature regimes both vacancies and interstitials annihilate, at least in part, at pre-existing or radiation-induced sinks, because even the slower defect diffuses fast enough to prevent the buildup of significant defect concentrations; therefore, recombination is not the dominant defect loss mechanism. In this intermediate temperature regime, typically 0.3 to 0.6 of the absolute melting temperature, major changes in microstructure on a scale above the resolution limit of conventional transmission electron microscopy (TEM) are induced or accelerated by irradiation. Generally, the high supersaturation of point defects at the lower end of this temperature regime causes profuse nucleation of interstitial loops and vacancy aggregates and, hence, leads to a high density of microstructural features. With increasing temperature, defect supersaturations decrease. Consequently, fewer defect aggregates nucleate and coarser microstructures result. In the following sections we will touch on several distinct types of microstructural changes that occur during irradiation.

Void Swelling

In contrast to other microstructural changes during irradiation, to be discussed later, void swelling can occur only when a partial separation of interstitial- and vacancy-type defects is produced by preferential annihilation of interstitials and/or vacancies at specific types of sinks. The stronger elastic interaction of interstitials with dislocations than that of vacancies with dislocations is recognized as the major cause of this bias. The development of the dislocation structure during irradiation of an annealed material usually starts with faulted loops. These loops grow, unfault, and eventually interact with other loops and existing dislocations to form dislocation networks as illustrated in Figure 1 (8). The

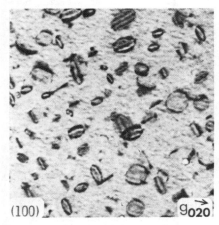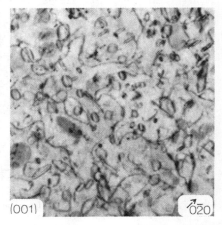

Fig. 1. Dislocation structure development in a Ni-12.8 at. % Al alloy during Ni-ion bombardment at 550°C. (a) Interstitial loops at 1.8 dpa. (b) Loops and network dislocations at 5.5 dpa (8).

development of the major microstructural features in an annealed austenitic steel during electron irradiation is shown in a quantitative fashion in Figure 2 (9). The number density of loops decreases as they interact and form dislocation networks, which in turn leads to an increase in total dislocation density, ρ_d. Saturation of ρ_d is approached with increasing dose presumably as a consequence of mutual dislocation annihilation. Generally, the saturation density achieved during irradiation is independent of the dislocation density present in the material prior to irradiation. In 316 stainless steel the saturation density is less than the dislocation density present in the 20% cold worked material. Brager et al. have shown that the total dislocation density in cold worked 316 stainless steel decreases initially during neutron irradiation and approaches the same saturation level as that of the annealed material (10). Generally, the nucleation of interstitial loops ceases after low doses, but nucleation may continue at low levels as careful observations in 316 stainless steel have shown (10). Nucleation of voids also appears to cease after low doses as can be deduced from the peak in the number densities, see Figure 2. Observed decreases in void number densities with dose are likely to be caused by a process similar to Ostwald ripening, i.e. by shrinkage of small voids in favor of the growth of larger voids, or, sometimes, by void coalescence. The growth of surviving voids more than compensates for the decrease in number density as evidenced by a monotonic increase of the fractional void volume, $\Delta V/V_o$.

Typical void microstructures at a fixed dose are illustrated as a function of irradiation temperature in Figure 3 (11). Consistent with the expectations outlined in the previous section, the void number density decreases with increasing temperature and the void volume attains a maximum at intermediate temperatures, where few defects are lost by recombination and thermal annealing is sufficiently slow to permit the nucleation of defect clusters at moderate number densities.

With some exceptions, voids and dislocation loops are usually distributed fairly randomly throughout the material. In a number of cases, many of the initial voids are observed in close proximity to dislocations, as for example in high purity Fe, Mo and Nb (12). Void and loop denuded zones occur near surface and grain boundary sinks. In some materials interstitial loops are aligned in strings or rafts. The most striking non-random defect cluster distributions are ordered void arrays in which the voids form a superlattice, usually with the same type of unit cell and orientation as that of the host metal lattice. Examples of ordered void

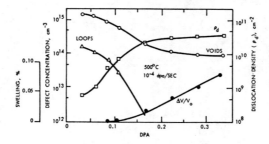

Fig. 2. Void and interstitial loop number densities, swelling ($\Delta V/V_o$) and network dislocation density ρ_d as function of dose for solution annealed type 316 stainless steel, irradiated with 1.0 MeV electron at 500°C (9).

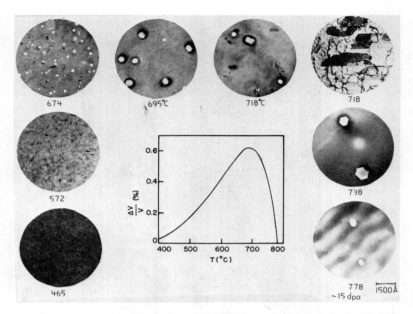

Fig. 3. Micrographs of a Ti-14.4 at. % Al alloy nickel-ion bombarded to
approximately 15 dpa at the temperatures indicated. The curve shows
the calculated swelling as function of temperature. Courtesy of
D. I. Potter.

arrays, shown in Figure 4, are taken from the work of Loomis et al. (13).
The figure shows that the perfection and temperature range of ordered void
arrays in Nb depend on the oxygen content.

Radiation-enhanced Microstructural Processes

Diffusion of substitutional elements in crystalline solids generally
involves point defects. It has long been recognized that the radiation-
produced excess point defects accelerate diffusion processes. Figure 5
illustrates the calculated enhancement of the diffusion coefficient by
excess vacancies for a species diffusing by a vacancy mechanism. In the
intermediate and low temperature ranges, diffusion is greatly enhanced by
irradiation. Values of the diffusion coefficient which can lead to
significant microstructural changes in reasonable times are maintained to
significantly lower temperatures during irradiation than without
irradiation. Note that interstitials contribute to the enhancement of
diffusion. However, increased recombination limits long range diffusion
and, hence, significant redistribution of alloy components on a
microstructural scale, at low temperature when one or both principle types
of defects are sufficiently immobile.

The calculations for Figure 5 are based on reaction rate theory which
yields spatially averaged defect concentrations and diffusion coefficients
(4). Complications arise near defect sinks where defect concentrations and,
therefore, diffusion coefficients have a strong spatial dependence.
Furthermore, defect-flux induced segregation, to be discussed in the next
section, may overshadow any effects of diffusion enhancement in the vicinity
of defect sinks. As an example of a radiation-enhanced, diffusion-
controlled microstructural process, we will discuss precipitate coarsening

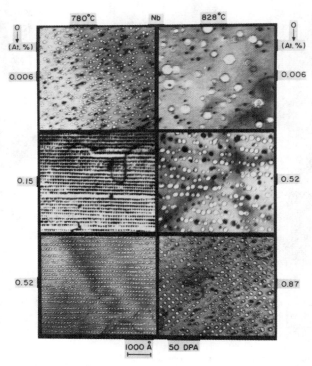

Fig. 4. Ordered Voids in Nb-O alloys after Ni-ion bombardment at 780°C and at 828°C to 50 dpa. Note that a higher oxygen concentration is required at the higher temperature in order to form a well-ordered void lattice (13).

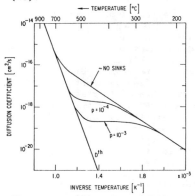

Fig. 5. Diffusion coefficient as function of inverse temperature calculated from rate theory for a nickel based substitutional alloy. Displacement rate used was 10^{-6} dpa/s. The sink density is characterized by the sink annihilation probability, p. p^{-1} is the average number of jumps of a defect between creation and annihilation at a sink. The thermal diffusion coefficient, D^{th}, is shown for comparison.

20

in γ/γ' alloys. The γ'-precipitates are ordered A_3B intermetallics (Ll_2) that form coherently with the fcc disordered matrix and often exhibit little if any lattice parameter misfit. Therefore, coarsening can be studied in these systems without the undue interference from defect sinks that was alluded to above, unless significant defect-precipitation interactions of the nature discussed below exist.

Coarsening or Ostwald ripening occurs by growth of large particles at the expense of smaller ones. The driving force is the reduction of interfacial energy. According to the Lifschitz-Slyozov-Wagner theory the average particle diameter, $\bar{a}$, raised to the third power increases linearly with the time, t, i.e., $\bar{a}^3 - \bar{a}_o^3 = kt$. The rate constant, k, is proportional to the diffusion coefficient. Figure 6, taken from the work of Potter and McCormick, illustrates that coarsening of a Ni-12.8 at. % Al alloy during irradiation at various temperatures follows this relation well and, hence, can be considered as Ostwald ripening (14). However, the rate constant (which is proportional to the ordinate value in Figure 6 at any fixed dose, e.g., 3 dpa) is greatly increased relative to that for thermal aging. In Figure 7 coarsening rate constants for a Ni-12.7 at. % Si alloy during irradiation are compared with those calculated on the basis of rate theory assuming that Si atoms act as immobile or mobile traps for interstitials, and for the case that Si atoms do not trap interstitials (15). The results indicate that precipitate coarsening under irradiation can be described as a radiation-enhanced diffusion process in the Ni-Al and

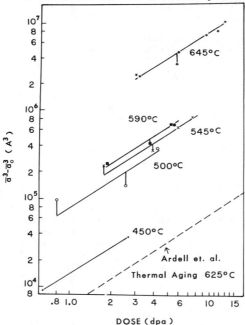

Fig. 6. Average diameter, raised to the third power, of γ' precipitates in Ni-12.8 at. % Al as function of dose during Ni-ion irradiation at the temperatures indicated. The dose rate was ~ 2.7 x 10^{-3} dpa/s or ~10 dpa/h. For comparison thermal aging data for 625°C is shown as a dashed line using a time scale in which 1 h corresponds to ~10 dpa (14).

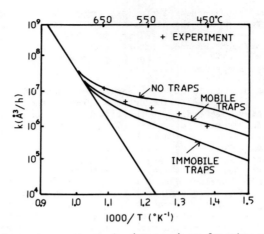

Fig. 7. Coarsening rate constant, k, (crosses) as function of inverse
temperature for a Ni-12.7 at. % Si alloy during Ni-ion bombardment
at a displacement rate of ~2.7 x 10^{-3} dpa/s. The curves were
calculated from rate theory assuming that interstitials form immoble
complexes (immobile traps), mobile complexes (mobile traps) with
silicon atoms, or do not become trapped at silicon atoms (15).

the Ni-Si systems. Some complications that arise from radiation-induced
segregation will be discussed later.

Radiation-Induced Phase Changes

Three principally different types of radiation-induced phase changes
can be distinguished. The changes may be caused by (a) the presence of
excess defects, (b) displacement disordering or (c) radiation-induced local
compositional changes.

The presence of excess interstitials and vacancies increases the free
energy of a phase. Maydet and Russell suggested that as a consequence,
shifts in phase boundaries should occur (16). However, the energy
contributions from excess defects are usually small compared to typical
internal energy differences between neighboring phases. Moreover, only the
difference between the energy contributions of the excess defects in
competing phases can lead to shifts in the phase boundaries. These shifts
are expected to be small and no unambiguous experimental evidence for such a
shift seems to exist. Some conceptual difficulties for phase boundary
shifts resulting from excess defects occur when the interface between the
competing phases acts as a defect sink, because for this situation the
defect concentrations are close to their thermal concentrations at the
locations where the phases could establish mutual equilibrium.

Atom displacements tend to disorder a solid and can change a
crystalline phase into an amorphous phase. Such transitions appear
especially prevalent in covalently bonded semiconductors but are by no means
restricted to this class of materials. For example, Rechtin et al. have
shown that incompletely crystallized, glassy metallic alloys of $Nb_{40}Ni_{60}$
revert completely to a glassy structure when irradiated below the
crystallization temperature (17). Similarly, irradiation can disorder an
ordered alloy without a severe reduction of the crystallinity. At

temperatures at which atomic rearrangements by diffusion can occur, with or without radiation, radiation-induced disorder and reordering by atomic rearrangements may set up a "steady state degree of order" during irradiation (18). Figure 8 shows some experimental evidence for this phenomenon taken from the work of Potter (19). At lower temperatures the Ni_3Si film tends towards a lesser degree of order during irradiation. At higher temperatures reordering is fast enough to maintain the degree of order close to that at thermal equilibrium. Since reordering requires only short diffusion distances, phase changes due to radiation-induced disordering are expected to be important mainly at relatively low temperatures.

The third type of radiation-induced phase change, i.e., that caused by segregation, appears to be of the most practical importance and is by far the best documented type. Radiation-induced segregation has been reviewed recently by Okamoto and Rehn (20). A simple theoretical treatment has been

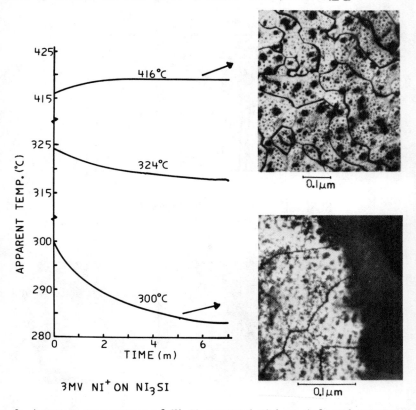

3MV NI⁺ ON NI₃SI

Fig. 8. Apparent temperature of Ni_3Si measured with an infrared pyrometer as function of bombardment time with Ni-ions. The constant actual temperatures are indicated at the curves. The apparent temperature changes are caused by emissivity differences resulting from the changing degree of order. TEM micrographs of the Ni_3Si surface films after bombardment at 416° and 300°C are shown on the right. Courtesy of D. I. Potter.

given by Wiedersich et al. (21). As previously mentioned, at intermediate
temperatures, significant fractions of the randomly produced defects
annihilate at sinks; thus, defect fluxes are induced. A preferential
association or exchange of specific alloy components with defects couples
net fluxes of alloy components to defect fluxes, which in turn alter the
local composition near defect sinks. Usually, the matrix near sinks becomes
depleted of the large atomic size components and enriched in the small
atomic size components of the alloy (20). Local enrichments of solutes may
be sufficiently large to exceed the solubility limit and precipitation can
then occur in nominally solid solution alloys. Examples are shown in Figure
9 for dilute Ni–Si alloys in which Ni_3Si precipitates form on surfaces (a),
on dislocation loops (b) and on grain boundaries (c) (22). A normally
single phase Ti–6Al–4V alloy forms a high density of β–phase precipitates
during elevated temperature irradiation as shown in Figure 10 (23). Energy
dispersive x-ray analysis on large precipitates has shown that V, a β–phase

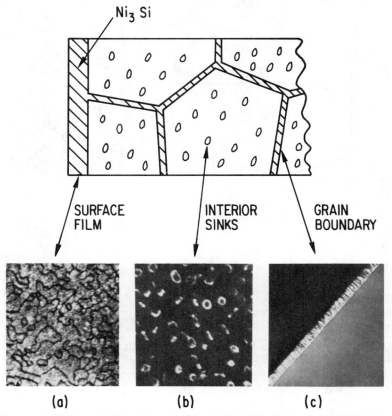

Fig. 9. Precipitation of Ni_3Si on defect sinks in solid solution Ni–Si
alloys during irradiation; (a) the domain structure of the
continuous surface film of Ni_3Si; (b) toroidal Ni_3Si precipitates
that form on interstitial dislocation loops; (c) a Ni_3Si film that
covers a grain boundary. Courtesy of P. R. Okamoto and
K.-H. Robrock.

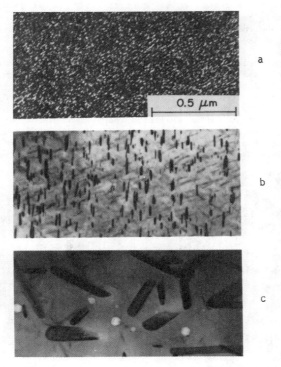

0.5 μm

Fig. 10. Formation of b.c.c. precipitates in an h.c.p. commercial purity
Ti-6 wt. % Al -4 wt. % V solution annealed alloy during V-ion
bombardment; (a) 465°C, 25 dpa, (b) 547°C, 12 dpa, (c) 660°C,
25 dpa. Control specimens with the same temperature history, but
shielded from the ion beam, remained single phase (23).

stabilizer, is enriched in the precipitates relative to the matrix
composition. No change in the Al concentration could be detected.

Cauvin and Martin have observed in a solid-solution Al-1.9 at. % Zn
alloy that homogeneous precipitation of Zn-rich Guiner-Preston zones and of
β-phase can be induced by electron irradiation, i.e., without the prior
presence of TEM-resolvable sinks at the location of the precipitates (24).
The authors suggest that an attractive solute-defect interaction in
conjunction with defect-induced solute fluxes in the same direction as that
of the defects stabilize solute clusters present as a consequence of
concentration fluctuations. The effect can be understood in a simple way.
Because of the defect-solute attraction, the defect concentration in solute
rich clusters is enhanced. This leads to higher recombination losses within
the clusters, which, thus, act as defect sinks. The clusters then grow by
radiation-induced segregation.

Solute depletion near sinks occurs in those alloys in which solvent
atoms are preferentially transported with the defect fluxes to sinks. A
corresponding enrichment of solute results in the matrix. However, in solid
solution alloys solute enrichment in the matrix is generally small because
the volume around sinks, in which significant concentration changes and

gradients occur, is a rather small fraction of the total volume of the material. Hence, generally no precipitation will occur for solid solution alloys which show solute depletion at sinks, except when the average composition is close to the solubility limit.

The solute depletion case is, however, important for redistribution of phases in multiphase alloys as will be shown by examples from the work of Potter and co-workers on a Ni-12.8 at. % Al alloy (25,26). In contrast to the uniform distribution of γ' precipitates which is observed in the thermally aged alloy, precipitate free zones develop during irradiation around defect sinks such as dislocation loops as illustrated in Figure 11. Nickel enrichment around sinks brings the local composition into the solid solution range; γ' precipitates dissolve near these sinks and are concentrated in sink-free regions. As dislocations penetrate, at high

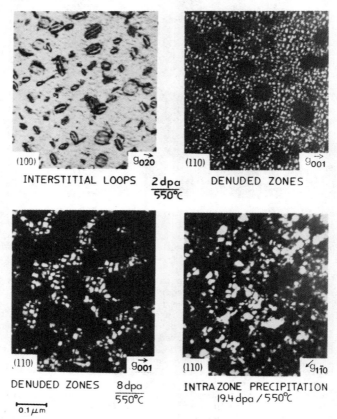

Fig. 11. Development of the precipitate microstructure during irradiation of a Ni-12.8 at. % Al two-phase alloy. Top: Precipitate-free zones form around interstitial loops. Bottom left: The zones grow and precipitates become highly concentrated between zones. Bottom right: At high doses dislocation-precipitate interactions partially dissolve existing precipitates and renucleation in precipitate-free zones occurs (26).

doses, into regions with large and dense precipitates, precipitates in the vicinity of the dislocations partially dissolve as shown in Figure 12. Fragmentation of precipitates occurs and renucleation takes place in previously precipitate free zones. The number density of precipitates eventually increases at high doses, so that the averge precipitate size goes through a maximum as a function of dose during irradiation, Figure 13, in contrast to the monotonic increase in size during thermal aging (25).

Precipitate redistribution due to radiation-induced segregation also occurs, of course, in two-phase alloys in which solute enrichment at sinks prevails. Precipitates then nucleate and grow preferentially at defect sinks, as discussed in conjunction with Figure 9 for solid solutions. The matrix loses solute and eventually all precipitates not associated with defect sinks are dissolved, as has been observed in a two-phase Ni-Si alloy by Potter and Wiedersich (15).

In two component alloys the composition of the precipitate phase should be close to that of the thermal equilibrium phase at the precipitate matrix interface. However, concentration gradients within the precipitate due to defect fluxes can be set up and, if segregation within the precipitate phase exceeds the width of the corresponding phase field, a second, solute-richer precipitate should form. No clear evidence of such a decomposition of a two-component alloy into three phases has been reported thus far. However, Rehn et al. find by Auger depth-profiling that a thin film which has a substantially higher silicon content than 25 at. % Si forms in Ni-Si alloys on top of the Ni_3Si surface layer during prolonged irradiations (27).

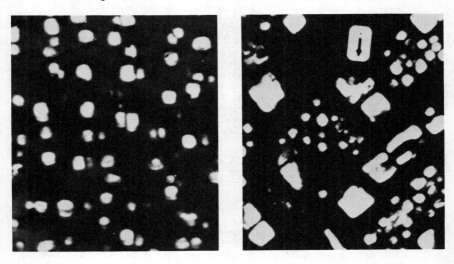

a   0.1 μm   b

Fig. 12. Dislocation-precipitate interaction as illustrated by dissolution of particles present prior to irradiation. (a) Dislocation loops nucleate almost exclusively at γ/γ' interfaces in a sample aged 6 h at 715°C, then irradiated to 0.1 dpa at 550°C. (b) Growth of dislocation loops at locations such as that marked with an arrow is accompanied by localized dissolution of particles, in a sample aged 38 h at 715°C, then irradiated to 5 dpa at 650°C (25).

27

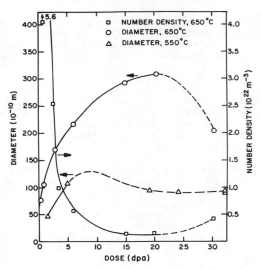

Fig. 13. Precipitate diameter and number density as function of dose in a
Ni-12.8 at. % Al alloy irradiated at the temperatures indicated
(26).

In multicomponent systems, the different coupling strengths between the
various components and the defect fluxes can lead to precipitate
compositions that differ markedly from their thermal equilibrium composition
in the alloy, because the matrix composition at the precipitate interface
may be altered significantly from the average matrix composition. An
example of this phenomenon may be the formation of the so called G-phase
that is found after neutron irradiation in silicon-containing Fe-Cr-Ni based
alloys (28). The G-phase is structurally similar to $M_{23}C_6$, but has a
significantly different composition, morphology and orientation relationship
to the matrix.

Problem Areas in Quantitative Characterization of Defect Clusters and Microstructural Features

As is apparent from the brief discussions of the primary and the
secondary damage microstructures presented in the two introductory papers of
the Symposium, a multitude of defects, defect clusters and other
microstructural features plays an important role in the development of the
damage structure in materials under irradiation. Although a good
qualitative understanding of the major processes and features has been
achieved during the past decade, quantitative and predictive descriptions of
the response of a material during elevated temperature irradiation require a
substantial increase of our knowledge about properties and structures of
defect clusters. Some of the most important areas in which the techniques
discussed in the Symposium could contribute are outlined here.

The structures, binding energies and kinetic properties of defect-
solute complexes are of great importance to the quantitative understanding
of alloying effects on nucleation of interstitial and vacancy aggregates as
well as to the quantitative description of radiation-induced segregation
processes. Quantitative information on structures, size-distributions,

compositions and stabilities of vacancy, interstitial and solute aggregates, especially those whose size is below the resolution limit of conventional transmission electron microscopy, would provide valuable checks on existing theories and would give guidance for necessary improvements. Measurements of chemical composition and phase identification of precipitates in irradiated materials are prerequisites for a thorough description of phase stability during irradiation. At present most quantitative microstructural investigations rely on quantitative transmission electron microscopy which is highly labor-intensive. Characterization of size-distributions of voids, loops and precipitates and measurements of dislocation densities by less laborious methods would be highly desirable.

Irradiation microstructures are frequently nonuniform. For example, void and dislocation loop depleted regions occur near surfaces and grain boundaries; massive precipitation can be associated with defect sinks and precipitates can form continuous grain-boundary and surface films; frequently preferential associations between different microstructural features are observed, e.g., voids attached to precipitates. It is easily surmized that such nonuniformities could affect, for example, crack propagation or the relative contributions of grain boundary sliding and intragranular slip to overall deformation. At present, no valid concepts seem to exist for useful, quantitative descriptions of these nonuniformities, nor are reliable models available for relating them to property changes.

Acknowledgement

The author expresses his appreciation for many fruitful discussions on the subjects covered in this paper with collegues presently and formerly at Argonne National Laboratory as well as for their generous contributions of illustrative figures, part of which are unpublished. Thanks are also extended to Ms. S. L. Ruffatto for careful preparation and layout of the manuscript for printing.

References

Note: The references given here are intentionally kept to minimum and an attempt was made to give only rather recent publications where possible.

(1)   Proc. Intl. Conf. on Radiation Effects in Breeder Reactor Structural Materials, eds. M. L. Bleiberg and J. W. Bennet, AIME, New York, 1977.

(2)   Proc. First Topical Meeting on Fusion Reactor Materials, eds. F. W. Wiffen, J. H. DeVan and J. O. Stiegler, J. Nucl. Mater. $\underline{85}$ & $\underline{86}$ (1979).

(3)   Workshop on Solute Segregation and Phase Stability During Irradiation, November 1-3, 1978, Gatlinburg, TN, J. Nucl. Mater. $\underline{83}$ (1) (1979).

(4)   H. Wiedersich in Radiation Damage in Metals, eds. N. L. Peterson and S. D. Harkness, ASM, Metal Park, OH, 1976, p. 157.

(5)   A. D. Brailsford and R. Bullough, J. Nucl. Mater. $\underline{69}$ & $\underline{70}$, (1978) 434.

(6)   L. K. Mansur, Proc. Workshop on Correlation of Neutron and Charged Particle Damage, CONF-760673, 1976, p. 61.

(7)   A. D. Brailsford, J. Nucl. Mater. $\underline{84}$ (1979) 245 and 269.

(8)   Private communication, D. I. Potter, University of Connecticut, Storrs, 1979.

(9)   J. J. Laidler, F. A. Garner and L. E. Thomas in Radiation Damage in Metals, eds. N. L. Peterson and S. D. Harkness, ASM, Metals Park, OH, 1976, p. 194.

(10)   H. R. Brager, F. A. Garner, E. R. Gilbert, J. E. Flinn, and W. G. Wolfer, in Ref. (1) pp. 727-755.

(11)   R. A. Erck, D. I. Potter and H. Wiedersich, J. Nucl. Mat. 80 (1979) (1979) 120.

(12)   K. Kitajima, K. Futagama and E. Kuramoto, in Ref. (2), p. 725.

(13)   B. A. Loomis, A. Taylor and S. B. Gerber, in Fundamental Aspects of Radiation Damage in Metals, eds. M. T. Robinson and F. W. Young, Jr., CONF-751006-P2, 1975, p. 1245.

(14)   D. I. Potter and A. W. McCormick, Acta Met. 27 (1979) 933.

(15)   D. I. Potter and H. Wiedersich, in Ref. (3), p. 208.

(16)   S. I. Maydet and K. C. Russell, J. Nucl. Mater. 64 (1977) 101.

(17)   M. D. Rechtin, J. Vander Sande and P. M. Baldo, Electron Microscopy 1978, Vol. 1, J. M. Sturgess, Ed., Microscopical Society of Canada, 1978, p. 388.

(18)   P. Wilkes in Ref. [3] p. 166, K-Y. Liou and P. Wilkes, J. Nucl. Mater. 87 (1979) 317.

(19)   Private communication, D. I. Potter, University of Conneticut, Storrs, 1979.

(20)   P. R. Okamoto and L. E. Rehn, in Ref. (3), p. 2.

(21)   H. Wiedersich, P. R. Okamoto and N. Q. Lam, in Ref. (3), p. 98.

(22)   K.-H. Robrock and P. R. Okamoto, in Comportement Sous Irradiation Des Materiaux Metalliques Et Des Composants Des Coeurs Des Reacteurs Rapides, J. Poiriers and J. M. Dupouy, eds., CEA, Gif-Sur-Yvette, France, 1979, p. 57.

(23)   S. C. Agarwal, G. Ayrault, D. I. Potter, A. Taylor and F. V. Nolfi, Jr., in Ref. (2), p. 653.

(24)   R. Cauvin and G. Martin, in Ref. (3), p. 67.

(25)   D. I. Potter and H. A. Hoff, Acta Met. 24 (1976) 1155.

(26)   D. I. Potter and D. G. Ryding, J. Nucl. Mater. 71 (1977) 14.

(27)   L. E. Rehn, R. Averback and P. R. Okamoto, this volume.

(28)   L. E. Thomas, Proc. of the 37th Annual Meeting of the Electron Microscopy Soc. of America, G. W. Bailey, Ed., Claitor's Publishing Div., Baton Rouge, 1979, p. 422.

IMPACT OF ADVANCED MICROSTRUCTURAL CHARACTERIZATION TECHNIQUES

ON MODELING AND ANALYSIS OF RADIATION DAMAGE*

F. A. Garner and G. R. Odette [1]

Hanford Engineering Development Laboratory
Richland, Washington

[1]University of California
Santa Barbara, California

Radiation-induced alterations of dimensional and mechanical properties have been shown to be a direct and often predictable consequence of radiation-induced microstructural changes. Recent advances in understanding of the nature and role of each microstructural component in determining the property of interest have led to a reappraisal of the type and priority of data needed for further model development.

This paper presents an overview of the types of modeling and analysis activities in progress, the insights that prompted these activities, and specific examples of successful and ongoing efforts. More importantly, however, a review is presented of some problem areas that in the authors' opinion are not yet receiving sufficient attention and which may benefit from the application of advanced techniques of microstructural characterization. Guidelines based on experience gained in previous studies are also provided for acquisition of data in a form most applicable to modeling needs.

*This research sponsored by the United States Department of Energy under contract DE-AC-14-76FF02170.

Introduction

The majority of the modeling and analysis activities concerned with the evolution of radiation-induced microstructure in metals derive their support from organizations dedicated to the commercial application of nuclear technology. These organizations also support extensive experimental programs to develop a data base on macroscopic properties necessary to design and support the operation of power generation facilities. Design correlations based only on empirical analysis of macroscopic data are frequently insufficient, however, to meet all design needs. The constraints imposed by leadtime and dollar requirements usually ensure that the data base seldom covers the entire spectrum of possible application, requiring interpolation and extrapolation to untested environments and operating histories. Therefore the impact of radiation-induced microstructural alterations on dimensional and mechanical properties is of prime importance.

The appearance of unanticipated phenomena can also require design changes or modification of operating procedures prior to accumulation of a relevant data field. Historical examples of significant unanticipated phenomena are void swelling and the magnitude of irradiation creep in fast breeder reactors, the magnitude of irradiation growth in zirconium alloys in CANDU reactors, and the effect of minor impurities on the embrittlement of pressure vessel steels in thermal reactors. Even more recently, the radiation effects community has become aware of a substantial alteration of phase stability that occurs during irradiation of many structural alloys.

Modeling and analysis activities have been employed to address both the extrapolation of data to untested environments and the anticipated impact of unexpected phenomena. A major tool in this effort is the accumulation of knowledge of the various radiation-induced microstructural components and the role of each component in determining the property of interest. Such modeling efforts have recently gone beyond a relatively narrow focus on void growth to treat the entire time history and correlated evolution of microstructural and "microchemical" features. This has led to a rethinking of both the type and priority of data needed for further model development.

This paper presents an overview of the types of modeling and analysis activities in progress, the insights that prompted these activities, and specific examples of successful and ongoing efforts. More importantly, however, a review is presented of some problem areas that in the authors' opinion are not yet receiving sufficient attention and which may benefit from the application of advanced techniques of microstructural characterization. Guidelines based on experience gained in previous studies are also provided for acquisition of data in a form most applicable to modeling needs.

Types of Modeling and Analysis Activities

Modeling and analysis activities can be conveniently divided into four categories. The first of these is the identification and description of the various roles played by each microstructural component in response to in-reactor or ex-reactor environments. Not only has the action of some important components previously escaped detection, but components may evolve either late in the irradiation or exist at such small sizes that they are effectively invisible, beneath the resolution limit of previously employed characterization techniques. Other visible microstructural components such as precipitates were once assigned relatively minor roles, particularly with respect to the assumed larger roles of dislocations and voids. Recent insight, however, has led to an upgrading of the importance of precipitates (1).

32

The second category of analysis involves the search for data sets which contain microstructural evidence that allows the validation of various proposed mechanisms for the action of a given component. Where more than one mechanism has been proposed, such data sets are also analyzed for clues which allow the various mechanisms to be ranked in order of their relative contribution to the property of interest.

The third category comprises the simulation, often computer-assisted, of the consequences of the competitive action of all microstructural components in each of their various roles. This type of modeling is particularly valuable for complex microstructures as well as for complicated operational histories.

The last category of modeling and analysis involves the application of the insight gained in the previous three categories toward the selection of the form of the empirical equation on which to condense the available data. This also provides a physically-based rationale for extrapolation of the equation beyond the boundaries of the current data base.

The various analysis activities are subject to some important limitations. In general the available data are not derived from single variable experiments, nor is there usually a complete understanding of the physical processes involved. Therefore the various data-fitting or modeling approaches require initial plausible guesses of the relative importance of competing phenomena. This often precludes a guarentee of uniqueness in the final models. The use of advanced microstructural characterization techniques can greatly aid in the testing of such models, by allowing tests of the predicted microscopic response of a material being tested for its macroscopic response.

Radiation Damage in Metals: Major Insights

It is possible in a greatly simplified fashion to outline the major insights which in the authors' perception form the basis of most of the current modeling efforts on radiation-induced property changes relevant to various fission and fusion reactor programs. Some of these insights derive their inspiration primarily from theoretical efforts and others primarily from experimental programs.

Current understanding of the interaction of a microstructural component with radiation-induced point defects is largely based on theoretical grounds. The first major insight involves the concept of a selective bias of each microstructural component toward acceptance of various point defects, with interstitials thought to be accepted at dislocations somewhat more readily than vacancies (2-3). This results in a net partition of point defects between various microstructural components, some of which, such as voids, were originally thought to be neutral sinks. It was later postulated that both dislocations and voids possess strain fields capable of producing biases in favor of interstitials (4). The biases of each individual sink in this model are much larger than that previously ascribed to dislocations alone, and it is the competitive interaction of many sinks with different biases that determines whether a particular microstructural component grows as a net sink for vacancies or interstitials (5). Major advances have been made in the description of these biases with respect to their dependence on component size, spacing, applied stress, segregation at sink surfaces and other variables (5-10). An area of current interest is the concept that the bias of each sink changes strongly with the composition of the matrix in which it is embedded (1,8).

The initial development of experimentally-derived insight on problems relevant to fast reactors was hampered by the unrecognized interaction of

the large number of variables which influence the development of microstructure during irradiation. Many of these variables were insufficiently controlled in early experiments. As each of these variables was recognized and
brought under better control, the reproducibility of the various phenomena
waš improved and many trends became evident. One major recent insight is
that some alloys evolve toward a saturation microstructure which appears to
be relatively independent of starting microstructure (11). Another is that
gradients in point defect concentration that develop near microstructural
components could produce a selective flow of various elements along the
gradients, leading to radiation-induced segregation of these elements (12).
The consequences of this phenomenon are sufficient to induce in some alloys
a microchemical evolution of the alloy matrix involving significant changes
in the concentrations of both solute and solvent atoms (1,13,14). This
process also causes a substantial alteration of phase stability and appears
in some alloys to proceed toward a saturation state that is relatively independent of starting microstructure (13-15).

Another largely experimentally-derived insight has been the concept that
microstructural records are sometimes impressed on the postirradiation microstructure by the action of physical mechanisms which cannot be observed
directly while in progress. Whereas it was once accepted that, unlike void
growth, irradiation creep left no record in the postirradiation microstructure, the expectation that such a record might exist has led to the discovery
of such records (16-18). Various radiation-induced diffusional processes have
also been found to leave interpretable records not only in the matrix regions
surrounding various sinks but also inside precipitates (15,19).

Examples of Current Modeling Activities

With the exception of irradiation of thin foils with electrons in a high
voltage microscope, it is generally impossible to observe the action of microstructural components during irradiation. After the irradiation has
ceased, processes such as dislocation climb and enhanced bulk and surface
diffusivity decrease sharply or terminate. Some processes such as radiation-
induced solute segregation may actually be reversed and some components such
as dislocation loops may shrink. In the most pessimistic sense, then, the
use of postirradiation microstructure in modeling efforts is best confined
to the description of ex-reactor properties. In general, however, it appears
that most microstructural components that develop in structural steels during
irradiation at low temperature and even in the range 300 to 700°C are relatively
stable during cool-down, extraction from reactor, and during subsequent storage
and handling. In the following sections, examples will be shown of the successful application of postirradiation microstructural data to the modeling
of both in-reactor and ex-reactor properties.

Ex-Reactor Mechanical Properties

Tensile test experiments yield a variety of data relevant to material
performance, two of which are shown in Figure 1, but the yield strength is
the best measurement of the influence of radiation-induced rather than de-
formation-influenced microstructure. As our understanding of the microstructural evolution of 300 series stainless steels has evolved, a number of
attempts (20-22) have been made to determine the relative contributions of
each microstructural component to the hardening or softening of the material
observed for a given set of irradiation conditions. These efforts have been
hampered by incomplete or incorrect descriptions of the microstructure and
some ambiguity concerning the exact nature of hardness models for each component. Much more complete characterizations now exist of the irradiation-
modified microstructure of both annealed and cold worked AISI 316 stainless

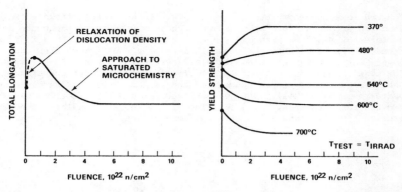

FIGURE 1. Schematic illustration of the fluence and temperature dependence
of several mechanical properties of 20% cold-worked AISI 316 (23-24).

steel. These descriptions have been recently used to successfully construct
a microstructurally-based yield stress correlation (23). In this effort
those microstructural components sensitive to flux, stress, and temperature
have been identified. An estimate has also been made of cavity contribution
to hardening.

The current empirically-derived design correlation for yield-stress of
AISI 316 is based on a flow model/state variable approach derived only from
cold work data obtained from irradiation in a fast breeder reactor (24). This
equation contains no guidance on how to extrapolate the data to other environments, particularly those that involve different displacement rates and high
helium/dpa ratios. It is anticipated, however, that microstructural information
developed from specimens irradiated in any new environment can be used with
the microstructurally-based correlation to aid in the extrapolation of the
empirical equation.

Correlations of microstructure with post-yield flow (i.e., work hardening)
and fracture properties will probably be more difficult to obtain. Not only
is the understanding of the "micromechanical" mechanisms of these phenomena
less well developed, but they generally involve a wider range of microstruc-
tural/microchemical processes. For example, it is well known that boundary
segregation of both major and minor alloy constituents can profoundly influence
brittle fracture parameters and even cause a ductile-brittle transition of
fracture mode (25). In ferritic steels this influence can be magnified by
irradiation-induced strengthening (26). Considerable progress has been made
recently in developing correlations between segregation and embrittlement
parameters in terms of critical stress models (26).

Ductile fracture processes usually involve the nucleation and growth of
cavities at second phase or inclusion particles. Nucleation, growth and
critical volume of cavities at the final fracture instability are very likely
to be sensitive to the inhomogeneous deformation (flow localization) characteristics found in some irradiated alloys. Hence, developing microstructural-
property correlations for ductile fracture will require microstructural
examination both before and after fracture testing; and also during the
development of the deformation microstructure prior to fracture. Emphasis
should be placed on quantifying the degree of flow localization and the
fracture parameters for regions of intense flow resulting in shear band
decohesion (27).

Considerable insight can be gained with microstructural correlations

which are less than complete. For example, the dominance of the network
dislocation component of hardening was recently used to validate and define
the limits of a simple property-property correlation (22). This is illustrated in Figure 2 where uniform ductility as a function of displacement
exposure for solution annealed stainless steels is compared to a simple
model based on the concept that irradiation-induced dislocations effectively
function as a "prestrain". The map of the regime where this simple model
appears reasonable (R∿1) versus where it breaks down also delineates major
microstructural regimes: void swelling at low and intermediate temperatures
and grain boundary cavitation at high temperatures.

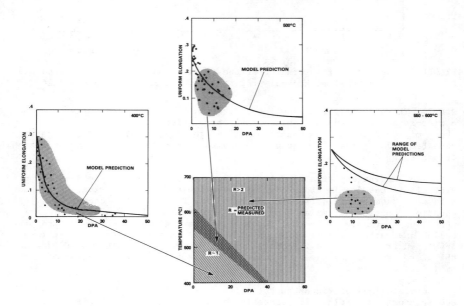

FIGURE 2.   Comparison of predictions of a ductility correlation based on
             yield to ultimate strength ratio with elongation data derived
             from irradiated solution-annealed 316 stainless steel (22).

     At high temperatures creep-rupture properties are of prime interest.
In this case, the concentration and distribution of helium at grain boundaries
is critically important. Figure 3 shows the predictions of an intergranular
creep rupture model based on stress-induced formation of creep cavities on
helium bubbles at grain boundaries, using assumed bubble size distributions
that are in broad agreement with the limited microstructural data (28).   The
creep rupture data are derived from both helium-injected and neutron-irradiated titanium-stabilized austenitic DIN 1.4970. Further development of such
models requires better microstructural data on precipitates, dislocation
networks, cavities and segregation profiles at grain boundaries.

In-Reactor Dimensional Changes

     As reviewed elsewhere, it is possible to empirically correlate the onset
and development of both swelling and irradiation creep to the details of the
microstructural and microchemical evolution (1,13). It was, in fact, the
inability of simple microstructural interpretations alone to account for the

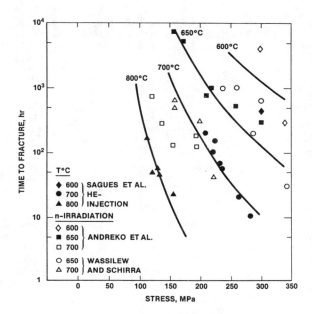

FIGURE 3. Comparison of data and microstructurally-based predictions of the postirradiation rupture life of DIN 1.4970 at elevated temperatures (28).

observed relative swelling behavior of cold-worked and annealed AISI 316 steel that led to the prediction and search for the microchemical evolution (14). This latter evolution could not have been identified and studied without the aid of analytical electron microscopy using energy-dispersive X-ray analysis. The techniques involved in these studies are in a state of rapid development and their application to irradiated materials is reviewed elsewhere (15,19).

A typical insight derived from such studies is shown in Figure 4 where the acceleration at high fluence of irradiation creep and the concurrent onset of swelling were found to occur long after the attainment of a

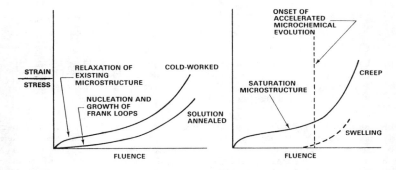

FIGURE 4. Schematic illustration of the evolution of swelling and irradiation creep in AISI 316 and the relationship of these to the evolution of microstructure and microchemistry.

37

saturation network of Frank loops and dislocations. However, the acceleration of creep and swelling were found to coincide with the onset and acceleration of nickel (and silicon) removal from the alloy matrix, as shown in Figure 5.

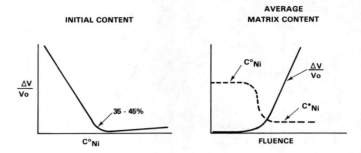

FIGURE 5. Schematic representation of the correlation observed between swelling $\Delta V/V_0$ versus the original alloy nickel content C°_{Ni} and swelling versus the average nickel content in the matrix of AISI 316 (13). C^*_{Ni} is the saturation level of nickel eventually reached by the matrix. For AISI 316 $C^\circ_{Ni} \cong 14\%$ and $C^*_{Ni} \cong 9\%$.

A similar situation was found for the response of both swelling (29–31) and irradiation creep (31) to changes in temperature during irradiation. Note in Figure 6 that small gradual decreases in temperature lead to decreased creep rates as expected from microstructural and diffusivity considerations.

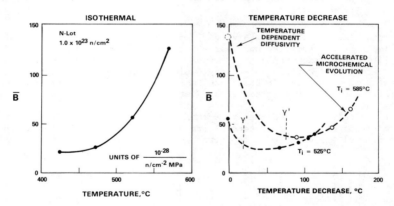

FIGURE 6. Comparison of average creep coefficient, $\overline{B}$, of 20% CW AISI 316 obtained in response to isothermal and gradually declining irradiation temperatures (31). The onset of γ' formation is shown for each starting temperature T_i.

This trend persists only so long as the microchemical evolution is not perturbed or accelerated. At some point, however, the further decline in temperature causes an accelerated formation of gamma prime precipitates, as confirmed by microstructural observations (31). An acceleration of both creep and swelling (Figure 7) was the direct result.

It is therefore necessary for modeling purposes to expand the definition of relevant microstructural features required to describe the radiation-induced

38

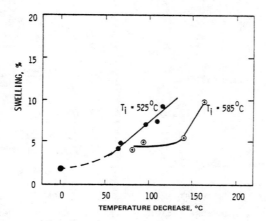

FIGURE 7. Enhancement of swelling by gradual temperature reductions during neutron irradiation of 20% cold-worked AISI 316 (replotted from Reference 29). T_i is the starting temperature.

evolution. In addition to the usual descriptions of component identity, size distribution and density, the experimenter should include the time-dependent and spatial details of the matrix and precipitate compositions that evolve during irradiation. The spatial details are particularly important as discussed in the following section.

The impact of the interaction between the microstructural and microchemical evolutions and other detailed examples of its consequences in alloy behavior have been discussed by Odette (32). In this latter paper a thorough review of the microstructural data base has been published.

Microstructural Records of Diffusional Processes

As discussed elsewhere, the precipitate phases that evolve in 300 series stainless steels become progressively richer in the elements nickel and silicon, and in some cases precipitates rich in these elements form which are only stable in the presence of irradiation (1,14,33,34). The study of this segregation process has been facilitated by microanalysis of the compositional gradients which develop in the vicinity of such precipitates. As shown schematically in Figure 8 and discussed in detail elsewhere, (15,34) the direction of solute flow can sometimes be inferred from the nature of the defect gradients near each precipitate, providing that the observation is made while the flow is occuring. While the onset of these segregation processes has been correlated empirically with the onset of swelling and accelerated irradiation creep, the termination of the segregation process is signaled by a reduction in the gradients, and appears to correspond to the attainment of steady state deformation rates.

Elemental segregation also appears to occur at dislocations, loops, grain boundaries and void surfaces (15,19,35,36). The segregation of nickel at void surfaces is primarily balanced by an outflow of chromium as would be expected from their respective diffusivities and the operation of the inverse-Kirkendall effect (37). As shown schematically in Figure 9, the effect is amplified in the region between two voids and demonstrates that at least at 650°C molybdenum and silicon are not directly visible as participants in the segregation process at voids (15,34). Data of this type are becoming very

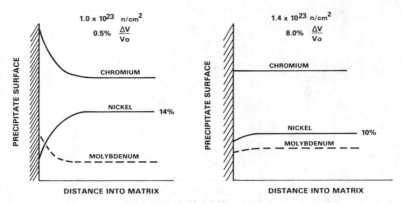

FIGURE 8. Schematic representation of compositional profiles measured near Laves precipitates in 20% cold-worked AISI 316 at 650°C (1,15). Fluence and swelling values are shown for each condition. When the swelling level is low the gradients are much more pronounced than when observed at higher swelling levels. Note the changes in matrix composition which result from the infiltration-exchange process across the precipitate-matrix interface.

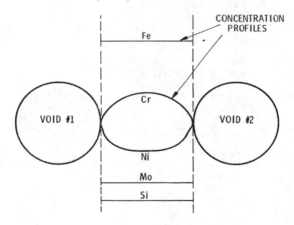

FIGURE 9. Schematic representation of the compositional profiles observed between two voids in 20% CW AISI 316 irradiated to 1.4×10^{23} n/cm^2 (E >0.1 MeV) at 650°C (15,34).

useful in the validation and ranking of various diffusional and solute-binding mechanisms.

Microstructural Records of Irradiation Creep

There is currently considerable controversy as to the parametric dependence of various proposed irradiation creep mechanisms, particularly that of stress-induced preferential absorption (SIPA) of interstitials (38). Indeed, the very existence of this creep mechanism has been questioned. The description of this and other creep mechanisms has been aided recently by the development of detailed studies of Frank loop and dislocation populations which evolve in response to anisotropic stress distributions. These studies were made possible by the development of weak-beam transmission electron microscopy techniques.

Okamoto and Harkness (39) first provided the evidence shown in Figure 10 that indicates that Frank loop populations might respond in a nonuniform way to the biaxial tensile stress state found in the walls of pressurized tubes

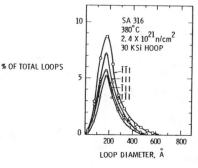

LOOP PLANE	ALIGNMENT FACTOR
$\bar{1}\bar{1}1$	1.30
111	1.07
$\bar{1}11$	0.86
$1\bar{1}1$	0.78

FIGURE 10.   Frank loop distributions observed by Okamoto and Harkness at low fluence in annealed AISI 316 irradiated under a biaxial tensile stress in a pressurized tube (39). The hoop stress on this tube was 207 MPa. The alignment factor is defined by the ratio of the loop density on a specific plane to the average density on all planes.

irradiated to low fluence. Fluctuations of as much as ±30% from the mean value were found, but unfortunately it was not possible in that experiment to relate the planar distributions to the stress state. Wolfer later showed that the most optimistic assessment of loop alignment models (based on rotation of di-interstitial clusters) could not account for this magnitude of fluctuation (40).

Brager and co-workers (41) showed (Figure 11) that at higher fluence and stress levels, more pronounced fluctuations about the mean planar density were possible and related these to the stresses acting on each plane. Subsequent reinterpretation of these data by Garner and co-workers (16,17) indicated that this evidence indeed appeared to validate the existence of the SIPA mechanism. It was also asserted that a relative ranking in importance of the SIPA mechanism over the stress-induced alignment mechanism was supported by these data (17). Later data for AISI 316 at higher fluence and in other steels has provided even more insight on the nature of and competition between various irradiation creep mechanisms (18). Validation of the existence of SIPA is important not only to fast reactor studies but also the extrapolation of low-fluence creep data on zirconium alloys to higher fluence in the CANDU reactors (42).

Problem Areas Requiring Attention

There are a number of areas in which the application of advanced microstructural characterization techniques might possibly aid in the modeling of material response to irradiation. Several of these problem areas are discussed and some examples of the application of advanced techniques are given. It is hoped that more attention will be brought to bear on each problem and provide the needed insight.

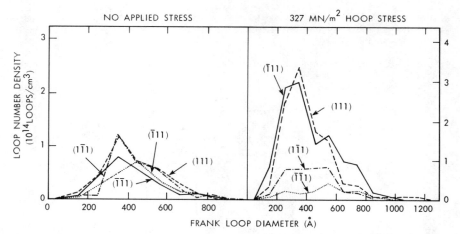

FIGURE 11. Comparison of Frank loops size distributions on each {111} plane
in two specimens of 20% cold-worked AISI 316 irradiated at different stress levels at 500°C to 3×10^{22} n/cm^2 (E >0.1 MeV) (41).

Superfine Microstructure

Irradiation-induced increases occur in the yield stress and the ductile-
brittle transition temperature in pressure vessel ferritic steels during
irradiation in light water reactors. These have been correlated primarily
to the copper and possibly the phosphorous content of these alloys (43-44).
In a simple Fe-0.3%Cu alloy irradiated to about 3×10^{19} n/cm^2 (E >0.1 MeV)
at 290°C, the possible role of copper in the degradation of mechanical properties has been identified. Brenner and co-workers (45) have demonstrated
by the use of field-ion microscope and atom probe techniques that 0.6 nm
copper-stabilized microvoids form during irradiation at densities of 8×10^{17}
cm^{-3}. This exceptionally high density of potential pinning centers could not
be resolved in transmission microscopy studies, in which only a much smaller
density of 3-5 nm dislocation loops was found. Gelles and co-workers (46)
later demonstrated in this same material that a dislocation pinning point
analysis conducted on specimens strained at room temperature confirmed the
existence and density of these clusters. The alloys actually employed in
pressure vessels are much more complex, i.e., tempered bainite which is
generally stable under light-water reactor irradiation conditions. Illumination of the mechanisms of embrittlement in such alloys will require a better
understanding of the governing microstructures, both visible and unresolvable.
Another method now being employed to study unresolvable microvoids and small
precipitates in pressure vessel steels is that of neutron scattering (47).

Several researchers have noted that in 300 series stainless steels large
scale radiation-induced segregation of nickel into precipitate phases leads
to a transformation of portions of the austenite matrix to ferrite (34,48).
There is a possibility that these ferrite phases may develop from superfine
ferromagnetic nucleii formed at very low fluences. Using advanced magnetic
measurements, Stanley (49) observed in AISI 316 $\sim 3 \times 10^{17}$ cm^{-3} of small magnetic particles after irradiation to fluences of $\sim 3 \times 10^{22}$ n/cm^2 (E >0.1 MeV)
in the range of 450 to 620°C. Microscopy did not detect resolvable ferrite
regions however. Although Stanley noted that nickel segregation to void
surfaces could be a possible source of the magnetization increases, the
magnetic particle density was about three orders of magnitude greater than
that of the voids. Baron and co-workers also found ferromagnetism in AISI 316L
in structural components irradiated in the Rapsodie fast reactor (50). The

relationship of the magnetic permeability profile relative to that of the temperature, flux and swelling profiles led these researchers to hypothesize that ferrite formation and trapping of transmutation-produced hydrogen in voids might account for the magnetism.

Other types of superfine microstructural components that might not be resolvable with currently applied techniques are small point defect cascade remnants in low temperature irradiations and very small helium aggregations that probably function as void embryos. Positron annihilation studies have demonstrated the existence of such microvoids and microbubbles (51).

Distribution and Influence of Helium

The large level of helium that will be generated in a fusion environment is expected to lead to significant degradation of mechanical properties relative to that developed in fast reactors at comparable displacement doses (52). It is also expected that radiation-induced dimensional changes will be sensitive to the helium/displacement ratio. In an effort to study this possibility, specimens are being irradiated with dual ion beams (53-54) and neutrons in mixed-spectrum reactors such as HFIR (55). It has been shown that the larger levels of helium in these environments lead to alterations in the temperature regime of swelling (56), as well as alterations in the Frank loop, void and precipitation response (56-59). It also appears that under some conditions the correlated development of voids and precipitates can lead to cavity/precipitate interactions that greatly accelerate swelling (54).

The modeling of these effects requires not only a description of the gas contribution to cavity growth (60-64) but also measurements of the initial distribution and subsequent mechanisms and rates of redistribution of helium and hydrogen. At the moment this area comprises the major uncertainty in modeling of the effect of transmutation-produced gases.

Another area requiring experimental input is the time-dependent distribution of helium generation. The use of mixed-spectrum irradiations and the nickel two-step transmutation reaction (55) will lead to a non-uniform helium generation in and around those areas where nickel segregates. Thus helium deposition will be greatest at the surface of large precipitates (65) and grain boundaries. Since voids and cavities tend to nucleate preferentially on these microstructural components, the mixed-spectrum simulation experiments will develop a distribution of voids and cavities that are possibly atypical of that obtained in both fast reactor and fusion device spectra.

Austenite → Ferrite Transformation

The question arises whether the γ→α transformation observed in several studies occurs by a nucleation and growth process during irradiation or by a martensitic inversion upon cooling. Porter (66) has deduced that the transformation is a continuous process involving nucleation and growth on stacking faults, citing as evidence the nature of the γ→α boundaries and the absence of twinning in the α-phase. Mazey and co-workers (67) arrive at the opposite conclusion from an ion irradiation study of a series of 12Cr-15Ni-Fe alloys. Their conclusion was based on the observation that the nickel content was reduced by precipitation of Ni_3Si and other phases, which should lead to an increase in the martensitic transformation temperature. They also note that X-ray analysis shows the remaining austenite and the ferrite have essentially identical compositions, whereas they would expect substantially different nickel or nickel-equivalent compositions for a diffusion and growth process.

Mazey also notes that voids were found in the ferrite regions which are known to resist swelling. Brager and Garner disagree with Mazey's conclusions while citing essentially the same type of evidence derived from neutron irradiations of silicon-modified AISI 316 alloys (34).

The issue here centers on whether the ∿2% volume change that accompanies the transformation occurs during irradiation or only on cool-down. The latter possibility allows the loss of this volume change upon reheating. Since this controversy has not been resolved by postirradiation analysis of microstructure, it appears that in-situ examination during irradiation is required. The best tool to observe such an effect is the high voltage electron microscope, allowing observation during irradiation and on subsequent cooling and heating.

Effect of Stress on Microstructural Development

The various roles of stress on radiation-induced microstructural development have been clearly demonstrated in a number of studies (11,12,16-18,68). It has been shown, however, that the operating stresses are not always determinate, particularly when the material experiences large (and possibly anisotropic) strains due to precipitate-related volume changes (17-18). In materials possessing low intrinsic swelling and creep rates this can lead to the influence of internally-generated stresses overwhelming the microstructural record of the externally-applied stresses (18). The acquisition of component and stress orientation data in complex microstructures may require the development of additional data collection and analysis techniques. In the dense microstructures produced by irradiation in complex steels it is often difficult to extract information on stress-dependent orientation. Efforts to date have concentrated on the "2½D" weak beam technique (41).

The development of preferred orientation of Frank loops during irradiation under stress (12,18,41) leads to the prediction that the resulting network dislocations will also exhibit a corresponding but slightly different anisotropy. This possibility has not yet been experimentally demonstrated and has large consequences in the modeling of irradiation creep, particularly in response to changes in stress magnitude or direction.

It also appears that the microstructural record of various irradiation creep processes must be interpreted very carefully in that the record changes as the number of competing components increase (18). This requires the experimenter to choose carefully the specimens with which the record is sought in order to maximize the record for the component under study.

The effect of tensile stress on swelling of AISI 316 and other alloys has recently been definitively determined to be related to changes in the duration of the transient regime of swelling rather than in the steady-state swelling regime behavior (68). The effect of stress is most pronounced at high temperatures and this sensitivity has been shown to be related to the stress-sensitivity of intermetallic phase formation, as shown in Figure 12. This insight requires that the swelling equation developed for non-tensile stress states might be different than was originally envisioned. The low-temperature radiation-stable precipitates do not appear to be stress-sensitive. This has been verified experimentally for the γ' phase in AISI 316 and has been inferred for the G-phase from the relative stress-insensitivity of swelling of annealed AISI 316 at low temperatures (69). Future modeling efforts require knowledge of lattice parameter mismatch, precipitate/matrix orientation and the degree of coherency for each precipitate phase.

The major problem with the above studies is that they were conducted in tensile biaxially-stressed specimens. In order to confirm the results of these studies and the prediction of microstructurally-based models for torsional and compressive stress states, specimens must be irradiated in non-tensile experiments.

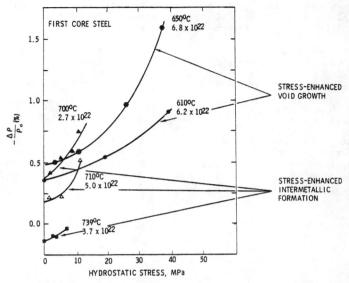

FIGURE 12.   Stress-enhanced density changes $(-\Delta\rho/\rho_o)$ observed in 20% cold worked AISI 316 (68). Note that dilational strains associated with formation of intermetallic phases can be observed at temperatures and fluences where void swelling has not commenced.

Guidelines for Data Acquisition

The preceeding discussion has spotlighted a number of principles which can be condensed as guidelines for experimenters who anticipate the further use of their data in modeling and analysis activities.

(1)   The choice of specimens to examine is often the critical step. Given the tremendous complexity and synergistic nature of the processes involved it is best to choose whenever possible specimens for comparison which clearly differ in only a single variable. Since many of the most important processes are transient by nature, it is best to examine specimens just before, during, and just after the transient. Examinations conducted long past the terminus of the transient will often suffer erasure of the microstructural or microchemical records.

(2)   After having chosen the specimen it is important that the experimenter anticipate the presence of fine structure and pursue it even in the absence of foreknowledge of its existence.

(3)   The microstructure should be characterized in its entirety, including associated microchemical gradients and concentrations. Wherever possible all data should be extracted from a single region of the specimen, recognizing the relatively heterogeneous nature of the microchemical evolution. The authors recognize of course the difficulty of obtaining fine and coarse structure in the same foil thickness.

(4) The experimenter should also recognize the need for mass balances in modeling efforts.

(5) In studies involving the effect of stress it is important to retain knowledge of the relative orientation of the microstructural components and the stress state. One should also be alert for signs that signal the large role of internally-generated stresses.

(6) Whenever possible, microscopy studies should be supplemented by other advanced techniques applied to the same specimens. Techniques such as X-ray analysis, and magnetic, resistivity and microhardness measurements can yield additional information not obtainable by microscopy.

Conclusions

It appears that the need is increasing for the coupled use of advanced microstructural characterization techniques and modeling/analysis activities. Substantial challenges are posed by recent insights on fundamental processes and the needs of materials people in development of descriptions of component response in fission and fusion environments.

REFERENCES

1. F. A. Garner, "The Microchemical Evolution of Irradiated Stainless Steels," Proceedings of the Fall AIME Symposium on Irradiation Phase Stability, Pittsburgh, PA, October 5-9, 1981, (to be published by the Metallurgical Society of AIME).

2. F. S. Ham, J. Appl. Phys., 30, (1959), p. 915.

3. R. Bullough and R. C. Perrin, in Irradiation Effects on Structural Alloys for Nuclear Reactor Application, ASTM STP 484, A. L. Bement (Ed.), (1970), p. 317.

4. W. G. Wolfer, M. Ashkin and A. Boltax, in Properties of Reactor Structural Alloys After Neutron or Particle Irradiation, C. J. Baroch (Ed.), ASTM STP 570, (1975), p. 233.

5. W. G. Wolfer, L. K. Mansur and J. A. Sprague, in Radiation Effects in Breeder Reactor Structural Materials, M. L. Bleiberg and J. W. Bennett (Eds.), The Metallurgical Society of AIME, p. 479.

6. W. G. Wolfer and A. Si-Ahmed, Physics Letters, 76A, (No. 3,4), (1980), p. 341.

7. A. Si-Ahmed and W. G. Wolfer, "On The Simultaneous Formation of Inter-stitial - and Vacancy - Type Loops During Irradiation," University of Wisconsin Report UWFDM-360, June 1980.

8. W. G. Wolfer and L. K. Mansur, J. Nucl. Mat., 91, (1980), p. 256.

9. A. D. Brailsford and R. Bullough, "The Theory of Sink Strengths," A.E.R.E. Harwell Report TP 854, August 1980.

10. L. K. Mansur and W. G. Wolfer, J. Nucl. Mat., 69 and 70, (1978), p. 825.

11. H. R. Brager, F. A. Garner, E. R. Gilbert, J. E. Flinn and W. G. Wolfer, Ref. 5, p. 727.

12. P. R. Okamoto and L. E. Rehn, J. Nucl. Mat., 83, (1979), p. 2.

13. H. R. Brager and F. A. Garner, in Effects of Radiation on Materials, ASTM STP 725, D. Kramer, H. R. Brager and J. S. Perrin (Eds.), ASTM, (1981), p. 470.

14. H. R. Brager and F. A. Garner, in Effects of Radiation on Structural Materials, ASTM STP 683, J. A. Sprague and D. Kramer (Eds.), ASTM, (1979), p. 207.

15. H. R. Brager and F. A. Garner, "Analysis of Radiation-Induced Microchemical Evolution in 300 Series Stainless Steels," this proceeding.

16. F. A. Garner and W. G. Wolfer, Trans. ANS, Vol. 28, (1978), p. 144.

17. F. A. Garner, W. G. Wolfer and H. R. Brager, Ref. 14, p. 202.

18. D. S. Gelles, F. A. Garner and H. R. Brager, Ref. 13, p. 735.

19. L. E. Thomas, "EDX Microanalysis of Neutron-Irradiated Alloys," this proceeding.

20. J. J. Holmes, R. E. Robbins, J. L. Brimhall and B. Mastel, Acta Met., 16, (1968), p. 955.

21. J. R. Matthews, Contemp. Phys., 18, (1977), p. 571.

22. G. R. Odette and D. Frey, J. Nucl. Mat., 85 and 86, (1979), p. 817.

23. G. D. Johnson, F. A. Garner, H. R. Brager, and R. L. Fish, Ref. 13, p. 393.

24. R. L. Fish, N. S. Cannon and G. L. Wire, Ref. 14, p. 450.

25. See for example Olefjord, I., Int. Met. Revs., (1978), p. 149.

26. R. O. Ritchie, W. L. Server and R. A. Wullaert, Met. Trans. A, (1979), p. 1557.

27. J. F. Knott, Met. Science, 14, (1980), p. 327.

28. S. S. Vagarali and G. R. Odette, "Influence of Irradiation on Creep Rupture of Stainless Steels-Comparison of Theory and Experiment," Proceedings of 2nd Topical Meeting on Fusion Reactor Materials, Seattle, WA, August 1981, (in press).

29. J. F. Bates and D. S. Gelles, J. Nucl. Mat., 71, (1978), p. 365.

30. J. P. Foster and A. Boltax, Nucl. Tech., 47, (1980), p. 181.

31. F. A. Garner, E. R. Gilbert, D. S. Gelles and J. P. Foster, Ref. 13, p. 698.

32. G. R. Odette, J. Nucl. Mat., 85 and 86, (1979), p. 533.

33. W. J. S. Yang, H. R. Brager and F. A. Garner, "Radiation-Induced Phase Development of AISI 316," Ref. 1, (in press).

34. H. R. Brager and F. A. Garner, "Radiation-Induced Evolution of the Austenite Matrix in Silicon-Modified AISI 316 Alloys," Ref. 1, (in press).

35. A. Kenik, Scripta Met., 10, (1976), p. 733.

36. D. S. Gelles, J. Nucl. Mat., 83, (1979), p. 200.

37. A. D. Marwick, R. C. Piller and P. M. Sivall, J. Nucl. Mat., 83, (1979), p. 35.

38. See for instance contributions by W. G. Wolfer, F. A. Nichols, R. Bullough and M. H. Wood in Proceedings of the International Conference on Fundamental Mechanisms of Radiation-Induced Creep and Growth, published in J. Nucl. Mat., 90, (1980), p. 1, 29, 44 and 175.

39. P. R. Okamoto and S. D. Harkness, J. Nucl. Mat., 48, (1973), p. 49.

40. W. G. Wolfer, "Correlation of Radiation Creep Theory with Experimental Evidence," University of Wisconsin Report UWFDM-312, also in Ref. 38, p. 175.

41. H. R. Brager, F. A. Garner and G. L. Guthrie, J. Nucl. Mat., 66, (1977), p. 301.

42. See for instance contributions of R. A. Holt, E. J. Savino, E. C. Laciana, H. Wiedersich and S. R. MacEwen in Ref. 38, p. 89, 108, 157 and 193.

43. L. E. Steele, Nuclear Safety, 17, (1976), p. 327.

44. F. A. Smidt and J. A. Sprague, in Effects of Irradiation on Substructure and Mechanical Properties of Metals and Alloys, ASTM STP 529, (1973), p. 78.

45. S. S. Brenner, R. Wagner and J. A. Spitznagel, Met. Trans. A, 9A, (1978), p. 1761.

46. D. S. Gelles, "HVEM In-Situ Deformation of Neutron Irradiated Fe-0.3% cu," DAFS Quarterly Progress Report DOE/ET-0065/8, (1980), p. 148.

47. Frisius and D. Buneman, "The Measurement of Radiation Defects in Iron by Means of Small Angle Neutron Scattering," Proceedings of International Conference on Irradiation Behavior of Metallic Materials for Fast Reactor Core Components, Ajaccio, Corsica, June 4-8, 1979.

48. D. L. Porter, J. Nucl. Mat., 83, (1979), p. 90.

49. J. T. Stanley, J. Nucl. Mat., 85 and 86, (1979), p. 787.

50. J. L. Baron, R. Cadalbert and J. Delaplace, J. Nucl. Mat., 51, (1974), p. 266.

51. M. Doyama, K. Hinode and S. Tanigawa, J. Nucl. Mat., 85 and 86, (1979), p. 781.

52. E. E. Bloom, J. Nucl. Mat., 85 and 86, (1979), p. 795.

53. N. H. Packan and K. Farrell, J. Nucl. Mat., 85 and 86, (1979), p. 677.

54. S. Wood, J. A. Spitznagel, W. J. Choyke, N. J. Doyle, J. N. McGruer and J. R. Townsend, Scripta Met., 14, p. 211.

55. M. L. Grossbeck and P. J. Maziasz, J. Nucl. Mat., 85 and 86, (1979), p. 883.

56. K. Farrell and N. H. Packan, J. Nucl. Mat., 85 and 86, (1979), p. 683.

57. E. A. Kenik, J. Nucl. Mat., 85 and 86, (1979), p. 659.

58. P. J. Maziasz, J. Nucl. Mat., 85 and 86, (1979), p. 713.

59. J. A. Spitznagel, F. W. Wiffen and F. V. Nolfi, J. Nucl. Mat., 85 and 86, (1979), p. 629.

60. W. G. Wolfer and R. W. Conn, "New Analysis of First Wall Lifetime Considerations for Fusion Reactors," University of Wisconsin Report UWFDM-212, March 1977.

61. M. H. Yoo and L. K. Mansur, J. Nucl. Mat., 85 and 86, (1979), p. 571.

62. R. Bullough, M. R. Hayns and M. H. Wood, J. Nucl. Mat., 85 and 86, (1979), p. 559.

63. G. R. Odette and M. W. Frei, Proceedings of First Topical Meeting on the Technology of Controlled Nuclear Fusion, CONF-740402-P2, 2, (1974), p. 485.

64. G. R. Odette and S. C. Langley, in <u>Proceedings of First International Conference on Radiation Effects and Tritium Technology for Fusion Reactors</u>, CONF-750989, <u>1</u>, (1976), p. 395.

65. D. S. Gelles and F. A. Garner, <u>J. Nucl. Mat.</u>, <u>85 and 86</u>, (1979), p. 689.

66. D. L. Porter, <u>J. Nucl. Mat.</u>, <u>79</u>, (1979), p. 406.

67. D. J. Mazey, D. R. Harris and J. A. Hudson, "The Effect of Silicon and Titanium on Void Swelling and Phase Stability in 12Cr-15Ni-Fe Alloys Irradiated with 46 MeV Nickel Ions," Ref. 47.

68. F. A. Garner, E. R. Gilbert and D. L. Porter, "Stress-Enhanced Swelling of Metals During Irradiation," Ref. 13, p. 680.

69. J. L. Boutard, G. Brun, J. Lehman, J. L. Seran and J. M. Dupouy, "Swelling of 316 Stainless Steel," Ref. 47.

Advanced Microscopy

AUTOMATED DATA REDUCTION FROM

TRANSMISSION ELECTRON MICROGRAPHS

J. A. Sprague

Naval Research Laboratory

Washington, D. C. 20375

Methods of extracting quantitative microstructural data from transmission electron micrographs are discussed. The relative advantages of manual, automated, and Fourier transform analyses of images are considered, with specific emphasis on the problems presented by diffraction contrast images of defects in crystalline solids. Manual methods have the obvious advantage of low capital cost, as well as the benefits of flexibility in adjusting for variations in defect contrast and recognizing artifacts and overlapped images, but problems of speed and operator fatigue can limit the quantity of good data that can be accumulated reasonably. A fully digital analysis system, in which an image is read directly via some type of image converter into a computer, requires a considerable capital outlay, but potentially offers large benefits in speed of defect analyses. Commercially available systems, however, have found their principal materials science applications in optical metallography, rather than transmission electron microscopy, mainly because of problems with defect contrast and image overlap. These problems, and some possible solutions to them, are discussed in terms of image digitization and segmentation processes. Fourier transform analysis of micrographs also has some potential advantages in certain cases, even though some information is lost in the process of recording the transform. Following the review of the various possibilities for micrograph data reduction, some useful future advances in this field are discussed.

*   This research was sponsored by the Office of Naval Research.

Introduction

Transmission electron microscopy (TEM) is widely recognized as a powerful tool for examining lattice defects in materials. Many defects, such as dislocation loops, voids, and precipitates, which significantly affect the physical and mechanical properties of materials, also produce characteristic electron diffraction and/or diffraction contrast effects that allow their identification and characterization by TEM of thin foils. By proper control of imaging conditions, it is often possible to obtain very detailed information on an individual defect, for example the shape, habit plane, and Burger's vector of a dislocation loop. This ability makes TEM a superb qualitative technique for microstructural characterization. When one desires quantitative correlation between observed microstructures and either theoretical models or macroscopic behavior, however, this level of detail can become something of a disadvantage, since large numbers of individual defect images must be counted and measured to obtain statistically significant average quantities or distribution functions. Since this data reduction process, typically performed by visual means, is extremely tedious, time-consuming, and expensive, successful automated image analysis would greatly improve both the amount and statistical quality of information generated.

Although automatic image analyzers have been commercially available for several years, their metallurgical use has been limited mostly to optical metallography. Although some success has been reported with relatively simple TEM images (1,2), the complexity of electron diffraction contrast images has frustrated most attempts to apply these instruments to TEM studies. Image processing and analysis, however, is a rapidly advancing field, spurred mainly by military and space-related applications, so that it is reasonable to assume that automated analysis will be applied to increasingly complex images. The aims of this paper are to examine a number of visual, digital, and coherent optical image analysis methods in relation to the special problems presented by TEM of defects in solids, and to indicate areas where additional development can provide useful advances in quantitative data extraction. The discussion will cover an overview of TEM defect analysis, an examination of visual, digital, and optical image analysis methods, and finally, a summary of the author's views on potentially useful advances in image processing techniques for these applications.

TEM Investigations of Lattice Defects

The study of defect clusters by TEM can be conveniently divided into four general categories: (1) defect identity and character; (2) size distribution; (3) number density; and (4) spatial distribution. The first category of investigation includes most of the qualitative aspects of the subject, and as noted by Frank (3), can be conveniently subdivided into deductive and inductive techniques. The deductive techniques assume that there is a simple relationship between the projected mass density distribution of the specimen and the image contrast. This approach is useful with weakly scattering biological objects and, in materials science, with relatively electron-opaque inclusions or precipitates in a more transparent matrix. The inductive methods, required for most lattice defects in crystalline solids, involve the solution of wave-optical or wave-mechanical equations for the diffraction of electrons in the vicinity of an assumed defect configuration and matching the calculated contrast profile and symmetry with observed images (4-6). This matching, sometimes performed for a number of diffraction conditions, is used to determine defect type, symmetry, strength, etc.

The second type of investigation, determining the size distribution of a set of features, requires the measurement of the distance(s) between equivalent points on each image to obtain a size or sizes that can be related to one or more characteristic dimensions of the physical defect producing that image. Imaging conditions must therefore be chosen to provide predictable relationships between contrast features and the defects producing them, and the contrast at the measuring points must be sufficiently localized (sharp) to allow accurate measurement. These requirements are fairly easily satisfied for many mass-thickness contrast features, but the image-defect relationship is often more complex and a strong function of imaging conditions for diffraction contrast features such as dislocation loops (6,7) and phase contrast features, such as small voids (8,9).

The final two categories, number density and spatial distribution, require information about the spatial dimension normal to the foil surface. This is commonly obtained by a stereo pair of micrographs, as illustrated in Fig. 1, which convert depth differences into parallax. By measuring the parallax of surface features, the thickness of a specimen region (and thus its volume) can be determined. Note that the equation given implies symmetric tilting of the foil normal about the electron beam direction. The parallax of each defect, along with its planar position, can be used to reproduce a complete three-dimensional model of the defect spatial distribution. Ideally, the contrast of each important feature should be identical in the two micrographs, but it must at least be sufficiently similar to allow matching. Furthermore, the relation between image contrast and physical feature must not be significantly different between the two micrographs, or false parallax could result. Finally, a proper accounting must be made of the fraction of each type of feature that will be visible for the imaging condition employed.

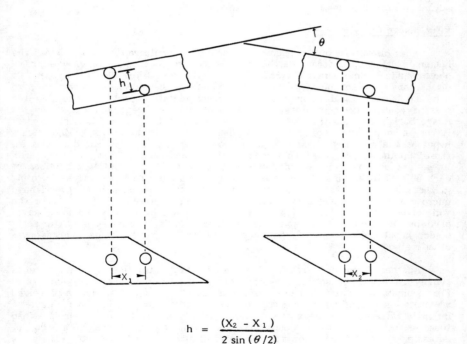

$$h = \frac{(X_2 - X_1)}{2 \sin(\theta/2)}$$

Fig. 1 - Stereo determination of depth position in electron microscopy. Two micrographs are taken of the same area at different angles of specimen tilt, and the resulting parallax of objects is used to determine relative depth.

A further consideration in TEM image analysis is the stereology of the situation - how the projection through a thin foil can be related to its three-dimensional structure. The stereological formulae available for analyzing projections through foils of finite thickness are not as general as those for planar sections, as encountered in optical metallography (10). To solve this problem, the microscopist normally resorts to one of two approaches: reducing the problem to a planar section case, as in determining dislocation density by counting intersections with random lines on the micrograph; or attempting to count and measure the individual defect projections, detecting overlapping images by their symmetry, or by working with a stereo pair. Rhines (11) correctly notes that, when determining number per unit volume, these methods are not satisfactory for all conceivable cases. In many cases of interest in microstructural studies, however, visually determining number density from transmission electron micrographs works quite well, although overlap compensation is a much more severe problem for computer image analysis, as will be discussed later in this paper. Two pathological cases that do present difficulties are very high number densities (12), and microstructural features of similar size to the foil thickness, although correction factors have been worked out for some similar biological electron microscopy problems (13).

Image Analysis Methods

Once suitable micrographs showing the features of interest have been obtained, a technique must be found to extract quantitative data from them. The available methods can be classified as visual, digital, and optical analysis, along with various combinations of these basic three. These techniques and their advantages and shortcomings will be discussed in the following sections.

Visual Image Analysis

In a discussion of techniques for analyzing photographs of any kind, the human visual system forms the baseline, since the most common purpose of a photographic or micrographic system is to create images for the human eye. To discuss any form of micrographic data reduction, therefore, it is necessary to consider human visual perception. The fundamental perceptive process in the present context is segmentation - the separation of image features from the image background and from each other. As discussed by Gregory (14), this process is very complex, the interpretation of visual data being based on an individual's hypotheses about how objects ought to look. In viewing the world around us, these hypotheses are based on the total of our experiences living in that world. If we are shown a picture of familiar objects in which significant portions of each object's image are obscured or missing, we usually have no difficulty filling in the missing features and perceiving the objects. Likewise, if an experienced microscopist is analyzing a micrograph of voids in an irradiated metal, he or she will often have little difficulty recognizing overlapping images and images partially obscured by background fluctuations. This inherent ability to apply object hypotheses is one of the major strengths of visual image analysis over other methods.

Another important aspect of visual feature extraction is that it relies heavily on edges, or spatially rapid changes in brightness, texture, color, etc. For example, Cornsweet (15) has shown that an image of a disk can be perceived when its only contrast is a sharp postive-negative one at its edge, with no intensity difference between the center of the disk and the background. When there is no sharp edge contrast however, and only a gradual lightening or darkening of the intensity toward the center of the disk, it is not seen. The same phenomenon is observed with void and bubble contrast in metals (8,9), for which the "best-focus" image is actually slightly underfocused, so that the phase contrast sharpens the image edges. In the context of quantitative micrograph measurement, this edge effect implies that an individual will tend to determine

56

the size of images by the distances between points with high local derivatives of intensity. These high-derivative points are thus the ones to be related to points on the physical objects being studied.

A variety of instrumentation is available to aid the visual analysis of micrographs. For measuring image sizes, a simple calibrated hand-lens offers good accuracy, but operator fatigue after measuring a large number of features can lead to unreliable results (as well as extreme boredom). Two types of instruments are commonly used to ease this problem. The first uses a variable-diameter light spot that is matched to some characteristic dimension on each feature. When the operator depresses a foot switch, the feature is marked, usually with a pin-hole, and the size of the light spot is recorded, either in one of a series of counters representing size ranges or directly in some form of computer memory. This system of size measurement is well suited to many radiation damage studies, in which large numbers of fairly regular objects, such as voids or dislocation loops, must be analyzed. The primary requirement for the use of this type of equipment is that a measurement strategy be worked out for each situation prior to the actual analysis, so that the image data can be related to the physical defect distribution in a consistent manner. With defects of regular shape, measuring a characteristic distance, such as edge length, diameter of inscribed circle, or diameter of circumscribed circle, will produce more usable data, with less variation from one operator to another than criteria such as estimated equivalent area of the measuring circle and the defect projection.

The second type of semi-automated visual analysis equipment consists of a digitizing tablet with a light pen or cursor, with which an operator can digitize points on a micrograph. The digitizing tablet is interfaced with a microcomputer, which is programmed to measure the positions of the digitized points and to perform a number of data reduction tasks. This type of equipment is not very convenient for measuring size distributions of individual objects, but it can be very useful for determining many stereological parameters, such as line intersections, boundary lengths, etc., The most commonly encountered TEM materials science problem that falls into this category is the determination of dislocation density by counting intersections with random test lines.

The second type of measurement required for quantitative TEM, that of the parallax between stereo pairs, can likewise be pursued by a variety of means. The simplest, and most tedious, of these methods, is the direct measurement of the distances of features, in the direction perpendicular to the tilt axis, from a common reference point. To more conveniently make these direct measurements, Minter and Piller (16) have recently described a technique using a graphic digitizer and a desk-top computer which automatically records these data and computes the three dimensional distribution of objects.

In addition to direct measurement of parallax, it is possible to present one image of the stereo pair to each eye in some form of stereo viewer so that the observer has a visual impression of the three-dimensional distribution. The stereo effect can be made quantitative by inserting a reference point into each half of the optical system. By providing one reference point with a calibrated movement along the direction of parallax, the operator can be given the impression of movement of the reference point in depth. The most common systems for introducing this floating reference point are the parallax bar, which is suspended slightly over the photographs, and the floating light spot stereom-eter, which projects two small lights into the optical system. By recording the relative parallaxes of a series of image features, their relative depths can be determined. The principal difference between this type of measurement and the method discussed in the previous paragraph is that the stereocope method takes advantage of visual depth perception, which is very accurate in many individuals.

While the parallax can be recorded manually as the readings on a micrometer barrel, this method becomes very tedious when the spatial distribution of a large number of features is to be determined. "Photogrammetric stereo compilation," which is the term for analyzing photographs to obtain full quantitative three dimensional descriptions of objects, has been highly developed in the fields of aerial reconnaissance and topographical mapping (17). A number of rather elaborate stereo plotters and stereo comparators have been commercially developed, but much of the complexity of these instruments is related to rectification of the photographs, correcting for variations in angular view over each photograph of a stereo pair. The TEM stereo situation, however, is much simpler in that each photo is at a constant angular view, and the angles are relatively small, approximately $\pm 6^\circ$ in the TEM situation, in contrast to the aerial mapping situation where acceptance angles for the lenses commonly used are 57 to 122°. Therefore somewhat less complicated equipment can be employed for TEM stereo compilation.

Simplified semi-automatic stereo analyzers suited to the TEM situation have been assembled in several laboratories and commonly take the form of a commercial sterescope with the addition of linear variable differential transformers (LVDT's) which measure the planar position of the stereo pair and the parallax distance representing the depth coordinate of the floating light spot (18,19). The outputs from these LVDT's are fed through an analog-digital converter into a small computer for reduction into x, y, and z coordinates of each measured point. Since the necessary calculations are very simple, and the speed is limited by the operator, a very simple desk-top computer suffices very well, provided it has a suitable data input port and a reasonable mass-storage medium, such as magnetic tape or disk. Another useful feature is a plotter or graphics display device to produce visual records of the measured distributions to determine, for example, the average tilt of foil surfaces (16).

A diagram of an automated visual analysis system constructed in the author's laboratory is shown in Fig. 2. The system consists of a mirror stereoscope. similar to the one described by Thomas (18) and a variable-light-spot particle size analyzer, modified with a precision potentiometer to give a voltage ouptut, with both instruments connected to a desk-top computer with output peripherals. The system has proven to be very flexible for analysis of voids, precipitates, and dislocation loops. A particular advantage realized by instrumenting the size analyzer is that, with an individual size stored for each defect, size distribution histograms are not arbitrarily dictated by the classes built into the analyzer. This gives the investigator greater flexibility in comparing size distributions from different micrographs and in setting the statistical significance of each size class. An additional benefit of the instrumented stereoscope has been its use for analyzing diffraction patterns. In this application, the plate or film is placed in one side of the stereoscope and the x-y movement of the table is used to measure diffraction ring, spot, and streak patterns rapidly, and without risk of damage to the negative.

Before leaving the subject of visual micrograph analysis, it is useful to consider the reliability and reproducibility of the visual techniques. As a part of the bcc ion correlation experiment (20), void size distributions were measured by eight laboratories on identical copies of a common micrograph, shown in Fig. 3. The micrograph was deliberately chosen for its good contrast and sharp focus. The void size measurements were fairly reproducible, varying by $\pm 4\%$ around the average. The void number density, on the other hand, varied from the average by +10% to -15%. This latter spread was especially interesting since the same specimen thickness was assumed by each laboratory, and the same formula was used to calculate number density. It implied that, among experienced microscopists working on a good micrograph, there was a significant difference in interpretation of what was and was not a void. This variability should be considered in evaluating the performance of any fully automatic system.

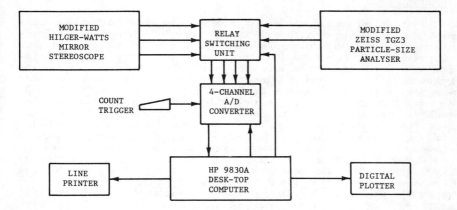

Fig. 2 - Semi-automatic visual micrograph analysis system, as assembled in the author's laboratory, incorporating computer aided recording and reduction of visually measured size and stereo position data.

Fig. 3 - Transmission electron micrograph of voids in ion-irradiated molybdenum. The micrograph was used in an interlaboratory comparison of size and number density measurement. The micrograph illustrates the problems of contrast variation and image overlap, commonly encountered in TEM analysis.

Digital Image Analysis

Even with automation of the data recording step, visual quantitative micro-graph analysis can become a time-consuming process. The situation is especially severe in that unskilled personnel cannot be used without risking large errors, since differences between defect images and artifacts are often subtle, and some exper-ience is required to separate the two. An appealing solution to this dilemma is to use the micrographs themselves as input to a computer, suitably programmed to recognize defect images, measure their sizes, and determine their positions. As mentioned previously, efforts to do this with diffraction contrast have been less than notably successful, so that it is of interest to examine digital image processing and analysis to determine the reasons for these failures and to look for developments that would improve the situation. The field of digital image processing has developed rapidly over the last decade, and several recent texts give good descriptions of its general aspects (21-23) and its application to transmission electron microscopy (24,25), although the microscopy-related literature is primarily concerned with image enhancement rather than the type of quantitative analysis being discussed here. The approach in the following section will be to discuss some of the fundamentals of digital image analysis as they relate to the specific problems of TEM of lattice defects, and then to briefly describe currently available equipment for these applications.

In any digital image processing or analysis scheme, the initial process is the conversion of an image to digital form, which is accomplished by dividing it into picture elements, or pixels, as shown schematically in Fig. 4. The area to be converted is divided into N lines, each of which is subdivided into n pixels. Each pixel is assigned an integer representing its gray level, which is stored as a byte, or portion of a digital word. Depending on the precision desired, the byte representing each pixel can be any number of bits long, although the most common lengths are 5 bits (32 levels), 6 bits (64 levels), and 8 bits (256 levels). Although 256 gray levels are far more than can be distinguished by the human eye, this high number is often useful in such processing operations as contrast stretching and various mathematical trans-formations. The common spatial divisions of digital images are 512 x 512 pixels and 1024 x 1024 pixels, which implies a considerable volume of data per picture, since a 512 x 512 image at 64 gray-level resolution consists of over 1.5×10^6 bits. This large amount of data per picture implies that data storage and transmission are important considerations.

In a photographic image, on the other hand, the information is stored at a higher spatial resolution (grains of emulsion) with fewer gray levels per element. If we consider a simplified case, with 1 micron grains, each of which is a binary storage element (black or white), then a 70 mm square negative would contain approximately 5×10^9 individual bits of data. The situation is slightly more complicated in the case of electron micrographs, since the effective "grain size" can be induced by local statistical fluctuations in electron exposure (26), but the fundamental difference between digital and photographic images remains, and can affect the relative results of visual and digital analysis methods. Figure 5, an enlargement of a small area of Fig. 3, illustrates the point. The photographic grain can be seen, along with a grid representing a 512 x 512 pixel division of the previous figure.

The next step in digital image analysis is known as segmentation, dividing the image into non-overlapping areas on the basis of one or more detectable properties. This step fulfills the same purpose as the segmentation process discussed under Visual Image Analysis - the separation of image features from each other and from the background. The two general approaches to digital segmentation are region methods, which identify the image segments as connected sets of pixels showing a given range of a property, such as gray level, texture, or color, and boundary methods which detect region boundaries as connected paths of high gradients in a given propoerty. In the absence of color information, the two properties that are useful for this initial segmentation are gray level and texture. Of these two, gray level is computationally

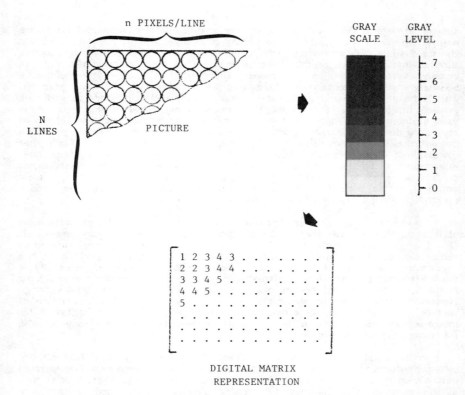

GRAY SCALE   GRAY LEVEL

7
6
5
4
3
2
1
0

n PIXELS/LINE

N LINES

PICTURE

```
⌈ 1 2 3 4 3 . . . . . . . ⌉
| 2 2 3 4 4 . . . . . . . |
| 3 3 4 5 . . . . . . . . |
| 4 4 5 . . . . . . . . . |
| 5 . . . . . . . . . . . |
| . . . . . . . . . . . . |
| . . . . . . . . . . . . |
| . . . . . . . . . . . . |
⌊ . . . . . . . . . . . . ⌋
```

DIGITAL MATRIX
REPRESENTATION

Fig. 4 – Schematic of the process of image digitization. The image is divided into picture elements, or pixels, each one of which is assigned an integer based on its average gray level. The picture is stored as a matrix of these integers.

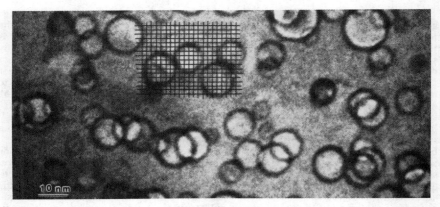

10 nm

Fig. 5 – Enlargement of an area of Fig. 3, illustrating the difference in digital and photographic image storage. The photographic grain can be seen, along with a grid representing a 512 x 512 pixel division of the micrograph in Fig. 3.

61

simpler to detect, since, in its simplest region-method form, individual pixels are assigned to one of two types of regions based on whether they fall above or below a threshold value. A somewhat more complex region method, known as adaptive thresholding, allows the threshold level to be a slowly varying function of position to compensate for changes in background. Differential or gradient thresholding, a boundary method, detects region boundaries as paths of high gray-level gradient. This technique can be best implemented by processing the input image to produce a gradient image and then thresholding that image. A boundary-following algorithm is used to detect closed curves of high gradient, and these boundaries are defined as separating image segments. Boundary methods of segmentation are, in general, more computationally complex than region-threshold methods. Referring to the earlier discussion of visual perception, however, it is apparent that the boundary techniques more nearly match the edge sensitivity of the human visual system. Furthermore, contrast sharpness is an important cue used for separating TEM defect images from artifacts, so that a well-designed boundary method would be more likely to produce useful digital segmentation of transmission micrographs.

Segmentation on the basis of texture separates a scene into regions on the basis of local patterns of gray level variation, rather than the absolute gray level itself (21). A simple example of this would be separating areas of grass from an area of concrete on a black and white photograph. Even if the average gray levels in the two regions are exactly the same, one would have no trouble separating them visually by their different textures. Although texture is used less than gray level in analyzing electron micrographs, one texture example related to diffraction contrast is fringe contrast of some types of planar precipitates. The mathematical description of texture is much more complex than that of gray level, and the use of texture for image segmentation is a recent development. It does, however, have some promise for future image analysis applications.

In the initial segmentation of an image, two or more types of features, for example defects and surface artifacts, may be detected, and these must be separated before counting and measurement. In such a case, some other characteristic(s), such as size or shape criteria, must be used to classify the detected features into appropriate groups. Any number of mathematically describable classifiers can be used in principal, although the computation time does increase rapidly with the number of classifiers. A good description of classifier training and operation is given by Castleman (23).

Following feature detection and classification, the final steps (which will often be performed concurrently with classification) are counting and measuring the features of interest. The only necessary comment on this phase of the analysis is that the measurement strategy should be worked out to use the mathematically simplest size description that can be reliably related to the physical defect dimensions. An obvious example of this principle is that the maximum caliper diameter (equivalent to the diameter of the circumscribed circle) is much simpler to measure digitally than the maximum diameter of an inscribed circle, which might be easy to measure visually.

To examine the above digital image processing fundamentals as they apply to a specific TEM case, consider the voids in Mo shown in Fig. 3 and enlarged in Fig. 5. First, the obvious characteristic to use for detection of the voids is the dark ring surrounding each void image, rather than the contrast in the center of each image, since that central contrast varies considerably from void to void. Second, either adaptive or differential thresholding would probably be required to allow for the variation in background. Third, looking at Fig. 5, it appears that a 512 x 512 pixel division of the area in Fig. 3 would not provide sufficient spatial resolution, since this division should be significantly smaller than the feature being detected, in this case the outer ring, rather than the entire void image. Fourth, the few dislocation images would have to be rejected on a shape criterion specific to the desired annular features. Finally, a technique would have to be devised to properly count and measure overlapped images. This last requirement is the most difficult one to fulfill with current equipment and software, since relatively little work has been done in digital analysis of overlapped images, and algorithms have only been developed for a few special cases (27). If a large number of micrographs similar to Fig. 3 had to be analyzed, however, a pattern recognition routine could be developed to separate these overlaps, since they are readily recognizable.

Turning to the subject of digital image analysis instrumentation, the currently available equipment falls roughly into two categories: commercially available packaged image analysis systems, exemplified by the Cambridge Instruments "Quantimet," Bausch and Lomb "Omnicon," Joyce-Loebel "Magicscan" and several others; and general-purpose image processing systems, primarily custom-built (23,24,28), although a few commercial systems are now being sold. Both types of systems operate on the general principles outlined above, the primary differences being complexity and flexibility. The basic components of a digital image analyzer are shown schematically in Fig. 6. The image is scanned by a video camera and displayed on a monitor while simultaneously being converted to digital form and fed sequentially to the controller which can be made up of hard-wired special function modules, a software-controlled microprocessor or mini-computer, or some combination of these. The total image is not stored at any one time in the controller, but rather is sampled by it. The controller also is able to superimpose its output on the monitor image to indicate the results of its operations: thresholding, shape classification, etc. In some systems a light pen is used by the operator to communicate with the processor for manual feature selection and other tasks.

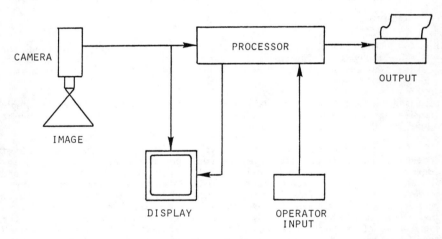

Fig. 6 -   The elements of a basic digital image analysis system for the extraction of quantitative information.

An example of a general-purpose digital image processing system is shown schematically in Fig. 7. Depending on the specific application, several devices might be used for input of image data, such as a video camera, video tape, scanning microdensitometer, etc. The image is commonly not displayed directly, but rather is recorded through the control computer onto a mass memory device such as a magnetic disk. For display, the image is again transferred from the disk through the central processor to a storage display device. Processing control information is supplied by the operator through a keyboard and a light pen or cursor. A normal processing sequence is to operate on an input image by loading pixels into the central processor serially as needed (not necessarily storing the entire image in the core memory at any one instant), performing the necessary operations, and storing the

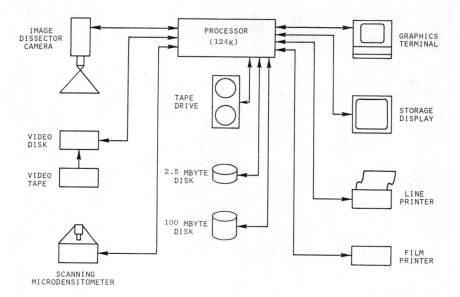

Fig. 7 -   The elements of a general-purpose digital image processing and analysis laboratory, having capibilities to mathematically transform images to enhance features of interest, as well as extract quantitative information.

results on an output image or data file. A processed image is viewed by simply transferring that file to the display. The final results of a processing session can be output by either a line printer (data) or a film printer (images).

From the above brief descriptions, it should be obvious that, barring cost considerations, the trade-off between the two types of systems is likely to be the convenience of a system specifically designed for data extraction vs the power and flexibility of a general-purpose image processing laboratory. In their present states of development, the commercial analyzers lack some important capabilities, such as overlap detection, and image differentation, and Fourier transformation and filtering, that would facilitate digital analysis of many diffraction contrast images. The shortcoming of the larger systems, however, is that specific analysis routines would have to be developed for each type of defect contrast before one of these advanced systems would be a cost-effective routine analysis tool. The commercial analyzers are continuing to become more sophisticated, however, and the advanced image processing systems are being assembled at an increasing number of laboratories, so that the necessary software for digital analysis of TEM diffraction contrast images is principally awaiting a sufficiently large single application to pay for its development.

Optical Image Processing and Analysis

Optical image processors make use of the Fourier transform properties of lenses to enhance images and extract information from them. The general principles of these systems have been well described in the text by Goodman (29), while Horne and Markham (30) have reviewed their applications to electron microscopy. Although the components of a basic optical image processor, equivalently called an optical diffractometer, can be arranged in several different ways, the main principles can be

64

discussed with reference to the layout shown in Fig. 8. First, light from a laser with beam expander uniformly illuminates a transparency of the input micrograph. The first lens forms a Fourier transform, or optical diffraction pattern, of the input image in the back focal plane of the lens. Either a film, to record the transform, or a filter, to modify it, is placed at this plane. When a filter is used, a second lens reconstructs the enhanced image, which is then recorded at the back focal plane of this lens.

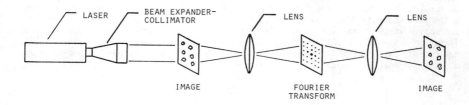

Fig. 8 -   The basic elements of a coherent optical image processor, or optical diffractometer. The first lens forms a Fourier transform of the input image, which can either be recorded itself or spatially filtered and retransformed by the second lens to form a reconstructed enhanced image.

While the Fourier transform contains the same information as the input image, the transform is organized in terms of spatial frequencies, rather than real space coordinates. This frequency form of information organization can be convenient for extracting average quantities from images, but when one records the transform as intensities on film, part of the information, carried as relative phases of different portions of the transform, is lost. This loss of information is exactly analogous to the phase problem faced in X-ray diffraction and scattering experiments. In certain cases, however, the remaining information is sufficient and in a convenient form to extract the data of interest. One example of such an application is determining the periodicities of objects such as lattice fringes (31) or periodic defect arrays. Some information can also be extracted from Fourier transforms of non-periodic objects (32), but these techniques have not been applied to many TEM image analysis problems. The optical transform has also proven valuable in high-resolution studies, to analyze the imaging conditions and resolution of the microscope (31). In addition to the optical methods, it is possible to record the Fourier transform directly in the microscope, as an electron diffraction pattern (33), from which much of the same information can be extracted. Depending on the characteristics of the specific electron microscope, however, it may not be possible to suitably control the analysis area and the diffraction camera length.

If a filter is placed in the back focal plane of the first lens, the amplitudes and/or phases of portions of the transform can be altered to enhance features of interest in the reconstructed micrograph. This type of processing has been applied to biological studies of low-contrast periodic objects (34), as well as lattice image studies in materials science (35,36), but the technique also has promise as a pre-processing step for general quantitative image analysis. By suitable filtering, such as d. c. blocking to remove the average background, desired features could be enhanced and artifacts removed to ease the problems of digital analysis. All of these transform steps can of course be done digitally, but where suitable filters can be readily constructed, optical techniques offer significant advantages in time and cost. An important consideration in any image processing operation involving spatial filtering of the Fourier transform is that care must be exercised to avoid introducing any spurious contrast that could invalidate the subsequent analysis of the reconstructed image (37).

A somewhat more complex type of optical/digital processor, which has some intriguing possibilities for TEM image analysis, is the optical stereo correlator, which is being developed to provide automatic stereo compilation for aerial mapping. (38). Conceptually, the principle behind these systems is to produce maps of the correlation between two images of a stereo pair as a function of displacement of one of these images along the direction of parallax. The displacement which produces the maximum local correlation of an image point represents the depth of that point. For a description of the operating principle of optical/digital correlators, the reader should refer to Balasubramanian's review (38). For the purposes of the present discussion, the important point is that at least two such experimental prototype systems, the Image Matched Filter Correlator and the Heterodyne Optical Correlator, have shown promising results in contour generation from aerial photographs. Their use for analysis of TEM stereo pairs should be examined to determine if such equipment can operate as successfully with transmission images of overlapping objects and with the types of image structure produced by diffraction contrast.

Summary

After reviewing the currently available methods for image analysis in relation to the special problems of TEM lattice defect studies, it appears that automated digital and optical systems do offer potentially large improvements in speed of data generation over presently used visual methods with computer-aided data recording. Because of the complexity of diffraction and phase contrast images, and the great capabilities of human visual perception, however, designing a digital system to out-perform visual micrograph interpretation is a very complex task. Given the impressive task that current-generation digital processes can successfully perform, it is likely that presently available hardware could be adequate for TEM applications. The need that has not yet been fulfilled is the development of a set of analysis software specifically oriented to the characteristics of TEM lattice defect images. With the continually improving capabilities of the commercial image analysers, the prospects that this development will occur are fairly good. A strong impetus for the further commercial development of both fully digital and computer-aided visual image analysis equipment should be provided by the growing use of scanning transmission electron microscopes, which produce their primary image information as electrical signals, which may be readily digitized and processed. Except in applications such as periodic images, the coherent optical diffraction techniques do not appear to offer many advantages for the direct extraction of detailed quantitative data in TEM investigations, but the optical processing techniques do have potential applications in efficiently pre-processing micrographs to remove artifacts and enhance features of interest for subsequent analysis. Finally, the optical correlation techniques, which are still at an early stage of development, have interesting future possibilities for automatic stereo defect analysis. Whatever visual, digital, or optical methods are used to extract detailed quantitative data from electron micrographs, this phase of microscopy investigation is likely to remain a time-consuming and expensive proposition for the near future.

References

1. B. Ralph and A. R. Jones, "Instrumentation for Quantitative Microscopy," pp. 206-209 in Proceedings of the Thirty-fifth Annual Meeting of the Electron Microscopy Society of America, G. W. Bailey, ed.; Claitor's Publishing Division, Baton Rouge, LA, 1977.

2. F. A. Heckman, E. Redman, and J. E. Connally, "Progress in the Practical Application of Automated Image Analysis to TEM Negatives," pp. 214-217 in Proceedings of the Thirty-fifth Annual Meeting of the Electron Microscopy Society of America, G. W. Bailey, ed.; Claitor's Publishing Division, Baton Rouge, LA, 1977.

3.   Joachim Frank, "Image Analysis in Electron Microscopy", Journal of Micros-
      copy, 117 (1) (1979) pp. 25-38

4.   A. K. Head, P. Humble, L. M. Clarebrough, A. J. Morton, and C. T. Forwood,
      Computed Electron Micrographs and Defect Identification; North Holland,
      Amsterdam and American Elsevier, Inc., New York, 1973.

5.   M. J. Whelan, "Dynamical Theory of Electron Diffraction," pp. 43-106 in Dif-
      fraction and Imaging Techniques in Materials Science, 2nd ed., S. Amelinckx,
      R. Gevers, and J. Van Landuyt, eds.; North Holland, New York, 1978.

6.   M. Wilkens, "Identification of Small Defect Clusters in Particle-Irradiated
      Crystals by Means of Transmission Electron Microscopy," pp. 185-215 in Dif-
      fraction and Imaging Techniques in Materials Science, 2nd ed., S. Amelinckx,
      R. Gevers, and J. Van Landuyt, eds.; North Holland, New York, 1978.

7.   D. J. H. Cockayne, "The Weak-Beam Method of Electron Microscopy," pp.
      153-184 in Diffraction and Imaging Conditions in Materials Science, 2nd ed., S.
      Amelinckx, R. Gevers, and J. Van Landuyt, eds.; North Holland, New York,
      1978.

8.   M. Ruhle and M. Wilkens "Defocusing Contrast of Cavities, I. Theory,"
      Crystal Lattice Defects, 6, (1975) pp. 129-140.

9.   J. A. Sprague, "Void and Bubble Size Measurement: The Effect of Focus
      Condition," pp. 324-325 in Proceedings of the Thirty-fifth Annual Meeting of
      the Electron Microscopy Society of America, G. W. Bailey, ed.; Claitor's
      Publishing Divison, Baton Rouge, LA, 1977.

10.   E. E. Underwood, "Quantitative Stereology and Image Analysis," pp. 202-205
      in Proceedings of the Thirty-fifth Annual Meeting of the Electron Microscopy
      Society of America, G. W. Bailey, ed.; Claitor's Publishing Division, Baton
      Rouge, LA, 1977.

11.   F. N. Rhines, "Microstructure-Property Relationships in Materials," Met.
      Trans. 8A (1977) pp. 127-133.

12.   J. H. Chute and J. G. Napier, "Image Overlap in Transmission Electron
      Microscopy," Philosophical Magazine 10 (1964) pp. 173-176.

13.   M. A. Williams, Chapter 2 of Quantitative Methods in Biology, Vol. 6, pt II of
      Practical Methods in Electron Microscopy, A. M. Glauert, ed.; North Holland,
      Amsterdam, 1977.

14.   R. L. Gregory, The Intelligent Eye; McGraw-Hill Inc., New York, N. Y., 1974

15.   T. N. Cornsweet, Visual Perception, p. 273; Academic Press, New York, N. Y.
      1970.

16.   F. J. Minter and R. C. Piller, "A Computerized Graphical Method for
      Analyzing Stereo Photomicrographs", Journal of Microscopy 117, Pt. 2 (1979)
      pp. 305-311.

17.   C. D. Burnside, Mapping from Aerial Photographs; John Wiley and Sons, New
      York, N. Y., 1979.

18.   L. E. Thomas "Quantitative Steroscopy in the HVEM," pp. 460-461 in
      Proceedings of the Thirtieth Annual Meeting of the Electron Microscopy
      Society of America, G. W. Bailey, ed.; Claitor's Publishing Division, Baton
      Rouge, LA, 1972.

19. L. E. Thomas and S. Lentz, "Stereoscopic Analysis of Irradiation-Induced Voids," pp. 362-363 in Proceedings of the Thirty-Second Annual Meeting of the Electron Microscopy Society of America, G. W. Bailey, ed.; Claitor's Publishing Division, Baton Rouge, LA, 1974.

20. J. L. Brimhall, "Summary Report of the BCC Ion Correlation Experiment," pp. 241-251 in Proceedings of the Workshop on Correlation of Neutron and Charged Particle Damage, Oak Ridge National Laboratory, 1976, J. O. Stiegler, ed.; CONF-760673, 1977.

21. A. Rosenfeld and A. C. Kak, Digital Picture Processing; Academic Press, New York, N. Y., 1976.

22. W. K. Pratt, Digital Image Processing; John Wiley and Sons, New York, N. Y. 1978.

23. K. R. Castleman, Digital Image Processing; Prentice-Hall, Inc. Englewood Cliffs, N. J., 1979.

24. W. O. Saxton, Computer Techniques for Image Processing in Electron Microscopy; Academic Press, New York, N. Y., 1978.

25. D. L. Misell, Image Analysis, Enhancement, and Interpretation; Elsevier/-North Holland Biomedical Press, Amsterdam, The Netherlands, 1978.

26. G. C. Farnell and R. B. Flint, "Exposure Levels and Image Quality in Electron Micrographs," Journal of Microscopy 3 (1975) pp. 319-332.

27. K. Preston, Jr. "Digital Picture Analysis in Cytology," pp. 209-294 in Digital Picture Analysis, A. Rosenfeld, ed.; Springer-Verlag, New York, N. Y., 1976.

28. R. M. Wilson, D. L. Teuber, J. R. Walkins, D. T. Thomas, and C. M. Cooper, "Image Data-Processing System for Solar Astronomy," Applied Optics 16, No. 4 (1977) pp. 944-949.

29. J. W. Goodman, Introduction to Fourier Optics; McGraw-Hill, Inc., New York, N.Y., 1968.

30. R. W. Horne, and R. Markham, "Application of Optical Diffraction and Image Reconstruction Techniques to Electron Micrographs," pp. 325-434 in Practical Methods in Electron Microscopy, Vol. 1, A. M. Glauert, ed.; North-Holland/-American Elsevier, New York, N. Y., 1972.

31. R. Gronsky, R. Sinclair, and G. Thomas, "Optical Microdiffraction in Lattice Image Analysis," pp. 494-495 in Proceedings of the Thirty-fourth Annual Meeting of the Electron Microscopy Society of America, G. W. Bailey, ed.; Claitor's Publishing Division, Baton Rouge, LA, 1977.

32. H. J. Pincus, "Optical Diffraction Analysis in Microscopy," pp. 19-71 in Advances in Optical and Electron Microscopy, Vol. 7, V. E. Cosslett and R. Barer, eds.; Academic Press, New York, N. Y., 1978.

33. R. W. Carpenter, J. Bentley, and E. A. Kenik, "Small-Angle Electron Scattering in the Transmission Electron Microscopy," Journal of Applied Crystallography 11 (1978) pp. 564-568.

34. H. P. Erickson, W. A. Voter, and K. Leonard, "Image Reconstruction in Electron Microscopy, Enhancement of Periodic Structure by Optical Filtering," p. 39 in Methods in Enzymology, Vol. XLIX, Enzyme Structure, Part G, C.H.W. Hirs and S. N. Timasheff, eds.; Academic Press, New York, N. Y., 1978.

35.   J. Desseaux, C. D'Anterroches, J. M. Penisson, And A. Renault, "Optical Filtering of Images of Dislocation Core," pp. 310-311 in Ninth International Congress of Electron Microscopy, J. M. Sturgess, ed.; Microscopical Society of Canada, Toronto, Canada, 1978.

36.   J. M. Penisson and A. Bourret, "High-Resolution Study of [011] Low-Angle Tilt Boundaries in Aluminum," Phil. Mag. A40 (6) (1979) pp. 811-824.

37.   C. A. Taylor and J. K. Ranniko, "Problems in the Use of Selective Optical Spatial Filtering to Obtain Enhanced Information from Electron Micrographs," Journal of Microscopy 100 (3) (1974) pp. 307-314.

38.   N. Balasubramanian, "Optical Processing in Photogrammetry," pp. 119-149 in Optical Data Processing, D. Casasent, ed.; Springer-Verlag, New York, N.Y., 1978.

HIGH SPATIAL RESOLUTION CHEMICAL MICROANALYSIS DAMAGE STUDIES

David C. Joy

Bell Laboratories
Murray Hill, New Jersey   07974

The addition of x-ray and electron energy loss spectrometers to a transmission electron microscope allows high sensitivity, quantitative chemical
microanalysis to be carried out at a spatial resolution approaching that of
the normal image mode. This paper describes the technique of x-ray microanalysis, its sensitivity, spatial resolution and quantitation. The complementary method of electron energy loss spectroscopy is then discussed.
It is shown that taken together these two techniques enable detailed quantitative information on the chemical and physical nature of the sample to
be obtained from specimen volumes of the order of 10^{-18} cm^3.

Introduction

A valuable adjunct to high resolution microstructural studies of damaged materials is the ability to determine the chemistry of the sample, if possible, at about the same level of spatial resolution. This chemical microanalysis should be able to both identify what elements are present in any given area of the sample, and measure the concentrations of these elements. Such an analysis could then be applied to either the bulk composition of the specimen, or to the identification of second phases, precipitates or inclusions in the matrix of the sample.

The techniques of imaging and microanalysis at high resolution can be coupled by using a transmission electron microscope equipped with spectrometers to detect and analyze the fluorescent x-rays and the characteristic energy losses which occur as a result of interactions between the electron beam and the sample in the microscope. Figure 1 shows the arrangement of the specimen incident electron beam, and the spectrometers in such a device which could be either a conventional 100 keV transmission electron microscope (TEM) equipped with a scanning attachment (STEM), or else a dedicated STEM. In both cases the electron probe is focused to a diameter of between 5 and 50 nm and is scanned over the sample so as to generate a high resolution transmission image of the sample. This probe can also be used to select any region of interest for analysis. With modern commercial instruments using tungsten thermionic cathodes, the current density into this probed area is of the order of 20 A/cm^2, corresponding to an incident beam current of about 10^{-10} A at the sample. On instruments using a field emission electron source current densities as high as 10^4 A/cm^2 can be obtained.

With the instrumental configuration shown in Fig. 1 either, or both, of the spectrometer systems can be operated while an image is being collected. The microanalytical data obtained can thus be placed in the context of the microstructure of the specimen. In general, any specimen that is thin enough to yield a good image will also be suitable for high spatial resolution microanalysis, but because the requirements for the energy dispersive x-ray (EDS) and electron energy loss (ELS) spectrometer are different, the

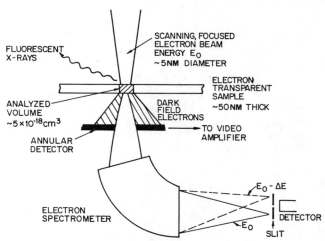

Fig. 1 – Schematic layout of the components of an analytical electron microscope equipped with X-Ray and Electron Energy Loss Spectrometers.

limitations placed on the sample in this respect will be discussed in more
detail later.

Energy Dispersive X-Ray Microanalysis

The high energy incident beam of electrons ionizes some of the atoms as
it passes through the sample. These ionized atoms subsequently decay back to
their ground state, and in so doing a fraction w produce characteristic x-
rays which can then be collected by a suitable detector. The energy carried
by any x-ray photon produced in this process represents the difference in
binding energy between two allowed states of the atom. This energy is uniquely characteristic to a particular element, and so a measurement of the
energy (or the equivalent wavelength) of the photon is sufficient to unambiguously identify the element from which it came. In the usual arrangement
the x-rays are detected by a lithium-doped silicon diode, placed in line of
sight of the sample (see the sketch in Fig. 2). The detector is typically of
about 20 mm² active area and placed at 15 to 20 mm from the incident beam
axis. The x-rays which enter the detector diode produce a charge pulse whose
magnitude is directly proportional to the energy of the photon. This pulse
can then be amplified, shaped and stored by a multichannel analyzer (MCA) to
yield an energy spectrum of the x-rays produced by the sample. Because as
many as 10^4 photons/s can be processed in this way, all the x-rays produced
by the specimen and accepted by the detector are effectively analyzed
simultaneously (1, 2).

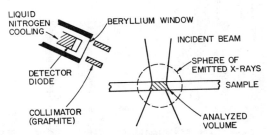

Fig. 2 - Geometrical arrangement of the x-ray detector and specimen in an electron microscope.
Typically, the detector only collects 1% of
the x-rays produced from the thin sample.

The number N_x of x rays per second *detected* from each atom of any element present in the volume of the sample irradiated by the beam is

$$N_x = J \cdot \sigma \cdot w \cdot \text{Efficiency} , \qquad (1)$$

where J is the incident electron flux (electrons/cm²/s),
   σ is the appropriate ionization cross section (cm²/atom), and
   w is the fluorescent yield for the x ray collected.

For all elements with an atomic number greater than three the product
(σ·w) is about constant, indicating that the number of x-rays produced per
atom is also constant. However, the efficiency with which these x-rays can
be detected varies considerably as we move through the periodic table. This
"efficiency" of collection and detection is controlled by two factors:

1. The x-rays are emitted uniformly into a spherical distribution from
the thin foil, so that only those traveling directly towards the detector

diode and its collimator are collected. Taking the dimensions quoted before as a guide, the x-ray detectors subtend at most 0.1 sterad at the sample, and the maximum collection efficiency is thus $0.1/4\pi$ (i.e., about 1%).

2. In order to keep out the high energy electrons (typically 100 keV) from the beam, as well as other photons in the visible range, the detector is usually protected by a beryllium window 8 to 10 μm thick. In addition, there are gold layers on the silicon to make connections to form a Schottky barrier diode. The effect of these metals is to absorb soft x-rays so that, for photons of 1 keV energy or less, the effective detection efficiency falls rapidly to zero. At the other end of the spectrum, high energy photons (10 keV or more) can pass right through the diode, again causing a fall in the detection efficiency. The overall response of the detector is sketched in Fig. 3. Within the optimum energy range (2 to 10 keV), x-rays from all elements above magnesium in the periodic table can be detected.

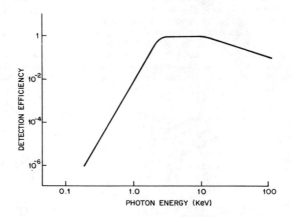

Fig. 3 - The efficiency of detection for x rays in the energy range 0 to 20 keV using a Si-Li detector equipped with the standard 8 μm beryllium window.

Assuming a conventional CTEM or CTEM/STEM microscope using a tungsten thermionic cathode, an incident flux of 20 A/cm^2 can be focused onto the specimen. If an analysis time of 100 s is assumed, then the minimum mass (MDM) of an element that can be detected can be calculated. Figure 4 shows the result of such a calculation assuming a detector accepting 0.1 sterad from the specimen (3) and a 100 keV incident beam. While this is an idealized calculation because it deals only with pure elements, it indicates the very high mass sensitivity available over much of the periodic table. It is clear, however, that for elements below Z of 11 or 12 the sensitivity is falling rapidly because of the absorption in the beryllium window. Recently "windowless" x-ray detectors have been made available to try and overcome this problem. Although these detectors do require very careful operation if satisfactory results are to be obtained, the removal of the beryllium window does make a dramatic difference in the sensitivity as can be seen from the dotted line on Fig. 4. Although some decrease in sensitivity still occurs (because of the absorption in the gold and silicon dead layers) a mass sensitivity of the order of 10^{-19} g (equivalent to the mass of a 3 nm sphere of carbon) is maintained down to about Z = 5.

In the presence of other elements, the sensitivity of x-ray analysis will be less because of the continuum (Bremsstrahlung) background contributed by the matrix in which the element of bremst is sitting. The amount by

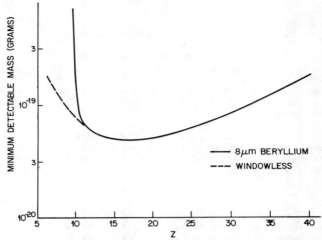

Fig. 4 – The minimum detectable mass of elements of atomic number Z
using energy dispersive x-ray analysis. The calculation
assumes a detector accepting 0.1 sterad from the specimen,
a 100 keV incident electron beam containing 200 A/cm^2 and
a recording time of 100 s. The solid line is for the conventional detector using an 8 µm window, the dotted line
for a "windowless" configuration.

which the MDM is increased will depend on the distribution of the elements
within the matrix, as well as on the composition of this matrix. The optimum
case is that in which the element of interest is present as an inclusion,
precipitate or second phase particle, and with a thickness close to that of
the foil. In such a case the electron beam can be focused down so as only to
irradiate the area required, and the MDM will then be close to the values
predicted from Fig. 4 because the conditions just described ensure that the
mass fraction of the element in the volume analyzed is close to unity. In
other less favorable conditions (such as a homogeneous distribution of the
element), where the mass fraction is substantially less than unity, detection
may only be possible if the beam is spread so as to irradiate the greatest
volume of material and maximize the x-ray production from the desired component. High spatial resolution microanalysis is thus not compatible with
the detection of trace amounts of an element, except when the trace is highly localized.

The electron beam is scattered as it passes through the sample and,
because a 100 keV electron has to lose a very high fraction of its energy
before it is no longer able to excite an x-ray, the volume within which x-
rays are produced is therefore generally larger than that defined by the
incident probe diameter and the thickness of the foil. While an exact calculation of the effective spatial resolution for x-ray microanalysis requires
a Monte-Carlo simulation (4) to a good approximation (5) the point-to-point
resolution $\underline{b}$ that is attainable in the EDS mode is given by the expression

$$\underline{b} = 6.24 \times 10^5 \left(\frac{Z}{E_0}\right)\left(\frac{\rho}{A}\right)^{1/2} t^{3/2} \qquad (2)$$

where E_0 is the accelerating voltage in eV,
    ρ is the density in g/cm^3,
    A is the atomic weight,
    Z is the atomic number, and

$\underline{t}$ is the thickness
$\underline{b}$ is the resolution ($\underline{t}$ and $\underline{b}$ are in centimeters)

Substituting typical values for 100 keV operation shows that a resolution of 5, 8, and 20 nm could be achieved in 100-nm-thick foils of carbon, aluminum, and copper, respectively. To achieve this level of performance, however, the background of stray electron and x-ray production contributed by the electron microscope itself must be completely suppressed. This topic is discussed in detail elsewhere (1, 6).

So long as the foil is "thin" the number of x-ray photons produced is proportional only to the variables defined in Eq. (1), and a quantitation can therefore readily be performed. The relative concentrations of two elements A and B will be given by

$$C_A/C_B = N_A/N_B \cdot k_B/k_A \qquad (3)$$

where N_A, N_B are the x-ray counts recorded from the elements A and B in the same time period and the terms k_A and k_B include the cross section, fluorescence yield and detection efficiency terms of Eq. (1). These "k-factors" (k_A, k_B, --- etc.) can either be determined experimentally (1) or calculated from first principles (3). In many practical situations, however, there is appreciable absorption of the emission of one element by another and this must be accounted for even though the foil is thin (1,6). In such cases the simple relationship of Eq. (3) is used to give approximate concentrations and an iterative method is then used to take account of absorption effects.

Absorption effects are, of course, more severe when the foil is thicker, but this effect must be balanced against the increased yield of x-rays from a thicker specimen, and the poorer spatial resolution obtained. Considering all these conflicting requirements, it is reasonable to suggest that the optimum sample thickness for EDS studies at 100 keV is about 50 to 100 nm. Increasing the accelerating voltage to 200 keV or higher will improve the spatial resolution attainable in a thick foil, but the problems caused by absorption will remain unchanged.

Electron Energy Loss Spectroscopy

While the x-ray technique is ideal for many applications, Electron Energy Loss Spectroscopy (ELS) is increasingly being used as a complementary technique for chemical microanalysis. ELS offers not only a very sensitive, quantitative technique for light element (Li to Cu) analysis, but in addition provides much other valuable information on the physical and electronic state of the sample.

ELS analysis is performed by passing the electron beam which has been transmitted through the specimen into an electron spectrometer which deflects the electrons by an amount proportional to their energy (see the sketch in Fig. 1). Typically a magnetic prism is used which produces a line focus dispersion of the incident beam. By placing a selecting slit and electron detector in the plane of the line focus, electrons of any given energy can be selected, or by scanning the spectrum perpendicular to the slits, the energy spectrum can be recorded sequentially and stored in a multichannel analyzer.

In practice, it is the energy loss of any electron that is important, rather than its absolute energy, so the spectrum has its zero at the incident beam energy E_0. The energy resolution of the spectrum is determined, to first order, by the slit spacing $\underline{d}$ and the dispersion $\underline{s}$ (in cm per eV) of the prism. Typically $\underline{s}$ is 4 µm per eV at 100 keV, so for a slit spacing of 20 µm an energy resolution of 5 eV can be obtained. Depending on the type of information required from the energy loss spectrum, resolutions of between 1 and 20 eV are useful (7).

Figure 5 shows a typical ELS display recorded from a thin foil of silicon at 100 keV. Each channel dot represents an increment of 1 eV of energy loss. The large peak at the left-hand side of the spectrum is the zero-loss peak containing those electrons which have passed through the sample without losing any energy. This peak is the most prominent feature of the specimen and typically contains 50 to 95% of the entire intensity of the spectrum. The width of this peak (full width at half height) is approximately equal to the energy resolution of the spectrometer. Other than defining the zero of the spectrum, this zero loss peak contains little useful information.

To the right-hand side of the zero-loss peak, that is at higher energy loss, three peaks (indicated 1, 2, 3) of decaying intensity can be seen. These are the "Plasmon Peaks" which measure the energy loss suffered by the incident electron as it passes through the sample and sets the free electrons into a collective excitation. The frequency W_p of these plasmon oscillations is proportional to $(n_E)^{1/2}$, where n_E is the number of free electrons/cm^3 involved in the oscillation. The energy lost E_{PL} in exciting such an oscillation is $E_{PL} = hW_p$ where h is Planck's constant. Since for most metals $W_p \sim 10^{16}$ rd/s, E_{PL} is about 20 eV. If the sample is thick enough the incident electron may generate a second plasmon giving it a total loss of $2E_{PL}$ and so on.

In many metals and alloys, such as the silicon shown here, the plasmon peaks are sharply defined. Because the plasmon loss E_{PL} depends on the free electron density, it can be used as a microanalytical technique (8). Unfortunately all the materials which give good plasmon spectra have plasmon peaks in the range 12 to 25 eV so the unambiguous identification of an unknown is

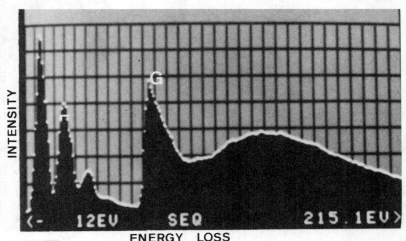

Fig. 5 -   The energy loss spectrum recorded from a thin foil of silicon
           showing the zero loss peak, plasmon peaks (labeled 1, 2, 3)
           and the Si L_{23} edge at 99 eV loss. The discontinuity at $\underline{G}$
           is a gain change in the recording system. Each channel repre-
           sents an energy increment of 1 eV.

77

impossible. But the plasmon loss technique can be very usefully applied to the study of alloying and segregation by making an initial calibration of plasmon shifts with concentration, and then using this data on an unknown composition. A precision of better than 0.5% is possible, with a spatial resolution of 5 to 10 nm. The actual value is limited because of the delocalized nature of the plasmon excitation.

The plasmon loss E_{PL} can also be affected by the geometrical nature of the sample. In the case of a very thin specimen, or one with a high density of voids, a surface plasmon can be detected. This has an energy loss ($E_{PL}/\sqrt{2}$). The appearance of a surface plasmon peak can therefore be used as a way of measuring the local concentration of voids, and by using the spectrometer as an energy filter tuned to this surface loss, high resolution in-focus images of the voids can be obtained (9).

Beyond the plasmon loss region the spectrum starts to fall very rapidly with increasing energy loss, and at some point a gain change (marked as $\underline{G}$ on Fig. 5) is required to maintain the signal level at an acceptable value. Shortly after this the smoothly falling background is seen to be interrupted by a discontinuity. This is the silicon-L_{23} ionization "edge." In order to ionize an atom a certain amount of energy is required, and this can only be obtained from the incident electron. The energy spectrum of the transmitted electrons therefore shows characteristic "edges" at losses corresponding to the classical binding energies. Since these binding energies are as unique a property of the element as its x-ray lines, they can equally well be used for the purposes of elemental identification.

The chemical identification is made by noting both the energy loss of the edge and the shape of the edge. The simplest to identify are K-edges which arise from the ionization of 1 s electron shells. Figure 6 shows the

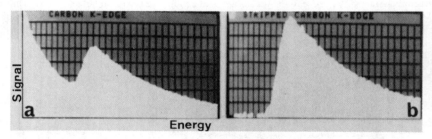

Fig. 6 – The carbon $\underline{K}$ edge (at 284 eV) before (a) and after (b) the removal of background to show the true edge shape.

appearance of the carbon K-edge at 284 eV loss before and after the characteristic background is subtracted away. The edge is seen to have a sharp leading edge followed by a smooth, rapid decay of intensity, and this "triangular" shape is found in all K-edges. With a typical electron spectrometer covering the energy loss range 0 to 2000 eV K-edges can be identified from the elements Li to Si. The range of elements that can be examined can be extended by using other ionization edges. The L_{23} edge (shown in Fig. 7 before and after background stripping) has a more gentle slope with the maximum intensity occurring well after the onset of the edge. Similar profiles can be obtained from the M_{45} and N_{45} edges (10). In practice it has been found safest to rely on K and L_{23} edges for identification, and this enables elements between Li and Cu in the periodic table to be detected.

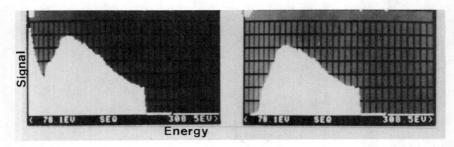

Fig. 7 – The silicon L_{23} edge (at 99 eV) before (a) and after (b) the removal of background.

The sensitivity of microanalysis using ELS is potentially very high especially for light elements. The detected signal I_{EG} is given as

$$I_{EG} = J \cdot \sigma \cdot \text{Efficiency (counts/s/atom)} \qquad (4)$$

where J is the incident electron flux (electons/cm^2/s) and σ is the ionization cross section. By comparison with the equivalent x-ray expression [Eq. (1)] the yield is seen to be improved by a least a factor of $\underline{w}$. Since for light elements $\underline{w}$ falls as Z^4, being only about 5×10^{-3} for carbon although close to 1 for iron, this means that ELS should be very sensitive for light elements compared to EDS. Secondly, the efficiency with which ELS events are collected and detected is much higher than in the EDS case. This is because the electrons which have suffered an inelastic collision are still traveling in a narrow forward cone. It is usually possible to detect between 25 and 75% of all these electrons, compared to the best 1% figure for x-rays. These advantages are balanced by the fact that ELS is a sequential analysis, so that only a small fraction of the total recording time is spent on each part of the spectrum, instead of recording the whole spectrum in parallel as is done in EDS.

Nevertheless, a high sensitivity can be achieved. Figure 8 shows the calculated miminum number of atoms of various elements between Li and Ca that can be detected in a 50 nm thick stainless-steel foil at 100 keV assuming a 30 nm diameter probe with a current density of 10 A/cm^2 and a total recording time of 100 s. As few as 10^3 atoms of lithium (about 0.06% by number) or 3×10^4 atoms of oxygen (about 1%) are seen to be detectable. The attainable spatial resolution is only limited by the probe diameter needed to obtain sufficient incident beam current, so in microscopes using field emission guns the same sensitivity can be achieved at resolutions of 1 to 3 nm. Unlike the EDS case there is no intrinsic "beam spreading" limit.

The thickness of the specimen is important in determining the quality of the energy loss spectrum. Figure 9 shows a K-edge from nitrogen ion-implanted into silicon to a concentration of 10%. In the first case, where the foil thickness was 10 nm the edge is prominent and of the expected shape, but in the second case, where the foil was 50 nm thick, the edge is both less pronounced and is visibly distorted. If the thickness of the sample becomes too great (typically 150 nm at 100 keV) the edge can disappear completely. Unlike the x-ray case, therefore, the best specimen for ELS is always the thinnest. Thicker samples can only be examined by increasing the accelerating voltage of the microscope.

Techniques for the quantitation of ELS data are now well developed (11). Because of the dominant background on which the edges rise, it is necessary to use computer modeling techniques to strip the edges, and to calculate the

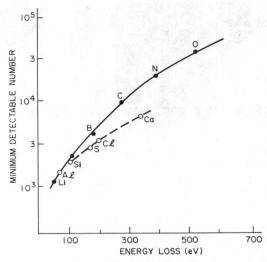

Fig. 8 – The calculated minimum number of atoms of elements between Li and Ca that can be detected in a 50 nm thick stainless-steel foil at 100 keV using ELS. A 30 nm diameter probe containing 10 A/cm^2 is assumed with a resolution of 10 eV and a recording time of 100 s.

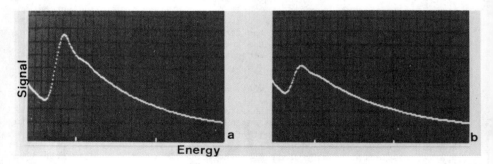

Fig. 9 – The appearance of the K-edge from nitrogen ion implanted into silicon for sample thicknesses (a) of 10 nm and (b) of 50 nm showing the effect of multiple scattering.

ionization cross-section required. Using a typical small computer interfaced to an MCA (i.e., PDP 11-03 and an ORTEC EDS II) a complete quantitation can be performed in less than a minute for all the K- and L-edges of interest using the QUEL program developed at Bell Laboratories (12).

The conventional ELS technique described above can only be used for elements above lithium in the periodic table, but information about the presence of hydrogen and helium can be obtained indirectly using the method sketched in Fig. 10. Here the electrons transmitted through the specimen are divided into two groups. The first comprises those that have been unscattered or elastically scattered — that is, those which have lost no energy. The majority of these elastic electrons form the wide-angle scattered "Dark Field" signal which can be collected with an annular detector as shown. The rest can be collected through the electron spectrometer. The total elastic signal

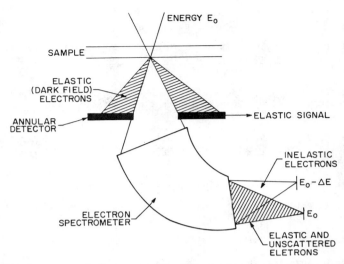

Fig. 10 – The arrangement for measuring the elastic to inelastic scattering
ratio of a material to determine the presence of hydrogen and
helium.

I_E is proportional to $Z^{3/2}$, where Z is the weighted average of the atomic
numbers of the elements in the irradiated volume.

The second group of electrons are the inelastically scattered component.
This is concentrated in the forward direction and can be separated from the
elastic and unscattered electrons by the electron spectrometer. The total
inelastic signal I_{IE} is proportional to $Z^{1/2}$. A ratio I_E/I_{IE} is thus
directly proportional to Z, the mean atomic number of the sample. This
simple expression has been found to work well in practice (13) and it has
been used to demonstrate the loss of hydrogen, due to radiation damage in a
polymer, at the 3×10^{-15} g level. In situations where the geometry of the
specimens is adequately characterized, this technique offers a possible way
of identifying and quantifying the presence of hydrogen or helium in voids,
etc., although the plasmon-loss technique may be more direct (17).

It is worth noting that much other information is also obtainable from
energy loss spectra. In addition to the chemical data discussed above, the
complex dielectric coefficient (14) can be determined and the electronic band
structure can be investigated (15). Finally, extended fine structure (EXAFS)
information can be obtained to give a detailed atomic model of the sampled
volume at a resolution of the order of 0.01 nm (16). Since all of this extra
information can be generated from the same volume of the sample, a highly
detailed model of any selected area of the specimen can be obtained.

Conclusion

Taken together, EDS and ELS techniques make it possible to characterize
the chemistry, as well as the electronic and physical detail, of specimen on
a scale of spatial resolution approaching that possible in a normal electron
micrograph. All elements in the periodic table can be detected, and for most
elements a sensitivity of the order of 10^{-19} g is possible. The development
of EXAFS in ELS will make it possible to understand the structure of disordered and amorphous regions in much greater detail.

References

1. J. Goldstein, "Principles of Thin Film X-Ray Microanalysis, pp. 83–117 in Introduction to Analytical Electron Microscopy, Hren, Goldstein, and Joy, eds.; Plenum Press, New York, 1979.

2. E. Lifshin, M. F. Ciccarelli, and R. B. Bolon, "X-Ray Spectral Measurements,"pp. 263–297 in Practical Scanning Electron Microscopy; Plenum Press, New York, 1975.

3. D. C. Joy and D. M. Maher, "Sensitivity Limits for X-Ray Analysis," Vol. 1, Proc. 10th Ann. SEM Symposium; IITRI, Chicago, pp. 325–335, 1977.

4. D. F. Kyser, Ref. 1, ibid., pp. 199–221.

5. J. I. Goldstein, J. L. Costly, G. W. Lorimer, and S.J.B. Reed, Ref. 3, ibid., pp. 315–324.

6. N. J. Zaluzec, Ref. 1, ibid., pp. 121–168.

7. D. C. Joy, Ref. 1, ibid., pp. 223–246.

8. D. B. Williams and J. W. Eddington, "High Resolution Microanalysis in Materials Science Using ELS," J. Microscopy 108 (1976) 113–45.

9. B. Jouffrey, "ELS with Special Reference to HVEM," pp. 29–46 in Short Wavelength Microscopy; New York Academy of Sciences, New York, 1978.

10. D. M. Maher, Ref. 1, ibid., pp. 259–291.

11. D. C. Joy, R. F. Egerton, and D. M. Maher, "Progress in the Quantitation of ELS," Vol. 2, Proc. 12th Ann. SEM Symposium, SEM, Inc., Chicago, 1979.

12. D. C. Joy and D. M. Maher, "The Quantitation of ELS Data," J. Microscopy, 124 (in press).

13. R. F. Egerton, "Measurements of Inelastic/Elastic Scattering Ratio," Phys. Stat. Sol. (a), 37 (1976), 663–71.

14. J. Daniels, C. V. Festenberg, H. Raether, and D. Zeppenfeld, Optical Constants of Solids; Springer Verlag, Berlin, 54 (1970) 77–135.

15. R. F. Egerton and M. J. Whelan, "ELS and the Band Structure of Diamond," Phil. Mag. 30 (1974) 739–46.

16. P. E. Batson and A. J. Craven, "EXAFS in ELS," Phys. Rev. Lett. 42 (1979) 893–95.

17. N. J. Zaluzec, T. Schober, and D. G. Westlake, "Application of EELS to the Study of Metal Hydrogen Systems," pp. 194–95 in Proc. 39th Ann. Meeting EMSA, G. W. Bailey, ed.; Claitor's Publishing Division, Baton Rouge, 1981.

EDX MICROANALYSIS OF NEUTRON-IRRADIATED ALLOYS*

L. E. Thomas

Hanford Engineering Development Laboratory
Richland, Washington

Energy-dispersive X-ray (EDX) spectrometry of 50 nm thick specimens in the scanning transmission electron microscope provides quantitative elemental analyses of selected regions as small as 20 nm in diameter. To analyze highly radioactive neutron-irradiated alloys it is necessary to reduce the high counting deadtimes caused by energetic γ-Compton scattering in the Si(Li) detector, and to account for spurious background contributions from γ-rays and characteristic X-ray emissions. Several simple methods for overcoming effects of specimen radioactivity are described, including use of a tungsten collimator to attenuate γ and X-rays coming from the thick edges of self-supporting disk specimens. These methods allow analyses of Fe-Cr-Ni based alloys with γ-activities up to 1000 μC_i. Techniques used to maintain high spatial resolution and accuracy in quantitative analysis are also described, and their use is illustrated with an example of solute concentration profiling at individual voids and dislocations.

*This work was sponsored by the Office of Fusion Energy, U. S. Department of Energy.

Introduction

In the few years since energy-dispersive X-ray spectrometers were first adapted to transmission/scanning transmission electron microscopes (TEM/STEMs), EDX microanalysis has developed rapidly as a means for chemical analysis at the microstructural level of dislocations, voids and small second-phase particles. EDX microanalysis as currently practiced provides quantitative analyses accurate to about 0.5 wt. % for all elements with Z >11 (sodium). The best spatial resolution is now about 20 nm, but improvements in electron source brightness and X-ray take-off geometry, reduced specimen contamination, and better control of specimen thickness should soon reduce this limit to about 5 nm. Also, "ultrathin window" (UTW) detectors developed recently for TEM/STEMs extend detectability limits to elements as light as carbon (1). EDX microanalysis is one of several complimentary microstructural characterization methods available in modern analytical transmission microscopes. Others include electron energy loss spectroscopy for microanalysis of light elements, convergent and microbeam diffraction for crystal structure determination from small crystal volumes, and conventional TEM imaging (2).

Several areas in which EDX microanalysis is used to characterize irradiated alloys are second-phase identification, measurement of local compositions, and measurement of concentration profiles at voids, dislocations, grain boundaries, and other point defect sinks. The techniques are similar to those developed for unirradiated materials. However, several special problems arise in analyzing highly radioactive specimens, such as Fe-Cr-Ni based alloys irradiated to high neutron fluences for nuclear energy applications.

This paper describes several methods developed to deal with the effects of specimen radioactivity in EDX microanalysis. It also describes techniques for quantitative microanalysis and recent applications to neutron irradiated alloys.

EDX Microanalysis

Microanalysis employs the simple strategy shown in Figure 1 to obtain characteristic X-rays from very small specimen volumes. The incident electron beam (probe) is focused to a small spot, typically less than 10 nm in diameter, on a thin region of the specimen selected by STEM mode imaging. Ideally, the probe diameter and beam spreading in the specimen determine the volume of material which generates X-rays. However, the resolution attained in practice also depends on hydrocarbon contamination buildup under the electron beam, relative drifts between specimen and beam, spurious radiation reaching the specimen and detector, and the specimen thickness along the incident beam direction. Beam spreading due to inelastic scattering of electrons in the specimen is the most important fundamental limitation, and must be minimized by selecting suitably thin specimen regions (3). In addition, the minimum probe diameter is limited by the need to provide enough electron signal for acceptable X-ray counting statistics over times set by the specimen drift rate. High electron intensity is achieved by using a convergent electron beam, using high brightness electron emitters such as LaB_6, and choosing the largest probe diameter consistent with the resolution sought. X-ray spectra produced in analytical TEM/STEMs have very low X-ray continuum backgrounds because of favorable X-ray take-off geometry. Also, the strong magnetic field of the objective lens in these microscopes prevents backscattered electrons and other charged particles from reaching an X-ray detector placed in the objective lens gap region.

The EDX detector is a lithium-drifted silicon crystal that converts the entire energy of most incident photons into electron-hole pairs; these are

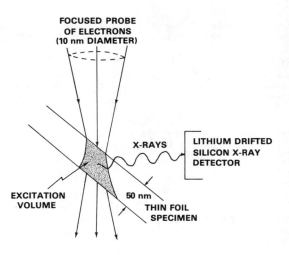

FOCUSED PROBE
OF ELECTRONS
(10 nm DIAMETER)

LITHIUM DRIFTED
SILICON X-RAY
DETECTOR

X-RAYS

EXCITATION
VOLUME

50 nm

THIN FOIL
SPECIMEN

Fig. 1 – EDX microanalysis geometry in a scanning transmission electron
microscope.

collected by a charge-sensitive preamplifier and amplified as voltage pulses
whose heights are proportional to the observed photon energies. The voltage
pulses are then digitized and counted by pulse height in a multichannel
analyzer to produce histogram displays of counts versus energy channel.

Radioactivity in Neutron-Irradiated Alloys and Its Effects on EDX Detection

When alloys are exposed to neutron bombardment, they form radioactive
nuclides that decay spontaneously by emitting charged α and β particles,
neutrons, and various forms of electromagnetic radiation including γ and
X-rays. Most of these nuclides decay to insignificant levels within a few
weeks after the material is removed from the neutron environment and, after
a year or so, one or two long-lived nuclides produce most of the radioactivity. For example, in AISI 316 stainless steel, an alloy commonly
used in nuclear reactor core components, the most prominent γ-emitters
three months after exposure in a fast reactor are ^{54}Mn (316 day half-life),
^{182}Ta (115 days), and ^{58}Co (70.8 days). After a year, ^{60}Co (5.27 years)
produces about 95% of the γ-activity and ^{54}Mn about 5% (A. T. Luksic,
Westinghouse Hanford Company, private communication).

For EDX microanalysis, the problems associated with specimen radioactivity arise from energetic γ-rays and characteristic X-rays reaching the
detector. The magnetic field around the specimen traps β and probably also
α particles; neutron and α emissions are not significant in the alloys
being studied; and the protective window between specimen and detector
stops light.

Prominent X-ray emitters in neutron-activated Fe-Ni-Cr based alloys
are ^{55}Fe (2.7 years), ^{54}Mn (312 days), and ^{58}Co (70.8 days). These decay
by capturing their own orbital electrons; the ionized daughter nuclides
subsequently emit characteristic X-rays. For example, ^{55}Fe produces
characteristic manganese X-rays. Another decay process that produces
characteristic X-rays, internal conversion, involves de-excitation of a
nucleus by ejecting an orbital electron. In this case, the X-rays are
characteristic of the original nuclide. However, internal conversion is not

an important source of X-rays in Fe-Cr-Ni alloys. Table I lists γ and X-ray emitters in activated Fe-Ni-Cr based alloys.

Table I. Important γ and X-Ray Emitters in Neutron Bombarded Fe-Ni-Cr Based Alloys

Nuclide	Half-Life	Origin	Principal γ-Energy (MeV)	X-Rays (by Electron Capture)
^{54}Mn	312 d	^{54}Fe (n,p)	0.835	Cr
^{58}Co	70.8 d	^{58}Ni (n,p)	0.811	Fe
^{60}Co	5.27 y	^{59}Co (n,γ) ^{60}Ni (n,p)	1.172 1.332	None
^{55}Fe	2.7 y	^{54}Fe (n,γ) ^{58}Ni (n,α)	No γ	Mn
^{182}Ta	115 d	^{181}Ta (n,γ)	1.289 1.374 1.453	None

Characteristic X-rays due to electron capture or internal conversion come from the entire specimen and typically produce high enough count rates in the EDX detector to require consideration. Since their relative contribution to an EDX spectrum increases for very thin specimens, one could, for example, be fooled into thinking that very small precipitates in irradiated stainless steel, which contains ^{55}Fe, are rich in manganese. In most EDX microanalyses, however, the X-ray contribution from radioactive decay is small compared to the X-ray signals produced by the electron beam.

The main problem in EDX microanalysis of radioactive alloys is high counting deadtimes caused by energetic γ-Compton events in the Si(Li) detector. At the energies of interest, 0.5 to 1.5 MeV, γ-rays interact with atoms mainly by Compton scattering, by which the incident photon energy is transferred to silicon orbital electrons. Energy transfers involving most of the incident energy are favored (4), and the Compton-scattered electrons convert most of their energy to electron-hole pairs in the detector crystal. The preamplifier in EDX detectors has as its first stage a charge-sensitive field-effect transistor (FET) from which the charge collected from successive X-ray detection events must be drained to maintain its performance (5, 6). The FET is allowed to accumulate charge up to a tolerable voltage shift, and is then discharged by momentarily triggering a light-emitting diode that causes photo-conduction by the FET. During the relatively long time required to reset the FET, the amplifier does not operate.

The normal "reset charge" for a pulsed-optical preamplifier is 1 MeV (expressed as energy collected as electron-hole pairs). Thus, only one or two γ-Compton events in the detector can cause a reset, and relatively low γ-count rates produce high counting deadtimes. For example, the "activity" spectrum shown in Figure 2 from a highly radioactive Fe-Cr-Ni alloy (about 500 μC) was taken (with no electron beam) at 70% deadtime. Yet the total X-ray count rate due to events with energies in the 0.2 to 10 keV region was only 350 cps. The deadtime in this case is caused by pulsed-optical reset of the preamplifier rather than by pulse rejection in the amplifier.

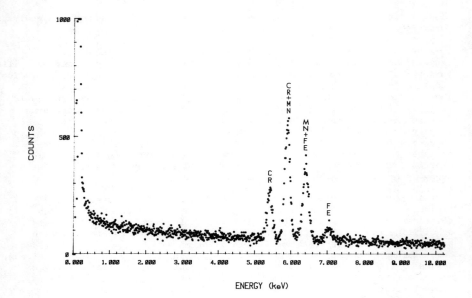

Fig. 2 – Activity spectrum obtained with no electron beam from a 500 µC$_i$
neutron-irradiated alloy specimen. The X-ray peaks result from
electron capture decay of ^{54}Mn, ^{55}Fe, and ^{58}Co.

Figure 2 also shows characteristic chromium, manganese, and iron X-ray
peaks due to electron capture nuclides, and a background "tail" which
increases at low energies. The background tail apparently occurs by escape
of Compton-scattered photons from the detector crystal. Continuum X-rays
due to radioactive decay are a negligible source of background in the
spectrum shown in Figure 2 but could be important for specimens with high β
emission.

Experimental Details

The analyses described in this paper were performed in a 100 kV TEM/STEM,
JEOL Model 100CX, equipped with a LaB$_6$ electron source and a minicomputer-
controlled electron beam scanning and EDX data acquisition system. Two
Si(Li) EDX detectors were used interchangeably. The standard detector had
a 30 mm^2 active area, active depth of 3.5 mm and a beryllium protective
window with nominal thickness of 7.5 µm. This detector was fitted with a
removable tungsten collimator designed to attenuate radiation from the
thick edges of self-supporting specimens. The ultra-thin window (UTW)
detector had a composite window of 0.1 µm thick parylene (C_8H_8) and 0.15 µm
thick aluminum to allow detection of elements as light as carbon. The
active area of the UTW detector crystal was 10 mm^2, its thickness was 2 mm,
and a knife-edge collimator was built-in (7). Neither detector was intentionally optimized for analysis of radioactive specimens.

High spatial resolution in EDX microanalysis is obtained by operating
the microscope in the STEM mode with a highly convergent, 8 to 10 nm diameter, electron beam. This range of probe diameters provides adequate image
resolution to position the beam probe on selected specimen features at
magnifications up to 150,000X. With emission currents of 20 to 30 µA, it

also yields the X-ray count rates needed for statistically acceptable analyses during 60 to 100 second data acquisition times.

Besides probe diameter, the factors limiting spatial resolution are beam spreading in the specimen, specimen contamination and instrument stability during analysis. Beam spreading calculations for 100 keV electrons in medium atomic weight alloys indicate that the specimen thickness along the incident beam direction must be less than 70 nm to restrict spreading to 10 nm (3). Since the specimen must be tilted 35 to 40° from normal incidence to allow X-rays to reach the detector (the detector is in the plane perpendicular to the beam, 45° to the specimen tilt axis), the actual specimen thickness required is less than 5 5 nm. The thin edges of jet electropolished alloy specimens usually include regions thinner than 5 5 nm; in practice, the specimen thickness is determined after analysis using stereoscopy or convergent-beam electron diffraction.

A spatial resolution of about 20 nm in EDX microanalysis has been achieved by using a 10 nm diameter electron beam, limiting the specimen thickness at the analyzed point to a maximum of 50 nm, interrupting data acquisition to correct image drift, and operating under conditions which produce no detectable specimen contamination mark during 60 seconds of data acquisition. Contamination is minimized by using freshly polished specimens, degassing the specimens in the microscope vacuum using a liquid nitrogen cooled shroud around the specimen, and flooding the specimen and holder with electrons for several minutes just before analysis in order to fix surface contaminants in a thin layer. The electron flooding technique prevents surface contamination from concentrating under the focused electron beam during EDX analysis. Specimen heating or cooling are more effective means of preventing contamination buildup but were not used in this work.

Since specimen thickness limitations imposed by the spatial resolution requirement usually preclude significant X-ray absorption and fluorescence in the specimen, quantitative analysis is very simple. Quantitative analysis involves (a) stripping (subtracting channel by channel) an "activity" spectrum acquired with the electron beam directed through a hole in the specimen, (b) determining the number of counts above continuum background in characteristic X-ray peaks, and (c) applying a correction to convert integrated peak intensities into weight fractions.

In practice it is always important to ascertain that absorption and fluorescence effects are indeed negligible, and that the spectra contain no significant contributions from "spurious" radiation sources. The effects of spurious instrumental radiation such as X-rays or strong electrons from the microscope illumination system which fluoresce the entire specimen are widely recognized (2). However, these and other instrumental effects have been virtually eliminated in most modern analytical TEM/STEMs including the one used in the present work. One notable spurious contribution occurred in the present work when the detector was withdrawn several centimeters from its normal working position. In this condition, characteristic copper X-rays from microscope components around the specimen were able to reach the detector because the detector collimator was rendered ineffective. Since the specimens being analyzed did not contain copper, the effect was ignored.

The correction procedure used for semi-quantitative analysis is based on a simple relationship between integrated peak intensities, I, and weight fractions, C, for the case where a characteristic peak is measured for each element present and where the region analyzed is sufficiently thin that absorption and fluorescence are negligible. For this case, the weight fraction of each element is given by the relationship:

$$C_i = k_i I_i / \Sigma k_i I_i \ , \tag{1}$$

where $\Sigma I_i = I_{total}$, $\Sigma C_i = 1$, and the constants k_i can be determined by analyzing large areas of thin specimens whose compositions are known by wet chemical analysis. This is essentially the method of Cliff and Lorimer (8), but differs in using fully normalized data to derive correction factors that do not depend on an arbitrary reference element. Table II gives the experimentally derived k-factors used in this work for the two detectors with a specimen/detector takeoff angle of 40° in the JEOL 100CX microscope. Element concentrations can be determined to within 0.5 to 1 wt. % in Fe-Cr-Ni based alloys by this method. This accuracy is possible because the elements having the greatest uncertainty in k-factor are present in relatively low concentrations in these alloys. K-factors can also be calculated using published data for fluorescent yields, ionization cross-sections, X-ray transmission coefficients, etc. However, the uncertainties in the published values, as well as in the effective thicknesses of the window, gold electrode and silicon dead-layer on the detector, make calculated k-factors much less reliable than those measured with the detector in use. The k-factors in Table II are reliable only for the detectors with which they were measured.

Table II. Experimentally Derived k-Factors

X-Ray Peak	k-Factor	
	Standard 30 mm² Detector	Ultrathin Window Detector
Al K_α	0.91 ±0.1	0.63 ±0.05
Si K_α	0.85 ±0.07	0.62 ±0.04
Ti K_α	0.80 ±0.02	0.76 ±0.03
Cr K_α	0.83 ±0.02	0.81 ±0.03
Fe K_α	0.96 ±0.02	1.93 ±0.02
Ni K_α	1.11 ±0.02	1.08 ±0.02
Mo L	1.71 ±0.20	1.26 ±0.08
Nb* L	1.62 ±0.10	1.32 ±0.04

*Note that the Mo and Nb L X-ray peaks overlap.

Methods for Analyzing Radioactive Alloys

Several methods have been used successfully in dealing with specimen activity in EDX microanalysis; most of these are simple but the techniques are improving as the effects of radiation on EDX detection become better understood and as detector manufacturers find it worthwhile to change standard designs.

The most obvious method (9) to reduce specimen activity is to minimize the specimen volume and to allow the activity to decay as long as possible

before analysis. It is often useful to extract precipitates from highly
radioactive specimens using extraction replica techniques, although extraction introduces its own limitations. Also, specimens irradiated to a low
fluence (2 x 10^{22} n/cm^2, E >0.1 MeV) and allowed to sit for a year or
so can often be analyzed with standard detectors. A second method is to
overpower activity effects by increasing the specimen to detector distance
to the point where system deadtime becomes tolerable, and then increasing
the electron beam current to obtain an acceptable X-ray count rate from the
specimen region of interest. This is often an effective approach; however,
increasing the probe current usually involves increasing the probe diameter
at the cost of spatial resolution and also increases spurious radiation –
i.e., X-rays from apertures, etc. – in the microscope.

An effective collimator between specimen and detector can greatly
reduce the flux of γ and X-rays at the detector by taking advantage of the
specimen geometry. Self-supporting specimens made by jet electropolishing
3 mm diameter disks to perforation at their centers may contain 90% of their
mass in the outer 1 mm periphery. Figure 3 shows the design of a collimator used on a standard 30 mm^2 detector to attenuate radiation from all
but the central 1 mm of centrally-perforated disk specimens. The collimator
is made of tungsten and is designed for a specimen to detector distance of
30 mm, a compromise between effective γ-ray attenuation and X-ray collection
by the detector. The unusual feature is a collimator tip to specimen
distance of only 2 mm. The collimator mounts on the end of the detector
housing tube and must be retracted before exchanging specimens via the
side-entry exchange mechanism on JEOL electron microscopes. Low magnification (10X) SEM mode imaging allows precise collimator alignment. Also, the
specimen holder is cut away to allow close proximity of the collimator to
the specimen.

An important modification not shown in Figure 3 is to shield the sides
of the detector from γ-rays scattered in the specimen chamber of the microscope. Although it might appear necessary only to shield the detector from
γ-rays coming directly from the specimen, about half of the γ-ray flux
reaching the detector from a specimen in the microscope comes from scattering within the specimen chamber. The use of high density collimator material

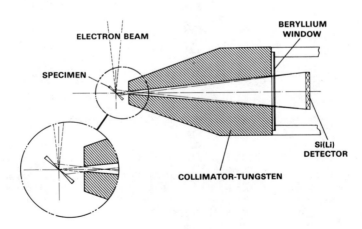

Fig. 3 – Tungsten collimator used to attenuate γ and X-rays emitted from the
outer 2 mm of a 3 mm diameter self-supporting disk specimen.

such as tungsten or platinum is also important; lead has only about half the density of these metals and is considered a poor choice for effective shielding.

Figure 4 shows activity spectra taken with and without the collimator, from a neutron irradiated Ni-Al specimen. Although the specimen received only 3 x 10^{22} n/cm^2 (E >0.1 MeV), it still contained 80 μC_i of ^{60}Co and considerable ^{55}Fe three years after neutron exposure. With the detector 30 mm from the specimen, the collimator reduced the counting deadtime from 41% to less than 10% (below detectability on the front-panel deadtime meter), and reduced the count rate for energies between 0.2 and 10 keV from 390 to 56 c/s. It also reduced the counts in the Mn-k X-ray peak from 10,260 to only 280. However, the background tail is still present.

A bonus from the collimator is that it prevents specimen radiation produced by the microscope illumination system from reaching the detector. Nevertheless, it is important in practice to strip (subtract channel by channel) an activity spectrum taken with the beam going through the hole in the specimen from the spectrum obtained during microanalysis.

Figure 5 illustrates stripping an activity spectrum to obtain a satisfactory EDX analysis from an Fe-Cr-Ni based alloy which had been irradiated to 14 x 10^{22} n/cm^2 (E >0.1 MeV) one year before analysis. In this example, the UTW detector was used. Because of the high specimen activity, 500 μC_i, the detector was withdrawn 44 mm from the specimen to obtain a deadtime of 70%. This procedure produced spurious copper X-ray peaks as noted previously. The electron probe diameter was also increased to 15 nm to produce an electron-excited X-ray signal (1400 c/s) twice that of the activity background. Quantitative analysis of the spectrum produced by the stripping procedure gave reasonable agreement with the known chemical composition of

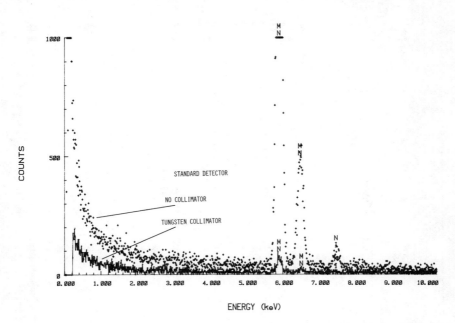

Fig. 4 – Activity spectra from a 80 μC_i neutron-irradiated Ni-Al specimen, showing effect of γ-ray collimator.

91

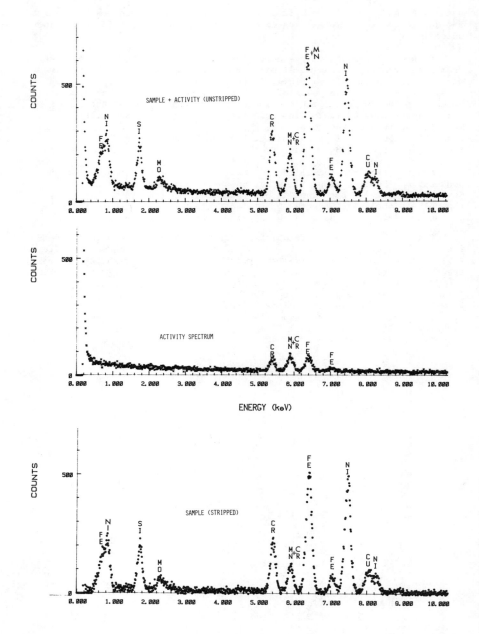

Fig. 5 – Microanalysis of a precipitate particle in a highly γ-active (500 μC$_i$) alloy specimen using ultrathin window detector at 70% deadtime.

the alloy. However, the stripped spectrum was clearly degraded by peak broadening, and attempts to use higher deadtimes produced severe peak broadening.

Modifications of detector electronics may also aid analysis of radio-active specimens. Recently we obtained, by request to the manufacturer, an EDX detector in which the reset charge of the pulsed optical preamplifier was set to 6 MeV instead of the usual 1 MeV. This modification was performed without warming the detector cryostat, and degraded the detector resolution by only 2 eV (from 158 to 160 eV) - a reasonable compromise for our pur-poses. Attempts to further increase the reset charge produced unacceptable peak broadening. Initial tests of this modified detector indicate a corresponding factor of six reduction in deadtime for cases where the deadtime is caused by pulsed-optical reset. Although paralyzation of the detector/preamplifier by energetic Compton events is the greatest problem in analyzing γ-active specimens, peak broadening may also occur in the amplifier/pulse processor. Recently-introduced "time variant" pulse processors can avoid amplifier saturation by detection and avoiding measure-ment of the oversized pulses from high energy photons.

The activity background tail or "Compton escape continuum" from the UTW detector is smaller than that from the standard 30 mm^2 detector when measured with the same specimen. This difference may be attributed to reduced γ-Compton interaction in the thinner crystal of the UTW detector. Accordingly, relatively thin (1 to 2 mm) detectors may be optimum for X-ray analysis of radioactive specimens if a decrease in efficiency for X-rays with energies above 10 keV can be tolerated. It is also possible that the background tail is partly due to light produced by scintillation of an internal detector component. Even without electronics modifications, an effective collimator extends EDX microanalysis to radioactive specimens with γ-activities up to 1000 μC_i.

Applications to Neutron-Irradiated Alloys

Recent applications of EDX microanalysis to neutron-irradiated alloys include identification and analyses of radiation-induced second-phases such as $Ni_3Si-\gamma'$, G-phase and complex carbosilicides in AISI 316, correlation of compositional fluctuations on a microscopic scale with local swelling and mechanical property changes, and studies of radiation-induced solute segre-gation effects at individual voids, dislocations and grain boundaries. A few years ago, most phase analyses were performed on chemically or electro-lytically extracted particles. However, the disadvantages of extraction.... loss of crystal orientation relationships, sensitivity to particle compo-sition and loss of compositional information about the alloy matrix.... can be avoided by analyzing particles in thin foils. On the other hand, extraction is a highly effective way to reduce specimen activity, and analysis of both extracted and combined particles is often desirable. The strength of analytical electron microscopy lies in the ability to combine microstructural, compositional and crystallographic information from micro-scopic specimen regions, and this approach may involve alternative specimen preparations.

The study of radiation-induced solute segregation at individual point defect sinks illustrates EDX microanalysis in a particularly important application. It is now recognized that nonequilibrium segregation of alloy elements in the point defect gradients around voids, dislocations and grain boundaries strongly influences void swelling, phase stability and mechanical properties in alloys being developed for nuclear energy applications. Initial work on the segregation phenomenon used surface analytical techniques such as Auger electron spectroscopy and ion scattering spectroscopy to determine concentration profiles at extended sinks such as ion-bombarded surfaces. However, pioneering work by Kenik (10) and others (11) who used EDX micro-analysis to analyze concentrations around individual defects in ion bom-barded alloys has led to similar work on neutron irradiated alloys. Figures 6 and 7 show examples of composition profiles around voids and dislocations in a neutron irradiated Fe-Cr-Ni based superalloy which had a γ-activity of 300 μC_i. By using the techniques described in this paper, compositions of selected regions 20 nm in diameter were analyzed with an accuracy near 0.5 wt. %.

It should be emphasized that these compositions are averages over the specimen volume excited by the electron beam. Rapidly varying composi-tion gradients at the voids and dislocations are not resolved at a spatial resolution of 20 nm. The voids and dislocations in this case had coatings of irradiation-induced Ni_3 (Al, Ti)-γ', and the resulting "composition profiles" reflect averaging over regions containing γ' as well as adjacent matrix regions with varying degrees of solute enrichment. The analysis of actual composition profiles at dislocations and voids requires careful measurement of the defect location in the specimen as well as mathematical deconvolution procedures. No such detailed analysis was attempted in this work.

The analysis results in Figure 7 indicate an important difference between segregation behavior of multicomponent Fe-Cr-Ni alloys and simple binary alloys. Aluminum, an oversized atom in these alloys, segregates at sinks to form Al-rich γ'-phase in the Fe-Cr-Ni alloy containing Ti and Al, but segregates away from sinks in Ni based binary alloys (12). This un-expected difference is not well understood and illustrates the need for further exploration of segregation effects by microanalytical techniques.

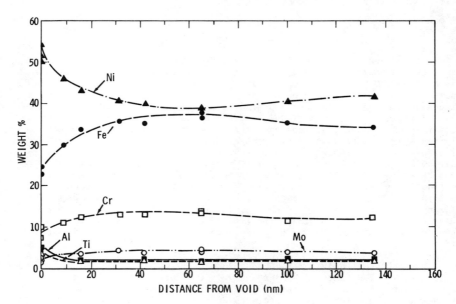

Fig. 6 - Concentration profiles at a γ'-coated void in a neutron-irradiated Fe-Cr-Ni based alloy.

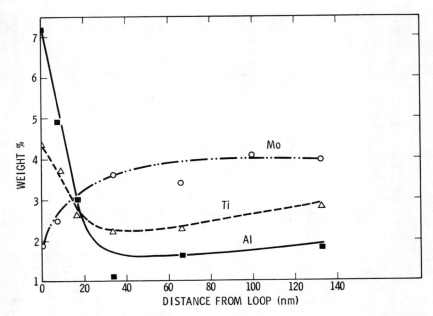

Fig. 7 - Concentration profiles at a γ'-coated dislocation loop in a neutron-
irradiated Fe-Cr-Ni based alloy.

Conclusions

Specimen radioactivity affects EDX microanalysis of neutron-irradiated
Fe-Cr-Ni based alloys in several ways. These include high counting deadtimes
due to energetic γ-Compton scattering in the EDX detector, low-energy
spectral background contributions from Compton escape in the detector and
from scintillation of detector components, and characteristic X-ray contributions from long-lived electron capture nuclides such as ^{55}Fe, ^{54}Mn and
^{58}Co. Specimens having γ-activities up to 1000 μC_i can be analyzed by
simple methods....using large specimen-dectector distances to obtain
acceptable deadtimes, increasing the electron probe current to improve X-ray
signal to activity background ratios, and stripping relatively large activity
spectra (obtained with the electron beam directed through a hole in the
specimen) from spectra obtained during microanalysis. However, these
techniques degrade X-ray spatial resolution and counting statistics. Effective collimation of γ and X-rays emitted form the thick edges of self-
supporting disk specimens is important to maintain high spatial resolution
and accuracy in quantitative analyses of highly radioactive specimens.

Although standard EDX detectors can thus be readily used with radioactive specimens, they are not optimum for this purpose. Recommended
detector modifications for analysis of neutron-irradiated specimens include
increasing the preamplifier reset charge to 6 MeV or as high as practicable
without degrading the energy resolution, using a relatively thin (eg.
1.5 to 2 mm) detector crystal to reduce γ-Compton interaction, adding on
appropriately designed heavy-metal collimator, and shielding the entire
detector end from penetrating radiation.

EDX microanalysis at high spatial resolution allows direct concentration profiling at point defect sinks in irradiated alloys....at individual voids, dislocations and grain boundaries. However, it must be recognized that the observed compositions are still averages over relatively large volumes excited by the electron beam. Increased use of this technique to study radiation-induced segregation phenomena is expected in the immediate future.

Acknowledgment

The author would like to acknowledge the contribution of G. M. Stevens (Westinghouse Hanford Company) in designing the γ-ray collimator.

References

(1) L. E. Thomas, "Microanalysis of Light Elements with an Ultrathin Window X-Ray Spectrometer," pp. 90-93 in Proceedings EMSA Thirty-eighth Annual Meeting, G. W. Bailey, ed.; Claiton's, Baton Rouge, 1980.

(2) For an up-to-date review of analytical electron microscopy techniques, see Introduction to Analytical Electron Microscopy, J. J. Hren, J. I. Goldstein and D. C. Joy (eds.), Plenum Press, New York (1979).

(3) D. F. Kyser, "Monte Carlo Simulation in Analytical Electron Microscopy," ibid. Ref. 1, p. 216; also, J. I. Goldstein, "Principles of Thin Film X-Ray Microanalysis," ibid. Ref. 1, pp. 101-103.

(4) R. D. Evans, The Atomic Nucleus, McGraw-Hill, New York (1959), pp. 672-693.

(5) Rolf Woldseth, X-Ray Energy Spectrometry, Kevex Corporation, Burlingame, California (1973).

(6) D. A. Landis, F. S. Goulding, R. S. Pehl and J. T. Walton, "Pulsed Feedback Techniques for Semiconductor Detector Radiation Spectrometers," IEEE Trans. Nucl. Soc. (1971), pp. 115-124.

(7) R. G. Musket, "Properties and Applications of Windowless Si(Li) Detectors," in Proceedings Workshop on E.D.S., NBS Special Technical Publication No. 604 (1980).

(8) G. Cliff and G. W. Lorimer, "The Quantitative Analysis of Thin Specimens," J. Microscopy, 103, 203-207 (March 1975).

(9) N. J. Zaluzec, "Quantitative X-Ray Microanalysis," ibid. Ref. 1, pp. 159-161.

(10) E. A. Kenik, "Radiation-Induced Solute Segregation in a Low Swelling 316 Stainless Steel," Scripta Met., 10, 733-738 (1976).

(11) A. D. Marwick, W. A. D. Kennedy, D. J. Mazey and J. A. Hudson, "Segregation of Nickel to Voids in an Irradiated High Nickel Alloy," Scripta Met., 12, 1015-1020 (1978).

(12) D. I. Potter and D. G. Ryding, "Precipitate Coarsening, Redistribution and Renucleation During Irradiation of Ni-6.35 Wt. % Al," J. Nucl. Mat., 14-24 (1977).

ANALYSIS OF RADIATION-INDUCED MICROCHEMICAL EVOLUTION IN

300 SERIES STAINLESS STEEL*

H. R. Brager and F. A. Garner

Hanford Engineering Development Laboratory
Richland, Washington

The irradiation of 300 series stainless steel by fast neutrons leads to an evolution of alloy microstructure that involves not only the formation of voids and dislocations, but also an extensive repartitioning of elements between various phases. This latter evolution has been shown to be a primary determinant of the alloy behavior in response to the large number of variables which influence void swelling and irradiation creep. The combined use of scanning transmission electron microscopy and energy-dispersive X-ray analysis has been the key element in the study of this phenomenon.

Problems associated with the analysis of radioactive specimens are resolved by minor equipment modifications. Problems associated with spatial resolution limitations and the complexity and heterogeneity of the microchemical evolution have been overcome by using several data acquisition techniques. These include the measurement of compositional profiles near sinks, the use of foil-edge analysis, and the statistical sampling of many matrix and precipitate volumes.

*This work was sponsored by the Office of Fusion Energy, U.S. Department of Energy.

Introduction

The irradiation of metastable alloys such as AISI 316 and 304 has recently been shown to lead to two concurrent and interactive evolutions, one involving the microstructural components associated with swelling and irradiation creep, and the other involving extensive repartitioning of elements between the matrix and various phases (1-5). This paper discusses the key role played by scanning transmission electron microscopy (STEM) and energy-dispersive X-ray (EDX) analysis in the identification and study of this latter phenomenon. The following sections describe the modifications of commercially-available STEM/EDX equipment necessary to overcome the difficulties associated with examination of radioactive specimens, and the types of data acquisition procedures used to determine the complex nature of the microchemical/microstructural evolution. Finally, a summary is presented of the radiation-induced evolution of AISI 316 as revealed by a combination of STEM/EDX and other techniques, and the impact of this evolution on dimensional and mechanical properties.

Modification of STEM/EDX Equipment

The majority of the measurements described in this study were performed on a JSEM-200 STEM equipped with a Kevex energy-dispersive X-ray detector system connected to a Tracor-Northern 880 multichannel analyzer. As described elsewhere (6), the analysis of radioactive specimens is complicated by the presence of both gamma radiation and characteristic X-radiation. Typical activities due to long-lived isotopes in AISI 316 specimens (nominally 3 mm in diameter and 0.25 mm thick) measure about 10 Rad/hr at 2.5 cm after irradiation to 1×10^{23} n/cm^2 (E >0.1 MeV). The decay-gamma rays contribute to increased deadtime in the X-ray detector system and a decreased spectral resolution, while the X-rays characteristic of various decay-daughter isotopes provide a false indication of elemental composition.

The majority of the decay radiation directed toward the detector is prevented from reaching it by placing a conical-shaped 16 mm long tungsten collimator between the specimen and the detector. The collimator also attenuates photons generated by other sources in the microscope. Shown in Figure 1, the collimator is mounted at a fixed location in the microscope column at the same elevation as the specimen. As discussed elsewhere (6), the attainment of an adequate signal-to-noise ratio at the detector requires a trade-off between the specimen-to-detector distance and the intensity of the beam on the specimen. The beam intensity and beam spreading determine the spatial resolution. The specimen/detector configuration used in this study requires a relatively intense electron beam and thus leads to a spatial resolution on the order of only 30 to 50 nm.

In order to determine the various elemental concentrations in an area under examination, the X-ray signal is analyzed by the detector circuitry and resolved as a function of X-ray energy. The characteristic decay X-rays are then measured for an equal amount of time when the electron beam passes through the hole in the center of the specimen. The decay spectra are comprised principally of mid-range (about 8 keV) X-rays which, for most cases, are less than 1% of the total signal. The resolved decay spectrum is then "stripped" or subtracted from each energy channel. Correlation factors between X-ray intensities and composition are determined from analysis of unirradiated homogeneous metal standards of known composition. The measurement uncertainties were determined to be about ±2% for major alloy components and about ±10% for minor solute components.

Two other modifications were made to the STEM. First, the specimen holder was replaced. The original holder, which was made of copper and which held

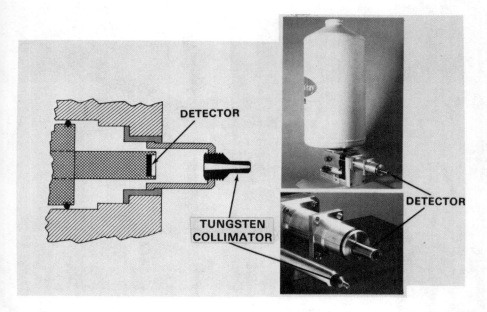

Fig. 1 – Cross section and location of fixed tungsten collimator. Note that the detector-specimen distance is variable.

the specimen in place by use of a stressed spring clip, was replaced by a holder made of beryllium. It secures the specimen by a threaded beryllium cup. Both the threaded cup and the holder, shown in Figure 2, were built to allow a larger solid angle for uninterrupted line-of-sight between the speci- men and detector. Beryllium was chosen to eliminate the generation of copper X-rays and to eliminate their second-order influence on the apparent concentration of other elements (7). The second modification was to alter one of the fixed resistors of the first condenser lens control to provide a continuous variation of lens current. This allows an optimization of the signal-to-noise ratio at the X-ray detector while also allowing maximum resolution of the STEM-generated image on the CRT screen.

Data Acquisition Strategy

There are a number of factors which complicate the analysis of radia- tion-induced evolution in 300 series stainless steel. First, the resolution limit of 30 to 50 nm is rather large compared to the scale of the micro- structure that develops at relatively low irradiation temperatures. It is also often large with respect to some of the concentration gradients found to develop near microstructural sinks during irradiation. Second, the microstructure under examination is often quite complex as shown in Figure 3, and the many phases are not only sensitive to irradiation temperature but are undergoing continuous evolution during irradiation. The evolution is quite heterogeneous in nature, however, progressing at different rates and sometimes by different paths in each small volume sampled. Therefore, con- clusions about the nature of the evolution cannot be safely deduced from a single or small number of determinations. Third, the major microstructural participants in the evolution are frequently small with respect to the usual foil thickness employed. This factor often does not allow determination

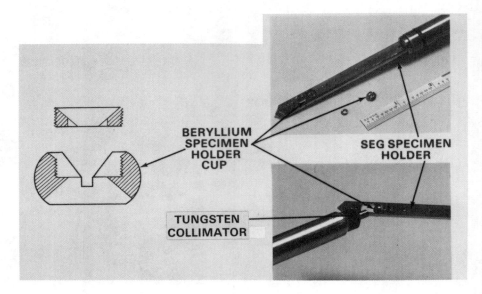

Fig. 2 - Cross section of modified specimen holder showing location in the
side entry goniometer (SEG) specimen holder assembly and its relationship to the collimator-detector assembly.

of the composition of a small precipitate, but yields an average composition
of the precipitate and surrounding matrix volume. Another problem is that
light elements such as carbon, nitrogen and oxygen cannot be detected with
the commercial EDX equipment available at the time this study was performed,
a problem which complicates the analysis of the various carbon-rich phases
that develop in these steels. Quantitative analyses of the light elements
have been obtained using either electron energy loss spectroscopy (8) or a
very thin window with an EDX detector on a TEM/STEM instrument (9).

A number of data extraction techniques have evolved in order to study
the nature of the radiation-induced evolution and to overcome some of the
complications discussed above. These involve different specimen
preparation methods, statistical sampling techniques and the use of bulk
extraction procedures which complement the microanalysis methods.

In-Foil Analysis

Figure 4(a) illustrates representative microstructural features which
can be studied using microanalysis techniques. Two of these features, voids
and grain boundaries, cannot be extracted and must be examined in-foil.
The nature of composition gradients adjacent to these features, as well as
to precipitates, is of primary interest but the X-ray signal generated by
the electron beam will represent a line-of-sight average over the composition
gradient as it extends through the matrix. It will also represent an average over the width of the excited volume, which increases with depth in
the foil due to scattering of electrons out of the incident beam. Figure 5
illustrates this point, showing the relationship of line-of-sight averages
to the actual profile for a highly idealized void, foil and electron beam.

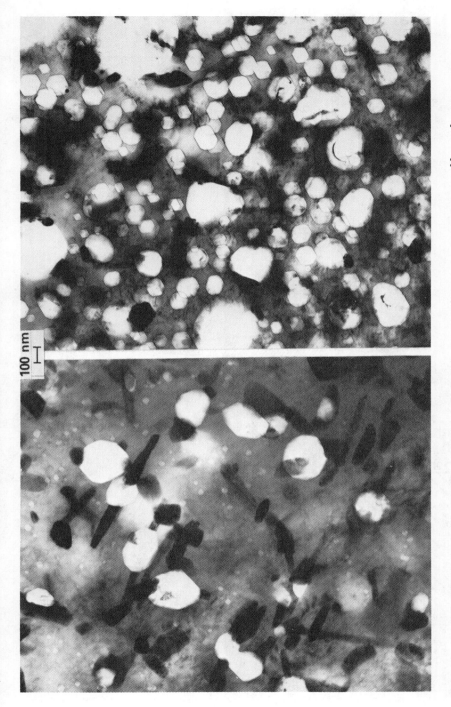

100 nm

Fig. 3 - Typical microstructures observed in irradiated AISI 316 at 600°C to 0.8 and 1.4×10^{23} n/cm^2 (E > 0.1 MeV), respectively.

101

(a) IN-FOIL ANALYSIS

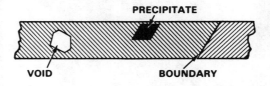

PRECIPITATE

VOID   BOUNDARY

(b) EXTRACTION REPLICA ANALYSIS

CARBON FILM

(c) FOIL - EDGE ANALYSIS

Fig. 4 – Schematic illustration of microanalysis techniques employed in this study. See text for a detailed description.

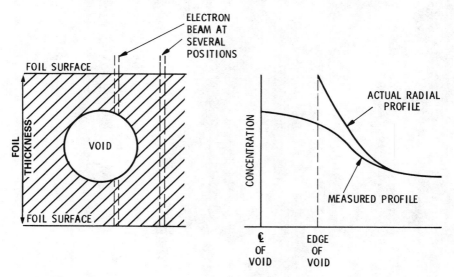

ELECTRON
BEAM AT
SEVERAL
POSITIONS

FOIL SURFACE

FOIL THICKNESS

VOID

FOIL SURFACE

CONCENTRATION

ACTUAL RADIAL PROFILE

MEASURED PROFILE

¢
OF
VOID

EDGE
OF
VOID

Fig. 5 – Schematic representation of idealized determination of line-of-sight concentration profile and its relationship to actual profile.

If the measured profile can be unfolded, it represents a microstructural record (10) of the diffusional processes in progress during irradiation. Figure 6 shows typical measured concentration profiles for nickel and chromium in the vicinity of void surfaces in AISI 316 at a high neutron fluence and temperature. The segregation of nickel near void surfaces is primarily balanced by an outflow of chromium. There is also a lesser outflow of iron, but no measurable gradients were found to exist in other major components such as molybdenum or silicon. This microstructural record therefore provides data which can be used to validate and rank the operation and relative contribution of various diffusional processes postulated to operate during irradiation. For instance, the segregation of the slow-diffusing nickel and the compensating outflow of the faster diffusing chromium offer substantial support for the proposed "inverse-Kirkendall" effect (11), wherein slow-diffusing species are concentrated by default at the bottom of vacancy gradients as a consequence of the selective outflow of faster-diffusing elements.

Similar microstructural profile records can be developed from regions adjacent to and within precipitates, but this subject will be developed in a later section. Grain boundaries also participate in the segregation process, but the microanalysis results can sometimes be misleading. Figure 7 shows the segregation profiles observed in annealed AISI 316 irradiated at 520°C. The apparent decline in nickel content near the boundary is illusory and reflects a problem peculiar to analysis of thin microstructural features. The relatively thick foils needed for boundary analysis and the 30 to 50 nm resolution limit tend to obscure the presence of small γ' precipitates within the foil. These precipitates are denuded near the boundary, however. The large decrease in nickel content in the vicinity of the boundary is due to the local absence of γ' while the small peak at the boundary represents the inverse-Kirkendall concentration of nickel.

Extraction Replica Analysis

Figure 4(b) illustrates the use of extraction replicas as a method of analyzing precipitates after separation from the surrounding matrix. An advantage of this technique is that the level of decay radiation is decreased substantially. The disadvantage of this technique is the partial or sometimes total loss of some kinds of precipitates due to selective dissolution. For instance, the normal procedure for thin foil preparation (electrolytic thinning in 10% perchloric and 90% glacial acetic acids) appears to dissolve the γ' or Ni_3Si phase while preserving most carbide and intermetallic phases. Since the irradiation-induced γ' phase is usually distributed as very small precipitates, it was necessary to extract this phase using another electrolyte (1). Cawthorne and Brown (12) were able to tentatively identify this phase in the foil by analysis of moiré fringe spacings and without specifying its exact composition, but extraction replication and EDX microanalysis were necessary to establish the elemental composition (1).

Foil-Edge Analysis

If a given precipitate is not selectively attacked by the electrolyte used to prepare the thin foil, then microanalysis can proceed on precipitates of reasonable size which are attached to the edge of the foil. Illustrated in Figure 4(c), analysis near the foil edge not only reduces the influence of the matrix but preserves a record of any compositional gradients in the adjacent matrix. Figure 8 illustrates the use of this technique in the examination of an irradiation-induced fcc precipitate (G-phase) formed in 20% cold worked AISI 316 irradiated to high fluence at 500°C.

103

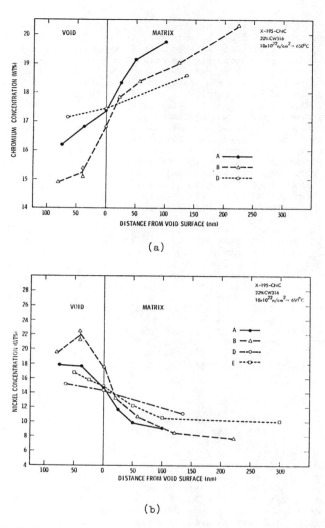

(a)

(b)

Fig. 6 – Typical line-of-sight profiles measured at void surfaces in 20% CW
AISI 316 irradiated to 1.0×10^{23} n/cm^2 (E >0.1 MeV) at 650°C.
Each letter (A, B, D, E) denotes a set of EDX measurements taken
with the electron beam incident on a different void and its adjacent matrix.

It is not unusual, however, to find several kinds of precipitates co-
existing in the same grain, with each type exhibiting a different solubility
in the electropolishing procedure. Figure 9 shows that both the γ' and
G-phases populate the grain located near the foil edge, but of the two only
the G-phase is both sufficiently large and stable during electropolishing
to allow the foil-edge analysis approach.

When the foil-edge technique was first employed in the study of the
elemental flow sequences in precipitates, several unexpected observations

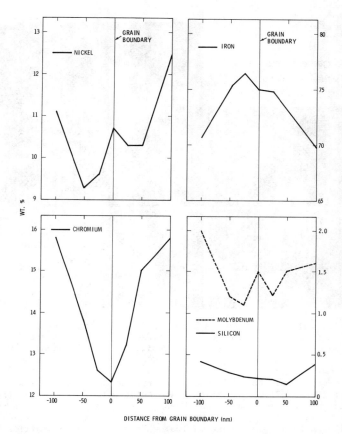

Fig. 7 – Line-of-sight measurement of elemental profiles in the vicinity of
a grain boundary in annealed AISI 316 irradiated at 520°C to
9×10^{22} n/cm^2 (E >0.1 MeV).

were made. First of all, Laves precipitates (nominally Fe$_2$Mo) appeared to
be changing in composition during irradiation. Although diffraction
analysis indicated the presence of Laves at all fluences above 5×10^{22} n/cm^2
(E >0.1 MeV) at temperatures over 600°C, foil-edge microanalysis showed a
steady reduction in iron and molybdenum in these precipitates with increasing fluence and a concurrent increase in nickel and silicon. It was
not known, however, whether Laves of traditional composition was continually
dissolving and nickel-rich Laves precipitating, or whether nickel and
silicon were infiltrating the precipitates and either diluting the precipitates or replacing iron and molybdenum. The first composition profiles
from the adjacent matrix indeed exhibited gradients in nickel composition
which were indicative of nickel flow into the precipitates. The profiles
shown in Figure 10 are not so illuminating on the direction of flow of
molybdenum and silicon, however. The lower level of these elements, their
higher diffusivity and the heterogeneity of the microchemical evolution all
combine to yield an ambiguous record of the elemental flow. The direction
of flow can be inferred from comparison of the mean matrix compositions of
these elements at different fluence levels, however. Note in Figure 11 that
for the two Laves phase particles analyzed, the molybdenum level in the

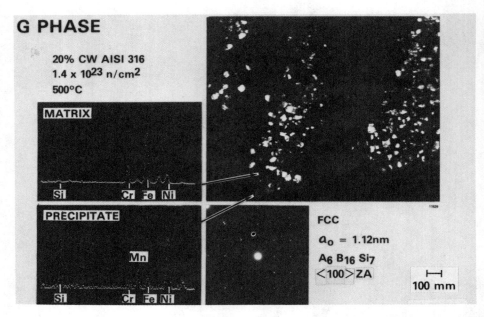

Fig. 8 – Determination of composition and crystal structure of G-phase found
in irradiated 20% cold worked AISI 316 using the edge-foil analysis. The adjacent matrix composition is also shown.

adjacent matrix areas have risen during irradiation to the average alloy
bulk composition. Since the precipitate molybdenum levels are rather low,
this indicates that molybdenum has been substantially depleted in the Laves
phase. The nickel level of the matrix has also fallen somewhat. It is
very difficult to draw a clear conclusion from a few such profiles since
the variability of the segregation process is expressed throughout the
irradiated volume, and the precipitate particle density is sufficiently
heterogeneous to preclude a simple determination of the total amount of
any one element in the precipitates.

Statistical Sampling Analysis

Additional information on the direction of elemental flows can be
gained from the random spot analysis of many small representative matrix
volumes. These volumes are chosen to be large enough to measure a substantial volume of material, but small enough to avoid including features such
as precipitates. As shown in Figure 12, the many areas sampled in one
specimen exhibit a large range of local concentration levels for each of
the elements nickel, silicon and iron. While these observations under-
score the heterogeneity of the evolution, it is rather obvious that nickel
and silicon flow in the same direction and toward the precipitates,
while molybdenum flows in the opposite direction.

When this information was obtained it provided an explanation for some
initially perplexing data published by Porter and McVay (5). Using bulk extraction techniques they reported that the precipitates formed in 20% cold
worked AISI 316 during irradiation at 550°C gradually increased in volume
but suffered substantial and progressive decreases in iron and molybdenum

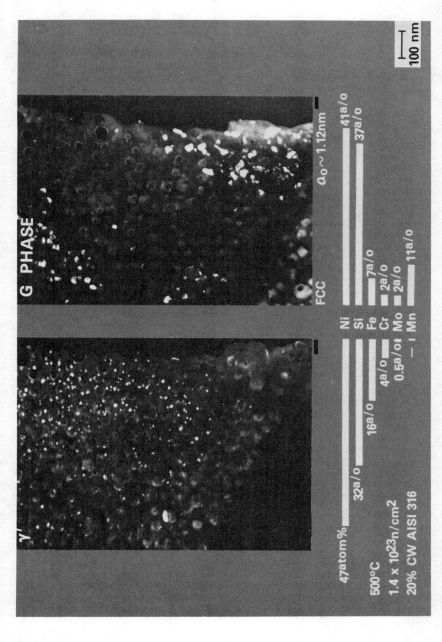

Fig. 9 - Coexistence of two precipitate phases, both rich in nickel and silicon, in one grain of 20% cold worked AISI 316 irradiated at 500°C to a fluence of 1.4×10^{23} n/cm^2 (E >0.1 MeV).

107

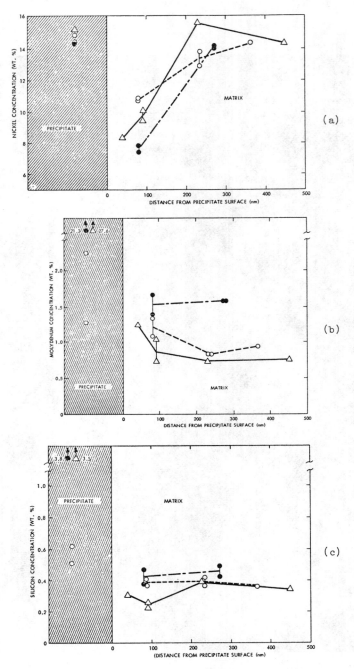

Fig. 10 - Concentration profiles of nickel, molybdenum and silicon observed near the surface of three Laves precipitate particles in 20% cold worked AISI 316 irradiated at 650°C to a fluence of 1.0 x 10^{23} n/cm^2 (E >0.1 MeV).

108

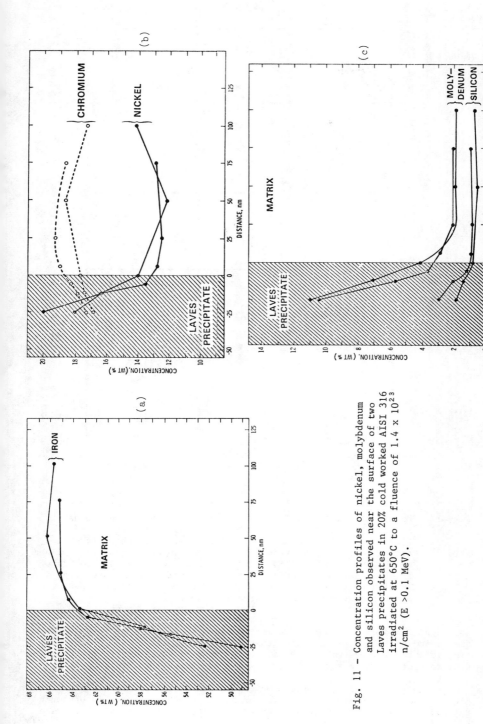

Fig. 11 – Concentration profiles of nickel, molybdenum and silicon observed near the surface of two Laves precipitates in 20% cold worked AISI 316 irradiated at 650°C to a fluence of 1.4×10^{23} n/cm^2 (E >0.1 MeV).

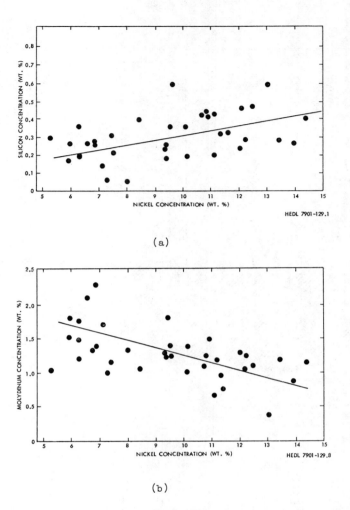

(a)

(b)

Fig. 12 - Local-spot determinations of elemental concentrations in many
regions in 20% cold worked AISI 316 irradiated at 650°C to a
fluence of 1.0 x 10^{23} n/cm^2 (E >0.1 MeV). Note that nickel and
silicon concentrations tend to increase together while nickel and
molybdenum do not. This indicates that nickel and silicon flow
together and against the flow of molybdenum. There are large
local variations in all elements, however. The trend lines were
established by a least squares analysis.

while gaining nickel. Whereas it was previously assumed that molybdenum,
nickel and silicon were all concentrated into precipitates during irradiation (3), the molybdenum concentration in the precipitates actually peaks at
some moderate fluence and declines thereafter during irradiation.

Another application of the sampling procedure is demonstrated in
Figure 13, in which the distribution of matrix nickel contents is presented
for two AISI 316 specimens of different starting conditions and which were

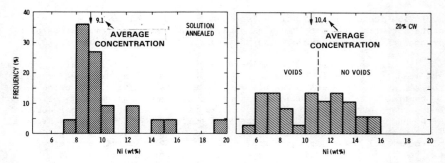

Fig. 13 – Distribution of nickel content and voidage at different locations
in the matrix of irradiated AISI 316 specimens irradiated at 650°C
to a fluence of 1.0×10^{23} n/cm^2 (E >0.1 MeV). Note the dependence
on cold working.

exposed to identical irradiation conditions. The solution annealed specimen
contained Laves precipitates which were substantially enriched in nickel and
silicon, with some indication that the mean levels of these elements were
higher than that of the 20% cold worked specimen. Note that the average
matrix nickel level of the annealed specimen has been reduced from the
original 13.5% level to about 9%. The cold worked specimen at these conditions has a matrix nickel content that has been reduced to an average of
only 10.4% and the distribution is much broader. Other studies (1,3,4) have
shown that the AISI 316 alloy in either the solution annealed or cold
worked condition is proceeding toward a matrix nickel content of about 9%,
a level which is independent of temperature and starting condition.

Another important observation was made in the data field shown in
Figure 13. The void swelling of the annealed specimen was about 8% and
quite homogeneous, while the void swelling of the cold worked specimen was
only 0.6% and quite heterogeneously distributed. In almost all cases,
however, large area (about 0.5 μm in diameter) analyses of voided regions
in the cold worked specimen showed the matrix nickel content to be in the ≤10
wt. % range while regions which had not yet begun to swell contained ≥11 wt. %
nickel. This information indicates that the onset of void nucleation is
determined by microchemical considerations as well as by dislocation density or helium level.

Summary of the Microchemical Evolution

The complexities and heterogeneity of the microchemical evolution of
AISI 316 have required that this study focus more on the techniques of data
extraction rather than the relatively minor problems and equipment modifications associated with the radioactivity of the specimens. This experience
has also been typical of that encountered in the study of AISI 304 (5) and
various modified AISI 316 alloys (3,4). Although the evidence compiled in
the course of this and other related studies is too large to present here,
a summary of the principles involved in the microchemical evolution of the
300 series alloys will be briefly presented to aid related studies by other
researchers. These principles were revealed by combined TEM, diffraction
and STEM/EDX techniques.

It appears that in the 300 series alloys the development of dislocation
and loop microstructure (and to a lesser extent voids) proceeds toward a
saturation state, composed of number densities and component identities
which are independent of starting microstructure. The microchemical

evolution induced by radiation also appears to proceed toward a saturation state but at a much more sluggish pace, although certain temperature paths or preirradiation treatments can short-circuit the sluggishness (3,13).

The various microstructural components interact with the defect fluxes and lead to point defect gradients which function as the driving force for the segregation process. The element nickel seems particularly prone to concentrate at defect sinks in response to the inverse-Kirkendall effect and perhaps other mechanisms. The matrix nickel content is then reduced, although not by much unless solutes such as silicon and carbon are present. These elements coprecipitate with nickel at microstructural sinks and form a microchemical reservoir capable of removing from the matrix approximately two to three atoms of nickel for each silicon (and possibly carbon) atom. In AISI 316 this leads to a reduction in the matrix of both nickel and silicon.

The most interesting feature of the precipitate-related segregation process is that the path by which the segregation process proceeds does not appear to influence the final outcome. A large variety of phases can develop, depending strongly on irradiation temperature and history, minor variations in composition and preirradiation thermomechanical treatment. But nearly all such phases either form as nickel and silicon-rich phases or become progressively richer in these elements as a result of the infiltration-exchange process. At fluences approaching 100 dpa even the Laves (Fe_2Mo) phase approaches a composition of Ni_3Si. If no other phase forms in the intermediate temperature range centered at about 500°C, the "phase of last resort" forms eventually. This is the γ' phase (nominally Ni_3Si) which not only requires irradiation to form but once formed will dissolve in the absence of displacive irradiation (3,4).

The consequences of the nickel segregation process are rather large. If the solute content is increased over the nominal specification for AISI 316, the matrix nickel content can be driven so low as to initiate the transformation of substantial amounts of matrix austenite to ferrite (3,4). In AISI 304 the initial nickel level is only 9 wt. % and the nominal 0.5 wt. % silicon of the alloy is sufficient to cause the transformation (5).

It also appears that the nickel content (at levels below 30%) is a major determinant of the swelling and irradiation creep rates, once the saturation microstructure has developed. Just as the initial nickel content (C_{Ni}^{o}) has been shown to be a major factor determining the magnitude of swelling (14), the declining localized instantaneous nickel content in a specific alloy has been correlated with the acceleration of the creep and swelling rates (3,4). This implies that when the average matrix nickel content reaches the level $C_{Ni}^{*} = C_{Ni}^{o} - 3$ (Si + C), then the steady state swelling and creep rate will have been established.

The many phases that can be involved in the segregation process all exhibit a different sensitivity to the host of variables which influence the evolution. Some are sensitive to cold work while others are not. In general, cold work decreases the rate of nickel removal from the matrix and its concentration into all precipitate phases in irradiated AISI 316. However, cold work accelerates the formation of some phases in the unirradiated steel (15,16). The sensitivity of the microchemical evolution to variables such as stress (17), temperature change (13) and neutron flux (4) has been shown to be related to the response of the various second phases to these variables.

112

Conclusions

The radiation-induced microchemical evolution of 300 series stainless steel has been shown to be a major determinant of alloy behavior in response to the host of operating variables known to influence the evolution of dimensional and mechanical properties. Conventional transmission microscopy techniques are not in themselves sufficient to characterize this evolution, however, and the combined STEM/EDX technique is an essential tool. The problems involved in examining a highly radioactive specimen can be resolved by equipment modifications, and the problems associated with instrument spatial resolution limitations and the heterogeneity of alloy composition can be overcome by developing a balance of data acquisition procedures which include the measurement of compositional profiles near sinks and statistical sampling of many matrix and precipitate volumes.

References

1. H. R. Brager and F. A. Garner, J. Nucl. Mat., 73, 9 (1978).

2. H. R. Brager, F. A. Garner, E. R. Gilbert, J. E. Flinn and W. G. Wolfer, "Stress-Affected Microstructural Development and the Creep-Swelling Interrelationship," in Radiation Effects in Breeder Reactor Structural Materials, M. L. Bleiberg and J. W. Bennett (Eds), The Metallurgical Society of the American Institute of Mining, Metallurgical and Petroleum Engineers (1977), pp. 727-755.

3. H. R. Brager and F. A. Garner, "Dependence of Void Formation on Phase Stability in Neutron-Irradiated Type 316 Stainless Steel," in Effects of Radiation on Structural Materials, ASTM-STP-683, J. A. Sprague and D. Kramer (Eds), ASTM (1979), pp. 207-232.

4. H. R. Brager and F. A. Garner, "Microchemical Evolution of Neutron-Irradiated 316 Stainless Steel," in Effects of Radiation on Materials, ASTM-STP-725, D. Kramer, H. R. Brager and J. S. Perrin (Eds), ASTM (1981), pp. 470-483.

5. D. L. Porter and E. L. Wood, J. Nucl. Mat., 83, 90 (1979).

6. L. E. Thomas, "EDX Microanalysis of Neutron-Irradiated Alloys," this proceedings.

7. N. J. Zaluzec, "Quantitative X-Ray Microanalysis: Microstructural Considerations and Applications to Materials Science," in Introduction to Analytical Electron Microscopy, J. J. Hren, J. I. Goldstein and D. C. Joy (Eds), Plenum Press (1979), pp. 121-167.

8. D. M. Maher, "Elemental Analysis Using Inner-Shell Excitation: A Microanalytical Technique for Materials Characterization," ibid Reference 7, pp. 259-294.

9. L. E. Thomas, "Microanalysis of Light Elements with an Ultrathin Window X-Ray Spectrometer," in Proceedings EMSA Thirty-eighth Annual Meeting, G. W. Bailey (ed); Claitor's, Baton Rouge (1980), pp. 90-93.

10. F. A. Garner and G. R. Odette, "Impact of Advanced Microstructural Characterization Techniques on Radiation Damage Modeling and Analysis," this proceedings.

11. A. D. Marwick, R. C. Pillar and P. M. Sivell, J. Nucl. Mat., 83, 35 (1979).

12. C. Cawthorne and C. Brown, J. Nucl. Mat., 66, (1977) pp. 201-202.

13. F. A. Garner, E. R. Gilbert, D. S. Gelles and J. P. Foster, "The Effect of Temperature Changes on Swelling and Creep of AISI 316 Stainless Steel," ibid. Reference 4, pp. 698-712.

14. J. F. Bates and W. G. Johnston, "Effects of Alloy Composition on Void Swelling," ibid. Reference 2, pp. 625-644.

15. B. Weiss and R. Stickler, Met. Trans., 3, (1972) pp. 851-866.

16. J. E. Spruiell, J. A. Scott, C. S. Ary and R. L. Hardin, Met. Trans., 4, (1973) pp. 1533-1543.

17. F. A. Garner, E. R. Gilbert and D. L. Porter, "Stress-Enhanced Swelling of Metals During Irradiation," ibid. Reference 4, pp. 680-697.

THE USE OF 2½-D ELECTRON MICROSCOPY TO STUDY

DISLOCATION CHANNELING EFFECTS IN IRRADIATED ZIRCALOY

W. L. Bell and R. B. Adamson

General Electric Company
Vallecitos Nuclear Center
Pleasanton, California 94566

    Through-focus dark-field images can be used as stereo images to study
post-irradiation deformation effects. Irradiation-hardened Zircaloy-2 will
often exhibit dislocation channeling in the gage sections of tensile specimens and these channeling effects are quite suitable for analyses using the
2½-D technique. It is observed that when grain boundary shearing is feasible, channel deformation is highly locallized. When the grain boundaries
resist shearing, the deformed regions become broadened. In both cases dislocations tend to move or aggregate with others of like sign, resulting in
closely spaced arrays and periodic reversals of Burgers vectors.

Introduction

The analyses of plastically deformed microstructures by conventional transmission electron microscopy are generally limited by dislocation density, diffracting conditions, foil thickness, etc.(1) Variabilities of dislocation features are hard to assess because of uncertainties in parameters such as specimen thickness, image magnification, invisibility criteria,(1) and so forth.

The introduction of the 2½-D imaging technique has allowed deformation features to be approached from an entirely different direction.(2) Changing crystal orientations can be detected by using asterism to produce dark-field image shifts during defocussing. Stereo analyses of such shifts can provide information about the manner of deformation and the level of dislocations and their distributions.(2) Such analyses have been used to analyse point-to-point variations in dislocation densities in post-irradiation tested Zircaloy.(3)

Objectives

The objective of this paper is to demonstrate that the 2½-D electron imaging techniques can be applied to the study of deformation which occurs in irradiated Zircaloy cladding during post-irradiation testing.

Basically the 2½-D technique consists of obtaining through-focus dark-field images and studying them with a stereo viewer.(2) In deformed materials the asterism, which results from arrangements of dislocations, will cause image shifts to occur when objective focussing conditions are changed. These image shifts constitute artificial parallaxes and, when viewed in stereo, different crystallographic orientations will appear to be at different spatial levels.

In single crystals or individual grains, orientation variations occur because of the presence of dislocations, which produce bending described by(4)

$$\rho = \frac{1}{rb} \tag{1}$$

where ρ = dislocation density
   r = radius of curvature of slip plane
   b = magnitude of the Burgers vector.

Crystal curvature, the reciprocal of r, is often a more convenient parameter(5) and is given by

$$\kappa = \rho b \tag{2}$$

There are two straightforward means of evaluating curvature using 2½-D images.(3) First, for a smoothly curved arc, the relationship between chord length, x, and subtended angle, α, for a smoothly curved circular arc is

$$x = 2r \sin (\alpha/2). \tag{3}$$

Hence the curvature can be expressed as

$$\kappa = \frac{2 \sin (\alpha/2)}{x} \tag{4}$$

The angular range of orientations will be manifest in a selected area electron diffraction pattern as asterism which is easily measured. The distance between points with orientation extremes can also be easily measured once the points are identified using 2½-D stereo pairs, i.e., the highest and lowest points.

The amount of parallax or image shift between points in 2½-D images with different reciprocal lattice vectors is given by (2)

$$\Delta \vec{y} = \lambda \Delta D \, \Delta \vec{g} \tag{5}$$

where λ is the electron wavelength, ΔD is the change in focus between the two images of the stereo pair and $\Delta \vec{g}$ is the change in reciprocal lattice vector. For two points on an arc of asterism, the situation relevant to deformed material,

$$\Delta g = |\Delta \vec{g}| = 2|\vec{g}|\sin (\alpha/2) \tag{6}$$

with $\vec{g}$ being the reciprocal lattice vector chosen for dark-field imaging and $|\vec{g}|$ its magnitude.

Combining expressions (4), (5) and (6), the relationship between curvature and parallax between two points separated by a distance Δx is

$$\kappa = \frac{\Delta y}{\Delta x} \cdot \frac{1}{\lambda \Delta Dg} \tag{7a}$$

The slope of a 2½-D image is, after rearranging terms,

$$\frac{\Delta y}{\Delta x} = \rho b \, (\lambda \Delta Dg) \tag{7b}$$

In these expressions the term $(\lambda \Delta Dg)$ is a constant for any given correct arrangement of 2½-D images viewed in stereo, hence any image which exhibits sloping terrain relates directly to lattice curvature. For the purpose of illustrating this report, the important features of these expressions are that dislocation density, which is proportional to the curvature, varies linearly with the slope of the terrain, $\Delta y/\Delta x$, in 2½-D images and the sense of slope is determined by the sign of the dislocations in the array. Other than these two aspects, the details of dislocation arrangements are beyond the scope of this report.

The actual curvature or dislocation density measurable will depend upon the numbers of excess dislocations of one sign present in an array. Hence, measurements of total dislocation density are not possible but only the excess dislocation density can be evaluated by 2½-D analyses.

Experimental Procedures

The materials used were Zircaloy-2 cladding, fully recrystallized prior to irradiation and/or deformation. Table I shows the details of the irradiation and post-irradiation deformation of the materials studied.

Table I. Irradiation and Deformation Details of Zr-2 Cladding

Fluence, n/cm² (E > 1 MeV)	Deformation
Zero (archive)	71% CW by tube reduction
1.1×10^{21} at 343°C	bulge tested at 343°C (closed end, internal pressure)
2.2×10^{21} at 323°C	tensile tested at 250°C (simple uniaxial tension)
1.2×10^{21} at 327°C	tensile tested at 250°C (simple uniaxial tension)

Transmission electron microscopy specimens were obtained from the bulk materials by microslicing and spark cutting. Prior to final polishing, specimen shapes were roughly 3 mm diameter discs approximately 0.3 mm thick. Thinning was done using a commercial twin-jet electrochemical polishing apparatus and the electrolyte was maintained at about -40°C. The polishing solution was 20% perchloric acid in ethyl alcohol.

All electron microscopy was done on a JEM-6A electron microscope operated at 100 kV. The high-resolution dark-field imaging conditions necessary for the 2½-D technique were obtained by manually tilting the electron gun-condensor lens assembly of the microscope until the desired diffracted beam passed down the optical axis of the lower microscope lenses. Standard objective apertures (20μm, 30μm and 50μm) were used for dark-field imaging. When prepared for stereo viewing, all 2½-D images were rotated, using the selected area diffraction pattern as a guide, until the diffracting planes were horizontal on the average. This results in an average vertical reciprocal lattice vector and produces the maximum visual effect from asterism.

Results

Figure 1 shows, for purposes of illustration and comparison, how deformation affects unirradiated Zircaloy-2. When viewed in stereo, the underfocus-overfocus image pair exhibits topographical features, in this case what is basically a ridge across the deformed grain. The importance of the ridge is that it separates areas where the excess dislocation densities are composed of dislocations of opposite signs. The eyes can immediately detect information, of a qualitative sort, which was inaccessible using standard transmission electron microscopy techniques. That is, there are variations in both the degree of slope (relatable to the magnitude of the excess dislocation density) and the sense of slope (related to the sign of the Burgers vector of the excess dislocations).

Figure 1 demonstrates what can be basically considered as a kink band.(4) That is, during deformation dislocation pairs are created, the individual dislocations of which move in opposite directions due to the applied stress on the slip plane. The presence of grain boundaries or other obstacles causes the dislocations to pile up and their densities to rise locally.

Caution must be used when attempting to relate the dislocation distributions observed in Zircaloy to those existing in other, disimilar metals with different crystal structures. For example, dislocation arrays in stainless steel cold-worked prior to irradiation are cellular, as are those in copper produced by post-irradiated deformation. Kink-like arrays consisting of large fields of dislocations of like sign have so far only been observed in the hcp Zircaloy materials.

Deformation in unirradiated cladding material tends to be relatively uniform with dislocation densities exhibiting some noticeable but small variations from point-to-point in a deformed grain. However, it is worth mentioning that the buildup of dislocation density as a function of the degree of cold work is non-uniform rather than gradual at the lower cold work levels of less than twenty percent. For material cold worked about 10%, dislocations pile up near the grain boundaries, leaving the grain centers relatively free of excess dislocations.

Irradiated Zircaloy cladding often exhibits locallized deformation features as a result of post-irradiation deformation. Dislocation channeling effects can usually be observed near the fractures of the radiation-hardened or embrittled material. What commonly constitutes a channel is an area (volume) wherein radiation damage has been removed by the passage of dislocations, thereby making the passage of more dislocations likely until work hardening occurs.

Fig. 1. Underfocus (left) and overfocus (right) dark-field images forming 2½-D stereo image pair. Dislocation arrays in unirradiated, 70% cold worked Zircaloy-2 consist of separated positive and negative fields as evidenced by the reversing slopes in such images.

Figure 2 demonstrates how dislocation channeling in irradiated Zircaloy can give rise to rather sharp grain boundary offsets under conditions where the buildup of stress can be relieved in the grain boundary regions. When the 2½-D image pair of Figure 2 is viewed in stereo, two remarkable features can be ascertained. First, consistent with the fact that large numbers of dislocations must have used the channels to produce the observed grain boundary offsets (400-700 Burgers vectors), the crystallographic orientations are different on either side of a channel (elevation changes in 2½-D). Second, dislocations of opposite sign have, in general, used alternate channels producing the step-up, step-down appearance of the 2½-D image corresponding to nearly regular, nearly reversible orientation variations for this example.

Here again we have encountered dislocation distributions which clearly exhibit separation of dislocations with opposite signs. The details of the distributions are greatly different for irradiated and unirradiated materials but each has definite separate distributions.

The grain boundaries in Figure 2 caused no great impediment to the passage of dislocations and thus grain boundary shearing has occurred. In a neighboring grain in the same specimen, shown in Figure 3, the conditions were not as favorable, presumably because of orientation differences between grains. Grain boundary offsets were not produced. Instead, masses of

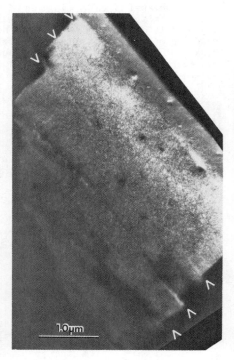

Fig. 2. 2½-D stereo image of dislocation channeling effects in Zircaloy-2 deformed in tension after neutron irradiation to 2.2 x 10^{21} n/cm^2 (E > 1 MeV). Dislocation channels are continuous grain to grain and remain relatively narrow.

Fig. 3. 2½-D stereo image of another grain in same specimen used for Figure 2. Dislocation channels are not able to pass through grain boundaries. Back stresses and dislocation walls produce large orientation reversals.

dislocations have aggregated in the grain boundary regions, their presence being responsible for the local orientation changes which produce depth variations in the 2½-D image. Again, it is the arrangements of dislocations which are of particular note. The dislocations are present in positive and negative arrays, producing a cyclic up-down 2½-D image near the grain boundary.

A further development of this behavior is illustrated in Figure 4, where an entire grain has been imaged in dark-field at low magnification. The grain boundaries still resist shearing and offsetting but do exhibit bowing. The buildup of stresses in this grain has been great enough to force the development of dense walls of dislocations extending nearly to the grain center from both sides of the grain. There are cyclic, periodic, regular lattice rotations which produce the depth in the 2½-D image. The grain has stored the energy of deformation by developing walls of dislocations and alternating the signs of the dislocations in the walls. What are seen as depth changes are, of course, orientation reversals. A Bragg contour running through such a microstructure is serrated as shown in Figure 5, taken in another grain of the same specimen.

Figure 6 illustrates the rather wide deformation bands which can occur during bulge testing. Dislocation channels were present in this specimen but these bands seemed to be somewhat of a modification of the normal channel. Besides being unusually broad, these bands are clearly associated with the corners of the upper grain. An unusual feature about this particular

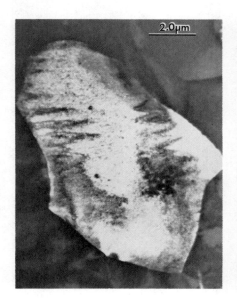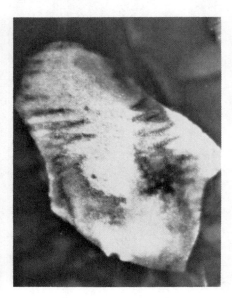

Fig. 4. 2½-D stereo image of grain in Zircaloy-2 deformed in tension after neutron irradiation to 1.2×10^{21} n/cm^2 (E > 1 MeV). Dense walls of dislocations with periodic reversals of sign produce the cyclic orientation variations which extend inward from the grain boundaries.

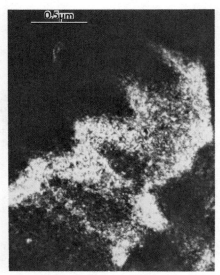

Fig. 5. 2½-D stereo image of a Bragg contour in the periodic deformation structure of sample used for Figure 4.

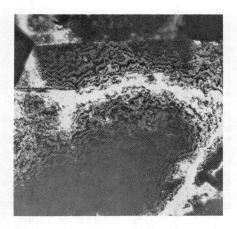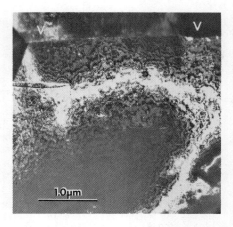

Fig. 6. 2½-D stereo image of dislocation bands in Zircaloy-2 bulge tested
after neutron irradiation to 1.1×10^{21} n/cm^2 (E > 1 MeV).

image is that all the dislocations (and most of the radiation damage, for
that matter) are invisible, the basal planes being used for dark-field imaging. This figure then serves to illustrate that the visible dislocation
densities and distributions are irrelevant to the analysis of 2½-D images.

Discussion

There is nothing new about observations of locallized deformation features of dislocation channeling in Zircaloy. The application of 2½-D imaging techniques and analyses, however, allows some of the details of the deformation process to be characterized. The tendency for neighboring channels, in specimens deformed in tension, to be used by dislocations of opposite signs is an interesting result which would be extremely difficult to
ascertain by other means.

The complicated deformation features in grains oriented so that grain
boundary shearing does not occur would be almost impossible to unravel without the aid of the 2½-D technique. Hindsight can be used, now that the 2½-D
images have provided descriptions of orientation variations, to determine
that contrast features are consistent with such variations, but the conclusion that, for example, wavy Bragg contours are indicative of periodic crystallographic orientation variations would be difficult to substantiate.

The main theme of this paper has been to emphasize the useful qualitative aspects of 2½-D imaging applied to deformation features in irradiated
Zircaloy. There are, additionally, opportunities for doing quantitative
analyses using such images. The excess dislocation densities of both the
positive and negative dislocation arrays in Figure 1, 70% cold-worked, unirradiated Zircaloy-2, have been analysed as 2.7×10^{11} cm^{-2} and 7.9×10^{10}
cm^{-2} respectively(3), the threefold difference in density being reflected in
the much steeper slope on the left side of the 2½-D stereo image. In a similar manner, by taking sufficient parallax measurements, the entire top half
of the grain of Figure 4 has been mapped to show point-to-point variations
in excess dislocation densities. The values range from a low of 1.3×10^{9}
cm^{-2} on the central ridge to a high of 2.4×10^{11} cm^{-2} in some of the canyon
walls near the sides of the grain(3).

Dislocation densities determined by analyses of 2½-D images are lower-bound estimates. There are numerous advantages in using such techniques compared to conventional counting techniques. Magnifications need not be high and large areas can be analysed for either point-to-point variations or average values. Neither the image magnification nor the foil thickness need be known to obtain dislocation densities. Very high densities of dislocations can be analysed and individual dislocation contrast or even visibility is not required(3).

Although not pursued in this study, the 2½-D technique may also provide quantitative means to investigate the mechanisms by which dislocation arrays determine the widths and spacings of channels in irradiated Zircaloy.

Conclusions

The through-focus dark-field imaging technique can be used to study deformation features in irradiated Zircaloy. Deformation effects tend to be localized as dislocation channels or bands if displacements can be transferred easily from grain to grain. When grain boundaries resist localized displacements, complicated deformation effects proceed inward from the boundary regions. 2½-D analyses allow the deformation patterns to be analysed as periodic orientation reversals resulting from arrays of dislocations in which the Burgers vectors change sign on a periodic basis.

References

1. P. B. Hirsch, A. Howie, R. B. Nicholson, D. W. Pashley and M. J. Whelan, Electron Microscopy of Thin Crystals, 4th ed., p. 422, Butterworths, London, 1971

2. W. L. Bell, "2½-D Electron Microscopy: Through-Focus Dark-Field Image Shifts", Journal of Applied Physics, 47 (4) (1976) pp. 1676-1682

3. W. L. Bell and H. S. Rosenbaum, "Microscopic Characterization of Deformed Polycrystalline Materials", to be submitted to Metallurgical Transactions

4. P. B. Hirsch, "Mosaic Structure", p. 249 in Progress in Metal Physics vol. 6, B. Chalmers and R. King, eds., Pergamon Press, London, 1956

5. J. F. Nye, "Some Geometric Relations in Dislocated Crystals", Acta Metallurgica, 1 (1953) pp. 153-162

THE STUDY OF DEFECTS, RADIATION DAMAGE AND IMPLANTED GASES IN SOLIDS BY FIELD-ION AND ATOM-PROBE MICROSCOPY†

David N. Seidman,* Jun Amano** and Alfred Wagner‡

Cornell University, Department of Materials Science and
Engineering and the Materials Science Center,
Bard Hall, Ithaca, New York 14853 U.S.A.

The ability of the field-ion microscope to image individual atoms has been applied, at Cornell University, to the study of fundamental properties of point defects in irradiated or quenched metals. The capability of the atom probe field-ion microscope to determine the chemistry — that is, the mass-to-charge ratio — of a *single* ion has been used to investigate the behavior of different implanted species in metals. A brief review is presented of: (1) the basic physical principles of the field-ion and atom-probe microscopes; (2) the many applications of these instruments to the study of defects and radiation damage in solids; and (3) the application of the atom-probe field-ion microscope to the study of the behavior of implanted ^{3}He and ^{4}He atoms in tungsten. The paper is heavily referenced so that the reader can pursue his specific research interests in detail.

† This work was supported by the U.S. Department of Energy. Additional support was received from the National Science Foundation through the use of the technical facilities of the Materials Science Center at Cornell University.

* John Simon Guggenheim Memorial Foundation Fellow 1980-81.

** Now at Hewlett-Packard Laboratories, 1501 Page Mill Rd., Palo Alto, CA 94304

‡ Now at Bell Laboratories, 600 Mountain Avenue, Murray Hill, NJ 07974

Introduction

In this paper an attempt is made to introduce the reader to some of the basic physical ideas involved in the field-ion and atom-probe field-ion microscope (FIM) techniques (see section on General Background Material), and to the applications of these techniques to the study of defects and radiation damage in solids (see section following the first one). The final section discusses, in précis form, the application of the atom-probe FIM to the study of the behavior of implanted ^{3}He and ^{4}He atoms in tungsten. The paper is heavily referenced so that the reader can pursue his specific research interests in detail.

General Background Material

The invention of the FIM and the atom-probe FIM by Müller (1,2) has provided the experimentalist with tools which allow both the direct observation of all the common defects (point, line, planar and precipitates) on an atomic scale and also the simultaneous determination of chemical effects on an atomic scale (the minimum detectable mass is equal to the mass of a *single atom*).

The atomic structure of the lattice is observed for those atoms which lie on the surface of a sharply pointed ($\sim$200 to 500Å in diameter) FIM specimen; the area imaged is $\sim10^{-10}$ to 10^{-11}cm^2. The information concerning the positions of the atoms is carried to a phosphor screen or a channel electron multiplier array (3,4) by an imaging gas which is typically helium or neon. The imaging gas atoms are ionized, by a tunneling mechanism, in the high local electric fields ($\sim$4.5 V Å^{-1} to ionize a helium atom) that exist at the site of individual atoms as a result of a positive potential applied to a sharply pointed FIM specimen (5). The positively-charged ions are repelled from the sharply-pointed specimen and then travel along the electric field lines to the phosphor screen which is at earth potential (see Figure 1); typically the phosphor screen is at a distance of between 4 to 10 cm from the FIM specimen. The image formed of the atoms on the surface of the FIM specimen, in the above manner, constitutes a point projection image with sufficient magnification to resolve individual atoms.

The interior of the specimen can be examined employing the field-evaporation process. The latter process consists of increasing the electric field to a value such that the potential energy curve for an ion on the surface of the specimen (this statement assumes that the state in which the metal atoms exist on the surface of the specimen is the ionic state) is deformed by the applied field to form a Schottky hump (6). The ions then evaporate (or sublime) by either jumping over this small Schottky hump as a result of a thermally-activated step or by tunneling through it; this process is called field evaporation or field desorption in the case of a solute atom. The field evaporation process can be controlled — with great precision — by applying the positive potential in the form of short (1 to 10 msec in width) high-voltage pulses. This latter technique is called pulse field evaporation; it is possible by this technique to dissect an atomic plane by removing one to two atoms per pulse. Thus, the atoms contained within the interior of the specimen can be imaged, albeit at the surface, at a rate which is determined by the experimentalist. In practice one can examine $\approx10^{-16}$ to 10^{-17}cm^3 of material, during the course of one afternoon, via the pulse field evaporation technique. At Cornell we have developed semi-automated techniques for the process of applying the field evaporation pulse in conjunction with the simultaneous recording of large numbers of frames of 35 ciné film [(15 to 30)$\times10^3$ frames per day] as well as developing techniques for the scanning of this film (7). It is clear, with the advantage of hindsight, that these two steps were essential to the successful application of the FIM technique to problems in the field of radiation damage.

The invention (2) of the time-of-flight (TOF) atom-probe FIM has provided the materials scientist with a unique instrument for the study of the interaction of impurity atoms or alloying elements with point, line or planar defects. The TOF atom-probe FIM (hereafter called an atom probe) consists of an FIM combined with a special TOF mass spectrometer (see Figure 1). This spectrometer allows the investigator to identify chemically any atom that appears in an FIM image. Thus, it is now possible to both image the microstructural features of a specimen and to measure the mass-to-charge ratio (m/n) of individual ions from preselected regions of a specimen with a lateral spatial resolution (i.e., within the surface) of a few angstroms and a depth spatial resolution that is determined by the interplanar spacing; the latter quantity can be tenths of an angstrom for a high index plane. An atom probe with a straight TOF tube has a mass-resolution (m/Δm) of ~200 while an atom probe with a Poschenrieder lens (8) has an m/Δm value of >1000 (9).

Figure 1 exhibits a schematic diagram that illustrates the main features of our straight TOF atom probe. A specimen with a radius of 50 to 500Å is maintained at a positive potential (3-20 kV) so that gas atoms surrounding the specimen are ionized over individual atomic sites and are projected radially outward to produce a visual image on the internal-image-intensification system. When a short high-voltage pulse is applied, atoms on the surface of the specimen are field evaporated in the form of ions. Those ions projected into the probe hole at the center of the internal-image-intensification system pass down the flight tube to the chevron ion detector. The TOFs of the ions and the voltages applied to the specimen are measured and the (m/n) ratios are calculated employing the equation:

$$m/n = 2e(V_{dc} + \alpha V_{pulse})\ (t-t_o)^2/d^2;$$

where e is the charge on an electron, V_{dc} the steady-state imaging voltage, V_{pulse} the pulse evaporation voltage, α the so-called pulse factor, d the flight distance, and $(t-t_o)$ the actual TOF of the ion. The quantity t is the observed TOF and t_o is the total delay time. The procedure we have developed to determine (m/n) is based on making V_{pulse} a constant fraction of V_{dc}; i.e., $V_{pulse} = fV_{dc}$ where f is a constant that is usually in the range 0.05 to 0.25 — the exact value depends on the specific alloy being analyzed. By the controlled pulse field evaporation of successive atomic layers it is possible to examine the bulk of the specimen and to reconstruct in three dimensions the correspondence between special microstructural features and chemical composition.

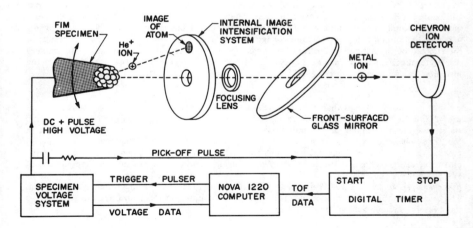

Fig. 1   Schematic diagram of the TOF atom-probe FIM. Shown at the top are the internal elements of the atom-probe including the FIM specimen, the internal-image-intensification system, the focusing lens, the 45° glass mirror, and the Chevron ion-detector. As indicated in the lower part of the figure, the specimen voltage system and the digital timer of the TOF mass spectrometer are operated automatically by a Nova 1220 minicomputer.

The details of the atom-probe are now summarized. The specimen is mounted on a liquid-helium-cooled goniometer stage which provides rotation about two orthogonal axes, thus allowing any portion of the specimen's surface to be projected into the probe hole for mass analysis. The goniometer stage is also translatable in three mutually orthogonal directions to facilitate alignment of the specimen with respect to the probe hole. The specimen is cooled by liquid helium in order to improve the quality of the FIM image and control the diffusivity of point defects; the temperature of the specimen is variable continuously from 13 to 450 K. The specimen is inserted into the goniometer stage via a high-vacuum ($<10^{-6}$ torr) specimen exchange device which allows rapid transfer of specimens without breaking the vacuum in the FIM. The specimen can also be irradiated *in-situ* with low energy gas ions (100 eV to $\sim$ 5 keV) employing a specially constructed ion-gun.

The internal-image-intensification system consists of a 75 mm diameter channel-electron-multiplier array (CEMA) and a phosphor screen with 5 mm diameter holes through their centers. The distance from the FIM specimen to the front surface of the internal-image-intensification system is continuously variable so that the magnification of the FIM image, as well as the size of the region projected onto the probe hole, can be varied by an areal magnification factor of $\sim$64X. An electrostatic lens immediately behind the internal-image-intensification system serves to focus those ions which pass through the probe hole down to a 1 mm diameter spot on the ion detector at the end of a 2.22 m long flight tube. The ion detector consists of two CEMAs placed in series in the Chevron configuration and a phosphor screen which provides a visual image of the ion beam. The atom probe was constructed to operate routinely in ultra-high vacuum ($\lesssim 5\times10^{-10}$ Torr) in order to minimize the interaction of residual gas atoms with the specimen.

The mass-spectrometer electronics, consisting of the specimen-voltage system and the digital timer are operated by a Nova 1220 minicomputer. This computerized system can automatically analyze up to 600 TOF events min^{-1} so that statistically significant results can be readily obtained even for small solute concentrations. As shown in the lower part of Figure 1 the computer triggers the V_{pulse} to the specimen which causes atoms on the surface of the specimen to be field-evaporated. A fraction of V_{pulse} is picked off and used to start an eight-channel digital timer which has a ± 10 nsec resolution. The pulses produced when ions strike the detector are used to stop the timer. A total of eight ion species from a single evaporation event can be identified. The controls of the dc and pulse power supplies, as well as the power supply for the focusing lens, are coupled so that the pulse and lens voltages are maintained at a constant fraction of the dc voltage. The values of V_{dc} and V_{pulse} applied to the specimen are measured by an analog-to-digital converter and read into the computer along with the TOF data. The (m/n) ratios are calculated by the computer and stored in the computer memory in the form of a histogram of the number of events versus m/n. In addition, the TOF and voltage data are stored on floppy discs so that the results of the run can be re-analyzed in the future. The computer is interfaced to a Tektronix 4010 graphics display terminal and Tektronix hard copy unit (not shown in Figure 1) so that a hard copy of the histogram can be obtained in $\sim$20 sec.

Three examples that illustrate the resolving power of our instrument are shown in Figures 2 and 3. Figure 2 exhibits the seven stable isotopes of molybdenum in the plus-two charge state; note that the isotopes are readily distinguished from one another. The peaks associated with the five stable isotopes of tungsten are also readily distinguished in the W^{+3} spectrum shown in Figure 3a. For comparison Figure 3b exhibits the stable isotopes of tungsten and rhenium in the plus-three charge state; this spectrum was obtained from a specimen with a nominal composition of W-25 at.% Re. The concentration of rhenium in the W-25 at.% Re alloy was determined by atom-probe analysis to be 22 ± 2 at.% and there was no apparent segregation or clustering of rhenium. The case of the tungsten-rhenium alloy is a worst case situation, for this instrument, as it requires a mass resolution approaching 200.

Additional material concerning the FIM and atom-probe techniques and their applications can be found in reference numbers (5, 10-17). The work performed at Cornell in the fields of defect physics and radiation damage has been summarized in detail in several review articles (18-21). For technical details concerning the computer controlled atom probe we have constructed at Cornell, see reference numbers (22-26).

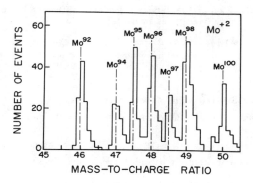

Fig. 2   Spectrum of Mo^{+2} obtained at a background pressure of 5×10^{-9} Torr, a specimen temperature of ~60 K and with the probe hole in the internal-image-intensification system near the (110) pole. The pulse fraction (f) was 0.025 and the calibration parameters used were $\alpha=1.482$, $t_o=0.56\mu$sec, and $d=2.213$ m. The total number of Mo^{+2} events in this histogram is 696.

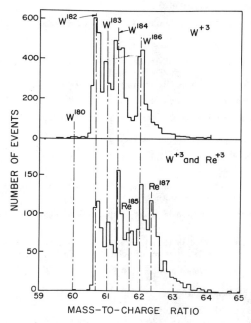

Fig. 3a   The W^{+3} spectrum of Westinghouse as-received tungsten pulse-field evaporated at 25 K with $f=0.05$ for V_{dc} varied continuously from 13 to 15 kV. The ions were collected from the (551) plane and the background pressure in the atom probe was 6×10^{-10} Torr. The calibration parameters used were $\alpha=2.0$, $t_o=0.56\mu$sec and $d=1.6003$ m. The total number of W^{+3} events in this histogram is 6045.

Fig. 3b   The W^{+3} and Re^{+3} spectrum of W-25 at.% Re thermocouple wire. The spectrum was recorded at a specimen temperature of ≈25 K with a $f=0.10$ at a pressure of 5×10^{-9} Torr. The calibration parameters used were $\alpha=1.5$, $t_o=0.56\mu$sec and $d=2.232$ m. The total number of W^{+3} and Re^{+3} events in this histogram is 1755.

A Catalogue of Applications to the Study of Defects, Radiation Damage and Implanted Gases in Solids

In this section we present, in catalogue form, a list of problems to which we have applied the FIM and atom probe techniques. The reader is referred to the references for the details concerning each problem.

Diffusive Properties of Self-Interstitial Atoms (SIAs)

a. Measured enthalpy change of migration (Δh_{ii}^m) of SIAs in pure metals, alloys and order-disorder alloys.

b. Measured the pre-exponential factor (D_{ii}^o) of the SIA self-diffusion coefficient.

c. Have studied SIAs in the recovery stages I, II and III of ion, electron or fast-neutron irradiated specimens.

d. The particular systems studied to date are W, W(Re), W(C), Mo, Pt, Pt(Au), Ni_4Mo and Pt_3Co.

e. For details see reference numbers (18, 19, 21, 27-40).

Volume Change of Migration (Δv_{ii}^m) of SIAs

a. Measured Δv_{ii}^m for the SIA in W, Pt and Mo in detail. Experiments were also performed on Ni_4Mo and Pt_3Co but in less detail.

b. For further details see reference numbers (18, 19, 27, 28, 31, 37-41).

Binding Enthalpy of an SIA to a Solute Atom (Δh^b)

a. Measured Δh^b by determining a dissociation enthalpy (Δh^d) and then determining Δh^b from the expression $\Delta h^d = \Delta h^b + \Delta h_{ii}^m$. The system Pt(Au) was studied in great detail and two thermally activated detrapping stages (II_B and II_C) were observed in Stage II.

b. The systems W(Re) and W(C) were also studied but in less detail.

c. For further details see reference numbers (19, 21, 36, 39, 40).

Diffusive Properties of Vacancies

a. Measured ratio of divacancy concentration to monovacancy concentration for one quench temperature in platinum specimens.

b. From (a) it was possible to determine the Gibbs free binding energy of a divacancy (Δg_{2v}^b) in platinum for one quench temperature.

c. Measured vacancy concentration in tungsten specimens which had been quenched from near the melting point.

d. The measurements discussed in (a) to (c) are important for the interpretation of the high-temperature self-diffusion data in terms of point-defect mechanisms.

e. For further details see reference numbers (18, 42-46).

Diffusive Properties of Gases in Metals

a. Diffusion of ^3He and ^4He in tungsten.

b. Diffusion of ^1H in tungsten.

c. For further details see reference numbers (47-51).

Range Profiles of Low-Energy Implanted Gases in Metals

a. Range profiles of ^3He and ^4He in tungsten (100 to 1500 eV singly-charged ions).

b. Range profiles of ^1H in tungsten.

c. For further details see reference numbers (47-53).

Point-Defect Structure of Depleted Zones: The Primary State of Radiation Damage

a. Depleted zones in ion-irradiated metals [W, Pt, Pt(Au)]

    (i)   Dimensions of depleted zones (DZs).

    (ii)   Number of vacancies per DZ.

    (iii)   Vacancy concentration within a DZ.

    (iv)   The distribution of first-nearest-neighbor vacancy clusters within a DZ.

    (v)   The radial distribution function for the vacancies within a DZ, out to ninth-nearest-neighbor.

b. Effect of projectile mass (M_1) on the vacancy structure of DZs at constant projectile energy (E_1).

c. Effect of E_1 — at constant M_1 — on the vacancy structure of DZs.

d. For further details see reference numbers (18, 20, 21, 49, 54-62).

Radiation Damage Profiles

a. Radiation damage profiles were measured in tungsten and platinum by determining the positions of all the vacancies, contained within DZs, as a function of distance from the irradiated surface.

b. Direct determination of radiation damage profiles in order-disorder alloys. After an irradiation each specimen was dissected on an atom-by-atom basis and the change in the Bragg-Williams long-range order parameter was determined as a function of distance from the irradiated surface. This approach was applied in great detail to Pt_3Co, which had been irradiated with 250 to 2500 eV Ne^+ ions, and in less detail to Ni_4Mo.

c. For further details see reference numbers (38, 49, 59, 60).

Sputtering of Surfaces

a. The sputtering of a metal surface is the result of the intersection of a collision cascade with the surface. In this work we compared the vacancy structure of DZs, produced by 30 keV W^+, Mo^+, Cr^+, Cu^+ or Ar^+ ions, that were found to have intersected the surface of a tungsten FIM specimen with those found in the bulk of the specimen.

b. For further details see reference numbers (64, 65).

Voids in Neutron-Irradiated Metals [Mo, Mo(Ti), Fe(Cu)]

a. Void number density: need a number density of $\approx 10^{17} cm^{-3}$ in order to be able to make measurements.

b. Void size distribution: same comment as above is applicable.

c. Direct observation of segregation of alloying elements.

d. For further details see reference numbers (66-68).

Distribution of SIAs in the Primary Damage State

a. The distribution of SIAs was determined in tungsten which had been ion irradiated at 10 K — below the Stage I recovery peaks — with 18 keV Au^+, 20 keV W^+ or 30 keV Cr^+ ions along high index crystallographic directions.

b. From the distribution of SIAs we were able to place upper limits on the ranges of focused replacement collision sequences in tungsten.

c. For further details see reference numbers (54,69).

Range Profiles of Low-Energy (100 to 1500 eV) Implanted
^{3}He and ^{4}He Atoms and the Diffusivity of ^{3}He and ^{4}He in Tungsten

General Background

Current interest in the fundamental properties of helium in metals has been generated by the materials problems associated with the development of the liquid-metal fast-breeder reactor (70) and the controlled thermonuclear reactor (71). However, because of a lack of appropriate experimental techniques the investigations of the range of low-energy (<1 keV) implanted helium atoms and the diffusivity of He in metals have been largely theoretical (72-74). Measurement of the range profiles of implanted He ions have been confined to energies (75) >1 keV; furthermore, the measurement of both the range profiles of implanted He and the diffusivity of He in metals have relied exclusively on the trapping of He at lattice defects introduced as a result of heavy-ion irradiation (76).

The accomplishments of our research on helium implanted in tungsten were: (1) the establishment of the ability of the atom-probe FIM to detect either implanted ^{3}He or ^{4}He atoms retained in a perfect (i.e., totally defect-free) lattice; (2) the detection of the presence of an isolated and immobile ^{3}He or ^{4}He atom in a perfect tungsten lattice; (3) the measurement of the range profiles of low-energy (100 to 1500 eV) implanted ^{3}He or ^{4}He atoms in a tungsten lattice; and (4) the measurement of the diffusivities of ^{3}He and ^{4}He in a perfect tungsten lattice.

The basic physical ideas involved in the experimental procedures are illustrated sequentially in Figure 4. A single-crystal tungsten FIM specimen, at an irradiation temperature (T_i), was irradiated **in situ** with ^{3}He$^+$ or ^{4}He$^+$ ions parallel to the [110] direction as shown in Figure 4(a). To study the diffusional behavior of either ^{3}He or ^{4}He in tungsten it was necessary to implant the helium under the condition of **no** radiation damage. For example, a 300-eV ^{4}He atom can transfer a maximum energy of $\sim$ 25 eV to a tungsten W atom in a head-on two-body elastic collision. Since the minimum displacement energy for the production of a stable Frenkel pair in tungsten is ≈42 eV (77), no self-interstitial atoms (SIAs) or vacancies were created at an implantation energy of 300 eV for either ^{3}He or ^{4}He. Thus for the diffusion experiments a standard implantation energy of 300 eV was employed. With no SIAs or vacancies present to act as trapping centers, implanted ^{3}He or ^{4}He atoms can remain in the specimen **only** if ^{3}He or ^{4}He is immobile at T_i. Thus, the state of the tungsten specimen after an implantation consisted of immobile interstitial ^{3}He or ^{4}He atoms implanted in a perfect tungsten lattice with a depth distribution that was determined solely by the range profile of the low-energy ions. Next the specimen was analyzed chemically, by the atom-probe technique, at a standard reference temperature (T_r), where $T_r \leq T_i$, and a ^{3}He or ^{4}He integral profile was plotted as shown in Figure 4(b); this was an integral profile since it measured the cumulative number of ^{3}He or ^{4}He atoms as a function of the cumulative number of tungsten atoms (depth) from the irradiated surface. The depth scale was converted from cumulative number of tungsten atoms to angstroms from the measured number of tungsten atoms per (110) plane contained within the cylindrical element sampled; see Figure 4(a). Finally the ^{3}He or ^{4}He range profile, Figure 4(c), can be constructed by taking the first derivative of the integral profile shown in Figure 4(b); or alternatively by plotting a frequency distribution diagram.

A novel technique for the determination of an absolute depth scale was developed; Figure 5 schematically illustrates the method. During the atom-probe analysis the specimen was oriented and the magnification adjusted so that only the central portion of the (110) plane of the tungsten specimen was chemically analyzed. The specimen was then pulse field evaporated through the repeated application of high-voltage pulses. Three successive stages in the pulse field evaporation of one (110) plane are indicated in Figure 5(a). As the specimen was pulsed, field-evaporated ions were detected as indicated by the positive slope in Figure 5(b). When a plane completely evaporated the slope of the curve in Figure 5(b) returned to zero. Therefore the removal of one (110) plane resulted in a single-step increase in the plot of the number of tungsten atoms detected versus the number of field-evaporation pulses applied to the specimen. Since the tungsten lattice was employed as a depth marker, the absolute depth of each implanted ^{3}He or ^{4}He atom from the initial irradiated surface was measured to within one (110) interplanar spacing ($\approx$ 2.24Å) **independent** of the total depth of analysis. Thus the spatial depth resolution of the atom-probe technique is limited solely by the interplanar spacing of the region being analyzed.

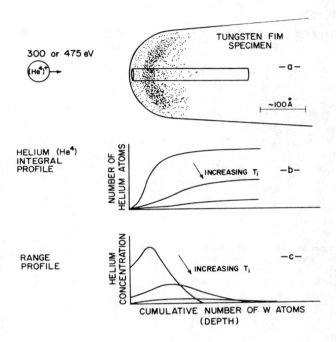

Fig. 4   (a) The *in situ* irradiation of a tungsten FIM specimen with 300 eV ^{4}He$^+$ ions at a T_i where the implanted ^{4}He atoms are immobile. The density of spots corresponds to the approximate range profile of ^{4}He in tungsten. The cylindrical volume element represents the volume chemically analyzed by the atom probe technique. (b) The cumulative number of ^{4}He atoms versus depth as a function of T_i. Note that the ^{4}He integral profile tends to flatten out as a T_i is increased. (c) The range profiles of ^{4}He in tungsten as a function of T_i. The same concepts illustrated here for ^{4}He apply, of course, to ^{3}He.

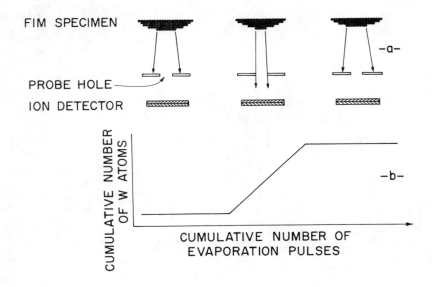

FIM SPECIMEN

PROBE HOLE

ION DETECTOR

-a-

CUMULATIVE NUMBER OF W ATOMS

CUMULATIVE NUMBER OF EVAPORATION PULSES

-b-

Fig. 5.   A schematic diagram illustrating the method employed to determine an absolute depth
scale. Three states in the field evaporation of one (110) plane of tungsten W are shown
in (a). The field-evaporation behavior of this plane is indicated in (b) by the steplike
increase in the rate at which tungsten atoms are detected.

Integral and range profiles of low-energy implanted ^{3}He and ^{4}He atoms

In this section we present a number of integral profiles and range profiles for both ^{3}He
and ^{4}He which had been implanted in tungsten at 60 K. The term integral profile reflects the
manner in which the data was recorded [see Figure 4(b)], whereas the range profile was constructed
by plotting a frequency distribution diagram from the integral profile [see Figure 4(c)]. The range
profile can also be obtained by drawing a smooth curve through the integral profile and taking the
first derivative of this curve. In all cases we have obtained the range profile by the former rather
than the latter technique.

Figure 6 exhibits ^{3}He integral profiles for the implantation energies 100, 500 and 1500
eV; the 100 eV profile is a composite of two integral profiles, each at a dose of 4.7×10^{15}ions cm^{-2};
both the 500 and 1500 eV integral profiles were obtained after implanting to a dose of
3×10^{15}ions cm^{-2}. In Figure 7 we show a composite range profile for 300 eV ^{3}He ions; this range
profile was constructed from seven integral profiles and includes a total of 385 ^{3}He events; the
values of the mean range ($\bar{x}$) and the straggling (Δx) are 54.9 and 41.5 Å, respectively.

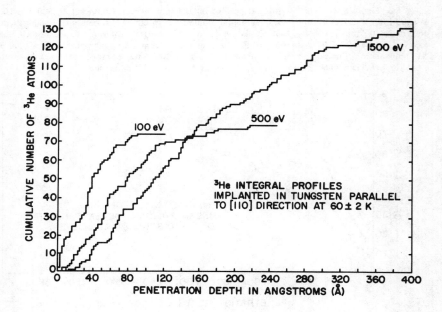

Fig. 6.   The ³He integral profiles for the implantation energies of 100, 500 and 1500 eV. The
tungsten specimens were implanted at 60 K parallel to the [110] direction.

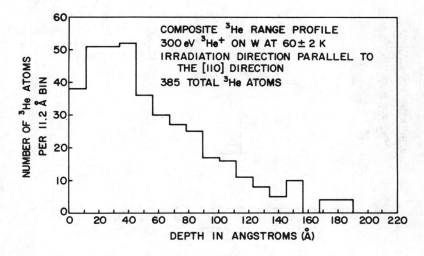

Fig. 7.   A composite ³He range profile for all the 300 eV implantations at 60 K. A total of 385
³He events were involved in the construction of this range profile.

Figure 8 exhibits ^{4}He integral profiles for the implantation energies 150, 500 and 1000 eV; the 150 eV data consists of a single integral profile for a specimen that had been implanted to a dose of 3×10^{15} ions cm^{-2}; the 1000 eV data is for a single integral profile for a specimen that had received a dose of 4×10^{15}ions cm^{-2}. In Figure 9 we show a composite range profile for a 1000 eV ^{4}He implantation; this range profile was constructed from three integral range profiles and includes a total of 147 ^{4}He events; the values of $\bar{x}$ and Δx are 133 and 104.2 Å, respectively.

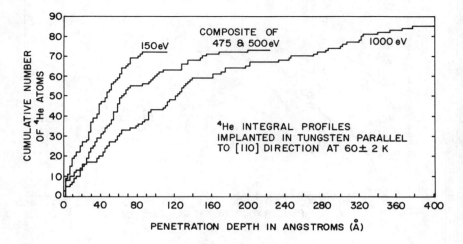

Fig. 8.   The ^{4}He integral profiles for the implantation energies of 150, (475 and 500), and 1000 eV. The tungsten specimens were implanted at 60 K parallel to the [110] direction.

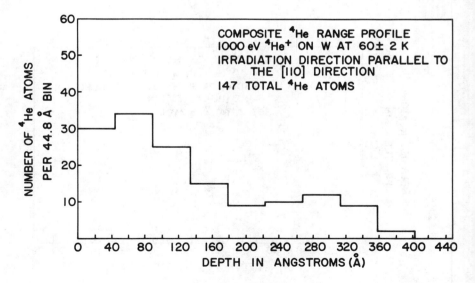

Fig. 9.   A composite ^{4}He range profile for all the 1000 eV implantations at 60 K. A total of 147 ^{4}He events were recorded in the construction of this range profile.

136

All of the ^{3}He and ^{4}He integral profiles exhibited positive skewness as expected for low-energy irradiations [Biersack and Haggmark (78)]; this implied that the mean range (7) was greater than the most probable range (or mode) and that the majority of the large deviations were to the right (positive) side of $\bar{x}$. The coefficient of skewness is related to the third moment about $\bar{x}$ and is given [Parratt (79)] by:

$$\text{Coefficient of Skewness} = \frac{\sum_{i=1}^{N}(x_i-\bar{x})^3}{N(\Delta x)^3}; \qquad (1)$$

where x_i is the measured depth of the ith detected helium atom from the initial irradiated surface, N is the total number of helium events detected and Δx is the standard deviation or straggling.

The values of $\bar{x}$ and Δx were calculated directly from the integral profiles for the ^{3}He and ^{4}He implantations. The values $\bar{x}$ and Δx were referred to as uncorrected quantities. The reason for this is that $\bar{x}$ and Δx must be corrected for the following systematic errors: (1) the random arrival of helium atoms at the surface of the specimen, from the residual partial pressure of helium, during the atom-probe analysis of the irradiated specimen; and (2) the effect of the finite curvature of the FIM specimen. A detailed analysis of the above effects is given elsewhere [Amano, Wagner and Seidman (52,53)], where it was shown that the corrections to $\bar{x}$ and Δx were very minimal in our experiments. Thus we shall not employ the word uncorrected any further in this paper.

Figure 10 exhibits $\bar{x}$ (in Å) versus the incident ion energy (in eV) for both the ^{3}He (solid black circles) and the ^{4}He (open circles) implantations. The total length of each error bar is two standard deviations in the mean (Δx_m), i.e., plus or minus one Δx_m. The quantity Δx_m is given by:

$$\Delta x_m = \Delta x/\sqrt{N}, \qquad (2)$$

where N was the total number of helium events detected at a particular incident ion energy for the composite profile. Figure 10 clearly shows that the quantity Δx_m was negligibly small when N exceeded 50 events. The smallest sample size was for the 100 eV ^{4}He implantation where N was equal to 21 events. In this case Δx_m was 3.5Å and the fractional standard error ($\Delta x_m/\bar{x}$) was

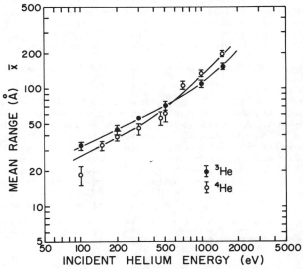

Fig. 10.   The mean range ($\bar{x}$) in angstroms versus the incident helium energy (eV) for all the ^{3}He and ^{4}He implantations in tungsten at 60 K.

137

~0.19. This was the only data point that laid slightly below the smooth line that passes through all the other data points for the ^{4}He implantations.

The results presented in Figure 10 show that $\bar{x}$ for both ^{3}He and ^{4}He increased monotonically, although not linearly, with increasing incident helium ion energy. Overall, for both ^{3}He and ^{4}He, the value of $\bar{x}$ increased from 18.7 to 194.9Å as the incident ion energy was increased from 100 to 1500 eV. For an incident helium ion energy of less than 600 eV the $\bar{x}$'s for ^{3}He were greater than the $\bar{x}$'s for ^{4}He; this indicated that ^{3}He penetrated more deeply into the lattice, on the average, than ^{4}He. At incident helium ion energies greater than 600 eV the $\bar{x}$'s for ^{4}He exceeded the $\bar{x}$'s for ^{3}He. A detailed discussion of these effects is given elsewhere [Amano, Wagner and Seidman (52,53)].

Figure 11 displays Δx as a function of the incident helium ion energy for both ^{3}He and ^{4}He; Δx is very commonly known as the straggling, since it determines the width of the range profile. The length of each error bar in figure 11 is equal to two universe standard deviations in the sample standard deviation (Δx_s), i.e., plus or minus one Δx_s. The quantity Δx_s, for a normal distribution, is given by [Parratt (79)]

$$\Delta x_s = \Delta x/\sqrt{2N}. \tag{3}$$

We have used Eqn. (3) to obtain approximate values of Δx_s for our range profiles, which are actually skewed from a normal distribution. It is seen from Figure 11 that for the sample sizes we employed the values of Δx_s were all rather small.

For both ^{3}He and ^{4}He the value of Δx increased monotonically, although **not** linearly, with increasing incident helium ion energy (see Figure 11). The quantity Δx ranged from 16 to 124 Å as the incident helium ion energy was increased from 100 to 1500 eV. At an incident helium ion energy of ~300 eV the two curves crossed one another and the Δx's for ^{4}He were greater than those for ^{3}He. This indicated that as the incident ion energy was increased the ^{4}He was distributed in space, both wider and deeper than ^{3}He.

Figure 12 exhibits the relative variance $[(\Delta x)^2/(\bar{x})^2]$ of the ^{3}He and ^{4}He range profiles, as a function of the incident helium ion energy (in eV). Within the scatter of the data the quantity $(\Delta x)^2/(\bar{x})^2$ for ^{3}He exhibited a constant value of ~0.47 and the same quantity for ^{4}He was ~0.61. Thus in the energy range 100 to 1500 eV the value of $(\Delta x)^2/(\bar{x})^2$ for ^{4}He was greater than for ^{3}He. This clearly indicated that the ^{4}He was distributed more broadly in space than was the ^{3}He.

Detection of possible radiation damage in the case of the 300 eV helium implantations

In order to establish that the ^{4}He detected in the case of the 300 eV implantation experiment was **not** trapped at structural defects in the tungsten lattice, the following isochronal recovery experiment was performed. A tungsten specimen was irradiated along the [110] direction with 300-eV ^{4}He$^+$ ions at ~30 K. After the irradiation $\approx$ two (110) planes, corresponding to $\approx$4.48 Å of material, were pulse field evaporated from the specimen. This procedure removed the sputtered surface and restored the surface to a nearly perfect state. The specimen was then warmed isochronally from $\approx$30 to 90 K at a rate of 1.5 K min^{-1}, while the FIM image was photographed at a rate of two 35-mm cine frames sec^{-1}. No SIA contrast effects were observed during this experiment, indicating that **no** SIA crossed the surface of the FIM specimen.† Our previous work (80) demonstrated that if·SIAs were present they would have appeared throughout the entire range of 38 to 90

---

† Helium atoms do not give rise to visible contrast effects in the FIM image, thus this type of experiment can not be used to detect the recovery behavior of interstitial helium atoms. Note, however, that the depth profiling experiments demonstrated that helium does **not** become mobile until 90 K.

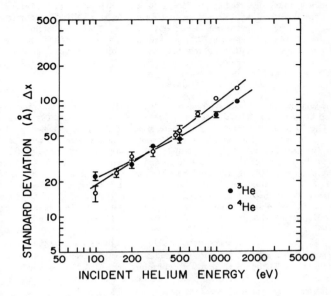

Fig. 11.   The standard deviation or straggling in angstroms (Δx) versus the incident helium energy (eV) for all the ^{3}He and ^{4}He implantations in tungsten at 60 K.

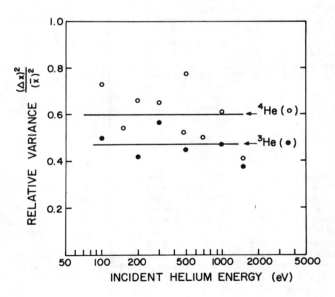

Fig. 12.   The relative variance $[(\Delta x)^2/(\bar{x})^2]$ versus the incident helium energy (eV) for all the ^{3}He and ^{4}He implantations in tungsten at 60 K.

K. The specimen was then dissected by the pulse field evaporation technique and was examined for point defects. The density of point defects was determined to be $<8\times10^{-4}$ (atomic fraction); their depth distribution was not related to the ^{4}He integral profiles. These results constitute conclusive evidence that the ^{4}He was **not** trapped at SIAs or vacancies. This indicated that the ^{4}He atoms were located in the interstices of the lattice and that they were immobile in tungsten at 60 K.

The diffusivities of ^{3}He and ^{4}He in tungsten

The temperature at which the interstitial ^{4}He (or ^{3}He) atoms became mobile in tungsten was determined by implanting ^{4}He (or ^{3}He) in an FIM specimen at different T_i's and then analyzing at T_r=60 K. The ^{4}He (or ^{3}He) integral profile determined at T_r was independent of T_i only if the ^{4}He (or ^{3}He) was immobile at all values of T_i. However, when T_i was above the temperature at which the ^{4}He (or ^{3}He) interstitials became mobile, the ^{4}He (or ^{3}He) implanted during the irradiation diffused to the surface of the FIM specimen and entered the gas phase. Therefore a sharp decrease in the measured ^{4}He (or ^{3}He) concentration was expected as T_i was increased [see Figure 4(c)]. Since only T_i was varied, significant changes in the integral profile could only be attributed to a sharp increase in the mobility of the interstitial ^{4}He (or ^{3}He) atoms at T_i. A dramatic change in the integral profile was observed upon increasing T_i from 90 to 110 K; thus indicating that interstitial ^{4}He or (^{3}He) atoms were immobile at 90 K but were highly mobile at 110 K. By employing a diffusion model, a value of the enthalpy change of migration ($\Delta h^m_{4_{He}}$) of 0.24 to 0.32 eV was estimated (47,48). The upper and lower limits on $\Delta h_{4_{He}}$ were determined by the values of the pre-exponential factor (D_0) chosen for the diffusion model and by the uncertainty in the diffusion temperature, i.e., 90 to 110 K; the lower limit was determined by a D_0 of 1×10^{-3} cm sec^{-1} and a T of 90 K and the upper limit by a D_0 of 1×10^{-2} cm^2 sec^{-1} and a T of 110 K. The uncertainty in $\Delta h^m_{4_{He}}$ was divided approximately equally between the uncertainty in D_0 and T.

The diffusivity of ^{3}He in tungsten was determined by actually following the isothermal recovery of 300 eV implantation profiles which had been implanted at 90, 95, 98, 100 and 110 K. The diffusion equation was solved with appropriate initial and boundary conditions, to describe the diffusion of ^{3}He out of an FIM tip under isothermal conditions. The fit of the experimental isothermal recovery data to the solution of the diffusion equation yielded the diffusivity of ^{3}He as a function of temperature. The results of this work are shown in Figure 13. It is seen that the data is best described by the expression

$$D(^3He) = (5.4 \pm \begin{smallmatrix}10.6\\3.8\end{smallmatrix})\times10^{-3} \exp[\frac{-0.28 \text{ eV}}{kT}]cm^2sec^{-1}.$$

Thus within the measured experimental uncertainties the Δh's for ^{3}He and ^{4}He in tungsten are identical. For further details on the diffusivity of ^{3}He in tungsten see Amano and Seidman (51).

Conclusions

The field-ion and atom-probe field-ion microscopes are ideally suited for studying a wide range of fundamental problems, concerning point defects in irradiated or quenched metals, that require information on an atomic scale. The atom-probe field-ion-microscope with its ability to measure the mass-to-charge ratio of a single ion can be brought to bear on problems that need chemical information on an angstrom scale. Since the atom-probe field-ion microscope is essentially a field-ion microscope coupled to a special time-of-flight mass spectrometer the main features of both instruments are contained in the atom probe. Thus, it is now possible to both image the microstructural features, with atomic resolution, of a specimen and to subsequently measure the mass-to-charge ratio of individual ions from these microstructural features, with a lateral spatial resolution (i.e., within the surface) of a few angstroms and a depth resolution that is determined by the interplanar spacing; the latter quantity can be **tenths** of an angstrom for a high index plane.

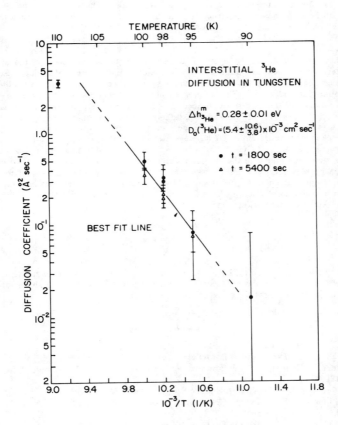

Fig. 13.   The diffusion coefficient of ^{3}He versus $(1/T)$ in the temperature range 90 to 110 K. The times indicated (1800 and 5400 sec) correspond to different recovery times at each temperature.   Note that $1\mathring{A}^2sec^{-1}$ is equal to $10^{-16}cm^2\ sec^{-1}$.

141

The principal shortcoming of the atom probe field-ion microscope is that only the more refractory metals and alloys have so far been routinely easily studied. Thus, to date, our own studies have focused heavily on the pure metals tungsten, molybdenum and platinum; the alloys tungsten (rhenium), tungsten (carbon), platinum (gold) and molybdenum (titanium), and the order-disorder alloys Ni_4Mo and Pt_3Co. This is not meant to imply that other pure metals or alloy systems are not amenable to the technique, but rather that they were the ones which we found easiest to work on in order to study certain specific physical problems. For example, it was possible to chemically analyze, on a quantitative basis, both an alloy as complex as stainless steel and a quaternary metallic glass (Metglas 2826) (24). The application of the atom-probe field-ion microscope to the study of clustering and precipitation in some reasonably complex alloys has been reviewed recently by Brenner (81) and the reader is referred to his paper for details. The atom-probe field-ion microscope could also be used to study the structure of metallic glasses as has been suggested recently by Jacobaeus et al. (82).

This paper, first, briefly reviewed the physical basis of the field-ion and atom-probe microscope techniques (see the section entitled **General Background Material**). Second we presented, in catalogue form, a list of the problems to which we have applied the field-ion microscope and atom probe techniques. This section contains many references to the original papers, so that the reader can follow his own research needs in detail. Finally, a presentation was made of our recent results on the behavior of ^3He and ^4He atoms implanted in tungsten, employing the atom-probe field-ion microscope technique. The range profiles of low-energy (100 to 1500 eV) ^3He and ^4He atoms implanted in tungsten, at 60 K, were measured with a depth resolution of one (110) interplanar spacing (2.24Å). At 60 K both ^3He and ^4He were found to be immobile. Thus, the range profiles were determined in the complete absence of any diffusional effects. All of the ^3He and ^4He range profiles exhibited positive skewness. Both ^3He and ^4He were found to be highly mobile in the temperature range 90 to 110 K. Experiments were described that measured, for the first time, the diffusivities of ^3He and ^4He atoms in the absence of any defects — i.e., in a perfect crystal lattice — in this temperature range.

Acknowledgements

We wish to thank Dr. S. S. Brenner for many useful discussions over the years concerning the field-ion and atom-probe microscope techniques and Mr. R. Whitmarsh for enthusiastic technical assistance in all phases of the experimental program.

References

1. E. W. Müller, Z. Physik *131*, 136 (1951).
2. E. W. Müller, J. A. Panitz and S. B. McLane, Rev. Sci. Instrum. *39*, 83 (1968).
3. P. J. Turner, P. Cartwright, M. J. Southon, A. van Oostrom and B. W. Manley, J. Sci. Instrum. *2*, 731 (1969).
4. S. S. Brenner and J. T. McKinney, Surface Sci. *23*, 88 (1970).
5. E. W. Müller and T. T. Tsong, *Field-Ion Microscopy* (American Elsevier, New York, 1969), Chapt. II.
6. E. W. Müller and T. T. Tsong, *Field-Ion Microscopy* (American Elsevier, New York, 1969), Chapt. III.
7. R. M. Scanlan, D. L. Styris, D. N. Seidman and D. G. Ast, Cornell Materials Science Center Report No. 1159 (1969).
8. W. P. Poschenrieder, Int. J. Mass Spectrom. and Ion Phys. *9*, 357 (1972).
9. E. W. Müller and S. V. Krishnaswamy, Rev. Sci. Instrum. *45*, 1053 (1974).
10. R. Gomer, *Field Emission and Field-Ionization* (Harvard University Press, Cambridge, Massachusetts, 1961).

11. *Field-Ion Microscopy*, edited by J. J. Hren and S. Ranganathan (Plenum Press, New York, 1968).
12. *Applications of Field-Ion Microscopy*, edited by R. F. Hochman, E. W. Müller and B. Ralph (Advanced Research Projects Agency, ARPA order No. 878 and the Georgia Institute of Technology, 1969).
13. K. M. Bowkett and D. A. Smith, *Field-Ion Microscopy* (American Elsevier, New York, 1970).
14. *Field-Ion, Field Emission Microscopy and Related Topics*, special issue of Surface Sci. *23*, 1 (1970).
15. E. W. Müller and T. T. Tsong, in *Progress in Surface Science*, edited by S. G. Davison (Pergamon Press, Oxford, 1973), Vol. 4, Part I, pp. 1-139.
16. J. A. Panitz, in *Progress in Surface Science*, edited by S. G. Davison (Pergamon Press, Oxford, 1978), Vol. 8, pp. 219-62.
17. *Proceedings of the Int. Symp. on Application of FIM to Metallurgy*, edited by R. R. Hasiguti, Y. Yashiro and N. Igata (Dept. of Metallurgy and Materials Science, Univ. of Tokyo, Tokyo, Japan, 1977); *Proceedings of the 27th International Field Emission Symposium*, edited by Y. Yashiro and N.. Igata (Dept. of Metallurgy and Materials Science, Univ. of Tokyo, Tokyo, Japan, 1980).
18. D. N. Seidman, J. Phys. F: Metal Phys. *3*, 393 (1973).
19. D. N. Seidman, K. L. Wilson and C. H. Nielsen, in *Proceedings of the Int. Conf. on Fundamental Aspects of Radiation Damage in Metals*, edited by M. T. Robinson and F. W. Young, Jr. (National Technical Information Service, U.S. Dept. of Commerce, Springfield, Virginia, 1975), pp. 373-96.
20. D. N. Seidman, in *Radiation Damage in Metals*, edited by N. L. Peterson and S. D. Harkness (American Society for Metals, Metals Park, Ohio, 1976), pp. 28-57.
21. D. N. Seidman, Surface Sci. *70*, 532 (1978).
22. A. Wagner, T. M. Hall and D. N. Seidman, Rev. Sci. Instrum. *4*, 1032 (1975); A. S. Berger, Rev. Sci. Instrum. *44*, 592 (1973).
23. T. M. Hall, A. Wagner, A. S. Berger and D. N. Seidman, Scripta Met. *10*, 485 (1976).
24. T. M. Hall, A. Wagner and D. N. Seidman, J. Phys. E: Sci. Instrum. *10*, 884 (1977).
25. A. Wagner, T. M. Hall and D. N. Seidman, J. Nuc. Mat. *69 & 70*, 532 (1978).
26. A. Wagner, T. M. Hall and D. N. Seidman, Vacuum *28*, 543 (1978).
27. R. M. Scanlan, D. L. Styris and D. N. Seidman, Phil. Mag. *23*, 1439 (1971).
28. R. M. Scanlan, D. L. Styris and D. N. Seidman, Phil. Mag. *23*, 1459 (1971).
29. P. Petroff and D. N. Seidman, Appl. Phys. Lett. *18*, 518 (1971).
30. D. N. Seidman and K. H. Lie, Acta Met. *20*, 1045 (1972).
31. P. Petroff and D. N. Seidman, Acta Met. *21*, 323 (1973).
32. J. T. Robertson, K. L. Wilson and D. N. Seidman, Phil. Mag. *27*, 1417 (1973).
33. D. N. Seidman, K. L. Wilson and C. H. Nielsen, Phys. Rev. Lett. *35*, 1041 (1975).
34. K. L. Wilson and D. N. Seidman, Radiat. Effects *33*, 149 (1977).
35. D. N. Seidman, Scripta Met. *13*, 251 (1979).
36. K. L. Wilson, M. I. Baskes and D. N.. Seidman, Acta Met. *28*, 89 (1980).
37. C. H. Nielsen, M. S. Thesis, Cornell University (1977).
38. J. Aidelberg, Ph.D. Thesis, Cornell University (1980).
39. C.-Y. Wei and D. N. Seidman, Radiat. Effects, *32*, 229 (1977).
40. C.-Y. Wei and D. N. Seidman, J. Nuc. Mat *69&70*, 693 (1978).
41. K. L. Wilson and D. N. Seidman, Rad. Effects *27*, 67 (1975).
42. A. S. Berger, D. N. Seidman and R. W. Balluffi, Acta Met *21*, 123 (1973).
43. A. S. Berger, D. N. Seidman and R. W. Balluffi, Acta Met. *21*, 323 (1973).
44. J. Y. Park, H.-C. W. Huang, A. S. Berger, and R. W. Balluffi, in *Defects and Defect Clusters in B.C.C. Metals and Their Alloys*, edited by R. J. Arsenault (University of Maryland, College Park, MD, 1973), Nuclear Metallurgy, Vol. 18, pp. 420-439.
45. J. Y. Park, Ph.D. Thesis, Cornell University (1975).
46. H.-C. W. Huang, Ph.D. Thesis, Cornell University (1975).
47. A. Wagner and D. N. Seidman, Phys. Rev. Lett. *42*, 515 (1979).
48. A. Wagner, Ph.D. Thesis, Cornell University (1978).
49. D. N. Seidman, U.S. Department of Energy Report No. COO-3158-77 (1979).
50. D. N. Seidman, U.S. Department of Energy Report No. COO-3158-87 (1980).

51. J. Amano and D. N. Seidman, Cornell Materials Science Center Report No. 4153 (1980); submitted for publication.
52. J. Amano, A. Wagner and D. N. Seidman, Cornell Materials Science Center Report No. 4107 (1980); submitted for publication.
53. J. Amano, A. Wagner and D. N. Seidman, Cornell Materials Science Center Report No. 4108 (1980); submitted for publication.
54. L. A. Beavan, R. M. Scanlan and D. N. Seidman, Acta Met. *19*, 1339 (1971).
55. K. L. Wilson and D. N. Seidman, in *Defects and Defect Clusters in B.C.C. Metals and Their Alloys*, edited by R. J. Arsenault (University of Maryland, College Park, MD, 1973), Nuclear Metallurgy, Vol. 18, pp. 216-239.
56. C.-Y. Wei and D. N. Seidman, Phil. Mag. A*37*, 257 (1978).
57. C.-Y. Wei, Ph.D. Thesis, Cornell University (1978).
58. C.-Y. Wei and D. N. Seidman, Appl. Phys. Lett *34*, 622 (1979).
59. D. Pramanik, Ph.D. Thesis, Cornell University (1980).
60. D. N. Seidman, M. I. Current, D. Pramanik and C.-Y. Wei, Cornell Materials Science Center Report No. 4278 (1980); accepted for publication in Nuclear Instruments and Methods.
61. C.-Y. Wei, M. I. Current and D. N. Seidman, Cornell Materials Science Center Report No. 4234 (1980); submitted for publication.
62. M. I. Current, C.-Y. Wei and D. N. Seidman, Cornell Materials Science Center Report No. 4309 (1980); submitted for publication.
63. J. Aidelberg and D. N. Seidman, Nucl. Instrum. and Meth *170*, 413 (1980).
64. M. I. Current, C.-Y. Wei and D. N. Seidman, Cornell Materials Science Center Report No. 4193 (1980); to appear in the Philosophical Magazine (1980).
65. M. I. Current and D. N. Seidman, Nucl. Instrum. Meth. *170*, 377 (1980).
66. S. S. Brenner and D. N. Seidman, Radiat. Effects *24*, 73 (1975).
67. S. S. Brenner, R. Wagner and J. Spitznagel, Met. Trans. A*9A*, 1761 (1978).
68. A. Wagner and D. N. Seidman, J. Nuc. Mat. *83*, 48 (1979).
69. C.-Y. Wei and D. N. Seidman, Cornell Materials Science Center Report No. 4088 (1980); accepted for publication in the Philosophical Magazine.
70. *Radiation-Induced Voids in Metals*, edited by J. W. Corbett and L. C. Ianniello (National Technical Information Service, Springfield, Va., 1972).
71. See, for example, J. Nuc. Mater. *53* (1974); *76* (1978); and *77* (1978).
72. W. D. Wilson and C. L. Bisson, Radiat. Effects *22*, 63 (1974).
73. O. S. Oen and M. T. Robinson, Nuc. Instrum. Meth. *132*, 647 (1976).
74. D. J. Reed, Radiat. Effects *31*, 129 (1977).
75. B. Terreault, R. G. St-Jacques, G. Veilleaux, J. G. Martel, J. L'Ecuyer, C. Brassard and C. Cardinal, Can. J. Phys. *56*, 235 (1978).
76. E. V. Kornelsen, Radiat. Effects *13*, 227 (1972); E. V. Kornelsen and A. A. van Gorkum, Nuc. Instrum. Meth *180*, 161 (1980).
77. F. Maury, M. Biget, P. Vajda, A. Lucasson and P. Lucasson, Radiat. Effects *38*, 53 (1978).
78. J. P. Biersack and L. D. Haggmark, submitted to Nuclear Instruments and Methods (1980).
79. L. G. Parratt, *Probability and Experimental Errors in Science* (John Wiley, New York, 1961), Chapt. 2.
80. R. M. Scanlan, D. L. Styris, and D. N. Seidman, Phil. Mag *23*, 1439 (1971); K. L. Wilson and D. N. Seidman, Radiat. Effects *27*, 67 (1975); D. N. Seidman, K. L. Wilson, and C. H. Nielsen, Phys. Rev. Lett. *35*, 1041 (1975); K. L. Wilson, M. I. Baskes, and D. N. Seidman, Acta. Met. *28*, 89 (1980).
81. S. S. Brenner, Surface Sci. *70*, 427 (1978).
82. P. Jacobaeus, J. U. Madsen, F. Kragh and R. M. J. Cotterill, Phil. Mag. B *42*, 11 (1980).

Profilometry, Resistivity, and Acoustical Measurements

STEP HEIGHT-ORIENTATION RELATIONSHIPS IN AN ION-BOMBARDED Fe-Cr-Ni ALLOY

A. L. Chang, R. Bajaj and S. Diamond

Westinghouse Electric Corporation
Advanced Reactors Division

Swelling induced by heavy ion bombardment can be measured in terms of either the voidage inside the specimen or the step height between the ion-bombarded region and the masked region. However, substantial grain-to-grain variation in step height has been observed which appears to be correlated with the grain orientation. A study of the step height-orientation relationships has been conducted in an Fe-15-Cr-25Ni pure ternary specimen bombarded by 3.5 MeV nickel ions to a dose of 95 dpa at 700°C. Selected step heights were measured by surface profilometry. Others were either visually estimated from the stereo scanning electron micrographs (SEM) using a comparative ranking procedure or they were calculated from the void distribution measured from separate specimens with identical bombardment conditions. Grain orientations were determined either from the SEM selected area channeling patterns or from TEM selected area diffraction patterns corresponding to void images. It was found that grains with orientations closest to the major poles, i.e., [111] , [001] and [011] showed a smaller step height than the grains oriented around the minor axes [113],[123] and [223].

Introduction

Ion bombardment of crystals can induce substantial volume changes in the damaged layer. In the case of heavy ion bombardment, the volume increase is due to void swelling. Johnston et al (1) have developed a direct method for measuring gross swelling in ion-bombarded specimens. The technique involves masking parts of the specimen during ion bombardment and determining the step height at the boundary between the masked and the bombarded region to provide a measure of the total swelling integrated along the ion path. They were able to quantitatively correlate the step height to the void swelling measured by transmission electron microscopy (TEM) in the peak damage region (2). However, differences in elevation of twins relative to the surrounding grains and grain-to-grain variation of the step height and the surface roughness were commonly observed on the surface of the ion-bombarded specimen (3). It is the purpose of the present investigation to relate the step height (and/or void swelling) variations with the grain orientation in a nickel ion-bombarded austenitic specimen.

Experimental Procedures

The Fe-15Cr-25Ni austenitic alloy was prepared by International Nickel Compnay by vacuum melting. The chemical composition of this alloy is listed in Table I. The material was swaged and drawn into a 3mm diameter wire with intermediate anneals. The wire was solution annealed at $1150°C$ for 1 hour and quenched by flowing argon gas. Disc samples about 0.38mm thick were cut from the wire for the ion bombardment experiment. The ion-entry surface was prepared using a multiple lapping technique with successively finer diamond abrasive down to 0.1µm. Each lapping step was designed to remove all of the cold-worked surface material from the original cutting of the disc and from the previous lapping step. The ion-entry surface was then subjected to ion-milling, electro-polishing and jetting over a 2mm diameter in the foil center to remove scratches introduced by handling. After uniform injection of 5 appm He to a depth of 25µm, an addition 6µm was jetted from the surface to remove fine scratches observed after He injection. The samples were then mounted into a tungsten alloy holder with the outer rim of the sample masked during ion-bombardment. The sample was bombarded by 3.5 MeV $^{60}Ni^+$ ions in the ANL Dynamitron at $700°C$ to a nominal peak dose of 95 dpa. The peak damage rate was about $5 \times 10^{-3} dpa/sec$. Details of the bombardment and the TEM examination have been reported elsewhere. (4)

Table I. Composition of Alloy

	Fe	Cr	Ni	O	N	C	Co	Cu	Si	Mn
wt. %	60.1	14.8	25.04	0.0154	0.0028	0.010	0.014	0.02	0.02	0.005

After ion bombardment, step heights of a few selected grains were measured by a surface profilometer to calibrate the range of variations. Stereo SEM micrographs showing step heights in neighboring grains were visually compared using a comparative ranking procedure. In an identical specimen, void distributions within the entire thickness of the damage layer were measured from several different grains by using the 1000 Kv. U. S. Steel HVEM. Equivalent step heights can be calculated by integrating the total void volume along the entire damage depth and by spreading this volume uniformly on the surface. The amount of material sputtered by the ion beam was estimated to be 10 ± 5 nm in thickness according to Johnston et al (2). On the other hand, the extra material injected by the ion beam also amounted to about 10 nm in thickness. Therefore, these two effects amost cancelled each other and the rather large grain-to-grain variation in step height can not be attributed to the differential sputtering. Grain orientations were determined either by matching the SEM selected area channeling patterns obtained from each grain to a standard map obtained from γ-Fe or by identifying the TEM selected area diffraction patterns and kikuchi bands corresponding to void images. The step height-orientation relationships can be expressed in the standard stereographic triangle.

Results

Individual grains that contained steps were identified and marked from A through P. Selected SEM micrographs of adjacent grains containing the step along the masked area are shown in Figure 1. Significant variation in step height between grains can be seen. They were visually compared to neighboring grains and were ranked as high, intermediate and low in terms of the relative step heights. Surface profilometry measurements on grain C and D gave ∿30 nm and ∿80 nm in step height, respectively. These two measurements represented the "intermediate" and the "high" on the relative step height scale. The "low" step height was estimated to be less than 20 nm. The equivalent step heights calculated from the HVEM void swelling data were found to be 34 nm along [112], 32-62nm* along [013] and 21nm along [011]. Some typical SEM selected area channeling patterns are shown in Figure 2. Although the surface material has been subjected to ion bombardment up to a dose of 95 dpa, these channeling patterns were still identifiable for orientation determination. Figure 3 summarizes the orientation dependence of the step height and void swelling data obtained. Grains with orientation closest to the major poles [111], [001] and [011] showed the smallest step heights, while largest step heights were observed around the minor axes like [113], [123] and [223]. This result is contrary to what Johnston et al (3) predicted from their investigation of nickel ion bombarded stainless steels. They speculated that grains orientated with [111] plane parallel to the ion-entry surface should swell most, since there is no constraint to the [111] type interstitial-platelet formation during ion bombardment. The degree of surface roughness was also found to vary from grain-to-grain in the bombarded region. Figure 4 shows a typical variation of the surface roughness. The high swelling Grain P shows a higher degree of surface roughness than the low swelling Grain O. There seemed to be a direct correlation between step height and surface roughness. Therefore, the surface roughness is lower around [001] and [111] and higher around minor axes in between. It should be noted that the step height at [001] axis was calculated from the HVEM void data in a separate specimen, surface roughness data close to the [011] orientation were not available.

---

* Due to truncation of the damaged layer by back-thinning, the foil thickness of this particular area was only ∿0.6µm instead of 1.0µm needed for void volume integration. The equivalent step height was estimated to be at least 32nm and could be as high as 62nm.

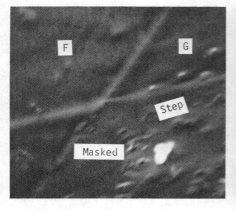

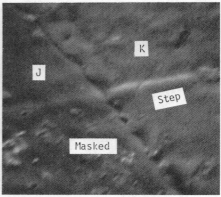

(a) Variation in Step Height
Between Grain F (~80nm)
and Grain G (~30nm).

(b) Variation in Step Height
Between Grain J (~30nm)
and Grain K (~80nm).

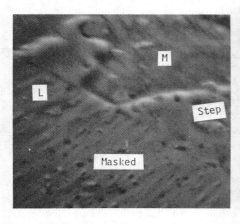

(c) Variation in Step Height
Between Grain L (<20nm)
and Grain M (~20nm).

(d) Variation in Step Height
Between Grain O (<20nm)
and Grain P (~80nm).

Fig. 1 - SEM Micrographs of Adjacent Grains Along the Masked Area of
Fe-15Cr-25Ni Alloy Bombarded With Ni Ions to 95 dpa at 700°C.

(a) (b)

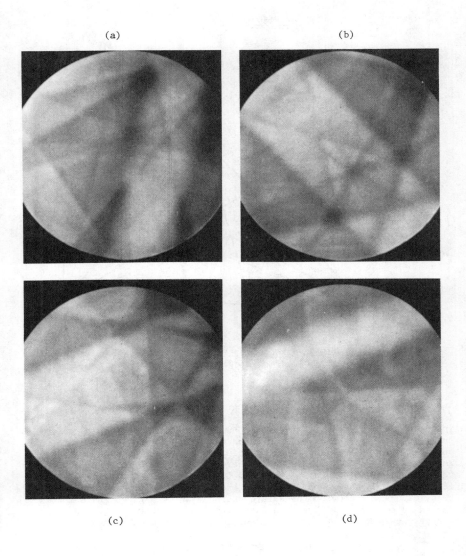

(c) (d)

Fig. 2 – Some typical SEM selected area channeling patterns (a) Grain
E (near [001]), (b) Grain H (near [111]) (c) Grain I (near
[113]), (d) Grain N (near [123]).

151

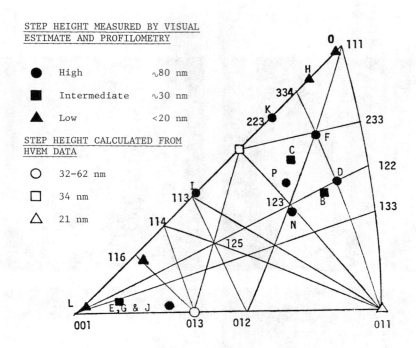

STEP HEIGHT MEASURED BY VISUAL
ESTIMATE AND PROFILOMETRY

●   High   ~80 nm

■   Intermediate   ~30 nm

▲   Low   <20 nm

STEP HEIGHT CALCULATED FROM
HVEM DATA

○   32-62 nm

□   34 nm

△   21 nm

Fig. 3 - Step Height-Orientation Dependence of Fe-15Cr-25Ni
Alloy Bombarded to 95 dpa at 700°C by Ni Ions.

Another feature worth noting is the grain boundary ridging effect shown
in Figure 5. These ridges were estimated to be ~2.5μm in width and ~100nm
in height with a grain size of ~2×10^{-2}cm. They are not grain boundary
precipitates. Furthermore, void denuded zone along grain boundaries were
observed at this temperature, indicating that these ridges could not be
due to excess local swelling. SEM energy dispersive x-ray analyses showed
that nickel was enriched while Fe and Cr were depleted at boundaries. The
matrix, on the other hand, exhibited the opposite solute segregation
behavior. The extent of the solute segregation in a given grain matrix
seemed to correlate with the surface roughness of that grain, i.e., rough
grains showed more severe solute segregation than smooth grains.

Discussion

In attempt to interpret the orientation dependence of the step height
(or void swelling), Garner et al (5) suggested that the undamaged and
undeformed bulk material beyond the depth of the damaged layer imposed
constraints to the swelling and the irradiation creep in the damaged layer
by setting up the swelling-induced anisotropic stresses. The irradiation
creep processes operating to relieve such stresses can give rise to creep
rates which are quite sensitive to the grain orientation. The orientation
dependence of the irradiation creep rate can cause the grain-to-grain
variation of the step height. Wolfer and Garner (6) recently calculated the
orientation dependence of the swelling induced stresses for three types of

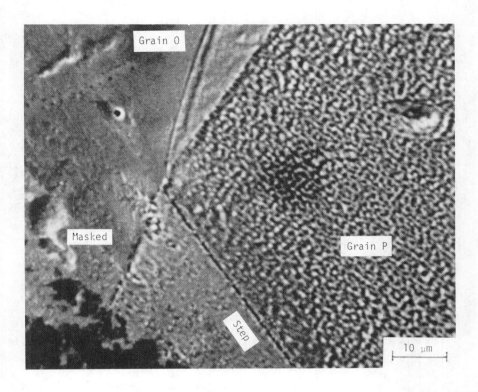

Fig. 4 – Surface Topography of Fe-15Cr-25Ni Alloy Bombarded to 95 dpa at 700°C with Ni Ions.

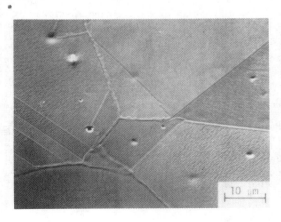

Fig. 5 – Grain Boundary Ridging Effect on the Surface of Fe-15Cr-25Ni Alloy Bombarded to 95 dpa at 700°C with 3.5 MeV Ni Ions.

of irradiation creep processes. The three processes considered were (a) the stress-induced preferential absorption of point defects on faulted dislocation loops (SIPAL), (b) the stress-induced preferential absorption of point defects on edge dislocations (SIPAD) or unfaulted loops, and (c) the climb-controlled glide (CCG) of dislocations. They found that the SIPAL process would give rise to the largest negative lateral stress for orientations along the [001]-[011] edge of the stereographic triangle, the SIPAD process would lead to the largest negative lateral stress near [001] orientation, intermediate stress near [011] and the least stress near [111]. On the other hand, the CCG process showed the largest negative lateral stresses along the [011]-[111] edge of the sterographic triangle and the least stress near [001]. Wolfer and Graner suggested that through the coupling of CCG to SIPAD, large compressive lateral stresses were expected for [001], [111] and [011] orientations, and lower stresses in between. Since compressive stresses suppress void nucleation (7), one should find smaller step heights around these orientations as compared to orientations in between. This is in qualitative agreement with the observation as shown in Figure 3. Closer examination of the void swelling-orientation data indicates that the grain which showed the largest void swelling also had the highest void number density, consistent with the stress-associated void nucleation (SAVN) theory (7).

The origin of the orientation dependence of the surface roughness has not yet been understood. However, the morphology of the surface very much resembles the blistering phenomenon on the light ion bombarded surface observed by Das and Kaminsky (8) in Nb and by Milacek and Daniels (9) in Al. The blister density in proton-irradiated polycrystalline Al was lower around [111] and [001] poles and higher around [011] and other minor axis in between. Since no surface roughness data are available at [011], it is difficult to assess the validity of the analogy at this point.

From the observed dimensions of the grain boundary ridges, it has been estimated that there are $\sim 2 \times 10^{16}$ atoms in these ridges per square centimeter of the ion-entry surface. This is roughly one-fifth of the number of Ni ions injected during the bombardment. Since nickel atoms tend to segregate to grain boundaries, a fraction of these injected nickel atoms has segregated as an excess interstitial flux toward boundaries and, subsequently, reached the surface. In other words, it is suggested that instead of depositing the injected atoms uniformly on the surface, a fraction of these atoms was preferentially deposited at grain boundaires to form ridges.

In quantitative comparison of step height or void swelling data among different ion-bombarded specimens, it is pertinent to take the orientation effect into consideration. In the case of the Alloy Development Intercorrelation Program (ADIP) (10) where Fe-Cr-Ni specimens, identical to the one used in the present investigation, were bombarded by Ni ions of various energies at various laboratories for quantitative swelling inter-correlation. The grain-to-grain variation of the void swelling was not properly taken into account in the process of inter-correlation. This orientation effect might have contributed its share of uncertainties to the discrepancies observed. More work is warranted in this area.

References

1. W. J. Johnston, J. H. Rosolowski, A. M. Turkalo and T. Lauritzen, "A Direct Measurement of Gross Swelling in Nickel Ion-Bombarded Stainless Steel", J. Nucl. Mater., 46, (1973), p. 273.

2. W. G. Johnston, J. H. Rosolowski, A. M. Turkao and T. Lauritzen, "An Experimental Survey of Swelling in Commercial Fe-Cr-Ni Alloys Bombarded with 5 MeV Ni Ions", J. Nucl. Mater., 54, (1974), p. 34.

3. W. G. Johnston, J. H. Rosolowski, A. M. Turkalo and K. D. Challenger, "Surface Observations on Nickel Ion-Bombarded Stainless Steels", Scripta Met., 6, (1972) p. 999.

4. S. Diamond, I. M. Baron, M. L. Bleiberg, R. Bajaj and R. W. Chickering, "HVEM Quantitative Stereoscopy Through the Full Damage Range of An Ion-Bombarded Fe-Ni-Cr Alloy", Proceedings of International Conference on Radiation Effect and Tritium Technology for Fusion Reactors, Gatlingburg (1975), CONF-750959, p. I-207.

5. F. A. Garner, G. L. Wire and E. R. Gilbert, "Stress Effects in Ion Bombardment Experiments", Proceedings of International Conference on Radiation Effect and Tritium Technology for Fusion Reactors, Gatlingburg (1975), CONF-750959, p. I-474.

6. W. G. Wolfer and F. A. Garner, "Swelling-Induced Stresses in Ion-Bombarded Surfaces; Effect on Crystalline Orientation", presented at the First Topical Meeting on Fusion Reactor Materials, Miami Beach, Florida (1979), UWFDM-286, J. Nucl. Mater. 85 & 86 (1979), pp. 583-589.

7. F. A. Garner, W. G. Wolfer and H. R. Brager, "A Reassessment of the Role of Stress in Development of Radiation-Induced Microstructure", ASTM STP 683, (1979), pp. 160-183.

8. S. K. Das and M. Kaminsky, in Radiation Effects on Solid Surfaces, ed. by M. Kaminsky, Advances in Chemistry Series, Vol. 158, Am. Chem. Soc., Washington (1976), Chapter 5.

9. L. H. Milacek and R. D. Daniels, "Orientation Dependence of Pitting and Blistering in Proton-Irradiated Aluminum Crystals", J. Appl. Phys., 39, (1968), pp. 5714-5717.

10. F. A. Garner, et al, "Summary Report on the Alloy Development Inter-correlation Program Experiment", Proceedings of the Workshop on Correlation of Neutron and Charged Particle Damage, held at Oak Ridge National Laboratory, June 8-9, 1976, CONF-760673, pp. 241-252.

Acknowledgement

We would like to thank C. W. Hughes of Westinghouse R&D Center for performing the SEM, S. Lally and the staff at the U. S. Steel Research Laboratories for HVEM, R. W. Chickering for surface profilometry measurements and A. Taylor of Argonne National Laboratory for the nickel ion bombardment. This work was performed within the National Alloy Development Program under the auspices of the U. S. Department of Energy.

STUDIES OF DEFECT PRODUCTION BY HIGH VOLTAGE ELECTRON

MICROSCOPY AND IN-SITU ELECTRICAL RESISTIVITY*

Wayne E. King[+][‡], K. L. Merkle[++], and M. Meshii[+]
Department of Materials Science and Engineering
Northwestern University
Evanston, IL 60201[+]

and

Materials Science Division
Argonne National Laboratory
Argonne, IL 60439[++]

We have developed a new technique to investigate point defect production and subsequent annealing by in-situ irradiation and measurement of the electrical resistivity at T < 10K in the high voltage electron microscope** (HVEM) at Argonne National Laboratory. This technique takes advantage of the high angular resolution of the electron beam in the HVEM, negligible multiple scattering of the irradiating electrons in thin specimens, accurate specimen alignment with respect to the electron beam using Kikuchi patterns, a wide range of precise irradiation energies which is essential for threshold energy studies, and high beam current densities that are available in the HVEM. The technique was first applied to the determination of the threshold energy surface of copper by making damage-rate measurements on single crystal films in the energy range from 0.4 to 1.1 MeV for forty different incident beam directions. The implications of these observations and the possible applications of this new technique are discussed.

*Work supported by the National Science Foundation and the U. S. Department of Energy.
‡Presently at Argonne National Laboratory.
**This microscope is interfaced with a 300 keV ion accelerator. A 2 MeV accelerator will become available later.

Introduction

Ever since the first in-situ observation of displacement damage, the high voltage electron microscope (HVEM) has been used extensively for quantitative defect production studies. (1) In the HVEM, it was possible to increase both the bore and gap sizes of the objective lens pole pieces compared to conventional electron microscopes. This provided a rather large specimen chamber that was conducive to in-situ experiments. With the introduction of side entry specimen stages, access to this specimen chamber with mechanical or electrical connections became very easy. In particular, cooling of the specimen became quite simple compared to the technique required by the top entry format, because of the straight line path from the specimen to the outside world (2).

Electrical-resistivity measurements have been used extensively in defect production studies at liquid helium temperatures for a variety of irradiating particles and samples. The great interest in studies of the damage function and ion damage led to the development of electrical-resistivity measurements on thin metal films, as reported by Averback et al. in these proceedings (3). Advances in the field of microelectronics enabled us to produce microminiature specimens from thin metal films in a configuration suitable for four-point dc electrical-resistivity measurements.

Of particular interest to experimenters using HVEM or electrical resistivity were experiments to investigate the anisotropy of the threshold energy for Frenkel pair production (4-7). Unfortunately, the results from HVEM and electrical-resistivity experiments could not be easily compared. The HVEM experiments were carried out mainly at room temperature or above where the analysis is complicated due to nucleation of visible defect clusters and sink effects which influenced both nucleation and growth of these clusters. At best, these experiments could only yield relative measurements of the total cross section for Frenkel pair production over a limited number of incident beam energies and directions. Therefore, as yet, no complete threshold energy surface has been derived using electron optical observations. In fact, many experiments reported only the apparent threshold energy for a given crystal direction. This apparent threshold energy may be quite different from the actual threshold energy for that direction. A number of electrical resistivity experiments were carried out at low temperature using conventional electron accelerators. By this method, derivations of threshold energy surfaces were obtained for several metals. Unfortunately due to the experimental setups, it was very difficult to precisely align these specimens and to obtain a parallel beam of electrons. Complex corrections for multiple scattering of the electrons in thick specimens further complicated the analyses. In most cases, data was obtained for only a few directions or for an insufficient number of directions and energies to obtain a detailed threshold energy surface because of the amount of experimental time that would be required.

Thus the development of a side entry helium temperature specimen stage that has the capability of in-situ electrical resistivity measurements for this work was a natural consequence of these earlier developments. This technique takes advantage of the high angular resolution of the electron beam in the HVEM to illuminate the specimen with a nearly parallel beam of electrons. Because we can use thin films, the effects of multiple scattering of the electron beam can be neglected (8). The specimens are, in fact, thin enough so that they can be precisely aligned with respect to the electron beam using Kikuchi patterns. The HVEM provides a wide range of well defined incident electron beam energies which are essential for studies of the threshold energy. Also, in contrast to conventional electron accelerators, very high electron beam current densities are available in the HVEM. This greatly shortens the time necessary to collect data which is particularly important at low energies where the total cross sections for Frenkel pair production are quite small. Since in first approximation the maximum temperature rise of the sample due to the incident electron beam is determined by the total electron beam current, the high electron beam current densities available in the HVEM can in fact be used on resistivity samples as long as the specimen dimensions are small enough. Here, microminiaturization offers a great potential for achieving extremely high displacement rates since the usable beam current densities increase quadratically with the inverse diameter of the irradiated area.

In the present work, in-situ damage-rate meausrements on thin, single-crystal copper films were carried out for electron irradiations in 40 crystallographic directions, at six energies ranging from 0.4 to 1.1 MeV. The measured total cross sections for Frenkel-pair production were compared to the total cross sections for Frenkel-pair production that were calculated from trial threshold energy surfaces composed of 41 ($5°$ x $5°$) blocks distributed over the unit triangle of a cubic crystal (defined by the directions $\langle 001 \rangle$, $\langle 011 \rangle$, and $\langle 111 \rangle$). An optimization procedure was used to find the best-fit threshold energy surface consistent with the present data. This threshold energy surface was used to derive the damage function $\nu(T)$, where T is the recoil energy, for $\nu(T) < 1$.

The results of this work represent just one of the many possible applications of the in-situ HVEM resistometry technique. Full utilization of this technique will include in-situ studies of atomic arrangements and defect structures through observations in the electron diffraction and electron microscopy modes in addition to the electrical resistivity measurements in the HVEM. A combination of these technqiues will certainly be of advantage in many investigations. Of major importance, as far as the applications to radiation effects are concerned, are the well defined electron beam conditions and the increased beam current densities relative to conventional resistivity methods. The ANL HVEM will also allow in-situ ion beam experiments. Thus ion beam implantation and ion-beam mixing effects can be studied and direct comparisons between electron- and ion-beam displacement effects are possible. In the following the emphasis will be on the application of the technique to a determination of the displacement threshold surface, however a few additional applications of interest will be mentioned.

Basic Principles

The coordinate system that will be used when referring to the incident electron-beam direction and the target-atom recoil angles and directions is the cubic [001] standard projection shown in Fig. 1.

The uppercase Greek letters (Θ, Φ) refer to angles measured with respect to the [001] pole while the lowercase Greek letters refer to angles measured with respect to the incident beam direction. The subscript 1 refers to the incident electrons and the subscript 2 refers to the target atoms. The direction of the incident electron beam is specified by the polar angle Θ_1 and azimuthal angle Φ_1. The incident electrons are scattered by angles (θ_1, ϕ_1) (not shown). As a result of this scattering event, the target atom recoils at angles (θ_2, ϕ_2) in the crystal direction specified by polar angle Θ_2 and azimuthal angle Φ_2. The angle ϕ_2 is the included angle defined by the intersection of the two great circles containing the poles ($\pi/2, \Phi_1$), (Θ_1, Φ_1) and (Θ_1, Φ_1), (Θ_2, Φ_2).

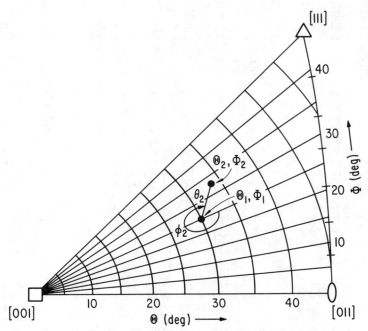

Fig. 1. The coordinate system, showing the electron—beam direction (Θ_1, Φ_1), recoil angle (θ_2, ϕ_2), and the recoil direction (Θ_2, Φ_2).

The probability for a recoil to create a stable Frenkel-pair is a function of both the recoil direction and energy and is assumed to be a step function going from a probability of zero to one when the recoil energy equals or exceeds the threshold energy in the recoil direction. This is the directionally dependent damage function,

$$\nu(\Theta_2, \Phi_2; T) = \begin{array}{l} 0, \text{ for } T < T_d(\Theta_2, \Phi_2); \\ 1, \text{ for } T \geqslant T_d(\Theta_2, \Phi_2). \end{array} \tag{1}$$

The total cross section for Frenkel-pair production, $\sigma_d(\Theta_1, \Phi_1; E_1)$, is given by:

$$\sigma_d(\Theta_1, \Phi_1; E_1) = \int_0^{2\pi} \int_0^{\pi/2} \frac{d\sigma(\theta_2; E_1)}{d\theta_2} \frac{d\phi_2}{2\pi} \nu(\Theta_2, \Phi_2; T) d\theta_2 \tag{2}$$

where $d\sigma(\theta_2; E_1)/d\theta_2$ is the differential cross section. This differential cross section weights the distribution of recoil directions to directions away from the incident beam direction, implying that most of the stable Frenkel pairs that are produced result from recoils in directions other than the incident electron beam direction. Consequently, any measured total cross section for Frenkel pair production may contain contributions from many directions and therefore does not provide direct directional information on the anisotropy of the threshold energy for Frenkel-pair production. The situation would be further complicated if multiple scattering was included in Eq. (2). Fortunately, in the case of thin specimens, multiple scattering may be neglected (8). To extract the threshold energy surface, values for the total cross section for Frenkel-pair production, based on a trial threshold energy surface, were calculated for directions and energies corresponding to the observed total cross sections for Frenkel-pair production. These calculated cross sections were compared to the measured cross sections using the χ^2-goodness-of-fit test. A derivative-type, unconstrained optimization scheme was used to find the best-fit threshold energy surface consistent with the present 0.502, 0.600, 0.701, 0.910, and 1.106 MeV data

Experimental Procedure

Equipment

1. The HVEM

The Kratos-AEI EM7 1200 keV high voltage electron microscope at Argonne National Laboratory, provides a unique miniature-laboratory environment for in-situ radiation damage experiments. The energy of this HVEM is continuously variable from 100 to 1200 keV. The microscope is equipped with a negative ion trap to insure that the beam is composed solely of electrons. In addition to producing atomic displacements with the electron beam, it is possible to study in-situ ion damage effects via an ion-beam interface. At this time, the HVEM is interfaced with a 300 keV ion accelerator. A 2 MeV accelerator will become available later.

2. The Specimen Stage

A side entry type, single tilt (± ~47°), cryogenic specimen stage was designed and developed for use in the Argonne HVEM (see Fig. 2) (9). With this stage, in-situ electrical-resistivity measurements as well as

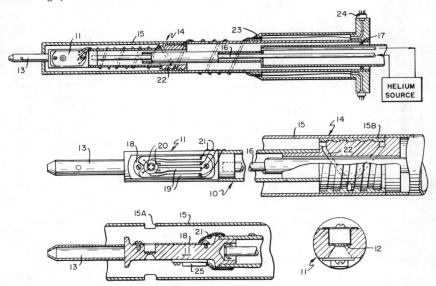

Fig. 2. Various schematic views and cross sections of the side entry helium temperature specimen stage.

irradiation and observation of the same specimen are possible. The specimen stage was designed as a flow cryostat that uses helium liquid or gas as a coolant to attain and maintain any temperature in the range < 10 to 300 K. The helium flow cools a heat exchanger [10] and thus the copper specimen block [11] which supports the sapphire specimen holder [12] with the specimen adhering tightly.

On the opposite end of the copper specimen block from the heat exchanger, a thin-walled locating tip [13] is attached which seats in the tilt cone of the specimen stage block of the HVEM. This locating tip reduces conductive heat flow from the specimen stage block (which is at ambient temperature) to the copper specimen block. The spent helium gas subsequently cools the cold shield heat exchanger [14]. The gold-plated cold shield [15] with apertures for the beam to pass [15A] is concentric to the copper specimen block, contacts the locating tip and is threaded onto the cold shield heat exchanger using a silver insert [15B]. The silver insert when cooled, contracts more than the copper of the cold shield heat exchanger, making very good thermal contact with the cold shield heat exchanger. Similarly, the cold shield contracts tightly on the locating tip effectively short circuiting heat flow from ambient through the cold shield to the cold shield heat exchanger rather than allowing the copper specimen block to be heated. The cold shield serves to reduce radiative heating of the copper specimen block and to reduce contamination at the specimen. The helium transfer tube [16] is inserted coincident with the axis of the specimen stage and touches the stage at one point only. This is achieved via an x, y, z positioning mechanism attached to the upstream part of the transfer tube which allows the tranfer tube to be tilted freely around a point concentric to the O-ring seal [17]. It is desirable to prevent contact between the transfer tube and the specimen stage to prevent vibration of the transfer tube from causing the specimen stage to vibrate thus degrading the electron-optical image.

The connections for the 4-point dc electrical-resistivity measurements are made via a sapphire junction plate [18] screwed to the top of the copper specimen block. Four gold pads [19] are silk-screened into the sapphire junction plate to facilitate the final electrical connection to the specimen via gold wires [20]. Four electrical leads [21] extend from the sapphire junction plate, which heat sinks the leads, to the exterior of the HVEM. They were wrapped spirally around the specimen stage, pass through the spiral groove [22] in the cold shield heat exchanger, through the vacuum feed through [23], and terminate at the miniature connectors [24]. A platinum resistance thermometer [25] is fastened to the bottom of the copper specimen block and the electrical leads are conducted to the miniature connectors in a manner identical to those used for the electrical resistivity measurement.

The specimen is attached to a sapphire specimen holder [12] with the portion to be irradiated left free standing over a small hole. The specimen holder is pressed into the copper specimen block using an indium gasket. Using this configuration, the path of the electron beam, from the accelerator to the specimen is unobstructed which insures that the beam divergence is defined only by the electron optics.

Dosimetry in the HVEM

The Argonne HVEM is equipped with a precision electron dosimetry system consisting of two Faraday cups, an electrometer and a current integrator shown in Fig. 3. The Faraday cup that is located in the top entry specimen chamber is of the Albany type (10) and is used to calibrate the Faraday cup

162

that is located in the viewing chamber. This lower Faraday cup is free to move and can be driven to any point in a plane relative to the projected image of the specimen. In this manner, the beam profile can be characterized and it is possible to have very accurate dosimetry in spite of the small beam spot sizes which are necessary for achieving a high rate of defect production. The collected current from the Faraday cup is conducted to an electrometer where the current is converted to a proportional voltage. This voltage is the input to a voltage-to-frequency converter, the output of which is counted by a scalar. The scalar controls a beam-stop and can thus be set to allow preset doses of electrons to hit the specimen.

Specimens

Single crystal, thin films (~0.4 μm) of copper were deposited epitaxially using vacuum evaporation (1.3 x 10^{-5} Pa) onto <100> sodium chloride substrates at 300-350°C. After removal from the vacuum system the substrates and films were cleaved into 3 x 3 mm squares. The film was removed from the substrate by dissolving the rock salt in water. The film was then picked up on a thin collodion film that had been stretched tightly across a brass ring. After drying, the film was as flat as the collodion. The collodion, with the specimen attached, was lowered onto a three millimeter diameter sapphire disk prepared with gold contact pads and a one millimeter hole through the center. The collodion film and the specimen were stretched taut on the sapphire disk to obtain the optimum flatness for the threshold anisotropy experiments. The collodion film was then partially dissolved in an acetone vapor, after which it was completely dissolved in a condensation washer using amyl acetate as a solvent. At this point, the flat single crystalline copper film was completely covering the 1 mm

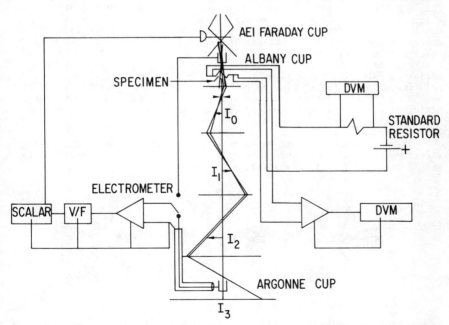

Fig. 3. Schematic diagram of the electron dosimetry system of the ANL HVEM and the electrical resistivity measuring circuit.

diameter hole in the sapphire specimen holder. The following procedure was used to generate the specimen shape needed for the 4-point electrical-resistivity measurement. The disk, with the specimen adhering tightly was mounted in an ion milling mask holder and the specimen was then ion milled using 2 keV argon ions to obtain the special configuration shown in Fig. 4. The specimen and disk were pressed into the copper specimen block of the specimen stage using an indium gasket and the final electrical connections were made.

All of the electrical resistivity measurements were carried out in-situ in the high voltage electron microscope with the objective lens deenergized. The electrical resistivity geometry factor was determined from the measurement of the room temperature resistance of the specimen. The specimen was cooled to <10 K where the electrical resistivity sensitivity was verified to be better than 1×10^{-11} Ω-cm.

Irradiation

The specimens were irradiated over their entire gauge length with a parallel beam of electrons. The current density at the specimen was ~1000 μA-cm^{-2}. The beam profile was defined by the condenser aperture and varied across the specimen by ~1%.

In this application of this technique, the specimens were aligned using Kikuchi patterns such that the major crystallographic direction that was perpendicular to the specimen surface was nearly parallel to the electron beam. This electron-beam direction was called the reference direction. After each irradiation step, damage-rate measurements were carried out. The specimen was then tilted by an angular increment that was typically a multiple of 5°, where another damage-rate measurement was carried out. Finally, the specimen was returned to the reference direction for a third damage-rate meausrement. This procedure was carried out cyclically, increasing the tilt angle with each cycle until the tilt limit was reached. After the procedure was completed, it was repeated at five additional energies and a Kikuchi map of the tilt path was photographically recorded with additional Kikuchi patterns recorded at the tilt angles that were used for irradiation. The entire procedure was repeated for four different tilt paths, three on one specimen and one on another, both of which had [001] surface normals.

The electron dose could be measured accurately to within 5%. The electron energy is known to an accuracy of 1%. The electron beam heating was less than 2 K during irradiation.

1 mm

Fig. 4. The configuration of the specimens used for the in-situ electrical-resistivity measurements showing the various stages of miniaturization.

Results

Experimental Observations

The measured damage rates were corrected for the electrical resistivity size effect using the Fuchs-Sondheimer theory (11). The probability of specular reflection of the conduction electrons at the specimen surface was taken to be zero and the ratio of the film thickness to the mean-free path of the conduction electrons was chosen in a self-consistent manner (12).

Saturation effects were corrected for by assuming that the bulk damage-rate curves (damage rate vs irradiation induced electrical resistivity) for copper are linear (13). Typically experimentally obtained total cross sections for Frenkel-pair production, $\sigma_d(\Theta_1, \Phi_1; E_1)$, are plotted in Fig. 5 as a function of the direction of the incident electron beam Θ_1, Φ_1 for several electron beam energies, E_1. The unit triangle in the upper left of the

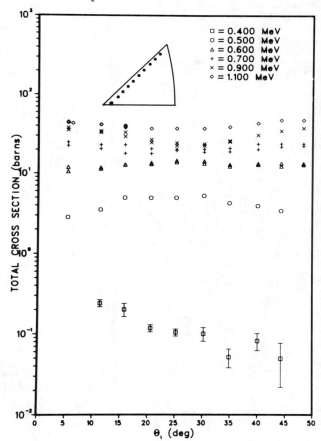

Fig. 5. Experimentally measured total cross sections for Frenkel pair production as a function of incident electron-beam direction and energy. Exact beam directions shown in the inset.

165

figure shows the exact beam directions plotted for that figure. The error
in the beam direction was ~1° with the largest contribution to this error
coming from slight intrinsic bending of the specimen. This plot represents
one scan across the unit triangle, that being a tilt path with approximately
constant Φ_1. A total of four scans were made to cover the unit triangle.
With increasing Θ_1 and Φ_1, differences appeared in the measured total cross
sections for Frenkel pair production due to the anisotropy of the threshold
energy. The detailed shape of these curves contained the necessary
information to extract the threshold energy surface.

Analysis

    The threshold energy surface as derived in the present study based on
the Frenkel pair resistivity ρ_F = 2.00 x 10^{-4} Ωcm (4) is shown in Fig. 6.
Figure 6 gives the best-fit threshold energy values and the sensitivity of
fit for each of the 41 (5° x 5°) regions (see ref. 12 for details). The
average threshold energy for the best-fit threshold energy surface is 28.5
eV; the minimum is 18 ± 2 eV and is located near <100>. A ring of very high
threshold energy (> 50 eV) surrounds the <111> direction. The threshold
energy surface generally varies smoothly from <100> toward <110> and
gradually increases toward <111>. The damage function as derived from this
threshold energy surface is shown in Fig. 7. This function rises rather
sharply from a value of zero at 17 eV to 0.8 at 42 eV. It has the value of
0.5 at 24.5 eV. Above 30 eV the curve begins to have a smaller slope

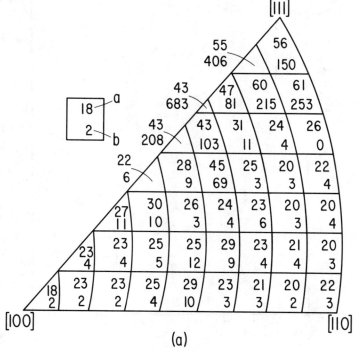

(a)

Fig. 6. The threshold-energy surface for copper, as derived in the present
        work. (a) Best fit values, and (b) sensitivities of fit for each of
        the 41 (5° x 5°) regions.

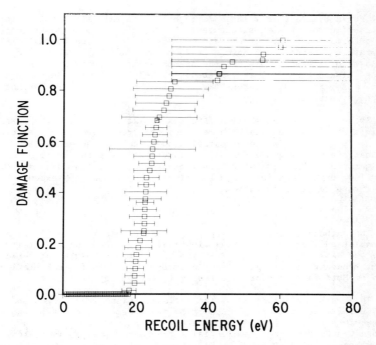

Fig. 7. The damage function for $\nu(T) < 1$ derived from the best-fit
threshold-energy surface of Fig. 6. The error bars indicate the
sensitivities of fit.

reflecting the high energy regions of the threshold energy surface. Details
of the interpretation of these results and comparison to previous work
(14,15,16) may be found in ref. 12 and 17.

Discussion and Conclusions

In this work, we have shown the successful application of this new
technique, which combined two well established techniques, to a rather
complicated problem. In the future, we plan to apply this technique to the
determination of the threshold energy surfaces for silver and aluminum along
with some bcc metals. In aluminum it should be possible to also investigate
the onset of multiple defect production with the electron and ion energies
available in the ANL HVEM. Direct comparison with ion damage rates will
also be possible. At the other end of the energy range, it should be
possible to carry out sub-threshold irradiations, that is, irradiations
where the maximum transferred energy is less than the minimum threshold
energy. This subthreshold regime is of interest for studies of stimulated
migration of point defects and impurity assisted defect production.

The defect production by electrons that travel in highly symmetric
lattice directions is modified by the so-called Bloch-wave channeling
effect. The resistometric technique could be used to study this phenomenon
if it is possible to produce quite flat and thin single crystals. Defect
production at temperatures where point defects are mobile have in the past
been studied separately by both direct observation and electrical
resistivity. It is expected that the analysis of such experiments will be

facilitated by the thin film geometry and the possibility of in-situ characterization of sink structures. The results of such experiments may give information on defect diffusion and sink properties as well as the temperature dependence of defect production. In concentrated alloys it should be of interest to study the displacement production and displacement induced disorder as well as radiation induced changes in short range order. Of particular interest may also be the study of systems that become amorphous under irradiation.

These few examples shall suffice as an illustration of the potential application of the in-situ HVEM resistometric technique. Much effort will have to be expended in the preparation of suitable specimens. Some of these applications will require further miniaturization of the specimen configuration. Further instrumental developments of the technique will have to include provisions for a double tilt goniometer and the extension of the temperature range of operation.

Acknowledgments

The authors gratefully acknowledge the technical support of the Argonne National Laboratory High Voltage Electron Microscope-Tandem Facility and crew, the assistance of Dr. I. Hashimoto, and the valuable suggestions of Dr. R. Benedek. Thanks are also due to Dr. O. S. Oen for calculating the ratio of Mott to Rutherford scattering in our energy regime.

This work was supported by the National Science Foundation and the U. S. Department of Energy. Support by a Walter P. Murphy Fellowship from Northwestern University and a Laboratory Graduate Participantship from the Argonne Center for Educational Affairs for Wayne E. King, is gratefully acknowledged.

References

(1)   M. J. Makin, in Proceedings of the Ninth Int. Cong. on Electron Microscopy, Toronto, Canada, edited by J. M. Sturgess (Microscopical Society of Canada, Toronto, 1978) p. 330.

(2)   P. R. Swann, ibid., p. 319.

(3)   R. S. Averback, K. L. Merkle and F. Dworschak, these proceedings.

(4)   P. Lucasson, in Proceedings of the Int. Conf. on Fundamental Aspects of Radiation Damage in Metals, edited by Mark T. Robinson and F. W. Young, Jr. (National Technical Information Services, U.S. Dept. of Commerce, Springfield, VA, 1975), p. 42.

(5)   P. Jung, in Atomic Collisions in Solids, edited by S. Datz, B. R. Appleton, and C. D. Moak, (Plenum, New York, 1975), Vol. I., p. 87.

(6)   K. L. Merkle, in Radiation Damage in Metals, edited by N. L. Peterson and S. D. Harkness (American Society for Metals, Metals Park, OH, 1975), p. 58.

(7)   P. Vajda, Rev. Mod. Phy. $\underline{49}$, 481 (1977).

(8)   M. Wilkens, Proceedings of the Fifth Int. Conf. on High Voltage Electron Microscopy, Kyoto, Japan, edited by T. Imura and H. Hashimoto (Japanese Society of Electron Microscopy, Tokyo, 1977), p. 475.

(9)   W. E. King and K. L. Merkle, United States Patent #4,162,40 (1979).

(10) J. N. Turner, J. of Phys. E: Scientific Instruments $\underline{8}$, 954 (1975).

(11) E. H. Sondheimer, Adv. in Phys. $\underline{1}$, 1 (1952).

(12) Wayne E. King, K. L. Merkle, and M. Meshii, Phys. Rev. B $\underline{23}$ 6319 (1981).

(13) G. Duesing, W. Sassin, W. Schilling, H. Hemmerich, Cryst. Lattice Def. $\underline{1}$, 55 (1969).

(14) J. B. Gibson, A. N. Goland, M. Milgram, and G. H. Vineyerd, Phys. Rev. $\underline{120}$, 1229 (1960).

(15) P. Jung, R. L. Chaplin, H. J. Fenzl, K. Reichelt, and P. Wombacher, Phys. Rev. B $\underline{8}$, 558 (1973).

(16) D. M. Schwartz, R. G. Ariyasu, J. O. Schiffgens, D. G. Doran, and G. R. Odette, in Proceedings of the 1976 International Conference on Computer Simulation for Materials Applications, Gaithersburg, Maryland, April 1976 (National Bureau of Standards, Washington, D. C., 1976), pp. 75–88.

(17) Wayne E. King and R. Benedek, Phys. Rev. B $\underline{23}$ 6335 (1981).

ELECTRICAL RESISTIVITY MEASUREMENTS ON ION-IRRADIATED

ULTRA-THIN METAL FILMS*

R. S. Averback, K. L. Merkle, F. Dworschak[+] and R. Benedek

Materials Science Division
Argonne National Laboratory
Argonne, IL 60439

A method has been developed to utilize electrical resistivity measurements (ERM) on ultra-thin metal films to study the properties of defects produced by ion irradiation. The procedures for producing ultra-thin resistivity specimens and for deducing the defect concentration from resistivity data, are described. Four studies which have employed this technique are presented to illustrate the scope of the method. These are studies of the damage function in fcc metals, electronic stopping powers, low-temperature hydrogen diffusion in gold and the interactions between self interstitials and defect clusters. This technique is complementary to such methods as Transmission Electron Microscopy for studying defects in irradiated materials as it is most sensitive to defects which have sizes below the visibility limit of electron microscopy.

*Work supported by the U. S. Department of Energy.
+Visiting Scientist from KFA, Jülich, W. Germany.

Introduction

Electrical resistivity measurements (ERM) have played an important role in investigations of radiation effects in metals [1,2]. Some of the advantages in using ERM are that they are sensitive to low defect concentrations < .001 ppm, and are rather easy to perform and with excellent accuracy. Moreover, it has been demonstrated that for many cases the electrical resistivity, $\Delta\rho$ is proportional to the defect concentration, c, i.e., $\Delta\rho = \rho_F c$ independent of the detailed configuration of the defects [3]. Therefore ERM can be used over a wide range [4], even when the defect configuration changes [3]. Although the proportionality constant, ρ_f, has been determined experimentally for only a few metals, reasonable estimates of ρ_f exist for most metals [5].

In the past, ERM were employed with electron, neutron, and very energetic proton and α-particle irradiations, for which damage is produced uniformly across specimens thicker than 1 μm. However, resistivity measurements did not seem compatible with heavy-ion and low-energy light-ion irradiations for which the ion range is $\lesssim$ 200 nm. Such ion irradiations, however, have many advantages for studying radiation effects. First, the primary recoil spectrum can be systematically shifted from low energy recoils, ~100 eV, to very energetic recoils, > 10 keV, by varying the mass and energy of the irradiation particle. Second, very high damage levels can be attained in a short time. Other advantages are that the cross sections for atomic collisions are relatively well known and ion doses can be measured accurately. Over the past several years we have developed experimental techniques and analytical methods involving thin films so that ERM could be used also in conjunction with ion irradiations at relatively low energy. In the following section we describe our experimental techniques for producing ultra-thin electrical resistivity specimens and our methods for analyzing the resistivity data. We then present four applications of the use of ERM on thin film specimens to illustrate the scope of problems for which these techniques have proven valuable.

Specimen Preparation

Considerable care is taken during specimen fabrication to avoid contamination. Up to now we have used only specimens which have been produced by vapor deposition. Electron beam heating is utilized to evaporate high purity materials onto rocksalt in a bakeable vacuum system which employs ion, titanium sublimation, and sorption pumping. The background pressure prior to evaporation is typically < 10^{-7} Pa; the measured pressure during evaporation is ~ 10^{-5} Pa. The specimens are either grown at 625K (to produce single crystal or large grain specimens), or at room temperature and immediately annealed to 575K. A quartz crystal oscillator is used during the evaporation to monitor the approximate film thickness. A more accurate determination ($\pm$ 5%) is made after the evaporation by weighing several monitor foils which are positioned adjacent to the specimens. The thicknesses of the specimens range from ~40 nm to 500 nm depending on the particular application. For specimens thicker than ~ 150 nm, the desired specimen geometry is obtained by placing masks over the rocksalt prior to evaporation. As shown in Fig. 1a, each specimen contains four segments on which different irradiations and four-probe dc resistivity measurements can be performed.

Before irradiation, these specimens are transferred from the rocksalt to sapphire plates on which gold film leads have been bonded [6]. The gold leads contact the specimen leads at one end and have fine gold wires bonded to them at the other end. The first step in transferring the specimens is

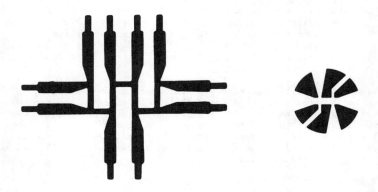

(a)   (b)

Fig. 1. Specimen geometries for specimens thicker than 150 nm (shaded area)
         (a), and thinner than 150 nm (unshaded area) (b).

to tape each specimen lead to a substrate lead with double-backed tape.
Small tabs at the end of each specimen lead are provided for taping. The
rocksalt is then dissolved in a 4:1 water to methanol solution. After the
salt is dissolved and the specimen has dried, the specimen adheres to the
substrate by surface tension and excellent thermal and electrical contact is
achieved. Finally the tabs are cut and the tape removed. Special platinum
and high resistance inks [6], which bond to sapphire, are employed to
produce a miniature thermometer and heater on the back of the specimen
holder. In addition, a gold foil which provides the thermal link to the
cryostat, is soldered to one end of the substrate. Because of to the low
mass of the specimen holder, the specimen can be rapidly cycled between the
irradiation or annealing temperature, and resistivity measuring temperature,
~6K. Heating and cooling times between 6K and 300K requires ~ 5 and 10
minutes repectively. Moreover, this technique maintains high vacuum
integrity and easy specimen changing.

     This procedure did not work well for specimens $\lesssim$ 150 nm as they
frequently broke during the transfer from the salt to the substrate;
therefore, a modification in the procedure was developed. These specimens
are produced by first evaporating onto the salt without masks. The salt,
with the metal film on it, is cleaved into 2.0 mm squares. The films are
floated off the salt and lifted onto sapphire annuli, 3.0 mm O.D., 1.0 mm
I.D. These annuli are metallized on one side and contain metallized leads
on the other. The square specimens are covered with masks as shown in Fig.
16, and ion-milled into the desired specimen geometry. The remaining
procedure is similar to that for the thick specimens. At present, we have
used the ultra-thin film procedure only for gold whereas we have
successfully applied the "thick" film technique to gold, silver, copper,
aluminum, and nickel. We are now preparing to produce specimens of dilute
alloys and iron, a bcc metal.

Electrical Resistivity

Performing electrical resistance measurements on thin films is essentially the same as on bulk specimens. Standard four-probe d.c. measurements are made using electronic preamplifiers, a digital voltmeter, and a computer. The system has an accuracy of ~.01% and a sensitivity of ~0.1 μV. With 30 mA measuring current and specimens which have a residual resistivity 1×10^{-7} Ω-cm Ω-cm, both typical values, the resolution of the system is ~1×10^{-11} Ω-cm. This corresponds to .01-.05 ppm Frenkel pairs. Cycling the specimen between 325K and 6K did not affect the electrical resistivity measured at helium temperature.

The determination in the defect concentration from the electrical resistivity data is more complicated for thin-film than for bulk specimens; however, it will be shown that the analysis is not as difficult as one might first expect. The first step in the analysis is to determine the "geometry factor" (GF) which relates measured resistance to resistivities. This is done by measuring the specimen resistance at room temperature, R(RT), and at 6K, R_o, and using the relationships

$$\text{GF} \cdot \text{R(RT)} = \rho(\text{RT}) = \rho_i(\text{RT}) + \rho_o = \rho_i(\text{RT}) + \text{GF} \cdot \text{R}_o \ . \tag{1}$$

Here $\rho_i(\text{RT})$ is the temperature dependent contribution to the resistivity at room temperature, and ρ_o is the residual resistivity. The relationships expressed in eq. (1) assume that there are no deviations from Matthiessen's Rule. This assumption is not usually valid; however, since ρ_o/ρ_i is typically < .05, any possible deviations in Matthiessen's Rule would not significantly affect the determination of GF [7].

The high residual resistivities of our specimens is due to surface scattering of electrons, not impurity scattering. Unfortunately, the surface scattering contribution to ρ is not simply additive with the radiation induced defect contribution, $\Delta\rho$, i.e.

$$\frac{d\rho}{d\Delta\rho} = 1 + \varepsilon(\Delta\rho) \tag{2}$$

where $\varepsilon(\Delta\rho)$ is the size effect correction. Two of the authors (F. Dworschak and K. L. Merkle) have discussed this problem previously [8,9] and have described the procedure for utilizing the Fuchs-Sondheimer theory [10] to determine $\varepsilon(\Delta\rho)$. The size effect correction can be made in a completely empirical manner for many cases. It has been shown [8,9] that $\varepsilon(\Delta\rho)$ becomes small at large values of $\Delta\rho$ because the electron mean free path is reduced by the radiation induced defects. Also, it has been observed during various irradiations of many metals, that the production rate, $d\Delta\rho/d\phi$, where ϕ is the dose, is a nearly linear function of $\Delta\rho$ as long as point defects are immobile [11,12]. Therefore $\varepsilon(\Delta\rho)$ can be approximated by plotting $d\Delta\rho/d\phi$ as a function of $\Delta\rho$ and extrapolating the linear portion of the production rate curve to low values of $\Delta\rho$. The difference between the measured and extrapolated curves yield $\varepsilon(\Delta\rho)$. Figure 2 illustrates an experimental production rate curve and the corresponding size-effect corrected curve for a 2 MeV α-particle irradiation of aluminum [13]. The size effect correction was made using the Fuchs-Sondheimer theory. It is clear from this figure that the empirical method can also be used for light ion irradiations of aluminum. Generally we have had excellent success in correcting for size effects. For any particular type of irradiation, we have found that the initial production rate may vary by as much as a factor of two from one specimen to another, but that the corrected values are reproducible to $\pm$ 10%.

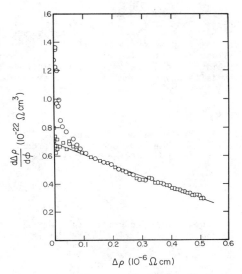

Fig. 2. Defect production rate, $d\Delta\rho/d\phi$, as a function of radiation-induced resistivity for 2 MeV α particle irradiation of aluminum [13]. The circles represent the measured production rates, the squares represent the data after correcting the size effects in the resistivity.

After the data has been corrected for size effects, the number of Frenkel pairs produced per incident ion can be deduced using the equation

$$\frac{d\Delta\rho}{d\phi} = \rho_f \frac{\nu}{N_o t} \qquad (3)$$

where ρ_f is the resistivity per atom fraction of Frenkel pairs
        N_o is the atomic density
        t is the specimen thickness
and ν is the number of Frenkel pairs produced per incident ion.

Eq. (3) is based on the assumption that the damage is produced homogeneously. In general, defect production by heavy ions or low-energy light ions is not homogeneous over the dimensions of our specimens. One might expect that nonhomogeneities in the damage distribution would invalidate the use of eq. (3) on the basis that the relatively undamaged regions of the specimen would tend to short-circuit the damaged portions. Our observations show, however, that this is not the case. We discuss these experimental observations in the next section, however, two qualitative remarks deserve mention here. First, $\Delta\rho$ is expected to be proportional to the total number of scattering centers (i.e. defects) but insensitive to their arrangement on a scale finer than the electronic mean free path. The electronic mean free path for our specimens is greater than the film thickness and therefore it is also larger than the scale of the damage nonhomogeneity. Second, even if we neglect the above consideration of the mean free path, and assume that the resistivity can be defined locally at a depth x as $\rho(x)$, as long as $\Delta\rho \ll \rho_o$, nonhomogeneous damage can be shown to have little effect on the value of ρ_f. Thus for studies in which the radiation induced resistivities, $\Delta\rho$, are $\lesssim 10^{-8}$ Ω-cm, the condition $\Delta\rho \ll \rho_o$ is satisfied.

175

Applications

The application of ERM on thin films is a useful technique to investigate essentially any phenomenon in which the defect concentration is altered. Four applications are cited in this section to demonstrate the versatility of the technique. In these examples the methods are emphasized, and not the results. Details of the results of each experiment can be found elsewhere.

Defect Production Studies [14]

A parameter fundamental to theories of radiation effects is the damage function, $\nu(T)$, which is the number of Frenkel pairs produced by a host-atom recoil of energy T. Generally the modified Kinchin-Pease expression [15]

$$\nu^{K-P}(T) = \frac{0.8 \ E_D(T)}{2 \ E_d} \qquad (4)$$

is employed to calculate the defect production rate during an irradiation. Here E_D represents the "damage energy" [16] and E_d is the average threshold energy for displacement. We have made studies of $\nu(T)$ for representative fcc metals and have found that in some cases, $\nu^{K-P}(T)$ is in error by as much as a factor of three.

The method we have used to investigate $\nu(T)$ can be understood by considering the equation for the number of Frenkel pairs produced by an ion coming to rest within a material,

$$\nu = \int \frac{dE}{S(E)} \int \frac{d\sigma(E,T)}{dT} \cdot \nu(T) \cdot dT \qquad (5)$$

where
	$S(E)$	is the energy dependent stopping power
	$d\sigma(E,T)/dT$	is the cross section for the ion of energy E to produce a host-atom recoil of energy T and
	$\nu(T)$	is the damage function

The functions $S(E)$ and $d\sigma(E,T)/dT$ are known reasonably well. Different regions of the damage function, $\nu(T)$, can be investigated by irradiating with different ion masses and energies. For example, for 15 keV proton irradiation of copper 80% of the defects are produced in recoils below 160 eV whereas for a 855 keV bismuth irradiation, 80% of the defects are produced in recoils above 7.0 keV. It is convenient to analyze the data by comparing experimental values of ν, ν^e, with values calculated using the Kinchin-Pease expression for $\nu(T)$ in eq. (5), ν^{K-P}. Accordingly, we define the defect production efficiency,

$$\xi = \frac{\nu^e}{\nu^{K-P}} \qquad (6)$$

Figure 3 shows the results for irradiations of copper. Here ξ is plotted as a function of the median recoil energy $T_{1/2}$; where half the Frenkel pair production results from primary recoil energies above $T_{1/2}$ and half below $T_{1/2}$. Silver and aluminum show a similar decrease of ξ with increasing $T_{1/2}$. A low but constant efficiency of $\nu(T) \sim 1/3 \ \nu^{K-P}(T)$ is observed at high values of $T_{1/2}$. An efficiency near unity at low values of $T_{1/2}$ indicates that $\nu(T) \sim \nu^{K-P}(T)$ at low recoil energies. Further

176

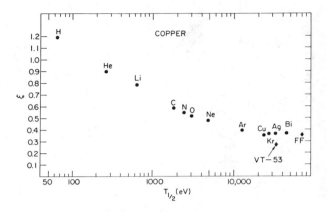

Fig. 3. Defect production efficiencies vs $T_{1/2}$ for several ion irradiations of copper. The irradiations were 17 keV H, 35 keV He, 54 keV Li, 100 keV C^+, 150 keV N^+, 160 keV Ne^+, 320 keV Ar^+, 500 keV Cu^{2+}, 520 keV Kr^{2+}, 560 keV Ag^{2+}, and 855 keV Bi^{3+}. Efficiencies for fast-neutron (VT-53) and for fission fragments (FF) are shown for comparison.

analysis reveals that the transition from high to low efficiency occurs for host-atom recoil energies between 1 keV and 10 keV.

During this investigation, we tested for possible effects of nonhomogeneous damage on the electrical resistivity. This was done by varying the damage distribution in the specimens using two methods. First the distribution relative to the specimen dimensions was varied by using several specimens of different thicknesses ranging from 220 nm to 460 nm. The values of ν^e were found to be independent of specimen thickness. Second, the distribution was varied by changing the projectile energy. For neon and argon irradiations of silver, it was found that changing the projected range of the ions from one-fourth to three-fourths of the specimen thickness caused no change in the values of ξ.

Stopping Powers of Low Energy Ions [17]

In the first example, thin films were employed because ion ranges are too short to use bulk specimens. This second example utilizes the specimen thickness as a length gauge for the study of damage distributions and stopping powers of low energy ions in metals. The idea of this experiment is as follows. The total number of Frenkel pairs produced along the range in bulk material increases as the energy of the incident projectile increases. In a thin film, however, a fraction of the beam may be transmitted through the back surface, which results in a reduced defect production at high energies. In addition, when all the ions pass through the specimen, the defect production decreases further with energy, since the displacement cross section is a decreasing function of energy. Thus the dependence of defect production on projectile energy in thin films contains information regarding the stopping of ions in metals. A quantitative determination of stopping powers, however, requires theoretical predictions of damage profiles. For this purpose we have used a computer simulation

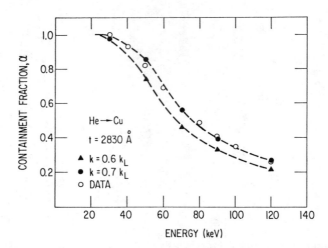

Fig. 4. The containment fraction versus energy for the irradiation of a
copper thin film with thickness = 238 nm. The points on the
theoretical curves indicate the energies at which the computer
simultation runs were performed.

code, (TRIM [18]). The low energy stopping power is expressed in the form

$$S = kE^{1/2}, \tag{7}$$

with the stopping coefficient, k, an adjustable parameter. Curves for the
defect production as a function of energy have been generated for different
values of k and compared with the experimental results. Rather than plot
the defect production directly, it is more illuminating to plot the
"containment fraction", α, the ratio of the defect production in the thin
film to that in a bulk specimen, i.e., α is the fractional damage energy
"contained", within the film. Figure 4 shows results for helium irradiation
of copper. The theoretical curves for α were obtained for two values of
k. The curve for $k = 0.7 \ k_L$, where k_L is the value of k obtained from the
Lindhard-Scharff formula [19], fits the data quite closely. This value of k
also gave best agreement in the damage production rates for helium
irradiations of copper specimens of different thicknesses.

Hydrogen Diffusion in Metals

A major difficulty in measuring low temperature diffusion of hydrogen
in fcc metals derives from the low solubility and high mobility of hydrogen
in these metals. Consequently, the technique of charging a material with
hydrogen at high temperature, and subsequently quenching, has not been very
successful in many cases [20]. Ion implantation provides an alternative
means of introducing hydrogen into materials. Often this method is not
satisfactory either, because many Frenkel pairs are produced during
implantation and the hydrogen can interact with these defects. However,
using films of ~ 50 nm, hydrogen can be implanted well into the specimen at
energies < 1 keV without the introduction of any Frenkel pair damage. For
heavy metals, such as gold, no displacement damage can be produced at these
energies since the maximum energy transfer between hydrogen and gold atoms
is less than the minimum displacement energy, ~30 eV. The electrical
resistivity increase is then associated entirely with interstitial

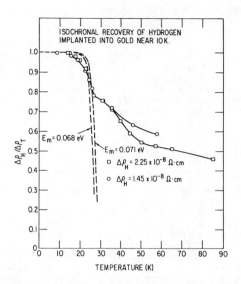

Fig. 5. Recovery of the electrical resistivity during isochronal annealing in gold after 1.0 keV H implantation to two concentrations, to $\Delta\rho = 1.45 \times 10^8$ Ω-cm, o, and $\Delta\rho = 2.25 \times 10^{-8}$ Ω-cm, $\square$.

hydrogen. Such experiments do in fact show a linear increase of $\Delta\rho$ with irradiation dose over a wide range of ϕ. From such an experiment it should be possible to also determine the specific resistivity of interstitial hydrogen, if the implanted dose is known with sufficient accuracy. For studies of hydrogen diffusion, an isochronal annealing program is performed after a low dose irradiation at 6 K and the recovery in the resistivity is measured. With the assumption that the surface is the dominant sink, the activation energy for migration can be determined. Figure 5 illustrates the recovery of a 60 nm gold specimen after 1.0 keV hydrogen implantation. Also shown are theoretical recovery curves which employ different values for the activation energy. A low temperature recovery stage with activation energy ~.07 eV is clearly evident. However, deviations occur from the recovery model based on a single activated process and it is also observed that the recovery is not complete after the low temperature stage. Additional recovery occurs at higher temperatures. We are presently trying to understand this recovery curve in more detail by studying the effects of hydrogen concentration, and of impurities on the recovery. An important advantage of using electrical resistivity measurements for this study is that the hydrogen concentration can be kept below ~20 ppm and the interaction between hydrogen atoms therefore minimized.

Interaction of Interstitials with Defect Clusters

Damage production rate measurements at temperatures at which interstitials are mobile but vacancies immobile have been used to study the interaction of interstitials with impurity atoms [21]. The diffusing interstitials may react with a number of different defects (e.g. impurity traps, dislocations, vacancies, interstitial- and vacancy clusters) but only those interstitials that do not recombine with vacancies produce measurable damage. Therefore, the defect state of a metal can be probed by damage rate measurements at temperatures at which interstitials are mobile. For maximum sensitivity in the method, the test irradiations should produce mainly

179

isolated Frenkel pairs and no defect clusters. Electron or proton
irradiations satisfy this requirement.

We have begun to investigate the structure of displacement cascades in
metals produced by ion irradiation in thin films and to study the dependence
of cascade structure on the energy density in the cascade and on annealing
temperature. The procedure is to dope the specimen with cascade damage and
subsequently probe it via 130 keV proton damage rate measurements at 55 K.
If the defect concentration produced by the test dose is small compared to
that of the defect cascades, the damage rate $d\Delta\rho/d\phi$ of the test irradiation
is given by

$$\frac{d\Delta\rho}{d\phi} = \rho_f \sigma_d^m \cdot (1 - f_{corr}) \frac{r_i^{cl} c_i^{cl} + r_t c_t + r_s c_s}{r_v^{cl} c_v^{cl} + r_i^{cl} c_i^{cl} + r_t c_t + r_s c_s} \tag{8}$$

where σ_d^m is the cross-section for producing migrating interstitials,
f_{corr} is the probability for interstitials to recombine
correlatedly, c_i^{cl}, c_v^{cl}, c_t and c_s are the concentrations and r_i^{cl}, r_v^{cl}, r_t
and r_s are the capture-radii for interstitial clusters, vacancy clusters,
impurity traps, and sinks (dislocations, etc.) respectively. Equation (8)
shows that interaction of interstitials with residual traps and sinks
reduces the sensitivity of the damage rates to changes in the cascade
structure. Consequently, the concentration of these residual traps and
sinks should be kept as small as possible, a condition which can be met by
careful sample preparation. The principle of this method shall be
demonstrated by some experimental results.

Figure 6 shows the damage rates in a 260 nm thick silver specimen
irradiated at 55 K with 130 keV protons versus defect concentration. Here
the defects are produced by the same irradiating particle as used for the
test dose. The damage rate decreases with increasing defect concentration
because the irradiation produces mainly immobile vacancies, which act as
recombination centers for the migrating interstitials, and no interstitial
traps. The defect state consists of single vacancies and interstitial

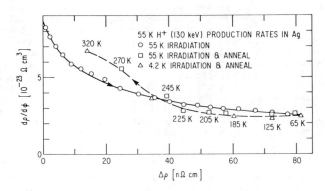

Fig. 6. Resistivity damage rate versus radiation-induced resistivity for
55 K proton irradiation in silver; o, during H^+ irrdiation at
55 K, □, after H^+ irradiation at 55 K and annealing at indicated
temperature; Δ, after H^+ irradiation at 4.2 K and annealing at
irradiated temperature.

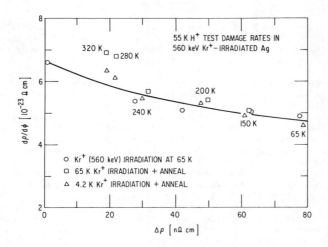

Fig. 7. Proton test damage rates in silver irradiated with 560 keV Kr$^+$; o, during Kr$^+$ irradiation at 65 K; □, after Kr irradiation at 65 and annealing at indicated temperature; Δ after Kr irradiation at 4.2 K and annealing to indicated temperature.

clusters nucleated at impurity traps. The damage rate of the test dose is then given by

$$\frac{\Delta\rho}{\Delta\phi} = \sigma_d (1 - f_{corr}) \cdot \rho_f \cdot \frac{r_{i,t}^{cl} c_{i,t}^{cl}}{r_{i,t}^{cl} c_{i,t}^{cl} + r_v c_v} ,$$

where $c_{i,t}^{cl}$ is the concentration of trapped interstitials and r_i^{cl} is the average capture radius per interstitial. During the subsequent anneal the damage rate of the test irradiation remains constant up to an annealing temperature of 225 K although nearly 50% of the damage has annealed out. This result indicates that the relative capture radii r_i^{cl} and r_v do not change although the interstitial clusters are known to grow in stage II [3]. At temperatures above 225 K migration of vacancies leads to vacancy cluster formation and hence to a decrease in the average capture radius r_v so that damage rate of the test dose increases.

Figure 7 shows the test dose damage rate in 260 nm thick silver irradiated at 65 K with 560 keV Kr$^+$ ions versus the ion-induced defect concentration. Damage rates of test irradiations performed on defect structures produced by heavy ion bombardment are expected to be constant as long as the cluster structure does not change, i.e. the defect clusters do not overlap, the damage rate being proportional to $r_i^{cl} c_i^{cl} / (r_v^{cl} c_v^{cl} + r_i^{cl} c_i^{cl})$. The decrease seen in fig. 7 indicates the influence of internal and external surfaces and of cascade overlap on the results; however, the decrease is considerably less than that observed for proton irradiation, Fig. 6. During the subsequent anneal (in addition the results of a sample irradiated at 4 K with 560 keV Kr$^+$ are also included) the damage rate of the test dose is practically identical to that measured during the production of the defect clusters at 65 K. Evidently, the cascade structure does not change significantly during this anneal. Further experiments are in progress.

Summary

It has been shown that the application of electrical resistivity measurements to ultra-thin metal films in conjunction with ion-beam irradiation has a wide range of application for studying defects and their properties as well as for studying ion-solid interaction processes.

We have demonstrated that, although the electrical resistivity of ultra-thin films is generally dominated at low temperature by surface scattering, the resistivity increments introduced by ion irradiation can be used for quanitative analyses of displacement and impurity defects. The method has considerable advantages over bulk resistivity measurements regarding, the achievement of very large damage rates, the possibility of automatically averaging over inhomogeneous damage distributions and the ultimate sensitivity. The method is most generally applicable to the study of point defects and small clusters thereof, providing thus a complementary tool to the transmission electron microscopy which is sensitive to point defect clusters above the TEM visibility limit.

Ion irradiation over a wide range of particle masses and energies in conjunction with the use of ultra-thin films provides the only practical procedure for systematically and continuously vary the displacement conditions, i.e. the recoil spectrum, over the whole range of interest in cascade studies. Such experiments have therefore been extremely valuable for the study of the damage function. Examples of the application of the technique to low energy ion stopping power, to the study of low-temperature hydrogen diffusion and a study of the interaction between self-interstitials and defect clusters have been given. Other applications have been concerned with the annealing behavior in cascades of varying energy density [22,23] and the degree of clustering in cascades [24].

In the future it is expected that the method will find further important areas of application in the study of defects and impurities in metals. However, for many problems, the development of suitable thin-film specimens may be a major obstacle.

References

[1]   W. Schilling, G. Burger, K. Isebeck, and H. Wenzl, in Vacancies and Interstitials in Metals, edited by A Seeger, et al. (North Holland, Amsterdam, 1970) p. 255.

[2]   H. Wollenberger, J. Nucl. Mater. 69-70, (1978) p. 362.

[3]   P. Ehrhart, H. G. Haubold, and W. Schilling, Festkoerperprobleme XIV, (1974) p. 87.

[4]   R. C. Birtcher and T. H. Blewitt, J. Nucl. Mater. 69-70, (1978) p. 783.

[5]   R. Benedek, J. Appl. Phys. 48, (1977) 3832.

[6]   Engelhard Industries Div., Hanover Thick Film, 1 West Central Ave., E. Newark, NJ 07029. Three inks were used, gold ink type A-3360, Platinum ink type A-3443, and Rely-Ohm type A-3000.

[7]   J. Bass, Adv. Phys. 21, 431 (1972).

[8]   F. Dworschak, H. Schuster, H. Wollenberger, and J. Wurm, Phys. Stat. Sol. 21, (1967) p. 741.

[9]   K. L. Merkle, in Proceedings of the International Conference on Solid State Physics Research with Accelerators, Brookhaven National Laboratory, 1967, Report No. BNL-50083, C52, p. 359.

[10]   E. H. Sondheimer, Adv. Phys. 1, (1952) p. 1.

[11]   G. Duesing, W. Sassin, W. Schilling, and H. Hemmerich, Crys. Lattice Defects 1, (1969) p. 55.

[12]   J. A. Horak and T. H. Blewitt, Phys. Stat. Sol. 9, (1972) p. 721.

[13]   K. L. Merkle and L. R. Singer, Appl. Phys. Letts. 11 (1967), p. 35.

[14]   R. S. Averback, R. Benedek and K. L. Merkle, Phys. Rev. B 18 (1978) p. 4156.

[15]   P. Sigmund, Radiat. Eff. 1, (1969) p. 15.

[16]   M. T. Robinson, in Radiation Induced Voids in Metals, edited by J. W. Corbett and L. C. Ianniello (U.S. AEC, Oak Ridge, Tenn., 1972) p. 392.

[17]   R. S. Averback, R. Benedek and K. L. Merkle, Phys. Rev. B 18 (1978) p. 4156.

[18]   J. P. Biersack and L. G. Haggmark, Nucl. Instrum. Meth. 174 (1980) 257.

[19]   J. Lindhard and M. Scharff, Phys. Rev. 124, (1961) p. 128.

[20]   See e.g. J. Völkl, G. Alefeld, in Diffusion in Solids, Recent Developments, edited by A. S. Nowick and J. J. Burton (Academic Press, New York, 1975) p. 231.

[21]   F. Dworschak, H. Schuster, H. Wollenberger, and J. Wurm, Phys. Stat. Sol. 29, 75(1968) and Phys. Stat. Sol. 29, 81 (1968).

[22]   R. S. Averback and K. L. Merkle, Phys. Rev. B 16, 3860 (1977) and R. S. Averback, L. J. Thompson, and K. L. Merkle, J. Nucl. Mat. 69/70, 714 (1978).

[23]   R. S. Averback, K. L. Merkle, and L. J. Thompson, Rad. Eff., 51, 91 (1980).

THE PIEZOELECTRIC ULTRASONIC COMPOSITE OSCILLATOR TECHNIQUE

(PUCOT) FOR MONITORING METALLURGICAL CHANGES [†]

Alan Wolfenden[*]

Westinghouse Research and Development Center
Pittsburgh, Pennsylvania 15235

The PUCOT has been used recently at 40-150 kHz to measure the temperature dependence of Young's modulus and the temperature and strain amplitude dependences of internal friction, $Q^{-1}(S)$, in a variety of materials. In addition to this traditional use, the technique has also been used to monitor order-disorder processes (e.g., Cu_3Au, Ni-25 at. % Co), phase changes in dental alloys (Hg-Sn-Ag), the melting and redistribution of grain boundary precipitates (Pb in α-brass), magnetic transformations near the Curie point (Fe and Ni), and precipitation processes (Mn-Cu alloys). In all cases the specimen size was relatively small (ca 2 mm dia x 20 mm long). Remote positioning of the specimen from the piezoelectric crystals during measurements was made possible by use of a quartz spacer rod of appropriate resonant length. This and other facets of the PUCOT could be of particular advantage for studying post-irradiation annealing behavior of irradiated metals.

[†]Research done at the Department of Scientific and Industrial Research, New Zealand.

[*]Formerly with Physics & Engineering Laboratory, DSIR, New Zealand.

Introduction

The ability of materials to dissipate vibrational energy is known as internal friction. Techniques for measuring internal friction have provided us with a wealth of information for understanding structural defects in crystalline solids. During the past thirty years there has been a mushrooming development of this field which is reflected in the books by Zener, (1) Mason, (2) Nowick and Berry (3) and De Batist (4). All the techniques are concerned with introducing elastic waves into the specimen so that they interact with point defects, dislocations and internal boundaries and then extracting information on the defects via the measuring system. The aims are to measure the internal friction or mechanical damping of the specimen $[Q^{-1}(S)]$, which corresponds to a change in the attenuation constant, and the modulus of elasticity of the specimen (E), which corresponds to a change in the wave propagation velocity. The importance of the latter term alone, E, can be assessed by its application in such wide-ranging areas as: load-deflection, thermoelastic stress, elastic instability, fracture mechanics, strength, interatomic potentials, lattice-vibration spectra, diffusion, lattice defects, thermodynamic equations of state, free energy, specific heat, thermal expansivity, and elastic Debye temperature. The combined knowledge of $Q^{-1}(S)$ and E is particularly effective for mechanistic studies of structural defects in solids.

In this paper one technique – the piezoelectric ultrasonic composite oscillator technique (PUCOT) – for measuring simultaneously the internal friction and Young's modulus in materials is described. The results and significance of the application of the technique to several materials demonstrating common physical metallurgical phenomena are outlined: order-disorder processes (Cu_3Au, Ni-25 at. % Co), phase changes (Hg-Sn-Ag), melting and redistribution of grain boundary precipitates (Pb in α-brass), magnetic transformations near the Curie point (Ni), and precipitation processes (Mn-Cu). Within the framework of this Symposium on techniques for characterizing the microstructure of metals, all of the metallurgical phenomena mentioned are of importance and hence the PUCOT has the potential of providing some of the information required for the microstructural characterization.

Experimental Technique

A schematic drawing of some of the PUCO components is shown in Fig. 1. The 4-component system has been described in detail previously (5,6) and only a brief description follows. The system consisted of two piezoelectric

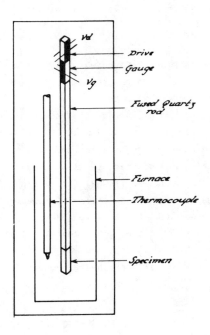

Fig. 1 – Schematic drawing of the piezoelectric ultrasonic composite
oscillator rig showing the four components: 2 piezoelectic
quartz crystals for drive and gauge, square cross-section
fused quartz spacer rod and square cross-section specimen.

quartz crystals (Drive D and Gage G) of square cross-section $3 \times 3 \text{ mm}^2$ tuned to resonate in the longitudinal mode at 40, 80 or 150 kHz, a fused quartz
spacer rod Q (of either square or circular cross-section) of appropriate
resonant length (usually about 370 mm), and the specimen (S). The joints
between the quartz crystals and the quartz rod and between the quartz rod
and the specimen were made with Loctite cyanoacrylate adhesive and
Sauereisen low expansion cement D-29, respectively. The PUCO system was
mounted vertically so that the quartz rod and specimen dangled into the
stainless steel furnace tube. The oscillator was under a vacuum of about
1 mPa for the experiments. For the studies at room temperature neither
the quartz spacer rod nor the furnace was required, i.e. the PUCO was then a

3-component system. Specimen lengths ℓ were cut to resonate at the appropriate frequency (f) and temperature (T) of interest according to:

$$\ell = (1/2f)\sqrt{(E/\rho)} = \lambda/2 , \tag{1}$$

where ρ is the density of the specimen and λ is the wavelength of the ultrasonic longitudinal wave in the specimen. During most of the experiments the
drive crystal was driven by up to 1000 V a.c. (Vd) through a closed loop
which held the gage crystal voltage (Vg) to 10 mV, corresponding to a speci-
men strain amplitude ε_{11} of about 3×10^{-7}. This strain amplitude was in
the amplitude independent range of internal friction. Amplitude dependence
of internal friction was studied by selecting various values of Vg in the
range 3 mV to 5 V. Values of τ(DGQS) (the resonant period of the composite),
Vd, Vg and T were recorded as functions of time on a millivolt chart recorder.
Polycrystalline specimens of initially ordered Cu_3Au, step-cooled Ni-25 at. %
Co, quenched Ni-25 at. % Co, Hg-Sn-Ag dental alloy Tytin, leaded α-brass,
lead-free α-brass, Ni, Cu-35 at. % Mn and Cu-55 at. % Mn were prepared for
the PUCOT investigations as described elsewhere (7-12).

The following standard equations for E, Q^{-1}(S) and ε_{11} were used (6):

$$\tau(S) = m(S)^{1/2}\tau(DGQ)\tau(DGQS)/A, \tag{2}$$

$$A = \left\{ \tau(DGQ)^2 m(DGQS) - \tau(DGQS)^2 m(DGQ) \right\}^{1/2}, \tag{3}$$

$$E = 4\rho\ell^2/\tau(S)^2, \tag{4}$$

$$Q^{-1}(S) = \left\{ (2/m(S)C_m)\left(N\tau(S)/\pi\right)^2 \right\} Vd/Vg, \tag{5}$$

$$\varepsilon_{11} = \left\{ (C_m\pi\sqrt{2})/N\lambda \right\} Vg, \tag{6}$$

where m(i) and τ(i) (for i = D,G,Q,S) are the masses and resonant periods,
C_m is the capacitance across the gage crystal (usually about 240 pF), and N
is the ideal transformer ratio for quartz (4.74×10^{-4} C/m). Other internal
friction relations that may be more familiar to workers in mechanical or
electrical fields are:

$$\tan \phi = \phi = Q^{-1}(S) = \delta/\pi = \Delta W/2\pi W, \quad (\phi \ll 1) \tag{7}$$

where ϕ is the loss angle (the angle by which the strain lags behind the stress), δ is the logarithmic decrement, ΔW is the energy dissipated in a full cycle (per unit volume), and W is the maximum stored energy per unit volume. The specific damping capacity is $\Delta W/W$.

Results and Discussion

Cu_3Au-Order-Disorder Process

This system is well known as exhibiting a classical long-range order-disorder transformation near 390°C and was used with the PUCOT to establish that the technique could readily monitor the order-disorder process as the initially ordered specimen was heated from room temperature to 425°C at a controlled rate and then cooled. Figure 2 shows the measured values of $Q^{-1}(S)$ as a function of temperature near the transformation temperature T_c.

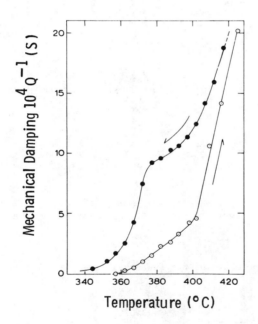

Fig. 2 – Mechanical damping $\left(Q^{-1}(S)\right)$ of the Cu_3Au specimen plotted as a function of temperature over the range 340–425°C for heating and cooling at 50°C/h.

Clearly the large rise in internal friction near T_c was associated with the disordering on heating. For cooling the internal friction decreased in more than one stage and a hysteresis loop was evident. The hysteresis loop was studied further in the form of τ vs. T for different heating/cooling rates. From Eq. (4) $E \propto \tau^{-2}$. Figure 3 shows the loop for a heating/cooling

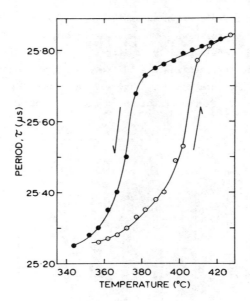

Fig. 3 - The resonant period of the composite (τ) plotted as a function of temperature over the range 340–425°C for heating and cooling Cu_3Au at 50°C/h.

rate of 50°C/h. The width of the hysteresis loop ΔT was found to fit the equation:

$$\Delta T = 240(\dot{T})^{1/2}, \qquad (8)$$

where $\dot{T}$ is the heating/cooling rate. Thus for $\Delta T \sim 390°C$, corresponding to the cooling rate to suppress ordering completely on cooling, $\dot{T} = 2.6°C/s$, i.e. 2.5 min to cool. This reflects the established observation that at

sufficiently high quenching speeds the disordered state obtained on anneal-
at T > T_c can be quenched in at room temperature.

Information on the mechanism of destruction of long-range order in
Cu_3Au on heating was obtained from an Arrhenius plot of the internal friction results (Fig. 4). For T > 370°C on heating the data points fitted a

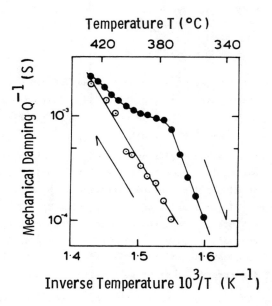

Fig. 4 - An Arrhenius plot of the mechanical damping of the Cu_3Au specimen
(Q^{-1}(S) vs. 10^3/T K) for heating and cooling at 50°C/h.

relation of the type:

$$Q^{-1}(S) \propto \exp(-F/kT), \qquad (9)$$

where F is an effective activation energy and k is Boltzmann's constant.
The value of F was 2.0 ± 0.2 eV which agreed with most of the values of
activation energy measured (13-19) for diffusion in Cu_3Au. Thus a possible
interpretation of the results for disordering could be based on atomic
diffusion via vacancies to produce disorder as the specimen temperature was
raised. Further support for this interpretation came from the consideration
that the value of F would be estimated as $h_f + h_m$, the sum of the enthalpies

of formation and motion of vacancies in Cu_3Au. We find that $h_f = 0.94 \pm 0.02$ eV (20) and

$$h_m = (0.71 + 0.31S^2) \text{ eV} \qquad \text{(Ref. 17)}, \qquad (10)$$

where S here is the Bragg-Williams parameter (22). For temperatures up to about 380°C S is in the range 1-0.8 (22). With S = 0.9 for our specimen, Eq. (10) yielded

$$h_m = 0.96 \text{ eV} \quad \text{and} \quad h_m + h_f = 1.9 \text{ eV}. \qquad (11)$$

This latter value was consistent with the F values of 2.0 ± 0.2 eV. Therefore the mechanism responsible for the change in $Q^{-1}(S)$ on heating in the range 340-425°C, which had an effective energy of 2.0 ± 0.2 eV, was probably the vacancy mechanism of atomic diffusion.

The situation on cooling (ordering) the Cu_3Au specimen was not identical with that on heating (disordering) (Figs. 2-4). Ordering seemed to be a more difficult process than disordering, agreeing with the concept of nucleation and growth of ordered domains on cooling below T_c. The complexity was reflected in Fig. 4 where at least three regions (and corresponding F values) on the cooling curve could be identified: 426°< T < ∿402°C (1.15 eV), ∿402°< T < ∿376°C (0.43 eV) and T< ∿376°C (3.3 eV). Further research is needed for a full interpretation of these results.

The PUCOT has the useful advantage that amplitude dependence studies of internal friction can be done by merely selecting various values of Vg. Figure 5 shows the amplitude dependence results for Cu_3Au obtained on heating

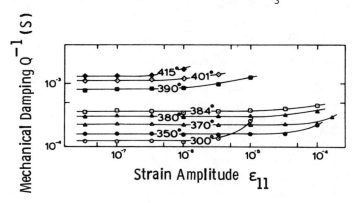

Fig. 5 — Amplitude dependence of the mechanical damping of a specimen of Cu_3Au $\left(Q^{-1}(S)\right)$ at several temperatures on heating.

and holding at several temperatures in the range 300-415°C. The curves
showed breakaway behavior which, on the basis of the Granato-Lücke (G-L)
theory (23), meant that the internal friction at high strain amplitude was
due to the motion of dislocations. The notation used is:

$$Q^{-1}(S) = Q_I^{-1}(S) + Q_H^{-1}(S) \tag{12}$$

where $Q_I^{-1}(S)$ is the amplitude independent internal friction (on the horizontal part of the $Q^{-1}(S)-\varepsilon_{11}$ curve) and $Q_H^{-1}(S)$ is the amplitude dependent part
of the internal friction. The breakaway strain amplitude $\varepsilon_{11}(b)$ is defined
as the strain amplitude at which $Q^{-1}(S)$ exceeds $Q_I^{-1}(S)$ by 10 or 20%. It was
found that $\varepsilon_{11}(b)$ decreased markedly near T_c, indicating that at T_c the
dislocations were able to break away from their pinning points more readily.
This could be caused by an increase in dislocation loop length or Burgers
vector, or a decrease in the strength of pinning points. It was likely that
the form of the dislocations (paired partials for $T<T_c$ on the one hand and
individual dislocations for $T>T_c$ on the other) and their arrangement made
some contribution to the internal friction behavior.

From Fig. 5 data were obtained for the G-L plot of $\varepsilon_{11}Q_H^{-1}(S)$ vs. ε_{11}^{-1}.
Such a plot for Cu_3Au (and also for brass) is given in Fig. 6. It should be
emphasized that since $Q_H^{-1}(S)$ is measured as the difference between $Q^{-1}(S)$
and $Q_I^{-1}(S)$ the errors in $Q_H^{-1}(S)$ increase tremendously for the smaller values
of ε_{11} (i.e. for larger values of ε_{11}^{-1}, say beyond 10^{-4} ε_{11}^{-1} = 3). For the
G-L model to hold for Cu_3Au the curves should be linear, which is seen to be
the case. The G-L analysis was applied to the Cu_3Au data to get information
(via the slope of the curve) on L_c, the minor loop length of the vibrating
dislocation lines. The values of L_c were found (24) to be about 200 nm at
300°C, falling to a minimum of 5 nm at 370°C, then rising to 1000 nm at
415°C. The L_c value at 350°C was within a factor of 4 of the values of the
separation of superlattice dislocations measured by Morris and Smallman (25)
and was smaller by a factor of 2 than the values found (26) for domain sizes
for isothermal annealing at 350°C for 100 min. Hence the internal friction
data suggested that during the order-disorder transformation the details of
the superlattice dislocations and domain sizes governed the minor loop
length. Most interestingly, when the values of $\varepsilon_{11}(b)$ for Cu_3Au were
plotted against the derived values of L_c (Fig. 7), the $\varepsilon_{11}(b) \propto L_c^{-1}$ depen-
dency expected from G-L theory was not obtained. Rather the curve was
horseshoe-shaped for Cu_3Au, being similar to that observed for Au-Ag (27)
for which short-range disordering above 400°C was suspected. The shape of
this curve could be rationalized by considering that the ordered Cu_3Au and
the disordered Cu_3Au behaved like different materials, each with its own

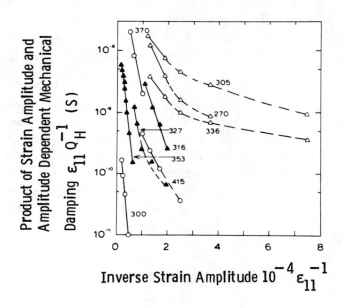

Fig. 6 – Granato-Lücke plots for Cu$_3$Au (circular data points), leaded brass (filled triangular points) and lead-free brass (open triangular points). The temperature (°C) for which each curve was determined is indicated.

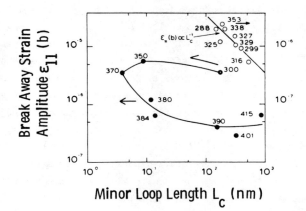

Figure 7 – Breakaway strain amplitude ε_{11}(b) as a function of derived values of minor loop length L_c for Cu$_3$Au (filled data points) and leaded brass (open points). The temperature (°C) for each datum and the expected dependence from G-L theory are indicated.

$\varepsilon_{11}(b)-L_c$ curve. As the temperature was increased we monitored the transition from one curve to the other. The turn-around or tip of the horseshoe was at 370°C, the temperature at which: (i) the rate of destruction of long-range order is a maximum, (ii) the L_c-T curve reached a minimum, and (iii) the relaxation time was postulated (28) to reach a minimum. Thus the functional dependence of $\varepsilon_{11}(b)$ on L_c may be regarded as a useful indication of some of the metallurgical processes occurring in the material under investigation by the PUCOT.

Ni-25 at. % Co-Order-Disorder Process (Possibly)

Over the past eight years the question of the existence of crystallographic order in Ni-25 at. % Co ("Ni_3Co") has been discussed and reexamined (8,29) several times. The direct methods of detecting long-range order - X-ray or neutron diffraction - have failed to demonstrate the existence of order in this system. Indirect techniques have yielded evidence that was inconclusive or contradictory since the effects could result from phenomena other than order (29). Many of the experimental problems stemmed from the proximity of the Curie temperature of ∿620°C to the reported or suspected T_c (critical temperature for ordering) values of 700-750°C. In light of the ease with which the PUCOT monitored the order-disorder transformation in Cu_3Au, an attempt was made to provide additional information on the possible existence of crystallographic order in Ni-25 at. % Co.

Specimens of step-cooled (possibly ordered) or quenched (presumably disordered) Ni-25 at. % Co were investigated with the PUCOT over the temperature range 25-840°C, particular attention being focussed on the temperature range (about 700-750°C) of interest for order-disorder reactions in this system. Figure 8 shows the internal friction in the form of an Arrhenius plot. For the step-cooled specimen two small peaks were found at 686 and 754°C for heating, while the cooling curve showed no peaks for this specimen. Therefore these peaks could be an indication of an order-disorder process. The effective activation energy associated with the cooling curve and also with the base-line heating curve was about 1.1 eV, this value not being equal to any of the literature values (approximately 2.5-2.9 eV) for bulk diffusion in Ni-25 at. % Co. The effective activation energy agreed with the activation energy value of about 1.03 eV of Million (30) who suggested that high-diffusivity paths occur for Co diffusion in Ni-25 at. % Co below 1000°C. This could be another internal friction phenomenon that was monitored here. These paths were most likely grain boundaries and the common impurities in Ni-Co alloys, C and S, may have been involved in the process

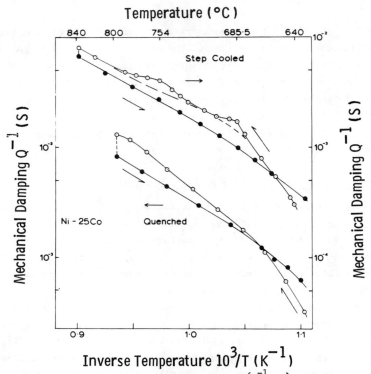

Fig. 8 - Arrhenius plot of the mechanical damping $\left(Q^{-1}(S)\right)$ of Ni-25 at. %
Co specimens heated and cooled at 50°C/h over the temperature
range 640°C to about 840°C.

too. Hence the observation of slight internal friction peaks at 686 and
754°C provides additional evidence for order-disorder effects in Ni-25 at. %
Co, though this evidence is not unambiguous. Further, the small peaks may
have been due to short-range disordering in that long-range disordering
effects would be expected to give much larger changes in internal friction
and resonant period. The possible reverse effect, disorder-order on cooling,
which was not observed, was absent because our cooling rate of 50°C/h was
too high. The internal friction curves for the quenched specimen showed
no peaks for heating or cooling (Fig. 8) at 640-800°C. This reinforces
the suggestion that the peaks observed for the step-cooled specimen on
heating may have been due to order-disorder effects.

So far the results from the PUCOT study of ordering in Ni-25 at. % Co
have been somewhat disappointing (in comparison with those for Cu_3Au). However, the experiments at temperatures below 640°C yielded spectacular
variations in internal friction in both the quenched and the step-cooled
specimens due to an effect other than crystallographic disordering. An
example is given in Fig. 9. The noteworthy feature was the large dip

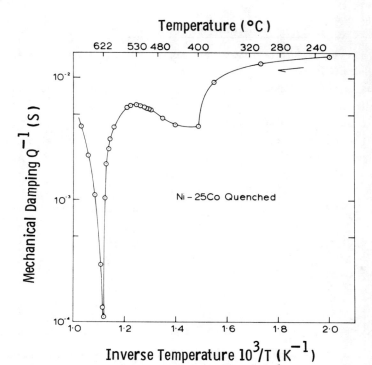

Fig. 9 – Arrhenius plot of the mechanical damping $\left(Q^{-1}(S)\right)$ of quenched Ni-25 at. % Co on heating at 50°C/h for temperatures above 227°C. The overnight anneal near 400°C reduced $Q^{-1}(S)$ by a factor of about 2. The Curie temperature of the alloy is near 622°C at which there is a large dip in $Q^{-1}(S)$ values.

(over an order of magnitude) in internal friction at 621 ± 2°C. This dip was attributed to a strong interaction between internal friction and the ferromagnetic-paramagnetic transformation in Ni-25 at. % Co at the Curie temperature reported as about 638°C (31). At temperatures below the Curie temperature it was found that magnetic fields (B) in the range 0.13-0.25 T applied to the specimen produced increases in internal friction for both types of specimen. This increase was proportional to B^2. At temperatures of 622-630°C the magnetic effect was progressively smaller and above 630°C it was absent. The magnetic field dependence of internal friction in ferromagnetic materials has been reviewed by Bozorth (32). The normal trend is for the internal friction to rise at a rate approximately proportional to B^2 and then at a linear rate, to reach a maximum at high values of B and then to drop to a low value at saturation. The Ni-Co results for low values of field are thus consistent with this general trend.

Phase Changes in Tytin, a Hg–Sn–Ag Dental Alloy

    Phase changes and mechanical properties of Hg–Sn–Ag amalgams are of
great concern in dentistry. The changes in Young's modulus after compaction
(condensing) of the dental amalgam are particularly significant. Since it
is difficult to obtain with standard tensile machines reliable E values for
such soft materials as amalgams just after their compaction, the PUCOT was
used (9) at 80 kHz to measure E and $Q^{-1}(S)$ in several dental alloys at small
strain amplitudes ($\varepsilon_{11} \sim 10^{-7}$) and hence also at low stresses (2–8 kPa).
Figures 10 and 11 show the time dependence of Young's modulus for aging a

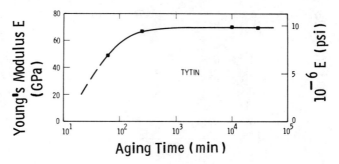

Fig. 10 – Measured values of Young's modulus E as a function of aging time
for the dental alloy Tytin.

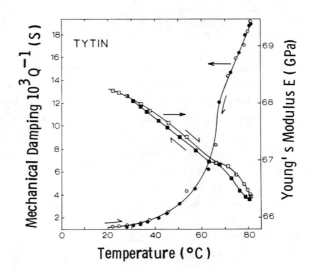

Fig. 11 – Temperature dependence of internal friction $Q^{-1}(S)$ and Young's
modulus E for the amalgam Tytin.

specimen of dental alloy Tytin at 37°C and the temperature dependence of E and Q^{-1}(S) over the range 20-80°C (heating/cooling rate of 20°C/h), respectively. At 37°C the Young's modulus increased by at least 43% over the time range used and saturated at a level of about 70 GPa after 10^4 min (1 week). The plateau value agreed with those from the literature (33). The Q^{-1}(S) values increased by a factor of 19 as the temperature was increased and the E values decreased only slightly (by 3%). The rapid increase in Q^{-1}(S) at 65-70°C was probably connected with the reported (34) $\gamma_1 \rightarrow \beta_1$ phase transformation in the alloy. The E-T curve had a corresponding change in slope at the same temperature. The values of Q^{-1}(S) measured at 80°C (about 2×10^{-2}) were high compared to those for common metals (except lead) and alloys. Thus the dental amalgam was very efficient at damping out vibrations at temperatures of 65-80°C. These observations are consistent with the fact that solid amalgams sweat mercury at temperatures near 70°C, the presence of the liquid raising Q^{-1}(S). Similar variations in Q^{-1}(S) are noted also in the following section for a melting metallic component (lead) within a ternary alloy system (Cu-Zn-Pb).

The Arrhenius plot (Fig. 12) of the internal friction results for Tytin yielded effective activation energies of 0.45 eV (T ≤ 65°C) and

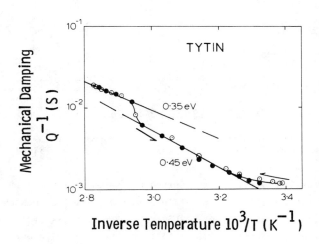

Fig. 12 - Arrhenius plot of internal friction Q^{-1}(S) for the alloy Tytin.

0.35 eV (T $\geq$ 65°C). The tail in the plot at $10^3/T \geq 3.25$ corresponded to
thermal inertia in the specimen-furnace system. Both energies were close
to the value of 0.42 eV for the diffusion of Hg in γ_1 (Ag_2Hg_3) (35). Thus it
seems likely that the diffusion of mercury in γ_1 influenced the internal
friction properties of this amalgam with a possible change in mechanism
near 65°C. Such complex behavior could be anticipated from a study of the
Ag-Hg-Sn phase diagram modified by small additions of copper, zinc, etc.

Leaded α-Brass–Melting and Redistribution of Grain Boundary Precipitates

Internal friction experiments were done on leaded (3.5 wt. % Pb) and on
lead-free (0.03 wt. % Pb) α-brass to investigate the internal friction peaks
found in the range 280–330°C for the former alloy only. The intriguing
results are shown in Fig. 13. On heating the specimen of leaded brass,

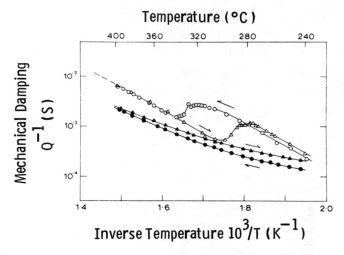

Fig. 13 - An Arrhenius plot of the mechanical damping of the specimen for
leaded (open data points) and lead-free brass (filled data points).

Q^{-1}(S) increased slowly and then started to decrease at 319 $\pm$ 1°C before
rising to a peak at 327 $\pm$ 0.5°C, while on cooling the specimen the internal
friction decreased smoothly until 305 $\pm$ 1°C when it increased to about the
original value for heating at 280 $\pm$ 1°C. The peak at 327°C was investigated
further by a series of experiments involving heating or cooling the specimen
over the range 310–330°C with or without pauses near 329°C (Fig. 14). It
became evident that prolonged annealing at 329°C eliminated the internal

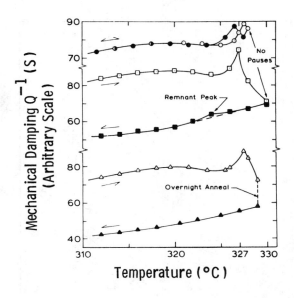

Fig. 14 – The mechanical damping of the leaded brass specimen plotted as a
function of temperature for experiments to follow the suppression
of the 327°C damping peak. Open data points: heating at 16°C/h;
filled data points: cooling at 16°C/h.

friction peak at 327°C on cooling and that heating to 330°C and then cooling
immediately almost removed the peak. Cooling the specimen to successively
lower temperatures made the 327°C peak reappear on reheating, as shown in
Fig. 15. The peak was fully re-established for lower holding temperatures
below 280°C with a characteristic temperature of 284 ± 1°C. For a possible
interpretation of these results the microstructure of the leaded alloy was
considered. The alloy Cu-Zn-3.5 wt. % Pb consists of polycrystalline α phase
with a distribution of lead in the form of precipitates, particularly at
grain boundaries, since the solubility of lead in brass is low (<0.29 wt. %
Pb in Cu above 600°C (36), <5 x 10^{-5} wt. % Pb in Zn (37)). Lead and zinc,
however, form a eutectic at 318.2°C and 99.5 wt. % Pb (38). Hence as the
specimen was heated above 318°C the eutectic would gradually melt until at
327°C the entire lead-based mixture would be completely molten. Thus the
observed decrease in internal friction starting at 319°C and the sharp peak
at 327°C may reflect the initial and then complete melting of the eutectic/
lead mixture. Indeed, globules of molten lead in brass have been observed
(39) on grain boundaries. The peak at 327°C may thus reflect the melting

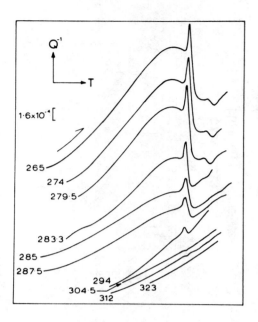

Fig. 15 – Schematic tracings of Q^{-1} vs. temperature for sequential experiments in which the temperature was oscillated between about 340°C and the temperature specified on each curve to monitor the reappearance of the peak at 327°C in leaded brass.

and redistribution of lead at or near grain boundaries. For cooling, the low temperature equilibrium distribution of the lead precipitates would take a longer time to occur and this process evidently did not give any internal friction peaks. The results suggested that the low temperature distribution reached equilibrium at temperatures below 280°C. Further, an amplitude dependence study of the internal friction of the leaded brass showed that the breakaway strain amplitude was a minimum at 316 ± 4°C (Fig. 16). This temperature almost coincided with the decrease in internal friction observed on heating at 319°C for this alloy. Thus the breakaway strain amplitude decreased with increasing temperature in keeping with the G-L theory until near 316°C there was a rapid increase in dislocation pinning possibly due to the release of lead atoms from the melting globules of lead in the grain boundaries.

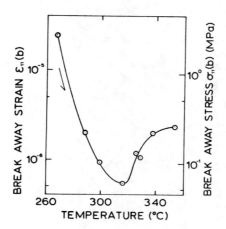

Fig. 16 - Temperature dependence of the break away strain amplitude and
break away stress amplitude for leaded alpha brass.

Thus the PUCOT experiments on leaded brass have led to the suggestions
that the decrease in internal friction at 319°C on heating may have been
associated with the onset of melting of the Pb-Zn eutectic while the sharp
peak at 327°C may have been due to the complete melting of the lead. The
internal friction rise starting at 305°C for cooling and finishing at
280°C was possibly caused by the redistribution of lead in or near grain
boundaries.

Ni-Magnetic Transformations Near the Curie Point

A brief survey (11) of the internal friction investigations in nickel
indicated that two points needed further study. These were the questions of
internal friction peaks or anomalies near the Curie temperature T_c and of
domain wall-dislocation interactions. The PUCOT study on nickel was done
over the temperature range 327 to 397°C (T_c ~360°C) to explore these questions. The amplitude independent internal friction of the polycrystalline
specimen for five values of magnetic field (B) for heating and cooling
through T_c is presented in Fig. 17. The $Q^{-1}(S)$ vs. T curves increased with
increasing temperature and displayed interesting maximum-minimum combinations near T_c. These combinations were readily resolved for B ≳ 51 mT and

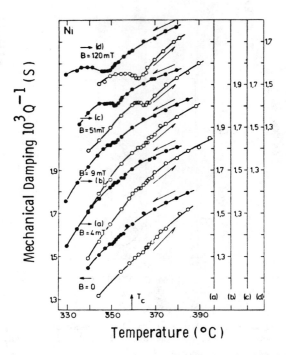

Fig. 17 - Temperature dependence of internal friction at 327–397 °C for
various applied fields. Heating (open data points) and cooling
(filled points) were at 100°C/h.

were analyzed in terms of a peak (with a corresponding T_{max}) or of a minimum
(with a T_{min}). For example, Fig. 18 shows the field dependence of the depth
of the minimum. Above 50 mT there was saturation of both the peak height
and the depth of the minimum.

If the interpretation of the $Q^{-1}(S)$ changes near T_c is in terms of an
internal friction peak, this peak cannot be due to domain wall motion since
for $B \gtrsim 25$ mT the specimen is magnetically saturated. The peak could arise,
therefore, from impurities such as C, O or Si. In Ni-0.5 wt. % C the carbon
peak was found near 467°C for 40 kHz (40). Thus our peak at $\sim$357°C was not
due to carbon alone but could have arisen from diffusion involving Ni, C,
O or Si as groups or clusters. Alternatively, the $Q^{-1}(S)$ vs. T behavior
near T_c can be assessed in terms of a minimum (Fig. 18). It was shown
earlier [Section "Ni-25 at. % Co-Order-Disorder Process (Possibly)"] that
an internal friction minimum occurred in Ni-25 at. % Co at T_c and was
attributed to a strong interaction between the operative internal friction

204

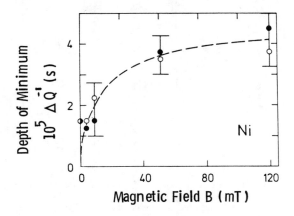

Fig. 18 - The field dependence of the depth of the minimum for heating
and for cooling.

mechanism and the ferromagnetic-paramagnetic transformation. Probably a
similar interaction occurred here for nickel. From the values of T_{min} at
$B = 0$ for heating and cooling T_c was estimated as 358°C, only 2°C lower
than the nominal T_c for nickel. The presence of the hysteresis in the
$Q^{-1}(S)$ vs. T curves indicated that a nucleation and growth mechanism
possibly involving clustering was operative near T_c.

The amplitude dependence of the internal friction was found to be
small but break away strain amplitudes $\varepsilon_{11}(b)$ could be measured. From
G-L plots values of the minor pinning length L_c were found to be in the
range 55 to 230 nm, much less than the usual domain sizes ($\sim 10^{-5}$ m). The
G-L theory predicts that $\varepsilon_{11}(b) \propto L_c^{-1}$. This dependence was found only for
$B \neq 0$ (in spite of the scatter, Fig. 19). For $B = 0$, the data suggested
a dependence of the form $\varepsilon_{11}(b) \propto L_c$, which is not in accord with the
G-L model. The rationalization of these PUCOT results is that for $B \neq 0$
there was no contribution to the internal friction from domain walls
(i.e., internal friction was due to dislocations pinned by impurities),
whereas with $B = 0$ the contribution from domain walls (magnetoelastic
internal friction) swamped the usual damping from pinned dislocations.

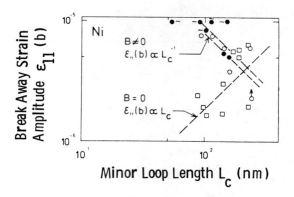

Fig. 19 - The break away strain amplitude $\varepsilon_{11}(b)$ plotted as a function of
dislocation loop length L_c. $B = 0$ □; $B = 73$ mT ●; $B = 126$ mT o.
The slope of -1 is the dependence expected from Granato-Lücke
theory (23).

Precipitation Processes in Mn-Cu

Excessive vibration and noise in machinery, piping, etc. can be com-
batted often with the use of Mn-Cu alloys in certain metallurgical conditions. The high internal friction that has been observed and measured at
or near room temperature arises from the tetragonality conferred on the
crystal lattice during aging and subsequent quenching of these alloys.
In this investigation the internal friction of Cu-35 at. % Mn and Cu-55
at. % Mn alloys was measured during (as opposed to after) aging of the
quenched alloys at 400 or 450°C with the PUCOT at 40 kHz. Further studies
were done with the former alloy to determine the amplitude dependence of
the internal friction.

The $Q^{-1}(S)$ values measured during aging are shown in Fig. 20. The
hardness data of Vitek and Warlimont (41) for Cu-40 at. % Mn and Cu-60 at. %
Mn alloys aged at 450°C and the lattice parameter results of Hedley (42) for
Cu-60 at. % Mn aged at 400°C are given also in the Figure. For the Cu-55
at. % Mn alloy the internal friction increased for aging times up to about
4 h, decreased rapidly at 4 to 6 h and then drastically at 6 to 18 h. The
rapid decreases occurred at aging times for which Hedley found rapidly
decreasing c/a values for a Cu-60 at. % Mn alloy. The increase in $Q^{-1}(S)$
(up to 4 h) was approximately proportional to the increase in hardness
measured (41) on a Cu-60 at. % Mn alloy. The internal friction variations
for the Cu-35 at. % Mn specimen were not so marked. There was a slight
increase in $Q^{-1}(S)$ for aging up to 1 h and then a decrease in two or three

206

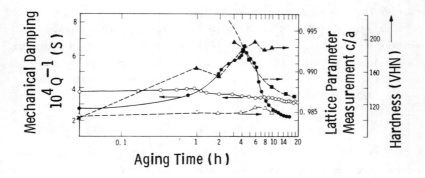

Fig. 20 - Mechanical damping $\left(Q^{-1}(S)\right)$ of Cu-35 at. % Mn (o) and Cu-55
at. % Mn (•) alloys measured at 40 kHz and a strain amplitude
of 3.3 x 10^{-7} <u>during</u> aging at 450 and 400°C, respectively.
The hardness results of Vitek and Warlimont (41) for Cu-40
at. % Mn (Δ) and Cu-60 at. % Mn (▲) alloys aged at 450°C and
the lattice parameter (c/a) measurements (■) of Hedley (42) for
Cu-60 at. % Mn aged at 450°C are shown for comparison purposes.

stages. This small response was similar to that found (41) in hardness
measurements on a Cu-40 at. % Mn alloy. As a general observation the $Q^{-1}(S)$
values ($\sim$5 x 10^{-4}) for the high temperatures (400 and 450°C) were much less
than those ($\sim 10^{-1}$) for low temperatures (100 to 200°C).

The internal friction behavior can be interpreted in terms of the continuous process that occurs during the decomposition of γ-Mn-Cu alloys in
the (α + γ) mixed phase field. First Mn-rich clusters form as a pre-
precipitation phenomenon. The internal friction peaks at 1 or 4 h for the
Cu-35 at. % Mn and Cu-55 at. % Mn specimens, respectively, could be related
to this. With the precipitation of α-Mn comes the eventual fall in hardness.
The drastic fall in $Q^{-1}(S)$ values at 6 to 8 h in the Cu-55 at. % Mn alloy
correlated well with this fall in hardness. Also, the slight decrease in
$Q^{-1}(S)$ values for the other alloy at 8 h correlated with the fall in hardness for the Cu-40 at. % Mn at 6 to 8 h. Thus it can be seen that the
PUCOT is particularly sensitive to the pre-precipitation process occurring
in this system.

The amplitude dependence investigation of internal friction in Cu-35
at. % Mn showed break away for all the curves and the G-L plots were
approximately linear, especially at low ε_{11}^{-1} values. The derived values of
the minor pinning length L_c of the vibrating dislocation lines were found
to be in the range 200 to 500 nm. These lengths were larger than the
spacings of 13 to 65 nm of the tweed structures found (41) in Cu-70 at. %
Mn aged at 500°C but were of the same order of size as some of the

207

boundaries between microtwins observed (42) in Cu-60 at. % Mn aged at 400°C. Thus it is possible that in addition to dislocation pinning by minor impurities such as C and N, the microtwin boundaries acted as dislocation pinning points. A confirmation of the dislocation damping mechanism of internal friction for Cu-35 at. % Mn was given by the good fit of the data to $\varepsilon_{11}(b) \propto L_c^{-1}$, the dependence expected from the G-L theory.

Summary

This paper has shown that the PUCOT can be used to investigate the mechanisms occurring in a variety of metallurgical systems. In Cu_3Au the order-disorder transformation was studied successfully, while in Ni-25 at. % Co some evidence for the postulated order-disorder processes was found. Both internal friction and Young's modulus were measured in a Hg-Sn-Ag dental alloy and related to the phases changes occurring in that system. In leaded α-brass the PUCOT results indicated possible melting and re-distribution of grain boundary precipitates, which were tentatively identified as lead or a Pb-Zn eutectic. The magnetic transformation at the Curie point in nickel and the pre-precipitation and precipitation processes occurring in the high damping Mn-Cu alloys were investigated. The technique has obvious application in the characterization of irradiated metals, particularly where a nondestructive method on a small specimen is required due to scarcity of specimen material. Further advantages of the PUCOT are that internal friction and Young's modulus measurements can be made readily and simultaneously over a range of temperatures (up to the melting point of the specimen or the devitrification temperature (∿1100°C) of the quartz spacer rod), with a range of frequencies (20-200 kHz), with a range of strain amplitudes (∿10^{-8} to 10^{-3}), at programmed heating/cooling rates or isothermally, in vacuum or inert gas, and with or without a magnetic field on the specimen. Additionally, the specimen is positioned relatively remotely from the drive and gage quartz crystals and their electrical connections, which is obviously an advantage with radioactive specimen materials. Of course, the technique has some limitations. Due to the sensitivity of the PUCOT the damping and elastic modulus curves often reflect complicated microstructural phenomena. To interpret an internal friction peak is not easy. Details of possible mechanisms and theoretical interpretations have to be examined. The technique is at its most powerful when used in conjunction with other physical metallurgical techniques such as metallography or X-ray diffraction. However, once a certain defect or microstructure has been identified, the PUCOT can be used most conveniently to monitor changes and to provide further details on defects and microstructure.

Acknowledgments

To a large extent this work was made possible by the cooperation of scientists on an international scale. Dr. A. W. Thompson (USA), Dr. W. H. Robinson and Mr. J. A. A. Hood (New Zealand) and Dr. S. R. Yeomans (Australia) contributed significantly. The experimental work was supported by the Department of Scientific and Industrial Research (New Zealand). Further support came from the Gesellschaft fuer Kernforschung mbH, Karlsruhe, West Germany. The assistance of Frl. E. Mittwoch and Mr. E. Simon (West Germany) is appreciated.

References

1. C. Zener, _Elasticity and Anelasticity of Metals_, University of Chicago Press, Chicago, IL, 1948.

2. W. P. Mason, _Physical Acoustics and the Properties of Solids_, Van Nostrand, Princeton, NJ, 1958.

3. A. S. Nowick and B. S. Berry, _Anelastic Relaxation in Crystalline Solids_, Academic Press, New York and London, 1972.

4. R. De Batist, _Internal Friction of Structural Defects in Crystalline Solids_, North-Holland Publishing Company, Amsterdam and London, and American Elsevier Publishing Company, Inc., New York, 1972.

5. J. Marx, _Rev. Sci. Instr._ 22 (1951) p. 503.

6. W. H. Robinson and A. Edgar, _IEEE Trans. on Sonics and Ultrasonics_ Vol. SU-21 (1974) p. 98.

7. A. Wolfenden and W. H. Robinson, _Scripta Met._ 10 (1976) p. 763.

8. A. Wolfenden, W. H. Robinson and A. W. Thompson, _Scripta Met._ 11 (1977) p. 71.

9. A. Wolfenden and J. A. A. Hood, _J. of Mater. Sci._ 15 (1980) pp. 2995-3002.

10. A. Wolfenden and W. H. Robinson, _Acta Met._ 25 (1977) p. 823.

11. A. Wolfenden, _Scripta Met._ 12 (1978) p. 103.

12. A. Wolfenden, _Phys. Stat. Sol._ (a) 44 (1977) p. K171.

13. A. B. Martin, R. D. Johnson and F. Asaro, _J. Appl. Phys._ 25 (1954) p. 364.

14. W. A. Goering and A. S. Nowick, _Trans. Met. Soc. A.I.M.E._ 212 (1958) p. 105.

15. R. Feder, M. Mooney and S. A. Nowick, _Acta Met._ 6 (1958) p. 266.

16. R. G. Davies, _Trans. Met. Soc. A.I.M.E._ 221 (1961) p. 1280.

17. S. Benci and G. Gasparrini, _J. Phys. Chem. Solids_ 27 (1966) p. 1035.

18. A. Chatterjee and D. J. Fabian, Acta Met. 17 (1969) p. 1141.

19. K. N. Tu and B. S. Berry, J. Appl. Phys. 43 (1972) p. 3283.

20. S. Benci, G. Gasparrini, E. Germagnoli and G. Schianchi, J. Phys. Chem. Solids 26 (1965) p. 2059.

21. W. L. Bragg and E. J. Williams, Proc. Roy. Soc. A 145 (1934) p. 699.

22. D. T. Keating and B. E. Warren, J. Appl. Phys. 22 (1951) p. 286.

23. A. V. Granato and K. Lücke, J. Appl. Phys. 27 (1956) p. 583.

24. A. Wolfenden and W. H. Robinson, Scripta Met. 12 (1978) p. 745.

25. D. G. Morris and R. E. Smallman, Acta Met. 23 (1975) p. 73.

26. D. G. Morris, F. M. C. Besag and R. E. Smallman, Phil. Mag. 29 (1974) p. 43.

27. A. Wolfenden and W. H. Robinson, Scripta Met. 11 (1977) p. 991.

28. S. Siegel, J. of Chem. Phys. 8 (1940) p. 860.

29. A. W. Thompson, Scripta Met. 8 (1974) p. 1167.

30. B. Million, Z. f. Metallk. 63 (1972) p. 484.

31. M. Hansen, Constitution of Binary Alloys, 2nd ed., p. 486; McGraw-Hill Book Co., Inc., New York, 1958.

32. R. M. Bozorth, Ferromagnetism, 2nd Printing, p. 699; D. Van Nostrand Company, Inc., New York, 1953.

33. G. Dickson, N.B.S. Special Publ. 354, Dental Materials Res., Proc. of the 50th Anniv. Symp., Oct. 6-8, 1969, Gaithersburg, Md., USA, p. 161, issued June 1972.

34. L. B. Johnson, Jr., J. Biomed. Mater. Res. 1 (1967) p. 285.

35. A. L. Hines, G. R. Lightsey et al., J. Biomed. Mat. Res. 10 (1976) pp. 371-5.

36. E. Raub and A. Engel, Z. f. Metallk. 37 (1946) p. 76.

37. I. S. Servi, M. Stern and W. W. Webb, Trans. AIME 212 (1958) p. 361.

38. J. Lumsden, Discuss. Farad. Soc. 4 (1948) pp. 60-68; Thermodynamics of Alloys, pp. 335-34; The Institute of Metals, London 1952.

39. R. M. Brick, R. B. Gordon and A. Phillips, Structure and Properties of Alloys, 3rd ed., pp. 134-146; McGraw-Hill, New York 1965.

40. S. Diamond and C. Wert, Trans. Met. Soc. AIME 239 (1967) p. 705.

41. J. M. Vitek and H. Warlimont, Metal Sci. J. 10 (1976) p. 1.

42. J. A. Hedley, Metal Sci. J. 2 (1968) p. 129.

Surface Techniques

SURFACE ANALYSIS TECHNIQUES*

M. G. Lagally**
Department of Metallurgical and Mineral Engineering
and Materials Science Center
University of Wisconsin-Madison
Madison, Wisconsin 53705

Major techniques for the chemical and elemental analysis of surfaces and interfaces are briefly reviewed. Four techniques, Auger electron spectroscopy, X-ray photoelectron spectroscopy, secondary ion mass spectroscopy, and ion scattering spectroscopy, are compared on the basis of a number of criteria useful in evaluating the applicability of a given technique to different surface and interface characterization problems.

*Supported in part by DOE
**H.I. Romnes Fellow

Introduction

There is no need to emphasize the importance of surface analysis techniques in the study of materials properties, or in fact in the development or improvement of technologically important processes. In the area of metallurgical phenomena alone there are applications in fracture, corrosion, oxidation and passivation, coatings technologies, crystal growth, bonding and adhesion, lubrication and wear, and radiation damage, to name just a few major ones. Applications exist in product development, process control, and failure analysis. Demand for increased reliability, new materials properties, new means of energy production, and resource conservation have led to an increasing emphasis on surface analysis techniques.

Reference to "surface" analysis techniques is actually insufficient, because in all cases an interface property is measured, be it only the solid-vacuum interface. Historically, "surface" analysis techniques have first been applied to the solid-vacuum interface, because this is clearly the most accessible. It is also the only one that can be classified as a "surface". Of all the interfaces that can exist, this one must surely be the lowest priority technologically, and it should not be thought that surface analysis techniques have not progressed beyond this stage. In fact, a large number of applications to the characterization of internal interfaces and thin films have been demonstrated. It should be borne in mind that the term "surface" analysis is used generally to include all these varied applications.

There are at present about fifty techniques that have been used for studying the properties of surfaces or interfaces at the microscopic level. Some are no more than laboratory curiosities, others have limited applicability to model systems in fundamental studies, while a few have more general applicability to technologically important problems as well as to fundamental studies. Attempts have been made to tabulate these techniques (1,2), but such tabulations are usually quickly obsolete because new methods or variations of old ones continue to appear. There are a few major surface analysis techniques. These have all been adequately reviewed, and a representative list of such reviews appears as a separate bibliography in the Appendix. Three others that are used for surface analysis, field ion microscopy (FIM), Rutherford ion backscattering (RIBS) or nuclear reactions profiling, and low-energy electron diffraction (LEED), are discussed elsewhere in this book. Two methods used for bulk analysis, electron energy loss spectroscopy (ELS) and extended X-ray absorption fine structure (EXAFS), have surface analogs that have recently become popular. Scanning electron microscopy is, of course, sensitive to surface morphology but it will not be considered here. We restrict our discussion to four of the ones that have had a major impact on microstructural characterization in systems of metallurgical interest, Auger electron spectroscopy (AES), X-ray photoelectron spectroscopy (XPS) (also known by the generic name ESCA: electron spectroscopy for chemical analysis), secondary ion mass spectroscopy (SIMS), and ion scattering spectroscopy (ISS).

The need to summarize four techniques dictates an approach somewhat different than that of the usual review based on the capabilities of a single one. No technique, of course, can give all the information that may be desired for any material under study. Hence it is necessary to define the type of surface or interface under investigation, to specify the properties of that surface or interface that are required to be known, and to evaluate the assets and liabilities of each technique that is potentially applicable. From this it is then possible to establish the suitability of a particular technique for the problem of interest, be it basic research, materials characterization, or process control. We therefore begin by illustrating commonly encountered "surface" structure/composition interrelations. We then outline the two general ways in which surface sensitivity is achieved by these four techniques, and then classify them according to the type of information they give. We list the major criteria that should be applied in evaluating the usefulness of a given surface analysis method. This is followed by a summary of the four techniques in terms of the criteria previously discussed.

Surface Structure/Composition Relationships

The type of information that one desires from a surface analysis technique is the microscopic characterization of the composition of the surface or interface, its structure, and the relationship between them. "Composition" includes both the elemental makeup as well as the actual chemical environment. Structure includes both the crystallography and defects in the surface or at the interface. The interrelation of these is obvious: chemical effects can affect structure (e.g. formation of an oxide at the surface) and structural effects can influence composition, e.g., defect-controlled diffusion. In Figure 1 is shown a summary (3) of the most commonly encountered combinations of surface or interface structure and composition. The most ideal situation, and the one frequently assumed for model systems (and unfortunately also for "real" systems), is shown in Fig. 1 a . Plane, homogeneous surfaces are not very common. Approximations are found in monocomponent systems that are single-crystal and whose surface can be prepared with a minimum of structural defects, e.g., by cleaving. Frequently, such surfaces serve as substrates for adsorption or epitaxial growth. To analyze such samples, surface sensitivity is not required. Figure 1b shows a somewhat more complicated situation, with inhomogeneities in the plane of the surface. These inhomogeneities could be purely

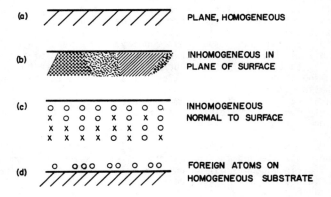

RELATION BETWEEN SURFACE STRUCTURE
AND SURFACE COMPOSITION

(a) PLANE, HOMOGENEOUS

(b) INHOMOGENEOUS IN PLANE OF SURFACE

(c) INHOMOGENEOUS NORMAL TO SURFACE

(d) FOREIGN ATOMS ON HOMOGENEOUS SUBSTRATE

(e) INTERFACE BETWEEN TWO HOMOGENEOUS MATERIALS

Fig. 1 - Structure/composition relationships for some simple surfaces and interfaces. (From Powell, Ref. 3).

structural, (e.g. a polycrystalline monocomponent material), or they could be chemical, (e.g. a multiphase material). For the former, the average chemical composition over a large surface area is the same as the chemical composition at any point, except for possible compositional inhomogeneities at the grain boundaries. For the latter, spatial resolution in the plane of the surface is required to measure the composition. There is no need for surface sensitivity in either case. Figure 1c shows a third possibility, a material that is homogeneous in the plane of the surface, but inhomogeneous perpendicular to the surface. Such situations arise, for example, for single crystals of multicomponent, single-phase materials in which segregation, either equilibrium or nonequilibrium, has occurred. Equilibrium segregation results from differences in the free energy or the size of the components (4) of the phase. Nonequilibrium segregation can be electrically (5), chemically (6), or defect induced (7). Techniques with good depth resolution are required for analysis of samples that have concentration gradients or inhomogeneities perpendicuar to the surface. Lateral resolution is not required. The most general situation for a free surface is a combination of Figures 1b and c. For a complete microscopic analysis of such surfaces both good depth and good lateral resolution are required.

The simplest kind of interface results with the adsorption of a mono-layer or fraction of a monolayer of foreign atoms on a homogeneous surface. "Adsorption" can refer either to adsorption from a vapor surrounding the surface or to segregation of a component dissolved in the bulk matrix. The two are completely analogous (8). If it is known that the foreign compo-nent is restricted to one layer at the surface such situations are easy to analyze. Lateral resolution is not required, and good depth resolution is required only in the sense that the technique must have enough sensitivity to detect a single layer. The analog to Figures 1b, laterally inhomogeneous equilibrium surface layers, might occur in chemisorption of a species in which the net adsorbed-atom interactions are attractive: the absorbate will form islands (9) at less than monolayer coverage. This is also one of the limiting cases for the initial stages of epitaxy and crystal growth (10). The same can apply to segregation. Kinetic phenomena, such as surface diffu-sion, can also result in lateral inhomogeneities (11). Clearly, good lateral resolution is required for such measurements, and depth resolution only in the sense just given. The difference between "surface" and "interface" be-comes blurred when attempting to consider an "interface" analog to Fig. 1c. A possible analog is a homogeneous monolayer or several-layer-thick surface phase in equilibrium with a bulk phase, with diffusion of the surface phase material into the bulk phase. Such situations are quite common and discon-certing and occur in most situations of surface contamination. However, the distribution shown in Figure 1c may itself refer to an interface, which one could arbitrarly draw at any atomic plane. Clearly, good depth resolution, preferably monolayer resolution, is required for the analysis of such dis-tributions.

More complex, but technologically quite important, situations arise if both the substrate and the surface phases are laterally inhomogeneous. For example, grain boundary diffusion in a multicomponent system can lead to lo-calized patches of diffusing species at the surface (12), which then may grow to form a complete layer as diffusion proceeds. Oxidation of a multi-component system may lead to phase separation as one of the species prefer-entially oxidizes and precipitates at grain boundaries and the surface. Good lateral and depth resolution are required for analysis.

In all the cases considered so far, the surface or interface is directly accessible with those techniques that are usually considered to be "surface-sensitive". A measurement will provide compositional analysis over the ex-cited sample volume, which ranges from one to perhaps ten atomic layers.

Figure 1e shows the simplest case of an interface that is buried, i.e. covered on both sides with sufficient material so that a "surface-sensitive"
technique can not see it directly. Such "internal surfaces" are technologically probably most important, e.g., in oxidation, corrosion, passive layers, epitaxial growth, fracture, deposition of protective coatings, adhesion,
and irradiation damage. Again all levels of complexity are possible. These
situations are more challenging to analyze by "surface-sensitive" techniques.
Some techniques, such as Rutherford ion backscattering (RIBS), can see such
interfaces nondestructively with relatively low resolution (as a simple measure, the depth that a technique can "see" into a material is inversely related to its ability to resolve the depth of any layer). To use truly
surface-sensitive techniques, however, destructive analysis is required,
either by fracturing, polishing, chemically etching, or sputter etching the
sample. In fact, the use of sputter etching in combination with surface
analysis techniques provides one of their main applications, thin film
compositional analysis with depth resolution of the order of one to several
atomic layers. Sputter etching is an inherent part of some techniques, e.g.,
secondary ion mass spectroscopy (SIMS).

Surface Sensitivity

The feature that distinguishes surface analysis techniques from bulk
probes such as transmission electron microscopy, x-ray microanalysis, or
backscattered electron detection in scanning electron microscopy, is the
very large sensitivity to a single layer of atoms. Surface sensitivity is
achieved in two different major ways. For electron spectroscopies, i.e.,
those techniques that involve the measurement of the kinetic energy or
momentum of an electron emitted from the solid, surface sensitivity is
achieved because of the very short mean free path of slow electrons in
solids. This is shown (13) in Figure 2. For an electron with E < 1000 eV,
the mean free path is smaller than about 2-3nm (20-30Å), with a minimum
of about 0.3- 0.5nm for most metals in the energy range 50-200 eV, giving
two-to-three monolayer sensitivity. This compares with a mean free path of

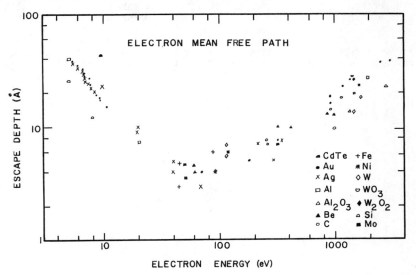

Fig. 2 - Kinetic energy dependence of the escape depth or mean
free path of low-energy electrons for various materials.

217

the order of 100nm for a 100keV beam. Major electron spectroscopies are Auger electron spectroscopy (AES), photoemission spectroscopy (PES) [also ESCA: electron spectroscopy for chemical analysis, XPS: X-ray photoelectron spectroscopy, UPS: ultraviolet photoelectron spectroscopy], and low-energy electron energy loss spectroscopy [ELS]. In addition low-energy electron diffraction (LEED) derives its surface sensitivity in this manner.

The second group of techniques depends on the transfer of momentum from incident ions to achieve surface sensitivity. This is shown in Figure 3. For any beam incident on a surface, a certain fraction of the ions will be reflected without neutralization. Because the neutralization cross section is very high at low ion energies (1-2 keV), reflected ions with these energies will come almost exclusively from the outer atomic layer. Low-energy ion scattering spectroscopy (ISS) derives its surface sensitivity from this feature. Because of the insignificantly small probability of backscattered ions coming from any but the outer layer, ISS has been proclaimed as the only "true monolayer" technique.

The incident ions that penetrate the surface will lose their momentum in a series of collisions, some of which will result in reversal of the momentum direction. This leads to the escape of particles by "sputtering". These particles come only from the surface region, because the momentum continues to be transferred to the least-tightly bound atom, analogous to the collision of a billiard ball with a row of balls, where only the last in the row moves. Cluster emission is possible, so that the sputtered ions may actually come from the first few atomic layers. The sputtering process is also shown in Figure 3. Secondary ion mass spectroscopy (SIMS), in which the sputtered ions are mass-analyzed, depends for its surface sensitivity on the surface selectivity of the sputtering process. Because of the possibility of cluster emission, SIMS is not as surface-sensitive as ISS.

Several other techniques, not discussed in detail here, derive their surface sensitivity in different ways. Field ion microscopy (FIM) derives its surface sensitivity from the ionization of gas atoms at a tip with a very high electric field gradient (14). Because subsurface atoms contribute little to the field outside the surface (14), they are effectively invisible in FIM. Rutherford ion backscattering (RIBS), the high-energy analog of ISS, is not a true surface-sensitive technique, although it has a number of extremely useful surface-related applications. Its "surface" sensitivity

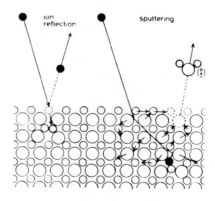

Fig. 3 - Interaction of ions with surfaces, demonstrating surface sensitivity for ion scattering and sputtering.

comes in these cases from its low detection limit, i.e., the ability to see a single layer or fraction of a layer of atoms in favorable cases (15). If it is known that this layer is at the surface (or at an interface) RIBS can determine the atomic concentration in the layer quantitatively.

Classification of Techniques

Surface analysis techniques can be classified according to the type of information they give. They can be broadly grouped into those that give chemical, electronic, or elemental information and those that give geometric or structural information. Most techniques provide only one or the other type, although a number of them offer limited information of both types.

Typically most important in the characterization of a sample is its composition; i.e., its elemental makeup. To describe the elemental distribution in a sample it is necessary to determine: 1) what species are present, 2) how much of each species is present, and 3) what the dispersion of a species is, e.g., does it occur as a precipitate, is it segregated to grain boundaries or surfaces, or is it uniformly dispersed. To determine the chemical makeup, it is additionally necessary to specify the chemical environment of a species, e.g., does it occur in elemental form or as a compound, and in which valence state. The electronic makeup will, of course, also be altered by compound formation.

The major techniques that are capable of giving elemental or chemical information are AES, XPS, SIMS, and ISS. Additionally, RIBS is frequently used for elemental analysis. The most successful one for chemical identification is X-ray photoelectron spectroscopy (XPS or ESCA). The species-specific chemical environment can be obtained from analysis of the line shapes in AES. The electronic environment, e.g. localized electronic surface or interface states, is most readily obtained with ultraviolet photoelectron spectroscopy (UPS). In addition to static chemical or elemental properties, all of these techniques can be used to follow the dynamics of a given elemental or chemical distribution, e.g. how much of what is moving from point A to point B in what amount of time, or the response time of localized electronic traps.

The most important geometric property of interest in surface analysis is, of course, the crystal structure of the surface phase, i.e., the size and shape of the unit mesh and the positions of atoms in it. Additionally it is desirable to obtain a quantitative measure of the density and distribution of point and extended defects on surfaces, and to study the thermodynamics of surface phases, e.g., phase transformations in two-dimensional or quasi-two-dimensional systems, as well as kinetics of surface structure formation, such as island or precipitate growth and surface diffusion. LEED is the major technique that provides such information. LEED has considerable potential for surface defect studies that is only now beginning to be realized (16).

Surface EXAFS provides in principle species-specific structural information, i.e., the short-range order (generally nearest neighbors only) around a given elemental species. The potential of this method has not yet been fully realized (17). RIBS channeling analysis can be used for surface structural analysis (18). In a similar way, ISS gives limited information on positions of surface atoms. Angular distributions in SIMS and PES may also give limited structural information.

219

Criteria for Evaluating the Capabilities of a Technique
for Surface Elemental or Chemical Analysis

Every technique has its own strengths and weaknesses, and no single technique can provide a total analysis. It is useful to establish a list of criteria to evaluate the applicability of a technique to a problem. To perform elemental or chemical analysis, such a list might include absolute sensitivity, bandwidth, resolution, element discrimination, sensitivity variation with atomic number, capability for quantitative analysis, ability to provide chemical bonding or valence state information, depth resolution, lateral resolution, and degree to which the technique causes damage to the surface being analyzed. These are each briefly discussed below.

Absolute Sensitivity. The sensitivity is properly defined as the minimum number of atoms required to give a meaningful signal. The sensitivity of every analytical technique is ultimately limited by a background that is in some way correlated with the desired signal.

Because of low-frequency or flicker noise it is not possible to measure an arbitrarily small signal sitting on a large background. Flicker noise is due to an instability of the measuring system, and increases with measurement time. It is described by a 1/f spectrum, where f is the frequency of the noise. A large background amplifies the flicker noise. Because background goes up proportionally with signal intensity, a given flicker noise is worse for a low signal-to-background measurement than a high one. The absolute sensitivity of a technique can thus be considered as some function of the cross section for creating a signal event, the signal-to-background ratio, and the stability of the measuring system.

Bandwidth. In the foregoing, there was no mention of a limited measurement time. This is as it should be, because the absolute sensitivity does not depend on measurement time. In actuality there are, of course, a variety of reasons why the measurement time must be limited. For surface analysis techniques this is so, if for no other reason, because a surface contaminates in a matter of a few hours at the best vacuum obtainable. The imposition of a finite time interval always results in a statistical uncertainty, called "shot noise". A measure of the quality of a signal is the signal-to-(shot) noise ratio, expressed by the following relation,

$$S/N = (\sigma I t)^{1/2} , \qquad (1)$$

where t is the measurement time, I is the rate at which the sample is probed (e.g. the intensity of the incident signal) and σ is the number of signal events counted per probe event.

The bandwidth is the inverse of the measurement time. Increased bandwidth (i.e., more rapid measurement to obtain the same signal-to-noise ratio) can be achieved by increasing the intensity of the radiation or by improving the collection and counting efficiency. Because measurement speed is obviously important, the bandwidth is a measure of the quality of a technique.

In principle, the frequently used term "detectability limit" is the same as the absolute sensitivity. In practice, detectability limit is used to refer to the minimum concentration of an element from which a meaningful measurement can be made by a technique in a given time. This time is never specifically defined, but is on the order of minutes, rather than hours. "Detectability limit" is thus not a well-defined concept, but as used, it is simply a rough measure of the bandwidth of a technique.

220

Resolution. Any instrument obscures the information it is trying to measure. This is in part useful, because it results in smoothing of high-frequency noise. Resolution is, of course, achieved at the expense of bandwidth, i.e., it takes longer to obtain high-resolution data with the same signal-to-noise ratio than low-resolution data. Resolution is important in various ways, e.g., mass resolution in SIMS and energy resolution for elemental analysis in ISS and for chemical analysis in ESCA and AES. A lucid discussion of resolution and its relation to sensitivity and bandwidth is given by Park (19).

Element Discrimination. The ability to discriminate elements of similar atomic number or atomic weight is an important attribute of an analytical technique. In SIMS this reduces to the problem of mass resolution. In ISS, it is a matter of resolving the energies of ions backscattered from similar-mass atoms on the surface. In the electron spectroscopies, the kinetic energies of electrons coming from different electron shells of surface atoms must be distinguished. An alternative means of providing element discrimination is through multiple spectral peaks for a given element. This reduces the need for high energy or mass resolution.

Sensitivity Variation with Atomic Number. Ideally, a technique should be equally sensitive to all elements in the periodic table. This is, of course, never achieved. Various of the techniques have special atomic-number-dependent sensitivity features. Thus SIMS is the only technique that can detect hydrogen. AES has high sensitivity to light elements and best sensitivity across the periodic table. ISS has good sensitivity to heavy elements. Figure 4 shows (20) a comparison of the sensitivity variation with Z for the four techniques discussed here.

Capability for Quantitative Analysis. It is clearly desirable to get absolute compositions. Various factors can affect the capability of a technique to provide quantitative information. These include unknown excitation or emission cross-sections, backscattering effects, surface topography effects, matrix effects, uncertain sample volume probed, etc. Some of these will arise later in the discussion of the individual techniques.

Chemical Information. It is desirable to know the chemical environment in addition to just the elemental composition. Some techniques, such as XPS, give quite reliable chemical information. In others, chemical (or "matrix") effects are so strong that they reduce the capability of the technique to give quantitative information (e.g. SIMS). Still other techniques, eg.. ISS, give no chemical information.

Depth Resolution. The different surface analysis techniques have somewhat different depth resolution, as already indicated. This ranges from single-layer sensitivity for ISS to a ~3nm mean free path for some XPS lines. In some cases, e.g. equilibrium segregation studies, the monolayer sensitivity of ISS is required. In many cases, especially when combined with sputter etching for depth profiling, XPS or SIMS depth resolution is quite sufficient.

Lateral Resolution. To observe surface phases such as shown in Figure 1b or, for example, to do compositional profiling of individual grains, good lateral resolution is required. Figure 5 shows a comparison of the lateral resolutions of the four techniques discussed below.

Surface Damage. All techniques damage to some degree the surface that is being probed. The damage can be of two kinds, physical mixing, induced by ion bombardment, and chemical decomposition, induced by electron bombardment. In AES and XPS, the damage is caused by electrons. XPS is believed to be the

221

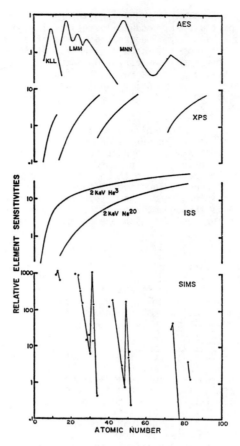

Fig. 4 - Variation of relative sensitivity with atomic number for AES, XPS, ISS, and SIMS. (After Holloway and McGuire, Ref. 20).

least damaging, because fewer electrons cross the surface. In part, this may be misleading: because electrons in the energy range of a few hundred electron volts or less have the highest cross-section for ionizing collisions with surface atoms, it is the emitted radiation that is causing most of the damage. The incident electrons in AES, having typically 3-5 keV energy, cause most of their damage much deeper in the sample. It is hard to quantify this, but it appears possible that the difference in observed surface chemical damage in XPS and electron-excited AES may in part be due to the more efficient method of signal detection in XPS.

In SIMS and ISS, the surface atoms are, of course, removed in the process of measurement. In addition, these techniques, as well as sputter etching for depth profiling in electron spectroscopies, cause intermixing of materials many atomic layers deep. The width of the damage zone gets larger the longer the material is sputtered. Thus buried interfaces can be observed only with poorer depth resolution than surface or near-surface layers.

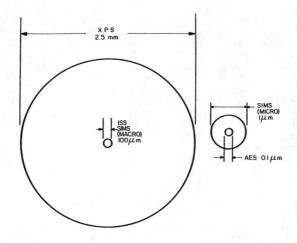

Fig. 5 - Lateral resolution capabilities of AES, XPS, ISS, and SIMS.

For any materials characterization problem some of these criteria will be more important than others. Because they can't all be achieved equally well by all techniques, a selection of the appropriate technique for a given problem can be made by considering each technique on the basis of these criteria. This is done in the following sections.

Auger Electron Spectroscopy

Probably the most universally applicable of the surface analysis techniques is Auger electron spectroscopy. It has been used in all the ways and for all the surface and interface combinations discussed in association with Figure 1.

The Auger effect is a deexcitation mechanism for ionized atoms that is competitive with X-ray emission. It yields a low-energy electron whose kinetic energy is characteristic of the element from which it came. Thus each element has its own signature. Excitation can be by electrons, X-rays, or ions. A series of Auger electrons results if a deep core level is originally ionized. For example, if a K shell is ionized, an Auger process can occur with an L shell electron dropping into the K shell, the excess energy being taken up by another L-shell electron (KLL process), an M-shell electron (KLM process), or an electron from a even higher level. This is then followed by LMM, etc. processes where the hole(s) on the L shell is(are) now being filled. The strongest Auger processes are generally those involving the valence shell and a core level close to it in energy; these are also the transitions that are the most surface sensitive, because the emitted electron has a kinetic energy near the minimum in mean free path. Because three electrons are required for an Auger process, hydrogen and helium do not have Auger spectra. Li has an Auger spectrum only in the condensed phase, because two L-shell electrons are required, which is achieved in the solid state because of interatomic transitions.

A schematic diagram of an Auger electron spectrometer with analog detection is shown in Figure 6. This figure also serves as a generic model for the other techniques, by simple appropriate modifications of the source (X-ray or ion instead of electron) and the detection scheme (pulse counting instead of analog detection in XPS and ISS, mass spectroscopy in SIMS). A typical Auger spectrum taken in the first-derivative analog detection mode, standard practice in AES to suppress background, is shown in Figure 7. It shows several things. A multiplicity of Auger lines exists for most elements (all but the first period). Thus elemental discrimination is very good, even when some of the lines of different elements overlap. Difficulties occasionally arise with light elements, such as carbon in ruthenium or nitrogen in TiN. In such cases, one of the other lines of the high-Z species is used to establish the line intensity of the overlapping high-Z line, which is then subtracted to give the low-Z line intensity. Light elements, such as oxygen and carbon, are visible in the Auger spectrum, even though their concentration may be quite low. The sensitivity of AES to light elements is very good, and the overall sensitivity variation with Z is only about one order of magnitude. This is shown in Figure 4.

The "detection limit" of AES for most elements is of the order of 1000 ppm of the analyzed sample volume (3-10 atomic layers) and 200 ppm for the best cases. This is limited by bandwidth considerations, with the applicable bandwidth chosen proportional to the patience of the experimentalist, which, as noted, is of the order of minutes rather than hours. For longer measurement times, assuming the instrumentation is stable and the surface remained clean, lower detection limits could be achieved. A 1000 ppm detection limit implies that if an impurity is uniformly distributed throughout the sample, the detection limit is 10^{-3} monolayer. If an impurity is not uniformly distributed but segregated to an interface, the detectability limit for bulk impurities is lower, of course. Thus if a 10 ppm impurity in a 1μm thick layer were all segregated to the surface layer, it would be easily observable.

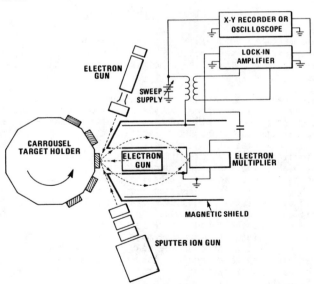

Fig. 6 - Schematic diagram of an Auger electron spectrometer.

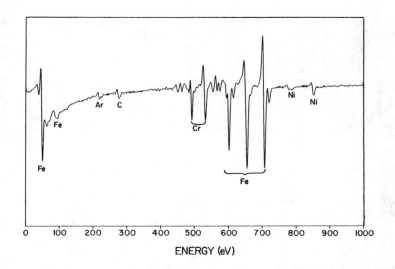

Fig. 7 - Auger spectrum of stainless steel. The derivative of the
secondary electron current is plotted vs the kinetic energy
of the electrons.

The resolution in Auger spectra can be made quite good. Resolution in
this case means the ability to separate lines lying close in energy, either
lines of different elements or multiple lines of the same element. The
latter can occur because there are close-lying energy levels of the atom
or, if only one energy level exists, because there are multiple chemical
states of the element, e.g. metal oxide or carbide precipitates in a metal
matrix. The electron transfer inherent in chemical bond formation (i.e.
change in valence state) causes a repositioning of the energy levels of the
affected atoms known as a "chemical shift" (21). The chemical shift is especially useful in XPS, giving it its chemical sensitivity. The chemical
shift also occurs in AES, but because of the involvement of several energy
levels and three electrons in the Auger process, the resulting line shift
is not easy to relate in a consistent way to chemical bond formation.
Thus very high resolution is generally not required in AES compositional
measurements.

The capability of AES for quantitative compositional analysis is limited somewhat by complicating factors, but can be quite good for homogeneous
materials of similar atomic number. For example, the line intensities for
Cr, Fe, and Ni in the stainless steel Auger spectrum shown in Figure 7
compare quite favorably with the bulk composition of stainless steel. Of
course, excellent relative concentrations with changes in sample processing
(e.g. segregation or diffusion phenomena) can be obtained. For quantification it is necessary to relate the current in the spectral feature of a
specific element to the number of atoms of this element that are present in
the analyzed volume of sample. The Auger current is given by (22)

$$I \quad \alpha \quad I_p (E,z) \; \sigma \; N(z) \; \gamma \; e^{-z/\lambda} \; d \, \Omega \; d \, E, \qquad (2)$$

where I_p is the primary excitation flux density at a given energy E and a
given depth z in the sample, σ is the ionization cross section for making a
core hole that will serve as the initial state for the Auger line in ques-

225

tion, $N(z)$ is the density at depth z of the element to be determined, γ is the probability of making the given Auger transition to be observed once a core hole has been made, and λ is the mean free path of the emitted Auger electrons under consideration. The difficulties in quantification are evident from the uncertainties in this equation. Sigma and gamma can be calculated with reasonable accuracy. Despite the apparent uniqueness of the mean free path curve in Figure 2, actual λ's for a variety of materials can vary by an order of magnitude at a given energy. This makes it difficult to extract the z-dependence of $N(z)$, which is, of course, the number to be determined. Additionally, I_p is not as simple as it seems, but includes two terms,

$$I_p(E,z) = I_{primary}(z) + I_{backscatt}(E,z). \qquad (3)$$

In addition to the attenuation of the primary beam itself (which is probably negligible over the escape depth of the emitted Auger electrons), there is a backscattering term that is energy dependent and may account for as much as one half of the flux density striking a given atom near the surface. This term is very hard to determine precisely. Finally surface roughness and lateral inhomogeneities can distort the Auger signal intensity, in addition to unresolved chemical shifts or line shape changes for transitions involving valence bands.

To quantify Auger spectra, external standards can be effectively used. Many standard spectra exist (23). If external standards with known composition are used, σ and γ are removed. If the composition of the standard is near to that of the test sample, λ and I_p are also removed. Then quantitative analysis reduces to comparing relative Auger peak heights. Of course, the effect of different surface roughness must still be considered, and the z-dependence of N can be deduced only by sputter profiling. If elemental standards are used, the possible differences in λ and I_p between the standard and test sample are simply ignored. Nevertheless, in such cases it is frequently possible to get absolute concentrations within several percent of the actual values. The lateral resolution of AES, shown in Figure 5, is very good, with 100nm beam diameter scanning Auger microscopes commercially available, and 30nm resolution demonstrated in a laboratory environment.

Auger electron spectroscopy can cause two forms of surface damage. For compounds, such as oxides, that are sensitive to electron beam decomposition, an elemental peak appears in the spectrum at the expense of the chemically shifted compound peak. The damage is total-dose dependent. For small beam sizes, e.g. in scanning Auger analysis, damage due to sample heating, which is flux-density dependent, can also occur.

Auger spectroscopy has found wide application in surface elemental analysis and depth concentration profiling. Its best applications have come where the good depth resolution, high sensitivity for almost all elements, and sputter profiling have been combined, such as in low-temperature bulk, grain boundary, and surface diffusion, oxide growth, equilibrium or nonequilibrium segregation, and contamination studies. Its greatest additional asset is the high obtainable lateral resolution, making possible studies on individual grains, epitaxial growth, and localized precipitation or chemical reaction, such as corrosion.

X-Ray Photoelectron Spectroscopy (XPS or ESCA)

Probably the second most widely used surface analysis technique is XPS, a companion to AES that is experimentally compatible and complementary in some of its capabilities. It has found especially great use in elemental analysis of beam-sensitive materials and whenever chemical analysis is required. It has been used less in such metallurgical studies as segregation, diffusion, or fracture analysis, but is finding increasing application in corrosion and passivation studies.

Photoelectron spectroscopy is simply the application of the photoelectric effect to elemental and chemical analysis. It is an excitation technique, where the incident photon is absorbed by an electron, which is ejected from the atom with a kinetic energy equal to the difference of the energy of the photon and the binding energy of the electron that absorbed it. The kinetic energy of the photoelectron is measured in the experiment. Because each element has a unique set of atomic energy levels, each element has its own signature of photoelectron emission lines for a given photon energy. Hydrogen and helium are in principle observable, although the cross-section for them is quite low. Surface sensitivity is again obtained because the kinetic energy of the emitted electrons is low. However, for typical commercially available sources (Al K_α and Mg K_α) the depth resolution is not as good as in AES, because many important photoelectron transitions occur at kinetic energies up to 1450 eV. Using synchrotron radiation, it is possible to do photoelectron spectroscopy at any energy, because the photon energy is tunable with such sources.

The XPS experiment can be visualized by referring to Figure 6. If the grazing-incidence ion gun is replaced by an x-ray source, XPS is possible, although in practice a somewhat more complicated analyzer and pulse counting electronics are required. It can be seen that XPS and AES can easily be combined, and this is quite common. A typical XPS spectrum, obtained with AlK_α radiation from a brass surface, is shown in Figure 8. Several of the

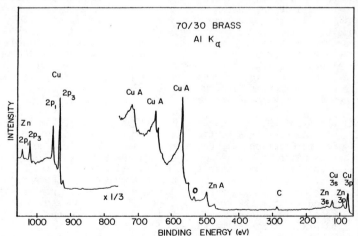

Fig. 8 - XPS spectrum of 70/30 brass, taken with Al K_α radiation. The secondary electron current is plotted vs. the binding energy of the photoelectrons. Note the Auger peaks in the spectrum, indicated by A.

core level lines of zinc and copper can be seen, in addition to the K levels of carbon and oxygen. Auger lines are also observed in an XPS spectrum, because the Auger process, as discussed earlier, is a deexcitation mechanism that fills the core holes left, in this case, by the photoelectric effect. However, because of the relatively low x-ray flux that can be achieved relative to the electron flux used in electron-excited AES, x-ray excited Auger analysis is much slower and thus used only when the circumstance demands it. The multiplicity of lines in the XPS spectrum again indicates that there is good elemental discrimination. Light elements are visible in XPS, but the sensitivity of XPS to them is low. The relative sensitivities are shown in Figure 4. It can be seen that the sensitivities in XPS vary somewhat more widely with Z than the AES sensitivities, especially for the light elements.

The average detection limit of XPS is somewhat worse than for AES, in the neighborhood of 0.3 to 0.5% of a monolayer for reasonable laboratory times. This can for the most part be attributed to a low source intensity. High-intensity soft x-ray sources are difficult to make. As a result, the signal-to-noise ratio, given in Equation (1), will be smaller for a chosen measurement time for XPS than for AES if σ is the same. It is not clear whether σ in Equation (1) is much different in the two spectroscopies; it depends on a number of factors such as cross sections for production of a photoelectron versus an Auger electron, detection geometry and detector gain, and penetration depth of the incident radiation. In the absence of other considerations, AES is preferred for simple elemental analysis or for depth profiling, because of its greater bandwidth.

The resolution of XPS is quite important because of the capability of XPS to characterize chemical states of an atom. As discussed earlier, chemical bonding manifests itself in an adjustment of the energy levels of the atoms, called the chemical shift, which may be from a fraction of an eV up to 10eV. Good energy resolution is requisite for accurate chemical-shift measurements. In addition to a high-resolution detector, a mono-chromatized source is useful. These additionally limit the bandwidth of the technique. Data deconvolution methods are being developed to extract chemical shift information even when the resolution is not high (24). These deconvolution techniques permit reasonably good chemical-shift measurements for the majority of systems that do not have a monochromatized x-ray source. The achievable energy resolution depends again on the bandwidth: for strong lines the signal-to-noise ratio at a given bandwidth is larger and greater resolution is possible. Some chemical shifts are not observable even with the best instruments. Chemical shifts for a large variety of compounds have been tabulated (26).

The capability of XPS for quantitative analysis is somewhat better than that of AES. The current in a particular spectral feature is given by a equation similar to Equation (2), but there is no factor γ and no back-scattering contribution to I_p. Also, since XPS spectra are usually taken in a direct pulse counting mode rather than a derivative mode as in AES, the area under a line in an XPS spectrum is a direct measure of the number of photoemitted electrons and thus directly proportional to the concentration $N(z)$ [with, of course, all the remaining uncertainties discussed under AES]. Standard XPS sensitivity factors exist for the two most common XPS sources (27), MgK_α and AlK_α. XPS spectra for simple multicomponent materials generally give better quantitative results than the corresponding Auger spectra.

There is no lateral resolution in XPS, as shown in Figure 5, because x-rays are difficult to focus. As a result there is also no scanning capability except through mechanical motion of the sample. XPS is less damaging to surfaces than AES. As already indicated, this is in part due to the greater detection efficiency in XPS, requiring fewer electrons at the sample. XPS, as a result, has major applications in surface elemental or chemical analysis of electron-beam-sensitive materials, such as polymers, oxides, or adsorbed molecular layers.

The other major application of XPS, as already indicated, is in surface chemical analysis. Used in conjunction with AES, it in addition becomes useful as a quantitative check on AES-determined compositions, e.g. in depth profiling. Its combination with scanning Auger microscopy provides a powerful tool for measurements of localized surface chemical phenomena, e.g. in corrosion or oxide formation.

Secondary Ion Mass Spectroscopy (SIMS)

The most widely used of the ion beam techniques is SIMS. It has some unique features not found in other techniques, notably a greater detectability limit than other techniques for many elements and the ability to detect hydrogen. SIMS depends simply on mass spectroscopic analysis of ions ejected from a surface by an incident ion beam. Surface sensitivity is obtained because the ejected ions come only from the top several layers of the solid, as already discussed. There is greater variability in the information zone than in AES or ESCA. The median escape depth can be from as little as one layer to greater than 20 Å, and depends on the collision cascade characteristics, the primary beam energy, the ion and target atomic numbers and masses, and the cohesive energy of the target.

A SIMS experiment can be visualized again with reference to Figure 6, if the detector is replaced with a mass spectrometer and the grazing electron gun by an ion gun. The mass spectrometer requires an energy filter to remove fast ions, because these can travel to the multiplier and cause a background signal. A typical SIMS spectrum is shown (27) in Figure 9. Both negative and positive SIMS spectra can be obtained. Generally a multiplicity of lines appears because of the possibility of emission of molecular fragments. This can complicate the elemental discrimination. On the other hand, mass spectrometers with mass resolutions $m/\Delta m$ up to 3000, as in the ion microprobe (28), can separate these overlaps, and even make isotopic separation possible.

The detection limit of SIMS can be made very good, but at the expense of lateral resolution. This is shown in Figure 10. Detection limits of $10^{-2}\%$ of a monolayer (100 ppm if uniformly distributed throughout the sample) are possible if the beam is made large and the sputter rate is high. This is again caused essentially by bandwidth considerations, this time for the mass spectrometer. A very sensitive mass spectrometer (or one operated at low resolution) will have a large bandwidth, i.e., to achieve a given signal-to-noise ratio requires a shorter time for such a mass spectrometer than for a poorer one or one operated at high resolution. Equation (2) can be used to calculate the signal-to-noise ratio, with σ containing the mass spectrometer sensitivity, the cross-section for producing a particular sputtered species, and the detection geometry.

The sensitivity variation with Z, shown in Figure 4, is greater than three orders of magnitude, which is much larger than the other surface analysis techniques.

229

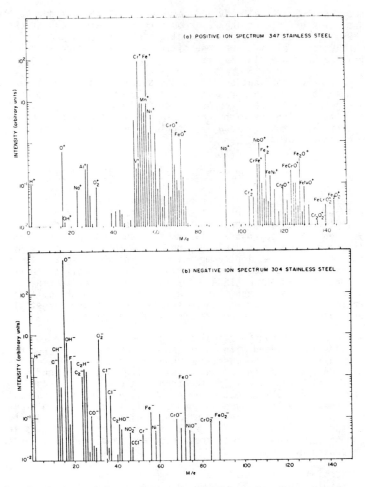

Fig. 9 - Typical positive and negative SIMS spectra from stainless
steel. The relative current in the mass spectrometer is
plotted vs. mass over charge ratio. (After McHugh, Ref. 27).

The capability of SIMS for quantitative analysis is limited by matrix
effects. The positive-ion yield of element A per incident ion, S_A^+, can be
written as (27)

$$S_A^+ = \gamma_A^+ \, C_A(z)S, \qquad (4)$$

where γ_A^+ is the ratio of positive secondary ions of A to the total number
of sputtered A atoms, $C_A(z)$ is the concentration of A atoms, and S is the
total sputter yield. An identical equation holds for negative ions. The
detailed process of producing a positive or negative ion is quite complex
(27), involving ground and excited-state interactions of atoms and molecules with surface electronic states. The process is as a result extremely

230

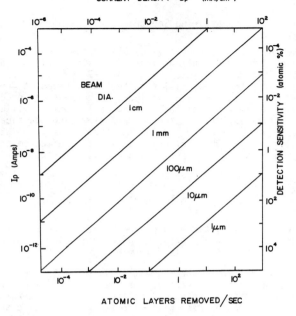

CURRENT DENSITY Dp (mA/cm²)

Fig. 10 - Relationship between ion current, current density, sputter etching rate, and detection limit for SIMS as a function of beam size.

matrix-dependent. The quantities γ_A and S are very sensitive to the matrix; γ as a result of the electronic configuration and S as a result of relative cohesive energies. Thus for a sample of unknown composition, quantitative information is very difficult to obtain. Only in cases where it is known that the matrix doesn't change, and where a concentration of a specific non-interacting species is desired is quantification possible. Standards can be used, but they suffer from the same problems as those for the other techniques. If surface roughness, composition, or matrix are different, standard spectra may not be comparable to spectra from the sample being analyzed. The technique has great potential for giving chemical information, but as yet the processes are too complicated to extract it reliably.

The technique is obviously destructive. In fact, the great sensitivity of SIMS is achieved only at the expense of rapid sample destruction. This points out a limitation of the method. If a low concentration of material on a surface or in a narrow interfacial layer is to be measured, bandwidth considerations require that the sputter rate be slow and the incident beam be large so that enough events can be accumulated to give a good signal-to-noise ratio. This is sometimes called the "static" mode of SIMS. Thus, in effect, a low detection limit and good depth resolution can be achieved only at the expense of lateral resolution.

The obvious applications of the technique are to hydrogen profiling and to situations where very high sensitivity for a particular element is required over a reasonable depth range, as in profiling a species implanted, for example, for semiconductor device doping or corrosion protection. It is thus more of a thin film probe than a surface or interface probe.

Ion Scattering Spectrometry (ISS)

ISS is the most surface-sensitive of all of the techniques described here. As already discussed, this great surface sensitivity is achieved because the neutralization cross-section of the 1-2 keV ions used in this technique is so large that essentially no ions are reflected from any but the outer layer of the sample being analyzed. The physical process of ISS is easy to visualize in terms of the elastic collision of billiard balls. If a light projectile is fired at the surface atoms and elastically reflected, the energy of the reflected ion is determined by momentum and energy conservation. For a 90° scattering angle, the reflected energy can be related to the masses of the projectile and target atoms by

$$E' = E_0 \left[\frac{M_T - M_0}{M_T + M_0}\right], \tag{5}$$

where M_T is the mass of the target atom and M_0 is the mass of the projectile ion. By measuring the kinetic energy E' of an inert ion of given mass M_0, the target atom mass can be readily determined.

The typical experimental arrangement is very similar to that shown for AES in Figure 6. All that is necessary is the replacement of the electron gun in the analyzer with an ion gun. A cylindrical mirror analyzer, the same as shown in Figure 6, can be used with reversed potentials to analyze positive ions. Alternatively a sector analyzer is used. The energy resolution of the analyzer can be quite good, as in AES. Pulse counting electronics are used to record the data, which are of the form shown in Figure 11. This figure shows (29) the ISS spectra for Mo contaminated Au/Ni alloy using both He$^+$ and Ne$^+$ as projectiles. This points out one of the difficulties with ISS. As the mass of the target atom increases, element discrimination

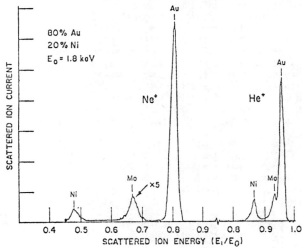

Fig. 11 - ISS spectrum of Mo-contaminated Au/Ni alloy, taken with He$^+$ and Ne$^+$ ions. The ion current is plotted vs. the reduced kinetic energy E'/E_0. (After McKinney, Ref. 29).

232

becomes poor, because according to Equation (5), $\Delta E'$ for two close-lying elements becomes too small to resolve. This can be alleviated by using a higher-mass projectile, but of course light masses can't be analyzed with heavy projectiles unless a grazing-incidence configuration is used. In this configuration hydrogen and helium can be observed with ISS. The instrumental resolution does not seem to be the limiting factor in the width of the peaks shown in Figure 11. The peaks in general have finite physical width as well as a low-energy tail that is due to beam attenuation or subsurface-layer scattering (30), which is especially prominent when He^+ is used as the projectile. In general, the sensitivity to light elements in the standard 90° scattering configuration falls off rapidly as Z decreases, as shown in Figure 4. The sensitivity is a smooth function of Z at high Z.

The detection limit for ISS is about the same as for AES, 0.1% of a monolayer for heavy elements, and less for low-Z elements. Little if any chemical information can be obtained with ISS, although there is some evidence that such information is contained in the ion yield as a function of primary energy (31).

The capability of ISS for quantitative analysis is probably about the same as AES, although much less work has been done with ISS in this area. The scattering yield depends on the differential cross-section for the ion-target atom collision, $d\sigma/d\Omega$, and on the probability γ, that the reflected ion will remain ionized (30), i.e.,

$$Y_A = \frac{d\sigma}{d\Omega} \gamma \ c_A , \qquad (6)$$

where c_A is the surface-layer concentration of the atoms being analyzed. Both $d\sigma/d\Omega$ and γ are not well known. Because the technique is sensitive to a single layer, c_A does not have a z-dependence. Of course, as with SIMS, the surface is sputtered as it is being measured, hence c_A in principle has to include deeper-lying levels. Generally the sputter rate of the primary ion beam can be made low enough that the experiment is done in the "static" mode. If information at greater depths is desired, generally an additional ion source is used. Sputtering, used in all four techniques, introduces a mixing effect that distorts the concentration profiles, and causes preferential sputtering in many multicomponent materials as well. Because of this, apparent relative concentrations can be in error by as much as a factor of 2 to 5. Using standards, ISS spectra can be quantified to ± 30%, with the same precautions that were addressed under AES applying here.

ISS is used for elemental analysis complementary to AES in many cases. Its distinct advantage of monolayer sensitivity is useful in studying such phenomena as equilibrium segregation, which is a monolayer or few-layer phenomenon. Additionally, the analysis of insulating samples is frequently simpler with ISS, because of lesser charging. It has another advantage that is not related to its capabilities as an elemental analysis tool. ISS can be used as a structural analysis tool for surfaces rough on an atomic scale, by doing shadowing experiments. In the most simple view of this experiment, an atom that can't be seen by line of sight will be invisible also to the ion beam. Thus an indication of the structural arrangement of atoms adsorbed on a surface can frequently be obtained by varying the angle of incidence of the ion beam with respect to the sample and observing the ion yield from the adsorbed species (32).

233

Summary

We have briefly compared the four most commonly used surface analysis techniques. An attempt has been made to cite the advantages and disadvantages of each technique, along with major applications, which, it should be evident, are not restricted to free surfaces.

In comparing these techniques to bulk elemental or chemical analysis techniques, reference should be made to energy (or wavelength) dispersive x-ray analysis, energy loss spectroscopy, and x-ray diffraction. Obviously the depth resolution is better in the surface analysis techniques. However, the lateral resolution is worse, at least when analysis using a scanning transmission electron microscope (STEM) is considered. This is simply a manifestation of absolute sensitivity. The total lower limit of sample volume analyzed is about the same in STEM x-ray analysis or ELS as in AES, and the detection limits are also similar, and there is simply a tradeoff between lateral resolution and depth resolution. There is generally better chemical information available from surface analysis techniques than from the bulk techniques. This information should, however, be contained in the bulk probes. It is a more difficult task to resolve the small kinetic-energy differences inherent in the chemical shift at the high energies used in bulk probes. Problems with quantification are similar. Factors such as surface roughness are traded for others, such as beam attenuation.

Surface analysis has grown tremendously in the last fifteen years, and, as is evident from this review, it is impossible to cover even one technique adequately in the limited space available. A number of books or extended reviews have been written on each of the techniques. These should be consulted for a more detailed description. They are listed for the interested reader in the Appendix.

Appendix
Representative Bibliography of Reviews on Surface Analytical Techniques

I. General or covering more than one technique

1. Methods of Surface Analysis, ed. A. W. Czanderna (Elsevier, 1975).

2. Characterization of Solid Surfaces, eds. P. F. Kane and G. B. Larrabee (Plenum 1974).

3. Systematic Materials Analysis, ed. J. H. Richardson and R. V. Peterson (Academic, 1974).

4. Treatise on Solid State Chemistry, Vol. 6A, ed. N. B. Hannay (Plenum 1976), Chapter 2.

5. Chemistry and Physics of Solid Surfaces, Vol. I, eds. R. Vanselow and S. Y. Tong, (CRC Press, 1977).

6. Chemistry and Physics of Solid Surfaces, Vol. II, ed. R. Vanselow (CRC Press, 1979).

7. R. L. Park, "Chemical Analysis of Surfaces" in Surface Physics of Materials, Vol. II, ed. J. M. Blakely (Academic 1975).

8. *Interfacial Segregation*, eds. W. C. Johnson and J. M. Blakely (ASM, 1979).

9. G. Ertl and J. Küppers, *Low-Energy Electrons and Surface Chemistry*, (Verlag Chemie, 1974).

10. C. J. Powell, *The National Measurement System for Surface Properties*, N.B.S. Report NBSIR 75-945. C. J. Powell, Appl. Surf. Sci. **2**, 143 (1979). A more complete list of references is given here.

11. *Quantitative Surface Analysis of Materials*, ed. N. S. McIntrye, ASTM STEP 643, (ASTM, 1978).

12. K. A. Sevier, *Low-Energy Electron Spectrometry*, (Interscience 1972).

13. A. W. Mullendore, G. C. Nelson, and P. H. Holloway, "Surface Sensitive Analytical Techniques: An Evaluation", in *Proc. Adv. Techniques Failure Analysis, Los Angeles* (IEEE 1977).

14. *Electron Spectroscopy for Surface Analysis*, ed. H. Ibach (Springer 1977).

15. *Surface Analysis Techniques for Metallurgical Applications*, eds. R. S. Carbonara and J. R. Cuthill (ASTM 1976).

16. R. L. Park, in *Experimental Methods in Catalytic Research*, Vol. III, eds. R. B. Anderson and P. T. Dawson (Academic, 1975).

17. T. A. Carlson, *Photoelectron and Auger Spectroscopy* (Plenum, 1974).

18. P. H. Holloway and G. E. McGuire, Thin Solid Films **53**, 3 (1978).

II. Specific to a Particular Technique

1. E. N. Sickafus, J. EMMSE **1**, 1 (1979) [AES].

2. P. H. Holloway, "Quantitative Auger Electron Spectroscopy: Problems and Prospects", in *Scanning Electron Microscopy, 1978* vol. 1, ed. O. Johari (SEM 1978).

3. K. Sieghahn, et al., *ESCA*, (Almqvist and Wiksell, Uppsala, Sweden, 1967).

4. R. S. Swingle and W. M. Riggs, Critical Reviews in Analytical Chemistry, **5**, 262 (1975) [ESCA].

5. *Secondary Ion Mass Spectrometry*, eds. K. F. J. Heinrich and D. E. Newbury, NBS Spec. Publ. 427 (1975) [SIMS].

6. J. T. McKinney and J. A. Leys, Eighth National Conference on Electron Probe Analysis (New Orleans, 1973) [ISS].

7. G. C. Nelson, J. Colloid Interface Sci. **55**, 289 (1976) [ISS].

References

1. C. J. Powell, Appl. Surface Science 1, 143 (1978). This paper also contains an extensive list of references on surface analytical techniques.

2. a. J. S. Murday, U. S. Naval Research Laboratory Memorandom Report #3062 (1975).

   b. J. P. Hobson, Japan J. Appl. Phys. Suppl. 2, 317 (1974).

3. C. J. Powell, American Laboratory 10, 17 (1978).

4. a. E. D. Hondros and M. P. Seah, Met. Trans. 8A, 1363 (1977).

   b. P. Wynblatt and R. C. Ku in Interfacial Segregation, eds. W. C. Johnson and J. M. Blakely ASM, Metals Park (1979).

5. D. Gupta, D. R. Campbell, and P. S. Ho, Thin Films - Interdiffusion and Reactions, eds. J. M. Poate, K. N. Tu, and J. W. Mayer, Wiley, New York (1978).

6. E. E. Latta and H. P. Bonzel, in Interfacial Segregation, eds. W. C. Johnson and J. M. Blakely, ASM, Metals Park, OH (1979).

7. H. Wiedersich and P. R. Okamoto, ibid; L. H. Rehn, S. Danyluk, and H. Wiedersich Phys. Rev. Letters 43, 1437, 1764 (1979).

8. J. M. Blakely and H. V. Thapliyal, ibid; J. M. Blakely and J. C. Shelton, in Surface Physics of Materials, ed. J. M. Blakely, Academic, New York (1975).

9. M. G. Lagally, T.-M. Lu, and G.-C. Wang, in Chemistry and Physics of Solid Surfaces, ed. R. Vanselow, CRC, Boca Raton, FL (1979).

10. J. H. van der Merwe, ibid.

11. D. A. King, J. Vac. Sci. Technol. 17, 241 (1980).

12. A. P. Janssen, J. A. Venables, J. C. M. Hwang, and R. W. Balluffi, Phil. Mag. 36, 1537 (1977).

13. Many such curves have been published. For a summary see, e.g., C. J. Powell, Surface Sci. 44, 29 (1974).

14. E. W. Mueller and T. T. Tsong, Field In Microscopy, Principles and Applications, Elsevier, NY (1969).

15. J. A. Borders, in Ion Beam Surface Layer Analysis, eds, O. Meyer, G. Linker, and F. Käppeler, Plenum Press (1976).

16. M. G. Lagally and D. G. Welkie, this volume.

17. J. Stöhr, Bull. Am. Phys. Sco. 25, 323 (1980).

18. L. C. Feldman, Appl. Surface Sci, in press.

19. R. L. Park, in Experimental Methods in Catalytic Research, Vol. III, eds. R. B. Anderson and P. T. Dawson Academic, New York (1976).

20. P. H. Holloway and G. E. McGuire, Thin Solid Films 53, 3 (1978). This paper lists the references for the original data.

21. K. Siegbahn, et al., ESCA, Almqvist and Wiksells, Uppsala (1967).

22. A. Joshi, L. E. Davis, and P. W. Palmberg, in Methods of Surface Analysis, ed. A. W. Czanderna, Elsevier, New York, NY (1975).

23. L. E. Davis, et al., Handbook of Auger Electron Spectroscopy, PEI, Eden Prairie, MN (1976).

24. For a list of references and several applications, see G. D. Davis, P. E. Viljoen, and M. G. Lagally, J. Electron Spectroscopy and Related Phenomena 20, 305 (1980); 21, 135 (1980). G. D. Davis and M. G. Lagally, J. Vac. Sci. Technol. 18, 727 (1981).

25. C. D. Wagner, et al., Handbook of X-Ray Photoelectron Spectroscopy, PEI, Eden Prairie, MN (1979).

26. These are tabulated in convenient form by W. M. Riggs and M. J. Parker, in Methods of Surface Analysis, ed. A. W. Czanderna, Elsevier, New York NY (1975).

27. J. A. McHugh, in Methods of Surface Analysis, ed. A. W. Czanderna, Elsevier, New York, NY (1975).

28. C. A. Anderson and J. R. Hinthorne, Science 123, 853 (1972).

29. Pricate communication, J. T. McKinney, 3M Company.

30. J. M. Buck, in Methods of Surface Analysis, ed. A. W. Czanderna, Elsevier, New York, NY (1975).

31. R. L. Erickson and D. P. Smith, Phys. Rev. Letters 34, 297 (1975).

32. W. Heiland, Appl. Surface Sci., in press.

USE OF AES AND RGA TO STUDY NEUTRON IRRADIATION ENHANCED

SEGREGATION TO INTERNAL SURFACES[*]

G. R. Gessel[†] and C. L. White

Oak Ridge National Laboratory, Oak Ridge, TN   37830

The high flux of point defects to internal interfaces during neutron irradiation can result in solute or impurity segregation at these interfaces. This segregation can strongly influence swelling and mechanical properties. Auger electron spectroscopy (AES), inert ion sputtering, and residual gas analysis (RGA) can be used to study the segregation phenomena if the internal interfaces can be exposed as free surfaces by fracture. RGA can monitor any helium release upon fracture. AES and inert ion sputtering can be used to study segregation in the vicinity of the fracture surface.

Results on an irradiated stainless steel, exhibiting intergranular failure at 610°C, are used to illustrate the application of these techniques. Significant segregation of phosphorus to the intergranular fracture surface was observed; however, no helium release upon fracture was detected. These techniques, coupled with scanning and transmission electron microscopy, provide insight into the microstructural factors affecting the irradiation-induced embrittlement.

[*]Research sponsored by the Division of Materials Sciences, U.S. Department of Energy, under contract No. W-7405-eng-26 with the Union Carbide Corporation.

[†]Present address: Department of Mechanical Engineering, University of Arkansas, Fayetteville, Arkansas 72701

Introduction

The effects of neutron irradiation on the properties of structural materials frequently appear to involve internal interfaces such as grain boundaries, precipitate-matrix interfaces, and internal-free surfaces (e.g., of voids and bubbles). The properties of these interfaces are likely to be strongly influenced by trace element segregation driven by both equilibrium and nonequilibrium mechanisms. The purpose of this paper is to describe the use of Auger electron spectroscopy (AES) and residual gas analysis (RGA) for the study of interfacial segregation in irradiated alloys. Experimental results on neutron-irradiated Inconel 706 (IN706) will be presented as an example of how these techniques can be applied.

Background

The correlation between interfacial segregation and changes in mechanical properties of polycrystalline materials is well documented. Reductions in both fracture stress and total elongation for tensile tests have been observed to coincide with metalloid segregation to metal interfaces. Such segregation-induced embrittlement has been reported for sulfur in nickel (1,2) and nickel base alloys (3); phosphorous in iron (4,5) and austenitic steels (6) and bismuth in copper (7). The data of Hondros and McLean (8) for the Bi-Cu system show both the ultimate tensile strength and grain boundary energy decreasing as bulk bismuth content is increased. A similar study by Joshi and Stein (7) showed that reductions in fracture stress, ultimate tensile strength, and total elongation correlate well with increases in bismuth concentration at grain boundaries in bismuth-doped copper.

The usual treatment of segregation-induced embrittlement begins with an expression for W, the work done in creating a grain boundary crack;

$$W = 2\gamma_s - \gamma_{Gb} + W_p + W_D \tag{1}$$

where γ_s is the surface energy, γ_{Gb} the grain boundary energy, W_p the plastic work, and W_D the miscellaneous dissipation. Thermodynamic arguments, as well as experimental observations, show that equilibrium segregation to an interface decreases interfacial energies. The $(2\,\gamma_s - \gamma_{Gb})$ term represents the energy expended in reversibly producing a brittle crack. This term could be increased or decreased as a result of segregation by some amount $\Delta\gamma$, which is the change in cohesive energy of the boundary. Because fracture ultimately means that bonds are broken, this is often converted to a change in average interfacial bond energy. This averaging approach, of course, neglects potentially important spatial variations in bond energy. Equation (1) is also restricted to the growth of an isolated crack and therefore neglects competing processes taking place throughout the specimen. Rather modest changes in surface energy can profoundly influence other material properties. Figure 1 shows the effect of surface energy on the energy for vacancy cluster (void) nucleation (9). Nucleation rates represented there (upper curve, 2.3×10^{-6} cm^{-3} s^{-1} lower curve, 3.1×10^9 cm^{-3} s^{-1}) span 15 orders of magnitude.

While much of the literature connecting impurity segregation with embrittlement addresses only equilibrium segregation, irradiated materials may also experience non-equilibrium segregation. Non-equilibrium segregation results from the flux of irradiation-induced point defects to internal sinks. Enrichment or depletion of impurities or alloy constituents at sinks can occur, depending upon the type of interaction that exists between these elements and the point defects.

240

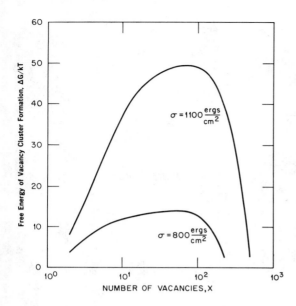

Fig. 1. Free energy of vacancy cluster formation versus cluster size for two values of surface energy (L. K. Mansur and W. G. Wolfer, ORNL/TM-5670).

Figure 2 schematically illustrates three mechanisms by which nonequilibrium segregation might occur. The first case shown in Figure 2 represents an enrichment of slower moving alloy components at a grain boundary (one type of sink) for elements diffusing by a vacancy mechanism. Case 2 addresses the case of self-interstitial diffusion and results in grain boundary enrichment of faster diffusing species. Defect-element binding is represented by case 3, which can result in grain boundary enrichment of the bound species. The current state of kinetic models for irradiation-produced segregation is discussed elsewhere (10).

It is important to keep in mind that nonequilibrium segregation (by definition) occurs in opposition to thermodynamic driving forces. At some point a steady-state condition will be reached when back-diffusion (driven by equilibrium thermodynamics) will balance the nonequilibrium flux due to defect production in the lattice. It is also important to note that, since equilibrium segregation tends to minimize γ_{Gb}, any nonequilibrium level of segregation will cause γ_{Gb} to be greater than its minimum value.

A second effect of irradiation on segregation could be increased kinetics of equilibrium segregation. The extent of segregation at equilibrium is generally expected to increase as temperature is decreased; however, the kinetics of segregation will decrease. This competition between kinetics and thermodynamics tends to limit the extent of segregation that actually occurs at low temperatures. By increasing atomic mobility, irradiation could permit thermodynamically favorable (but kinetically limited) segregation to occur in shorter periods of time or at lower temperatures than would be possible in the absence of irradiation.

Finally, significant quantities of helium can be produced via (n,α) reactions on many alloys, especially those containing nickel. Segregation of helium to internal interfaces could have significant effects on alloy behavior. If bubbles are formed at internal boundaries, this of course would lower the load bearing cross section and facilitate localized (and hence low ductility) failure in the vicinity of the boundaries.

In all cases where solute enrichment at internal interfaces occurs, the region of enrichment is expected to be rather narrow. For the case of equilibrium segregation, solute-enriched regions are typically only a few atom layers (< 1 nm) thick. Somewhat thicker regions are often anticipated for nonequilibrium segregation, sometimes extending several tens of nanometers from the interface plane.

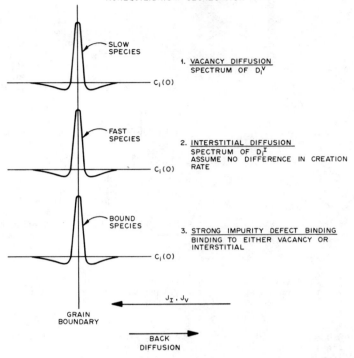

NONEQUILIBRIUM SEGREGATION

SLOW
SPECIES

$c_i(0)$

1. VACANCY DIFFUSION
   SPECTRUM OF D_i^V

FAST
SPECIES

$c_i(0)$

2. INTERSTITIAL DIFFUSION
   SPECTRUM OF D_i^I
   ASSUME NO DIFFERENCE IN CREATION
   RATE

BOUND
SPECIES

$c_i(0)$

3. STRONG IMPURITY DEFECT BINDING
   BINDING TO EITHER VACANCY OR
   INTERSTITIAL

J_I, J_V

GRAIN
BOUNDARY

BACK
DIFFUSION

Fig. 2. Schematic illustrating qualitatively three mechanisms for irradiation-induced solute segregation.

When segregation occurs in a very thin region at an interface, it can be difficult to detect using conventional analytical techniques. Auger electron spectroscopy (AES) is a well-established tool for examining thin interfacial regions when they can be exposed as external surfaces. AES is sensitive to only the top few atom layers on an exposed surface because it involves detection of characteristic low-energy Auger electrons. All elements except hydrogen and helium can, in principle, be detected using AES. Subsurface regions can be exposed for analysis by sputtering with inert gas ions. Alternate AES and sputtering permit information about the depth distribution of segregated species to be obtained.

An important requirement for the use of AES to study interfacial segregation is that the interface in question must be exposed as a free surface. In the case of internal interfaces, this is normally achieved by fracturing the specimen along those interfaces. The conditions under which this fracture takes place are important and can significantly affect the interpretation of AES results. One requirement is that the fracture surface be created under ultrahigh vacuum (UHV) conditions. At pressures greater than $\sim 10^{-7}$ Pa, contamination of clean surfaces occurs so rapidly that AES results are significantly affected.

One benefit of having to fracture specimens in a UHV environment is that any gas release upon fracture can be easily monitored. If a residual gas analyzer (RGA) is used, the pressure increase due to a specific gaseous species can be determined. This approach has proven useful in the past for

242

observing helium release (which cannot be detected by AES) from irradiated specimens; however, other gaseous transmutation products could also be detected.

In the remainder of this paper we will present and discuss experimental results on neutron-irradiated Inconel 706. This alloy exhibits poor elevated-temperature ductility in postirradiation tensile tests. The failure mode in these tests is largely intergranular, suggesting the possibility that irradiation-induced or -enhanced segregation might be important.

Experimental Techniques

Figure 3 is a top view of an Auger system that has been developed over a period of years at Oak Ridge National Laboratory. Experimental studies using this system have been previously reported by Clausing and Bloom (11), Sklad et al. (12) and White et al. (13). This system uses a cylindrical mirror analyzer with coaxial electron gun and electronics. The electron gun has a 5 μm minimum spot size and the electron beam can be rastered to provide images and elemental maps of the surface being analyzed. A normal incidence 1 kV inert ion sputter gun with movable sputter shield and a quadrupole mass analyzer are also part of the system. Not shown in Figure 3 is a transfer-lock device into which specimens are introduced prior to entry through the specimen entry port into the main chamber.

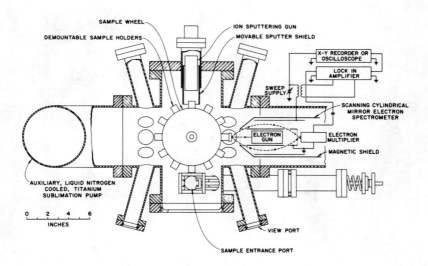

Fig. 3. Top view of the Auger system used at ORNL.

Figure 3 also shows the location of a tensile fracture stage, described in greater detail in Figure 4. Electrical leads are used to pass current through the specimen for elevated-temperature tests. The hollow threaded pull rod and ceramic standoffs limit the specimen load to approximately 400 N. The specimen holder (shown down-side up) together with a specimen 1.016 cm long with a 1.016 mm cross section are shown in Figure 5. All specimens used in this work were electropolished. Temperature measurements were made by sighting through the front viewing port (Figure 3) onto the notched portion of the Auger bar with an infrared pyrometer.

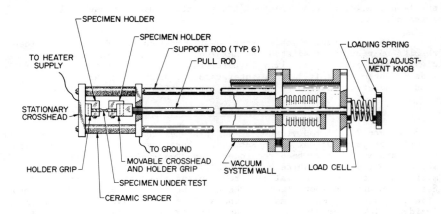

Fig. 4. Elevation view of the Auger straining stage with a specimen in place.

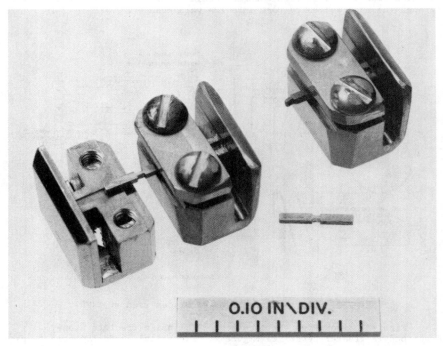

Fig. 5. Specimen, specimen holder, and one-half of specimen holder containing fractured specimen for Auger analysis of fracture surface.

Table I gives the composition of the IN706 used in this study. Tensile specimens of this alloy were given the following thermal treatment (955°C, 1 h/water quench/843°C, 3 h/air cool/720°C, 8 h/furnace cool to 620°C/10 h air cool), which we designate the STA (solution treated and aged) condition.

Table I. Composition of Inconel 706 (wt %)

Fe	Ni	Cr	Nb	Ti	Si	Mn	Al	C	Co	Cu	S	P
Bal	41	16	3	2	0.2	0.2	0.2	0.03	0.05	0.04	ND	ND

ND = not determined.
955°C/1 h WQ + 843°C/3h AC + 720°C/8hFC to 620°C/10 h AC.

These tensile specimens were then irradiated to a dose of 5×10^{22} n/cm^2 at
500°C. The results of tensile tests on these specimens are summarized in
Table II. When tested at 610°C, these specimens show almost no ductility, in
sharp contrast to the unirradiated alloy. There also appears to be some
increase in yield strength and decrease in ultimate tensile strength associated with the irradiation.

Table II. Tensile Results for Inconel 706 (STA)

Conditions	Designa-tion	Test Temper-ature (°C)	Strain Rate (s^{-1})	Yield Strength (MPa)	Uniform Elonga-tion (%)	Ultimate Tensile Strength (MPa)	Total Elonga-tion (%)
Unirradiated; Irra-diated @ 500°C	T_1	600	4×10^{-4}	903	7.7	1116	9.9
$\phi t = 5 \times 10^{22}$ n/cm^2 Irradiated @ 500°C	T_2	232	4×10^{-4}	1096	5.4	1224	5.8
$\phi t = 5 \times 10^{22}$ n/cm^2 Irradiated @ 500°C	T_3	610	4×10^{-4}	1032	0.4	1075	0.5
$\phi_t = 7.5 \times 10^{22}$ n/cm^2	T_4	610	4×10^{-4}	918	0	918	0

Figures 6 and 7 compare TEM micrographs of unirradiated and irradiated
IN706-STA, respectively. The unirradiated alloy (Figure 6) is strengthened
by γ' and γ'' and has a precipitate-free zone adjacent to grain boundaries,
which contains η phase precipitates. Figure 7 shows the microstructure of a
specimen from the head of the irradiated tensile specimen T3 (Table II). The
irradiated specimen has no detectable precipitate-free zone and does have
voids or bubbles at precipitate-matrix interfaces along the grain boundaries.

A notched Auger tensile specimen, A3, was cut from the gage section of
the failed tensile specimen, T3. The Auger specimen was fractured at approximately 610°C in the AES system and the resulting fracture surface was examined using AES. The helium partial pressure was monitored during fracture,
and no release of helium was detected.

A derivative type Auger spectrum from the fresh fracture surface of specimen A3 is shown in Figure 8. This spectrum was obtained by rastering the primary electron beam over roughly one-half of the fracture surface; and as such
represents an average surface composition for that surface. In addition to
the intentionally added metallic alloying elements (see Table I), P, C and O

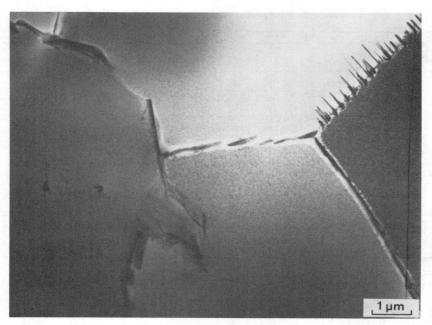

Fig. 6. Microstructure of unirradiated Inconel 706 in the STA condition.

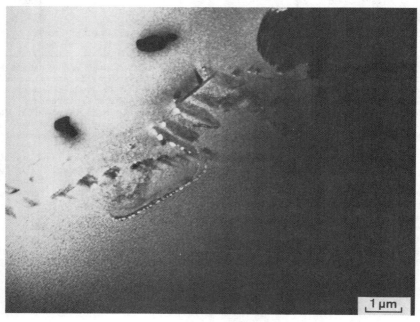

Fig. 7. Microstructure of Inconel 706 STA irradiated to 5×10^{22} n cm^{-2} at 500°C and tested at 610°C.

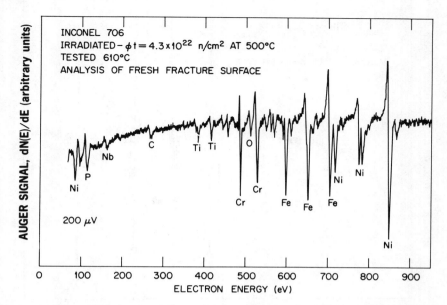

Fig. 8. Auger spectra obtained by rastering 4 keV electron beam over one-half A3 fresh fracture surface.

are present in significant concentrations. An approximate* quantitative analysis of this spectrum indicates P, C, and O levels of 5, 5, and 1 at. %, respectively. Spot analyses on this fracture surface indicated that the phosphorous concentration varied significantly, and was as high as 13 at. % in certain areas. This indicates phosphorous concentrations on the fracture surface over 600 times the average bulk concentration.

Carbon and oxygen levels indicated in Figure 8 are also significantly greater than bulk levels. The observed carbon could arise from at least three sources: (1) elemental carbon segregated to the fracture surface, (2) carbon-rich precipitates on the fracture surface, and (3) adsorption of carbon containing residual gases (e.g., CO and CH4) from the analysis chamber. Similar possibilities also exist for oxygen; however, it is believed that most of the oxygen observed in Figure 8 results from contamination by residual gases.

Following the analysis of the as-fractured surface, specimen A3 was positioned in front of the inert ion sputter gun and sputter etched for 5 min. The ion current density and ion energy were adjusted to allow an etching rate of approximately one atom layer per minute. Figure 9 shows an Auger spectrum of specimen A3 after the 5-min sputter etch. Comparison of the spectra in Figures 8 and 9 reveals that the phosphorous peak is gone. This indicates that the phosphorous-enriched zone near the fracture surface is only a few atom layers thick. This comparison will also reveal that the carbon and oxygen signals have increased. The increased carbon could be due to the "smearing" or uncovering of carbon-rich precipitates. Increases in

---

*These values are obtained by measuring the peak-to-peak intensity for each element, dividing that intensity by an elemental sensitivity factor, and normalizing these values to 100%. This does not correct for matrix effects, distribution effects, and differences in electron mean free path.

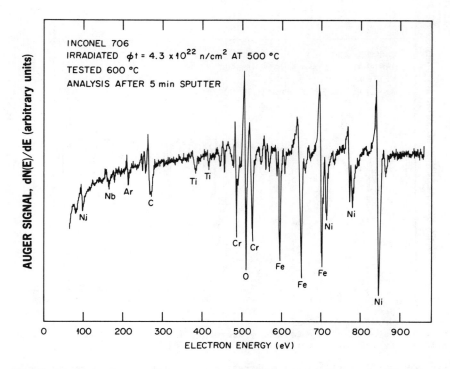

Fig. 9. Auger spectra obtained by rastering 4 keV electron beam over the same area as in Fig. 8 following 5 min sputtering with argon ions at 1 kV.

carbon and oxygen could also result from the presence of these impurities in the sputtering gas.

Figures 10, 11, and 12 show scanning electron micrographs of the fracture surface that was analyzed using AES. The failure mode in this specimen was a mixture of intergranular and transgranular modes. Figure 12 shows the small dimples that covered many of the grain boundary facets. The spacing of these dimples and the cavities in Figure 7 are similar.

In order to determine what effect, if any, the postirradiation heating at 610°C might have had on these results, in situ heating of unirradiated IN706 foils was carried out. AES analysis of external surfaces indicated that 2–4% phosphorous can segregate to the external surface of this material when heated for 10–30 min in the temperature range 600–650°C. These observations indicate that the grain boundary fracture surfaces created by high-temperature fracture may not have the same composition as the unfractured grain boundaries in "as-irradiated" material.

In an effort to expose grain boundaries (for AES analysis) without any postirradiation heating, electrolytic hydrogen charging has been exploited. Hydrogen charging was conducted in a round bottom flask outfitted with a heating mantle, reflux condenser, and two platinum electrodes. The specimen was held in a platinum wire mesh envelope and suspended between the electrodes using platinum wire. A 4% H_2SO_4 solution containing 0.25 g/L of As_2O_3 was used for charging, and a current density of 0.1 A/cm^2 at 4 V was maintained at the specimen. In order to maximize the thickness of the hydrogen diffusion

248

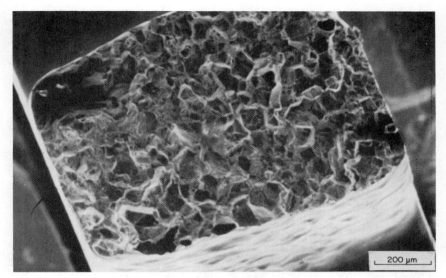

Fig. 10. Fracture surface of A₃.

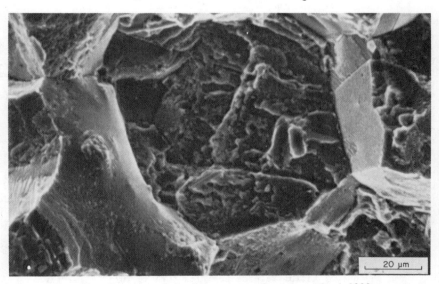

Fig. 11. Portion of A₃ fracture surface magnified 1000×.

zone, temperatures close to boiling are desirable. Our work has successfully used temperatures of 85—90°C and charging times up to 72 h. Longer charging times are desirable; however, during lengthy charges a surface film periodically forms on the specimen which must be removed by electrolytic polishing to ensure free entry of hydrogen. Preliminary AES results from this effort indicate that as-irradiated boundaries contain 1—2 at. % phosphorous. Helium partial pressure was again monitored, but no helium release was detected. Future experiments on hydrogen charged specimens are intended to compare grain boundary compositions for unirradiated and irradiated alloys.

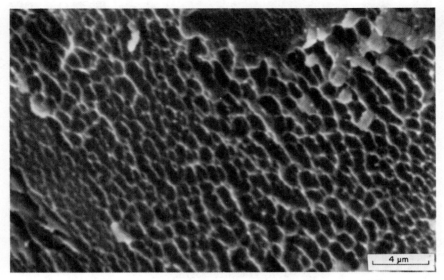

Fig. 12. Portion of fracture surface shown in Fig. 11.

Discussion

    A partially intergranular fracture surface of irradiated IN706, created
by in situ straining at elevated temperatures, has been analyzed using AES.
Large phosphorous enrichment (nearly 600 times the bulk concentrations) was
observed on this fracture surface. Scanning and transmission electron
microscopy indicate that the fracture path probably follows grain boundaries
and precipitate-matrix interfaces, cutting through voids on both types of
boundaries. Internal cracks, created during the initial tensile testing,
could also have been present in the AES specimen. The resulting fracture
surface, therefore, consists of grain boundaries, precipitate-matrix interfaces and void surfaces, as well as some regions of transgranular fracture.

    All of these interfaces (except internal cracks from tensile testing)
were exposed to neutron irradiation at 500°C. They were also exposed to
postirradiation thermal cycling during the tensile testing, and during fracture in the AES system. The phosphorous segregation indicated in Figures 8
and 9 could result from either the irradiation or the thermal exposure, or
some combination of the two. As we previously mentioned, AES analysis of
"as-irradiated" IN706 shows lower, but still significant, phosphorous enrichment on partially intergranular fracture surfaces.

    The absence of helium release upon fracture is puzzling. Previous
studies by Clausing and Bloom (11) and Sklad et al. (12) on alloys irradiated
to similar fluences showed definite evidence of helium release. The possibility exists that any helium present in the alloy could have been lost during
the tensile test; however, recent tests on hydrogen charged "as-irradiated"
IN706 also did not show any helium release. This seems to leave only the
possibility that the helium is trapped elsewhere in the microstructure, and
not released upon fracture. Determination of bulk helium levels in
"as-irradiated" alloys should help to clarify this point.

Conclusions

The use of RGA and AES to study solute segregation to interfaces has been discussed, and experimental results on irradiated IN706 have been presented. AES results indicate that phosphorous segregation to internal interfaces takes place during irradiation at 500°C; however, the mechanism (irradiation-induced nonequilibrium segregation, irradiation-enhanced equilibrium segregation, or purely thermal segregation) cannot be deduced without appropriate control experiments.

The process of exposing internal interfaces for AES analysis must be given careful consideration in these experiments. The composition of intergranular (or interfacial) fracture surfaces produced during fracture at elevated temperature may not be entirely representative of "as-irradiated" interfaces. To the extent that changes occur after irradiation, but prior to separation of the interface, they, of course, may still be relevant to the fracture process. To the extent that segregation occurs to a fracture surface after it is exposed, it cannot be considered to have influenced the fracture process. For this reason, the ability to expose "as-irradiated" interface without elevated-temperature fracture seems highly desirable. Hydrogen charging followed by low-temperature fracture is one promising way of doing this.

The complexity of the alloys, and the variety of interface types exposed for AES analysis, will also affect the extent to which experimental data can be interpreted. Clearly, experiments intended to explain mechanisms of fracture should concentrate initially on simple alloys where only grain boundaries, and possibly void surfaces, are exposed for analysis.

While no helium release has been observed in IN706, the use of RGA for this purpose seems well established. Much work seems needed to determine the factors that influence whether or not helium will segregate to interfaces and influence mechanical behavior.

Finally, extrapolation of our knowledge about unirradiated alloys suggests that segregation may strongly affect the fracture behavior of irradiated alloys; however, clear demonstration of a cause-and-effect relationship is far from established. In general, other irradiation-induced effects such as hardening, void formation at interfaces, and alterations of precipitate-free zones can be expected to act in combination with segregation to produce a net loss in ductility.

Acknowledgments

The authors wish to thank colleagues at ORNL for their contributions to various phases of this work. These include L. Heatherly and R. Padgett for assistance in conducting Auger analysis, L. Schrader for scanning microscopy of fracture surfaces, P. S. Sklad for transmission electron microscopy, and A. F. Rowcliffe for useful discussions. The assistance of F. A. Scarboro and S. P. Buhl in preparation of the manuscript is also gratefully acknowledged.

References

1. J. H. Westbrook and S. Floreen, "Grain Boundary Segregation and the Grain Size Dependence of Strength of Nickel-Sulfur Alloys," Acta Met. 17 (1969) 1175.

2. M. G. Lozinskiy, G. M. Volkogon, and N. Z. Pertsovskiy, "Investigation of the Influence of Zirconium Additions on the Ductility and Deformation Structure of Nickel Over a Wide Temperature Range," Russian Metallography 5 (1967) 65.

3. C. L. White and D. F. Stein, "Sulfur Segregation to Grain Boundaries in in Ni_3Al and $Ni_3(Al,Ti)$ Alloys," Met. Trans. 9A (1978) 13.

4. E. D. Hondros, "The Influence of Phosphorous in Dilute Solid Solution on the Absolute Surface and Grain Boundary Energies of Iron," Proc. Roy. Soc. London A 286 (1965) 479.

5. M. P. Seah, "Segregation and the Strength of Grain Boundaries," Proc. Roy. Soc. London A 349 (1976) 535.

6. W. Losch, "Temper Embrittlement and Surface Segregation, an AES and ILS Study," Acta Met. 27 (1979) 567.

7. A. Joshi and D. F. Stein, "An Auger Spectroscopic Analysis of Bismuth Segregated to Grain Boundaries in Copper," J. Inst. Metals 99 (1971) 178.

8. E. D. Hondros and D. McLean, "Cohesion Margin of Copper," Phil. Mag. 29 (1974) 771.

9. L. K. Mansur and W. G. Wolfer, A Study of the Effect of Void Surface Coatings on Radiation-Induced Swelling, ORNL/TM-5670 (1977). Available from NTIS, U.S. Department of Commerce, Springfield, VA 22161

10. P. R. Okamoto and L. E. Rehn, "Radiation-Induced Segregation in Binary and Ternary Alloys," J. Nucl. Mater. 83(1)(1979) 2.

11. R. E. Clausing and E. E. Bloom, "Auger Electron Spectroscopy of Fracture Surfaces in Irradiated Type 304 Stainless Steel," Proceedings Fourth Bolton Landing Conference, J. Walters et al., eds. (1974), p. 491.

12. P. S. Sklad, R. E. Clausing, and E. E. Bloom, "Effects of Neutron Irradiation on Microstructure and Mechanical Properties of Nimonic PE-16," ASTM-STP 61 (1976), p. 139.

13. Calvin L. White, Robert E. Clausing, and Lee Heatherly, "The Effect of Trace Element Additions on the Grain Boundary Composition of Ir + 0.3 Pct W Alloys," Met. Trans. A, 10A (1979) 683.

STUDIES OF EXTENDED DEFECTS ON SURFACES BY

LOW-ENERGY ELECTRON DIFFRACTION*

M. G. Lagally[†] and D. G. Welkie[††]

Department of Metallurgical and Mineral Engineering
and Materials Science Center
University of Wisconsin
Madison, Wisconsin 53706

A brief review is given of the use of low-energy electron diffraction
(LEED) for studying structural defects at surfaces and in layers adsorbed
on or segregated to surfaces. It is demonstrated that a variety of surface
extended defects can be distinguished and quantified by measuring the
dependence of diffracted-beam angular profiles on diffraction geometry.
The discussion is illustrated with examples of an epitaxially grown Ag(111)
film, a cleaved and ion-beam damaged GaAs(110) surface, and a partial monolayer of O adsorbed on W(110).

*Supported in Part by NSF Grant #DMR 78-25754
[†]H. I. Romnes Fellow
[††]Present Address: Perkin-Elmer, Physical Electronics Division,
Eden Prairie, MN

Introduction

Because of the interrelation between surface microstructure and surface chemical, electronic, and transport properties, there is considerable impetus for studying the crystallography and defect structure of surface phases and adsorbed layers. Whereas a large effort has been invested in the last decade in determination of the equilibrium positions of atoms in the "perfect" surface, little work has been done in analyzing deviations from perfect crystallinity. Perfect surfaces, just as perfect bulk crystals, are, however, thermodynamically not possible: there always exists a finite concentration of defects even at a single-crystal surface. In addition there are nonequilibrium defects that may be introduced in preparing or working the surface, as well as those purposefully introduced by chemical reactions or beam interactions at the surface. Although the importance of defects on other surface properties has been realized (1-3), there have been, until recently, few actual structural-defect studies.

The ideal probe of the geometric structure of any object is radiation with wavelength of the order of the dimensions one is trying to resolve. This radiation is then diffracted by the periodic arrangement of scatterers in the object. In cases where the radiation can be focussed, it is possible to recombine diffracted beams to form a real-space image of the object and its geometric structure. Forms of radiation that are suitable for analysis of atomic structure are, of course, most commonly x-rays and electrons. The transmission electron microscope uses high-energy electrons for the analysis of bulk defect structure, and, as is well known, produces an image by decomposing a particular diffracted beam back into its various Fourier components. Analysis of surface structure proceeds in an analogous manner by a quite similar technique, low-energy electron diffraction (LEED). There are, however, two major differences: 1) the diffraction beams in LEED are observed in reflection rather than in transmission, and 2) it has not yet been possible to form a real-space LEED image. The first difference makes LEED useful as a technique and is a natural consequence of the very strong interaction with the solid of the slow (10-1000 eV) electrons used, which limits their mean free path to several atomic layers. Because crystals as thin as this obviously can't be prepared, only backscattered beams can be observed. On the other hand, as a result of this strong interaction, the backscattered beams are relatively strong, each about 0.1 to 1% of the primary-beam intensity.

The second difference is a practical one. Imaging capability in LEED has been limited by lack of availability of sufficiently chromatic lenses at these energies, where diffraction angles, rather than being fractions of degrees, are in the tens of degrees. Thus all the analysis is in terms of the diffracted beams, very similar to x-ray diffraction, where imaging is, of course, limited by the difficulty in focussing x-rays.

In x-ray diffraction (as in high-energy electron diffraction) the equilibrium positions of the atoms are deduced from measurements of the positions and integrated intensities of the diffracted beams. It was realized early that deviations from long-range order result in distortions of the diffraction line profiles. This concept has been developed to such a degree that quite detailed information on bulk defect structures can now be obtained from the quantitative analysis of diffracted intensity distributions (4). In order to investigate the effects of defects on processes and properties at surfaces, to find ways to control the defects by varying sample preparation and crystal growth procedures, and to study the creation and decay of defects induced by for example, irradiation or sputtering, it is important to develop similar techniques with which different types of defects at a surface can be distinguished and quantitatively analyzed.

Although it is recognized that LEED beam intensity angular profiles must contain the same information on surface defects as x-ray diffraction does for the bulk ("sharp" diffraction spots are taken as an indication of a "well-ordered" surface, "fuzzy" spots or streaks, a "poorly-ordered" surface), there has been until recently little effort to make quantitative measurements. This has been a result of the belief that instrumental limitations make LEED relatively insensitive to surface defects. However, several studies (5-9) have demonstrated the ability of LEED to extract quantitative information on surface defect structures, given a proper consideration of instrumental effects (10-12).

The object of this paper is to review the extent to which LEED can presently be used in a quantitative investigation of the surface microstructure. In the next section a brief review of the principles of LEED is given. This is followed by an analysis of different types of extended defects that have been identified on surfaces, including mosaic structures, random strain, finite-size overlayer islands, steps, and antiphase domains in continuous layers. The discussion is illustrated with results of measurements on sputter-damaged and annealed GaAs surfaces, on epitaxially grown oriented Ag(111), and on the W(110)-O chemisorption system.

Observation of Defects by LEED

A LEED experiment entails the measurement, as a function of electron wavelength and diffraction geometry, of a current of slow electrons that are scattered elastically from a surface. A schematic diagram of a state-of-the-art LEED instrument is shown in Figure 1. Traditionally, measurements have been made either with a Faraday cup that measures diffracted currents directly or with a spot photometer focussed on a fluorescent screen on which the diffraction pattern is displayed. Two kinds of measurements can be performed, an integrated-intensity measurement and an angular distribution. The former can be analyzed in terms of equilibrium positions, while the latter gives information on structural defects. The simplest example of such an analysis is given in Figure 2, which shows the diffraction pattern expected from an infinite set and a finite set of slits. It is noted that both sets of slits give diffraction lines spaced equally far apart (reflecting the slit spacing), but that the lines in the finite set are broadened to reflect the limited number of slits (domain or "finite-size" broadening). An integrated-intensity measurement will give only the slit spacing, while an angular distribution measurement will indicate that the slit distribution is finite.

A generally useful representation of diffraction from any lattice is in terms of its reciprocal lattice and the Ewald construction. For a three-dimensional infinite crystal, the reciprocal lattice is a three-dimensional array of points given by the reciprocal-lattice vectors $\underline{G}$ (hkl) = $n2\pi/\underline{d}$(hkl) where $\underline{d}$ is the distance between (hkl) planes. The Ewald construction gives simply the conservation of energy for elastic scattering, i.e., $\lambda_{in} = \lambda_{out}$ or $|\underline{k}_0| = |\underline{k}|$, where λ and $\underline{k}$ are respectively the electron wavelength and momentum. The superposition of the Ewald sphere onto the reciprocal lattice forces conservation of momentum in the form of the Laue conditions $\underline{S} = \underline{G}_{hkl}$, where $\underline{S}$ is the diffraction vector or momentum transfer. The diffracted-intensity distribution is then given by the intersection of the Ewald sphere with the reciprocal lattice.

For an adsorbed monolayer (or in any case where a phase is only one atomic layer thick) it is easy to demonstrate that the reciprocal lattice becomes a set of rods normal to the surface of the layer. The strong attenuation of slow electrons in solids means that the crystal contributing to the scattering is effectively only a few atomic layers thick, so that the

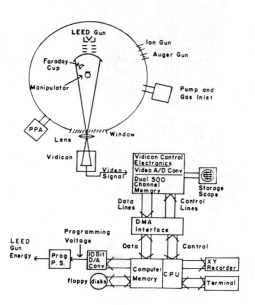

Fig. 1. Schematic diagram of a LEED diffraction system. The main features are an electron gun, a sample mounted on a goniometer, and a detector system, which in this case consists of a fluorescent screen on which the diffracted beams are displayed and a vidicon camera for recording the diffracted intensity distribution automatically. A dedicated computer is used to collect, reduce, and analyze the data.

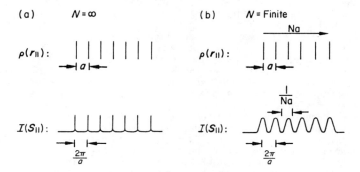

Fig. 2. Schematic illustration of the effect of finite size on the diffracted-intensity distribution. a) A perfect and infinite ordered structure, leading to a delta-function array of diffracted beams. b) Finite ordered regions of average size Na, leading to broadened reflections with width proportional to 1/Na.

approximation of reciprocal-lattice rods is useful for low-energy electron diffraction from any solid.

The reciprocal-lattice rods will have zero width if the surface or overlayer is laterally perfect and infinite, and the diffraction spots will be sharp. The effect of limitations to the long-range order is to modify the distribution of Fourier components that make up the diffracted amplitude. In the presence of defects the reciprocal-lattice rods have a finite width, which results phenomenologically in a broadening of the LEED beam angular profiles. It should be evident that different types of surface defects can cause different broadening of the intensity distribution, and that this fact can be used, as in x-ray diffraction, to distinguish and quantify different surface defects. This capability has, however, been exploited in LEED only in a very limited way (5-9,13). The types of surface defects that can presently be analyzed by consideration of the dependence of the LEED angular profile on diffraction variables will be illustrated in the next section.

Analysis of Surface Defects

The simplest type of defect that can exist is, of course, a point defect, such as a surface vacancy or an adsorbed atom. Such defects add a diffuse "gas scattering" background to the intensity that is featureless except that it is distributed with the weak angular dependence of the atomic scattering factor of the atom in question (14). Although surface thermal diffuse scattering has been investigated in a few cases (15,16), there have been no attempts so far to quantify point defect concentrations.

A number of different types of extended defects can exist on surfaces. Of these, the simplest one to picture is finite-size broadening. Mosaic structure (i.e., the existence of subgrain boundaries) in a crystal will appear at the surface as incoherently scattering finite-size domains with a small angular misorientation. The finite size of the domains manifests itself as a broadening of the reciprocal-lattice rods, just as finite crystallite size does in broadening reciprocal-lattice points (17). The angular misorientation leads to an additional broadening because each domain scatters the incident beam in a slightly different direction (8). This will be discussed in greater detail below, after consideration of an additional broadening mechanism, random incoherent strain.

Strain in a lattice can be either uniform or random. A uniform strain leads to a uniform expansion or contraction of the lattice, and thus causes a uniform shift in the diffraction features (17). Random strain, on the other hand, introduces new long-wavelength Fourier coefficients into the lattice, resulting in spot broadening (18). The same is true for surfaces or overlayers; for example, strain parallel to the surface causes a unique broadening of the angular distribution of intensity, as will be discussed below.

All these effects cause broadening in the reflections but each has a different dependence on diffraction parameters. Different extended defects can thus be distinguished. This can be quantified in the following way, using the approach of Warren and Averbach for x-ray diffraction intensities (18). Consider diffraction from a domain of finite size that contains distortions in a given direction a given by small random static displacements of the atom from their equilibrium positions. The intensity function for such a domain is given by

$$\mathcal{I}(\underline{S}_{||}) = \sum_{i,j=1}^{N-1} e^{i\underline{S}_{||} \cdot (\underline{R}_i - \underline{R}_j)}, \tag{1}$$

where $\underline{R}_i$ denotes the position vector of the i'th atom, $\underline{S}_{||}$ is the component of the momentum transfer parallel to the surface, and N is the number of atoms in the domain in the direction of $\underline{S}_{||}$. The $\underline{R}$'s can be written

$$\underline{R}_i = \underline{R}_{o_i} + \underline{\delta}_i , \tag{2}$$

where $\underline{R}_{o_i}$ is the equilibrium position vector of the i'th atom and is just ia, where $\underline{a}$ is the equilibrium interatomic spacing. δ_i describes the displacement of the i'th atom from its equilibrium position. The intensity function can then be separated into two sums

$$\mathcal{I}_o (\underline{S}_{||}) = \sum_{i,j=1}^{N-1} e^{i\underline{S}_{||} \cdot [(\underline{R}_{o_i} - \underline{R}_{o_j}) + (\underline{\delta}_i - \underline{\delta}_j)]}$$

$$= \sum_{i,j=1}^{N-1} e^{i\underline{S}_{||} \cdot (\underline{R}_{o_i} - \underline{R}_{o_j})} \sum_{i,j=1}^{N-1} e^{i\underline{S}_{||} \cdot (\underline{\delta}_i - \underline{\delta}_j)}. \tag{3}$$

The first factor is the diffracted intensity from a perfect finite-size domain of N atoms, and reduces to

$$\mathcal{I}_o(\underline{S}_{||}) = \frac{\sin^2 \frac{1}{2} N \underline{S}_{||} \cdot \underline{a}}{\sin^2 \frac{1}{2} \underline{S}_{||} \cdot \underline{a}} . \tag{4}$$

At the Bragg conditions, $|\underline{S} \cdot \underline{a}| = 2n_{||}\pi$, this function always has the same value and the function is periodic with $n_{||}$, the order of diffraction from rows perpendicular to $\underline{a}$. Hence the broadening introduced by finite N will repeat from zone to zone.

Because $\delta_i \neq \delta_j$ and the δ's are generally not rationally related, the second factor in Eq. (3) is not a periodic function and does not reduce to an analytic expression like Eq. (4). The phase factors represented by this sum continue to increase, causing an increasingly out-of-phase condition as the order number $n_{||}$ increases at constant $S_\perp$, the momentum transfer perpendicular to the surface. Thus reciprocal-lattice rods will get increasingly broad with increasing $\underline{S}_{||}$, or equivalently, if diffraction beam widths normalized to the separation between beams are measured, they will be increasingly broad if random strain is present at the surface. As can be seen from Eq. (3), strain broadening will be zero for the specular ($\underline{S}_{||} = 0$) reflection, because this beam carries no information on parallel components of displacement. These results are indicated schematically in Figure 3.

The angular misorientation alluded to earlier can also be described in terms of the reciprocal lattice. Each domain can be considered to have its own reciprocal lattice. These will all start at a common origin (000) but, because of the slight misorientations, will increasingly deviate, with an "opening angle" = 2α, where α is the mean domain misorientation angle. This is shown in Figure 4. The factor of 2 arises because for a misorientation α the diffracted beams will be displaced by 2α. This indicates that misorientation effects can be distinguished by measuring a particular diffraction spot (constant $\underline{S}_{||}$) as a function of energy, or equivalently $S_\perp$.

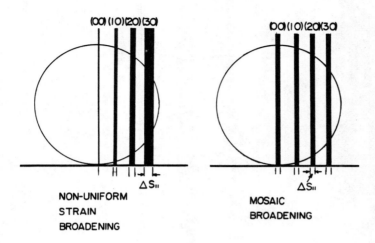

(OO) (IO) (2O) (3O)   (OO)(IO)(2O)(3O)

ΔS₁₁   ΔS₁₁

NON-UNIFORM   MOSAIC
STRAIN   BROADENING
BROADENING

Fig. 3. Schematic diagrams of the reciprocal lattice and Ewald construc-
         tion for a surface layer contains finite-size domains or random
         strain.

Misorientation out of the plane of the surface can be distinguished by the
diffraction spot shape in colatitude and azimuth. At $S_\perp = 0$, there will
be no misorientation broadening (at least for the (00) beam). Thus extrapolation of the broadening to $S_\perp = 0$ gives the pure domain-size broadening.

     The occurrence, separation, and quantification of mosaic and strain
broadening can be illustrated by the results of the measurements from a Ag
film grown epitaxially on a mica substrate (8). This film grows with a
preferred (111) orientation. Plotted in Figure 5 are the FWHM of the
instrument-corrected (i.e., after deconvoluting the instrument response
function (10-12)) angular profiles of three beams from the Ag(111) film,
expressed as a fraction of the two-dimensional Brillouin zone size, versus
$S_\perp$, the component of the diffraction vector perpendicular to the surface.
There is a significant amount of broadening for all beams, even for the
specular or (00) beam. Because this beam is not broadened by random strain
(8), the broadening must be due to finite-size and misorientation effects.
The fact that the widths broaden with $S_\perp$ indicates that there is a contribution from misorientation in the form of an out-of-plane or "tilt" distribution. The average degree of tilt can be determined by considering the
slope of the plot of beam width versus $S_\perp$, while the domain size contribution
is obtained by extrapolating such a plot to $S_\perp = 0$. Making quite simple
assumptions for the domain size distribution and the distribution of tilt
angles (8) gives for this particular film a mean tilt angle of ∼0.25°
between the surface planes of the domains, and an average projected domain
size of ∼75A (7.5 nm).

     Figure 5 shows that the broadening of all the beams at any $S_\perp$ is inde-
pendent of the order number to the accuracy of these data points. Hence,

259

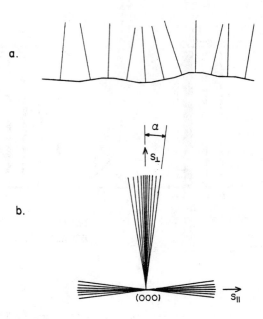

a.

b.

α

$S_\perp$

(000) $\overrightarrow{S_\parallel}$

Fig. 4.   Schematic diagram of tilt misorientation at the surface of a
          crystal and the resulting reciprocal lattices. The individual
          crystallites are assumed to be large enough to each give a delta-
          function intensity profile. α is the mean tilt angle.

there appears to be no strain at the surface of this film. A more careful
treatment using the method of Warren and Averbach (18) indicates that, in
fact, the random rms strain at this surface may be as much as 0.25%.

    A different defect, that is unique to surfaces and is the only one
that has been extensively studied, is atomic-height steps (7,12,19-24).
These are present on most surfaces as a result of surface preparation,
cleavage, or bombardment by ion beams. Two limiting types of step structures can be distinguished - regular step arrays or random step arrays.
Regular step arrays occur when a single crystal is cut at a small angle to
a singular direction, and are characterized by nearly constant terrace width,
regular step direction, and a definite "riser" orientation. Random step
arrays occur in most situations of surface damage, in most cases of cleavage,
and sometimes as a result of surface treatment, e.g. adsorption or chemical
reaction. The effect of steps on the angular profile of diffracted beams
is shown in Figure 6. Because the different terraces are at different
heights, a wave scattered from one terrace will alternately interfere con-
structively and destructively with that from adjacent terraces as the wavelength is changed. This causes the spot to be alternately narrow or broad.
The period of the broadening with change in energy gives directly the step
height (7), while the broadening itself gives the average terrace width, or
equivalently, the step density (20-22). At the positions of minimum width
in energy, the diffraction does not recognize the existence of steps, and
any broadening here will be due to the other mechanisms considered earlier.

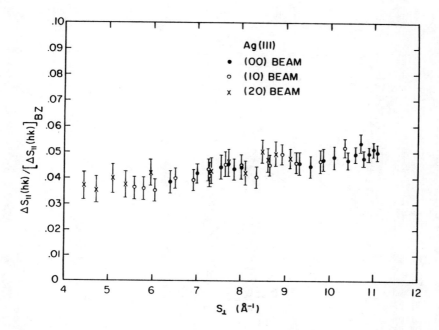

Fig. 5. Physical broadening of three reflections from a Ag(111) film
epitaxially grown on mica. The full width at half-maximum of
the reflections normalized to the width of the Brillouin zone
is plotted versus the normal component of the diffraction vector
$S_{\perp}$.

At the positions of maximum width, the diffraction observes each terrace as
an independent finite-size domain, and the broadening reflects the average
size of the terraces.

   If the terrace sizes are regular rather than random there will be only
discrete Fourier coefficients that make up the intensity sum, and the diffracted beams will split rather than broaden (7). The degree of splitting
indicates the average terrace size, while the width of each of the split
beams will give the distribution of terrace sizes about this average (20).

   Although the above is qualitatively correct, it has proven difficult so
far to devise models that quantitatively match the oscillatory behavior of
diffraction beam shapes (20-22). A simple model of random step distributions
(22) indicates that the diffraction spot shape may be a quite sensitive function of step density even for step densities as small as a fraction of a
percent of surface sites.

   An illustration of the existence of surface steps is shown in Figure 7
for GaAs(110) that has been damaged by a low-energy ion beam and then partially reannealed at various temperatures (24). It is noted that the
oscillations in beam width decrease with heat treatment, indicating steps
are being annihilated more rapidly at higher temperatures. From the observed
energies at which the maxima and minima in the widths of the (10) beam occur,
the step height is determined to be 2.0 ± 0.1Å (0.2 ± 0.01nm), corresponding
to the bulk interplanar distance. From the observed maxima in the widths,

261

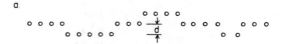

a.

b.

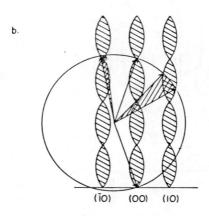

(Ī0) (00) (I0)

Fig. 6.   Schematic illustration of the effect of (a) a random distribution
          of up and down steps on a nominally singular surface on (b) the
          angular distribution of intensity, depicted in terms of the
          reciprocal lattice and Ewald construction.

step densities (21) of 33%, 21%, and 9% are obtained (24) respectively after
150°C, 350°C, and 500°C anneals. These results also suggest the possibility
of measuring the kinetics of step annihilation at surfaces or in overlayers,
information that can be valuable in understanding epitaxy, crystal growth,
and chemical interactions at surfaces, in addition to the energetics of the
extended defects themselves. Such kinetics studies should be possible for
other defects also.

     The defects considered so far are associated with the termination of a
bulk crystal. It is possible to consider in a similar way defects in a
single layer or fraction of a layer. Such situations are important both in
adsorption or, more importantly for metallurgical phenomena, in segregation
or precipitation at surfaces or interfaces. Adsorption and segregation are
totally analogous (25). In adsorption the chemical potentials are controlled
through the temperature and pressure of the adsorbing molecules, while in
segregation the chemical potentials are controlled through temperature and
concentration of the segregating species. Both equilibrium segregation
(25,26) and nonequilibrium segregation, either chemically driven (27) or
induced by bulk defect motion (28), have been observed. Preferential precipitation of one component of a multicomponent system can occur at a surface
even though the bulk remains single-phase (26). Nonequilibrium precipitation can, of course, also occur (28).

     In addition to point defects, some of the extended defects considered
so far may also be present in the single-layer "adsorbed" phase, particularly

262

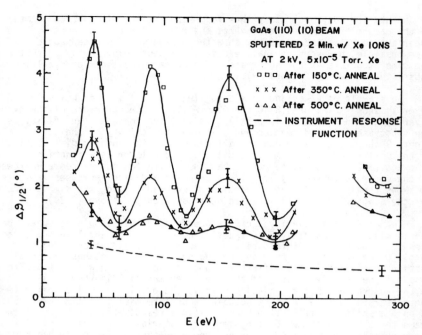

Fig. 7.   Step densities on ion-bombarded cleaved GaAs(110) as a function of
          annealing temperature.   The FWHM of the (10) reflection measured at
          room temperature is plotted vs. energy of the incident electrons
          after 10 minute anneals at 150°C, 350°C, and 500°C.

the equivalent of mosaic structure and strain.   Additionally, antiphase
domains are quite important in many overlayers.   Mosaic structure in a single
layer clearly can involve only translational randomness and misorientation
in the plane, but not out-of-plane misorientation.   One can think of "islands"
of adsorbed phase that meet at island boundaries that may have (9) a definite
phase relationship with each other.   If they do not, as, for example, for an
incommensurately "adsorbed" layer consisting of a number of islands randomly
nucleated, the analysis is the same as for mosaic structure.   Diffracted
beams will be broad because of the finite size of the islands, and spot
smearing occurs in azimuth if there is an in-plane misorientation.

       If the layer is commensurately adsorbed, i.e., in lattice sites of the
substrate, then a definite phase relationship exists between different islands.
Because of the existence of regular sites, only specific angular misorientations and translational position differences are allowed, giving antiphase
domains for all but (1x1) overlayers.   The resulting phase relationships can
be included in the analysis in a straightforward manner.   The broadening
caused by antiphase boundaries is qualitatively the same as for mosaic structure, but differs quantitatively (9).   It is thus possible to differentiate
between true "island formation", i.e. condensation in random patches of
finite size, and continuous phases with antiphase domain boundaries.

       To measure such broadening effects, the contribution of the overlayer
must be isolated.   This is easily possible for the large number of overlayers that form a structure with a unit mesh different from the substrate on

263

which they are adsorbed. They thus form superlattice diffraction beams, or in the case of incommensurate layers, their own diffraction pattern. The superlattice beams then directly show the broadening discussed above. It is also possible, however, to analyze defect structures in overlayers that have the same structure as the substrate under proper conditions of overlayer and substrate scattering factors (9).

An example of an island size determination (29) in a fractional mono-layer is shown in Figures 8-10 for 0 on W(110). The structure of the over-layer, shown in Figure 8, is p(2x1), i.e., it has the same periodicity as the substrate in one direction and twice the periodicity in the other. Figure 9 shows the full width at half-maximum of the angular distribution of the diffracted beam from the equilibrated overlayer as a function of coverage (9,29). It can be noted that the diffracted beams become narrower as the coverage increases, i.e. the average island size increases. Figure 10 shows the island size as a function of coverage, using the simple model for the scattering from finite-size domains discussed above (9,29).

There are other overlayer or surface defects that should be observable. One is slip lines on cleaved surfaces. These should behave like multiatomic steps. A second, very interesting one, is misfit dislocations (30) in a monolayer of material (either segregated or adsorbed) that has a natural lattice constant slightly different from the substrate on which it is grow-ing. If the cohesive energy of the overlayer material is very weak relative to the adsorbate-substrate binding energy, the overlayer will accommodate to the substrate lattice constant, causing a uniform misfit strain that results in only a shift in diffraction beam positions relative to positions of the beams for a bulk specimen of the overlayer material. Conversely, if the over-layer cohesive energy is very strong, the overlayer will grow incommensur-ately. In the intermediate range, where the overlayer cohesive energy is similar to the substrate-overlayer binding energy, regions where the layer accommodates to the substrate lattice will be separated by a dislocation and regions of changing lattice constant to make the average lattice constant equal to that of the overlayer (30). The range of the associated strain field will depend on the total amount of misfit and the degree of misfit that can be accommodated by uniform strain. The broadening effect of the LEED beams should be easily observable and quantifiable using the random strain model discussed earlier.

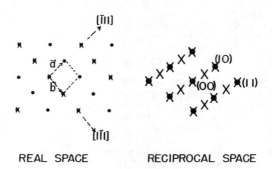

REAL SPACE   RECIPROCAL SPACE

Fig. 8.   Structure of the overlayer (a) and the corresponding diffraction
          pattern (b) for 0 chemisorbed on W(110) into a p(2x1) overlayer
          unit mesh. X's correspond to overlayer atoms and superlattice
          reflections.

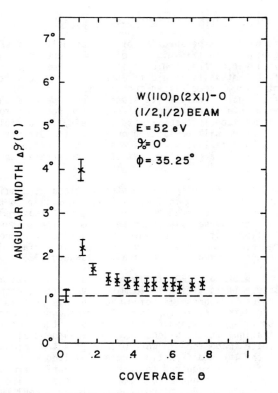

Fig. 9. Size effect in W(110) p(2x1)-O as a function of coverage. The minimum attainable FWHM of the (1/2 1/2) superlattice reflection is plotted vs. coverage. The dotted line represents the instrumental limit.

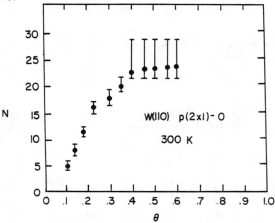

Fig. 10. Size of O overlayer islands chemisorbed on W(110) as a function of coverage, derived from the data of Figure 9 and using Eq. (4).

Discussion

In order to gain some perspective of the capability of LEED, we briefly compare the technique with transmission electron microscopy in this section. In TEM, a single layer or fraction of a layer of atoms generally contributes less than 1% of the intensity in a diffracted beam. Hence, it is not easily possible, except through specialized techniques, (31) to image a single layer of the substrate, e.g. to analyze steps and other substrate surface defects. For an overlayer with a superlattice, it should be possible to image a diffraction spot of the superlattice, if its intensity is sufficiently strong. In general, these beams will be sitting on top of the diffuse intensity of several hundred substrate layers. Thus, unless the scattering factor of the overlayer atoms is much larger than that of the substrate, the contrast will be very poor.

A more practical difficulty with most traditional transmission electron microscopes is that the vacuum is insufficient. A vacuum in the low 10^{-10} torr ($\sim 10^{-8}$ Pa) range is required to do reliable LEED measurements. This type of vacuum is, of course, also required to make TEM measurements free of surface contamination, but such contamination is less of a problem for bulk defect measurements. Advances in the use of TEM for surface measurements have recently been discussed in more detail elsewhere (32).

The major drawbacks of LEED at the moment are beam size and the lack of imaging capability. It is difficult to make a very fine, high-intensity beam at low energies. Progress is being made on this problem by putting more gain into the detector, in the form of microchannel electron multiplier arrays, making small - area scanning LEED likely in the near future (33). As regards the imaging, it is, of course, always conceptually much simpler to have a real-space image than to work with diffracted intensities. However, because of the large diffraction angles involved in LEED, considerable electron optics effort is required to make sufficiently chromatic electron lenses at low energies to image LEED beams. Progress is also being made in this field (34), so that a LEED microscope should be a reality in the near future.

References

1. W. K. Burton, H. Cabrera, and F. C. Frank, Phil. Trans. Roy. Soc. 243, 299 (1951).

2. Interfacial Segregation, eds. W. C. Johnson and J. M. Blakely, ASM, Metals Park, OH (1977).

3. Surface Physics of Materials, Vols. 1 and 2, ed. J. M. Blakely, Academic, New York (1975).

4. See papers in "Local Atomic Arrangements Studied by X-ray Diffraction", 36, ed. J. B. Cohen and J. R. Hilliard, Gordon and Breach, New York (1966).

5. R. L. Park, J. Appl. Phys. 37, 295 (1966).

6. R. L. Park, in The Structure and Chemistry of Solid Surfaces, ed. G. A. Somorjai, Wiley, New York (1969).

7. M. Henzler, Surface Sci. 19, 159 (1970); Surface Sci. 22, 12 (1970); in Electron Spectroscopy for Surface Analysis, ed. H. Ibach, Springer-Berlin (1977).

266

8. a) D.G. Welkie, M.G. Lagally, and R.L. Palmer, J. Vac. Sci. Technol. 17, 453 (1980); b) D.G. Welkie, Ph.D. Dissertation, University of Wisconsin-Madison (1981, unpublished).

9. G.-C. Wang and M.G. Lagally, Surface Sci. 81, 69 (1979); T.-M. Lu, G.-C. Wang, and M.G. Lagally, Surface Sci. 108, 494 (1981).

10. R.L. Park, J.E. Houston, and D.G. Schreiner, Rev. Sci. Instr. 42, 60 (1971).

11. D.G. Welkie and M.G. Lagally, Appl. Surface Sci. 3, 272 (1979).

12. T.-M. Lu and M.G. Lagally, Surface Sci. 99, 695 (1980).

13. R.L. Park and J.E. Houston, Surface Sci. 18, 43 (1969); J.E. Houston and R.L. Park, Surface Sci. 21, 209 (1970).

14. M.G. Lagally and M.B. Webb, in The Structure and Chemistry of Solid Surfaces, ed. G.A. Somorjai, Wiley, New York (1969).

15. J.T. McKinney, E.R. Jones, and M.B. Webb, Phys. Rev. 160, 523 (1967); R.F. Barnes, M.G. Lagally, and M.B. Webb, Phys. Rev. 171, 192 (1973).

16. R.L. Dennis and M.B. Webb, J. Vac. Sci. Technol. 10, 192 (1973).

17. L.H. Schwartz and J.B. Cohen, Diffraction from Materials, Academic, New York (1977).

18. B.E. Warren and B.L. Averbach, J. Appl. Phys. 21, 595 (1956).

19. M. Henzler, Appl. Phys. 9, 11 (1976).

20. J.E. Houston and R.L. Park, Surface Sci. 26, 269 (1971).

21. M. Henzler, Surface Sci. 73, 240 (1978).

22. T.-M. Lu, S.R. Anderson, M.G. Lagally, and G.-C. Wang, J. Vac. Sci. Technol. 17, 207 (1980).

23. G. Schulze and M. Henzler, Surface Sci. 73, 553 (1978).

24. D.G. Welkie and M.G. Lagally, J. Vac. Sci. Technol. 15, 784 (1979).

25. P. Wynblatt and R.C. Ku, in Interfacial Segregation, eds. W.C. Johnson and J.M. Blakely, ASM, Metals Park, OH (1977).

26. J.M. Blakely and H.V. Thapliyal, ibid.

27. E.E. Latta and H.P. Bonzel, ibid.

28. H. Wiedersich and P.R. Okamoto, ibid.

29. M.G. Lagally, T.-M. Lu and G.-C. Wang, in Chemistry and Physics of Solid Surfaces, Vol. II, ed. R. Vanselow, CRC Press, Boca Raton, FL (1979); M.G. Lagally, T.-M. Lu, and G.-C. Wang, in Ordering in Two Dimensions, ed. S. Sinha, Elsevier (1980).

30. J.H. van der Merwe, in Chemistry and Physics of Solid Surfaces, Vol. II ed. R. Vanselow, CRC Press, Boca Raton, FL (1979).

31. G. Lehmpfuhl and Y. Uchida, Ultramicroscopy $\underline{4}$, 275 (1979).

32. J. A. Venables, Ultramicroscopy $\underline{7}$ (1981), in press.

33. M. G. Lagally and J. A. Martin, Revs. Sci. Instr. (1982), to be published.

34. E. Bauer, private communication.

Characterization of Near Surface Regions in Irradiated

Ni(Si) Alloys

L. E. Rehn, R. S. Averback and P. R. Okamoto[*]
Materials Science Division
Argonne National Laboratory
Argonne, IL 60439

Transmission electron microscopy, Rutherford backscattering spectrometry, Auger electron spectroscopy, and infra-red pyrometry have been used to characterize the radiation-induced growth of Ni_3Si coatings on the surfaces of irradiated Ni(Si) alloys. Results from each of the four techniques are presented. The advantages and limitations of the techniques with regard to obtaining quantitative microstructural information are compared. Finally, AES measurements are reported which show that silicon concentrations significantly in excess of stoichiometric Ni_3Si are induced in regions very near to the surfaces of Ni-12.7 at. % Si specimens during irradiation.

[*]Work supported by the U. S. Department of Energy.

Introduction

Irradiation at elevated temperatures is known to cause nonequilibrium segregation of many alloying components (1). A particularly striking example of this radiation-induced segregation is the formation of ordered Ni_3Si coatings on the external surfaces of irradiated Ni(Si) alloys (2). Strong coupling between silicon and the defect fluxes produced during irradiation causes silicon to be swept toward defect sinks, such as the external surface. When the local enrichment of silicon in the near-surface region exceeds the solubility limit, precipitation of Ni_3Si occurs. As additional silicon is transported from the specimen interior, the Ni_3Si coating grows.

We have employed four experimental techniques to characterize the growth of these γ', Ni_3Si surface coatings during irradiation. In the following four sections of this paper, we present our results obtained via transmission-electron-microscopy (TEM), Rutherford backscattering spectrometry (RBS), Auger electron spectroscopy (AES), and infra-red pyrometry (IRP). Because of the variety of techniques which were used, only a brief description of each is given. However, appropriate references are cited. The discussion emphasizes the advantages and disadvantages of each technique with regard to understanding various microstructural aspects of the coatings.

Transmission Electron Microscopy

Figures 1-3 are TEM micrographs of Ni-12.7 at. % Si specimens which were annealed for 6 hours at 950°C, water quenched to room temperature and irradiated at 590°C with 3 MeV Ni^+ ions. Details of the irradiation procedures can be found elsewhere (3). After irradiation, specimens thinned from the unirradiated side were examined by TEM. Care was taken to minimize material removal from the irradiated surface during thinning. A contiguous surface coating of the Ll_2 ordered phase, Ni_3Si, was found on specimens irradiated to peak doses between 0.7 and 15 dpa. Figure 1b is an <001> diffraction pattern taken from the surface coating of a sample irradiated to 3 dpa. The superlattice reflections of the Ll_2 ordered structure are clearly visible. The corresponding many-beam bright field image is shown in Fig. 1a. The γ' coating

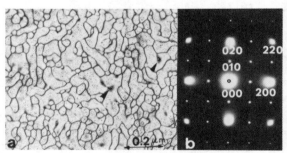

Fig. 1. Ni_3Si surface coating formed on a Ni-12.7 at. % Si specimen irradiated at 590°C to a peak dose of 3 dpa with 3 MeV Ni^+ ions. (a) Bright field image showing the complete cellular network of the (a/2)<110> type APB's. (b) Diffraction pattern showing γ' superlattice reflections. This pattern was used to form the bright field image in (a).

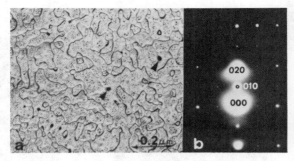

Fig. 2 (a) Dark field image of the same area as in Fig. la, formed
with the [010] superlattice reflection. (b) Diffraction
pattern for Fig. 2a.

contains a cellular network of (a/2)<110> type antiphase boundaries
(APB's) and is in epitaxy with the underlying substrate. The cellular
APB structure is complete since all three unique (a/2)<110> APB's are
visible under many-beam bright field imaging conditions. Only the
(a/2)<110> and (a/2)<011> APB's are visible in the g<010> dark field
image of the same area shown in Fig. 2a. The local deviations in the
background intensities of these micrographs indicate that the coating
thickness is not perfectly uniform. Except for the abrupt termination
of the APB's at the back edge of the coating, no strong feature
delineates the coherent interface between the coating and substrate.
Estimates of the coating thickness which were made using
stereomicroscopy and the APB's as markers range from 10 to 20 nm.

Coatings formed on specimens irradiated to lower dose levels are
highly nonuniform in thickness. Figure 3 is a dark field image, taken
with an <010> superlattice reflection, of the coating on a specimen
irradiated to a peak dose of 0.7 dpa. Stereomicroscopy shows that the
very bright particles are parts of the APB structure which are
considerably thicker than the rest of the coating. The APB structure in
Fig. 3 is much finer than that in Fig. 2a. The dark regions in Fig. 3
are areas of the coating that were etched away during thinning, exposing
the underlying alloy substrate. The absence of γ' particles in these
areas indicates the existence of a γ'-denuded zone beneath the γ'
surface coating. Figure 3 also shows that the APB contrast almost
disappears in the center of the etched out area, A. Therefore a lower
bound for the coating thickness can be obtained from the depth of this
area. Stereomicroscopy measurement yields a value of ~5 nm. The
brighter particles may be considerably thicker than 5 nm (4).

Additional characteristics of the Ni_3Si surface coatings were
noticed during the course of this work. The bright and dark field
images shown in Figs. 1 and 2 reveal a finely dispersed background of
tiny spherical particles. These precipitates always appear darker than
the background in dark field images taken with <100> and <110> type
superlattice reflections and hence they are not γ' particles. They are
barely visible in bright or dark field images formed under well defined
two-beam conditions using fundamental reflections. Stereomicroscopy
shows that these precipitates are not dispersed evenly throughout the γ'
surface film, but tend to cluster near the back of the coating. A
tentative interpretation is that these precipitates are particles of
solid-solution γ in an ordered γ' matrix.

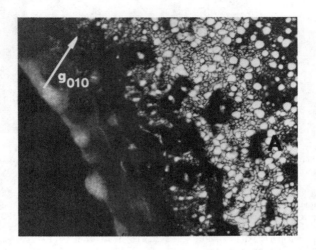

0.2 μm

Fig. 3. Ni_3Si surface coating formed on Ni-12.7 at. % Si irradiated to
0.7 dpa. The dark regions are areas of the coating that have
been etched away during thinning. The absence of γ' particles
in these regions indicates the existence of a γ' denuded zone
behind the γ' surface film. Note that the sizes of the
antiphase domains are much smaller than in Fig. 1b.

<u>Rutherford Backscattering Spectrometry</u>

RBS is used to profile the near-surface composition of materials as
a function of depth. The fundamental and practical aspects of RBS have
been well described elsewhere (5,6). Basically, RBS involves
irradiating a material with energetic light ions, typically 1-3 MeV
helium ions, and energy analyzing a fraction of the backscattered
particles. The incident particles lose energy via two channels,
electronic excitations and elastic collisions. The first channel, which
is characterized by the stopping power, provides depth analysis
information; the second, which is characterized by the kinematics of the
collision, provides mass analysis information. Quantitative analysis is
based on the equation (5),

$$E = K_A(E_o - \frac{\Delta x}{\cos\theta_1} \frac{dE}{dx}|_{in}) - \frac{\Delta x}{\cos\theta_2} \frac{dE}{dx}|_{out} \quad . \qquad (1)$$

Here E is the energy of the backscattered particle, E_o is the energy of
the incident particle, $\theta_1(\theta_2)$ is the angle between the surface normal
and the direction of the incident (backscattered) particle, and $dE/dx|_{in}$
and $dE/dx|_{out}$ are the average stopping powers at the incident and
backscattered energies, respectively. K_A, the kinematic factor, is the
fraction of the incident energy which is retained by the particle after
backscattering from target atom A, and Δx is the depth at which the
backscattering collision occurs. The stopping powers for helium in most
elements are accurately known and K_A can be calculated exactly.

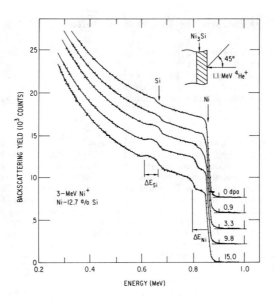

Fig. 4. RBS spectra for 1.1 MeV ^{4}He ions incident on five different Ni–12.7 at. % Si specimens which had been irradiated at 590°C to the indicated doses.

Figure 4 shows backscattering spectra for five Ni–12.7 at. % Si specimens obtained after irradiation at 590°C with 3 MeV Ni$^+$ ions to the indicated doses. The zero of the vertical axis of each RBS spectrum in Fig. 4 has been shifted for clarity. A beam of 1.1 MeV He$^+$ particles and a scattering angle of 135° were used to obtain the spectra. The composition of the unirradiated (0 dpa) alloy can be determined from its RBS spectrum using the expression,

$$\frac{H_{Si}^{\gamma}}{H_{Ni}^{\gamma}} = \frac{X}{(1-X)} \frac{\sigma_{Si}}{\sigma_{Ni}} . \tag{2}$$

The quantity on the left is the ratio of the measured backscattered yields from silicon and nickel atoms at a given depth; σ_{Si} and σ_{Ni} are the total cross sections for scattering of an incident He ion from a silicon and a nickel atom, respectively, into the solid angle defined by the detector; X is the atomic fraction of silicon in the material. In this way, the near-surface region of the unirradiated alloy was found to contain 12±3 % Si. The spectra from the irradiated specimens all show enhanced silicon concentrations in the near-surface region. The silicon enhancement is evident both in the increased backscattering yield from silicon atoms located at the surface (E = ~0.7 MeV), and the decreased yield from nickel atoms at the surface (E = ~0.85 MeV). Analysis (7) of the spectra based on the decreased yield from nickel atoms at the surface yields a surface layer composition of 25±4 % Si. The thickness of the layer can be determined directly from eq. (1) which can be recast in the form

$$\Delta E = E_2 - E_3 = \left[\frac{K_{Si}}{\cos\theta_1} \frac{dE}{dx}\Big|_{in} + \frac{1}{\cos\theta_2} \cdot \frac{dE}{dx}\Big|_{out} \right] \cdot \Delta x \; . \tag{3}$$

Here, ΔE is the difference in the energies of particles scattered from the front and back edges of the layer. The stopping powers in eq. (3) are for the compound. Generally the stopping powers of compounds have not been determined and must be approximated using Bragg's rule, which assumes the linear additivity of stopping cross sections (5).

$$S_{Ni-Si} = (1-X)S_{Ni} + X \; S_{Si} \tag{4}$$

Here, $S = (1/n)dE/dx$, and n is the atomic density. By rotating the specimen to the geometry indicated in Fig. 5, the depth resolution can be significantly improved. Spectra in this high resolution geometry for the same set of specimens analyzed in Fig. 4, are shown in Fig. 5. Again, the vertical axis zero of each spectrum has been shifted. The enhanced depth resolution is a result of the increased pathlength of the helium ions in the material. Limitations on the resolution that can be obtained by RBS are discussed in reference 6. Employing eq. (3), the thickness of the layer is found to increase approximately as the square root of the dose, the thickest layer being 44 nm.

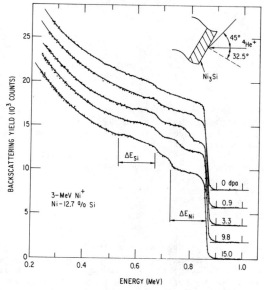

Fig. 5. RBS spectra from the same specimen (Fig. 4) in a higher resolution geometry.

Auger Electron Spectroscopy

AES yields a derivative with respect to energy of the number of Auger electrons ejected from a thin (0.5-2 nm) surface layer by a primary excitation beam (8,9). As a first approximation, the concentration of a particular element may be assumed to be proportional to the peak-to-peak height of this derivative at a characteristic transition energy of the element. A concentration versus depth profile is obtained by sputtering and measuring the Auger spectra at various time intervals. The accuracy of the depth measurement is determined by the accuracy to which the sputtering rate can be determined. Preferential sputtering of individual alloying components can produce erroneous concentration profiles. It can further complicate the analysis by causing the sputtering rate to vary with alloy composition. However, previous experience has shown that preferential sputtering is not a significant problem for the Ni(Si) alloy system (10).

The ratios of the AES, Si to Ni, peak-to-peak amplitudes which were measured for three irradiated Ni-12.7 at. % Si specimens are shown in Fig. 6 as a function of sputtering time. Irradiation parameters are included in the figure. A primary electron beam of 5 kV and 10 μA was used to excite the Auger transitions. The transitions at 848 eV for nickel and 1619 eV for silicon were monitored. The diameter of the analysis beam at the specimen surface was ~100 μm and a 1 kV Ar^+ beam, ~2 mm^2 in area, was used for sputtering. Peak-to-peak ratios of 30.5×10^{-3} were measured in all three specimens after sputtering for long times. This ratio is shown by the dotted line in Fig. 6, and represents the bulk concentration of 12.7 at. % Si. Larger ratios indicate regions of silicon enrichment, smaller ratios, silicon depletion.

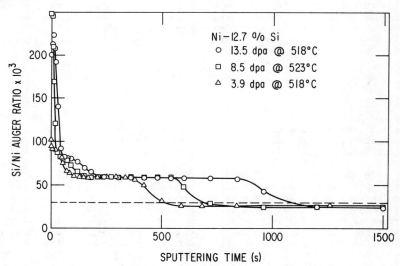

Fig. 6   Si/Ni AES ratios as a function of sputtering time for three Ni-12.7 at. % Si specimens irradiated to different doses at a nominal temperature of 520°C. The dotted line is the AES ratio obtained from the bulk alloy.

Each of the profiles in Fig. 6 shows an enriched layer of constant composition which begins slightly beneath the surface, and which extends to increasingly greater depths as the irradiation dose is increased. Assuming that the ratio of Si to Ni peak-to-peak heights varies linearly with the ratio of the Si to Ni concentrations, and using the ratio of 30.5×10^{-3} measured for the bulk alloy, a concentration of

22±1 at. % Si is found for the layer of constant composition. The reported stoichiometry range for Ni_3Si is 22 to 24 at. % Si. Beneath this silicon enriched layer lies a region which has been depleted to ~10 at. % Si and which also extends to increasingly greater depths (not shown in Fig. 6) as the irradiation dose increases. Because the return of the silicon concentration from ~10 at. % to the bulk value of 12.7 at. % is quite gradual, it is not possible to define precisely the extent of the depleted region. Very near to the surface, silicon concentrations considerably in excess of the 22 at. % found for the Ni_3Si layer are observed. Subsequent to the AES analysis, measurements were made of the total sputtered depth at the analyzed area. Dividing this depth by the total sputtering time yields a sputtering rate of 0.05±.005 nm/s. Assuming this same sputtering rate for the entire enriched region yields coating thicknesses of 22, 31 and 48 nm for the samples irradiated to doses of 3.9, 8.5 and 13.5 dpa, respectively. Again, we see that the coating thickness increases approximately linear with the square root of the irradiation dose.

Infra-Red Pyrometry

Significant changes in the infra-red emissivity of Ni(Si) alloys due to the formation and growth of the radiation-induced Ni_3Si surface coatings have been reported previously by Potter et al. (4). As a coating nucleates and grows, the emissivity of the specimen increases from that characteristic of the solid-solution, γ phase to that of the ordered, γ' phase. Neglecting multiple reflections within the coating, the intensity, $E(\ell)$, of the infra-red radiation which is emitted at a given temperature by a sample with a surface film of thickness ℓ, is given by (11)

$$E(\ell) = E_\gamma e^{-K\ell} + E_{\gamma'}(1-e^{-K\ell}) \ .$$

E_γ and $E_{\gamma'}$ are the infra-red intensities emitted from a bulk γ and a bulk γ' sample, respectively. Knowledge of the optical constants of ordered Ni_3Si is necessary for calculation of the absorption coefficient, K.

In the present experiments, an infra-red pyrometer sensitive to wavelengths near 2 μm and temperature changes of ~0.1°C was used to monitor the emmisivity. During the irradiation of a Ni-6 at. % Si specimen with 3 MeV Ni^+ ions to a peak dose of 14 dpa at 537°C, the pyrometer registered an apparent increase in the specimen temperature (ΔT_a) of 21±2°C. A thermocouple attached to the sample showed that the actual sample temperature remained constant during the irradiation. The increase in apparent temperature recorded by the IRP reflected the fact that the emissivity of the specimen had increased as a result of Ni_3Si formation near the irradiated surface.

Immediately after the irradiation, the specimen was cooled to room temperature. It was then annealed for about one hour at each of the following temperatures, 525, 530, 542, 550 and 556°C. During these isothermal anneals, the increase in IRP temperature which occured during irradiation decreased in an approximately exponential fashion. These results are shown in Fig. 7, where the decrease in apparent temperature is plotted as a function of time for the five anneals. The total decrease in apparent temperature during the anneals was 20°C, which agrees well with the 21°C increase observed during irradiation. The

276

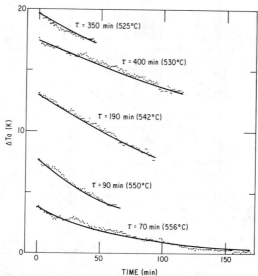

TIME (min)

Fig. 7. IRP apparent temperature change (ΔT_a) as a function of annealing time at the indicated temperatures. The total apparent temperature change induced in this Ni-6 at. % Si specimen during irradiation was $21\pm2^\circ$C.

time constant of the observed exponential decay at each temperature is included on the figure. Since the solubility of Si in Ni is ~10 at. % at these temperatures, the γ' coating on this 6 at. % Si specimen dissolves in the absence of irradiation, and the emissivity decreases. The measured time constants for the decay indicate an activation energy for dissolution of the Ni_3Si film which is close to the self-diffusion energy of nickel (12).

Discussion

In this section, the results from the four techniques are compared on the basis of the compositional and structural information they provide. In addition, the effects of certain types of measurements on subsequent analysis by other techniques, and the advantages of doing multiple kinds of measurements are discussed.

Three techniques, TEM, RBS and AES, show directly that well defined coatings are formed at the specimen surface during irradiation. IRP demonstrates that the specimen emissivity increases during irradiation, but in the absence of the additional information from the other techniques, the cause would remain speculative. The TEM diffraction pattern provides quite conclusive identification of the coating as ordered (γ') Ni_3Si. The AES and RBS results do not contain information about the degree of order of the γ' phase. However, the chemical compositions determined by AES and RBS are consistent with this interpretation. The complexity of the Ni(Si) binary phase diagram prevents a conclusive identification of the coating solely on the basis of the measured composition. Of the present results, only the AES profiles exhibit sufficient sensitivity to concentration changes to detect the Si depleted region which lies beneath the coating.

The best lateral spatial resolution, 2-5 nm, is obtained from
TEM. The present RBS, AES and IRP results are averages over analyzed
areas of ~1 mm^2, 100 µm^2 and 1 mm^2, respectively. RBS and AES, however,
are superior for determination of the coating thickness. In fact, since
no strong contrast was evident for the back edge of the coating, only a
qualitative estimate of the coating thickness could be obtained by
TEM. Since the coating absorption coefficient, K, is unknown, a
thickness determination by IRP was not possible. The depth resolution
of the combined AES-sputtering procedure decreases with increasing depth
from the surface. A commonly quoted estimate (13) of the resolution of
techniques which employ sputter-profiling is 5-10% of the sputtered
depth for situations with no significant preferential sputtering
effects. The accuracy of the present AES measurement of coating
thickness is also affected by the error (±10%) in the sputtering rate
measurement, and the fact that the analysis ignores any difference in
sputtering rates for the γ and γ' phases. Errors in the stopping power
values limit the accuracy of the RBS thickness determination to about
±5%. Because of energy straggling, the depth resolution of RBS also
decreases with increasing depth. A further complication which is
inherent in the thickness determination by any technique is a matter of
definition, since three techniques, TEM, RBS and AES indicate that the
inner edge of the coating is not a perfect plane. We performed normal
resolution RBS measurements on the same specimens which were
subsequently used to obtain the AES results in Fig. 6. RBS yielded
thicknesses of 18, 28 and 34 nm, which can be compared to the AES values
of 22, 31 and 48 nm. Hence, the average film thicknesses determined by
the two techniques on these three specimens are 20, 29.5 and 41 nm. The
average deviation of each measured value from these averages is ±11%.
Since the assumptions which are required for thickness determinations by
the two techniques are independent of each other, ±11% is considered to
be a reasonable measure of the accuracy of the present AES and RBS
results.

The IRP results are clearly inferior to those obtained from the
other three techniques for characterization of the coatings. However,
IRP does have the definite advantage that it can be conveniently
performed at temperature, during irradiation. This capability is
particularly useful for studying coatings near the high temperature end
of the growth regime or for studying the kinetics of the coating
growth. The TEM, RBS and AES measurements reported in this paper were
all made on specimens which had been irradiated at elevated temperature,
then cooled to room temperature. During cooling, an interval of 1-2
minutes elapses before the specimen temperature drops by more than
100°C. The IRP results in Fig. 7 show that significant changes in the
coatings can result from annealing during cooling from temperatures
above ~600°C. Since IRP can be used to make measurements during
irradiation, the results are free of any post-irradiation annealing
effects. The annealing results in Fig. 7 also demonstrates the time-
saving advantage of IRP measurements. Although similar information
might have been obtained by either RBS or AES, the requirement for doing
room temperature measurements would have greatly increased the number of
specimens and/or the analysis time. An example of the use of IRP to
determine the response of coatings to sudden changes in irradiation
conditions can be found elsewhere (14).

The destructive nature of certain techniques, e.g., sputter-
profiling, which prevent subsequent analysis by any other technique is
evident. However, it may not be so evident that certain techniques are
only selectively destructive. A selectively destructive technique

278

prevents further analysis by some but not all of the other techniques. For example RBS is often considered to be nondestructive. In the present set of experiments, AES profiles were obtained on some specimens after RBS measurements, demonstrating that RBS is indeed not destructive relative to subsequent AES analysis. However, another specimen which had been examined previously by RBS was thinned for examination by TEM. No superlattice γ' reflections were found in the diffraction pattern from this specimen. The dose of 1.1 MeV He ions used for the RBS analysis was sufficient to completely disorder the γ' at room temperature. Since the concentration profiles were not significantly altered by the disordering process, it was still possible to obtain AES results. However, the specimen was no longer useful for techniques which monitor the state of order of the γ', such as IRP or dark-field TEM.

A summary of the information obtained from the different techniques is given in table I. The values shown in the table are not state-of-the-art limiations, but rather are intended to be representative of the results which are reported in this paper.

Obtaining results from a variety of experimental techniques significantly enhances the degree to which the radiation-induced coatings can be characterized. One advantage of a multiple technique approach is that results obtained on the basis of certain assumptions can be tested by use of another technique based on an independent set of assumptions. An example of this is the agreement in thickness measurements between the present AES results which require certain assumptions about the sputtering process, and the RBS results, which are independent of sputtering effects but which do require an assumption about electronic stopping in a concentrated alloy. Another advantage of multiple techniques is that certain characteristics are revealed by only one technique. As example of this is the present AES result showing that silicon concentrations considerably in excess of the 22–24 at. % range of stoichiometric Ni_3Si are induced very near to the irradiated surface. We digress to emphasize this latter result, since it has important implications for the phase stability of alloys during irradiation (15). In the two highest dose specimens, silicon concentrations of roughly 50 at. % were observed for the first few atom layers. For all three irradiated specimens in Fig. 6, silicon concentrations of ~28–30 at. % are indicated in 2 to 5 nm wide layers immediately on top of the γ' coating. Some influence of surface

Table I. Characteristics of the four experimental techniques.

Technique	Sensitive to Degree of Order	Composition Determination	Thickness Determination	Lateral Resolution	Destructive	Performed In-situ
TEM	Yes	Indirect	Poor	2–5 nm	No	No
RBS	No	20%	10%	1 mm	Selective	No
AES	No	10%	10%	100 μm	Yes	No
IRP	Yes	Indirect	No	1 mm	No	Yes

contaminants, e.g., C and O, is to be expected in the AES analysis. However, the regions for which the 28-30 at. % Si concentrations are indicated extend beyond the depth where significant surface contamination is observed.

Reference

1. P. R. Okamoto and L. E. Rehn, J. Nucl. Mater. 83 (1979) 2.

2. D. I. Potter, L. E. Rehn, P. R. Okamoto and H. Wiedersich, Scripta Met. 11 (1977) 1095.

3. A. Taylor, J. R. Wallace, D. I. Potter, D. G. Ryding and B. Okray Hall, Radiation Effects and Tritium Technology for Fusion Reactors, CONF-750989, Vol. I (1975) 158.

4. D. I. Potter, P. R. Okamoto, H. Wiedersich, J. R. Wallace and A. W. McCormick, Acta Met. 27 (1979) 1175.

5. W.-K. Chu, J. W. Mayer and M.-A. Nicolet, Backscattering Spectrometry, Academic Press, New York, 1978.

6. J. S. Williams and W. Möller, Nucl. Instr. and Meth. 157 (1978) 213.

7. Ref. 5, page 130.

8. C. C. Chang, in Characterization of Solid Surfaces, eds. P. F. Kane and G. B. Larrabee (Plenum Press, New York, 1974) 509.

9. L. E. Davis, N. C. McDonald, P. W. Palmberg, G. E. Riach and R. W. Weber, Handbook of Auger Electron Spectroscopy, 2nd ed. (Physical Electronics Inc., Eden Prairie, MN, 1976).

10. L. E. Rehn, P. R. Okamoto, D. I. Potter and H. Wiedersich, J. Nucl. Mater. 74 (1978) 242.

11. O. S. Heavens, in Physics of Thin Films, eds. G. Hass and R. Thun, Vol. 2 (1964) 208.

12. L. E. Rehn, unpublished.

13. J. W. Mayer and A. Turos, Thin Solid Films, 19 (1973) 3.

14. L. E. Rehn, P. R. Okamoto, D. I. Potter and H. Wiedersich, J. Nucl. Mater. 85 and 86 (1979) 1139.

15. H. Wiedersich, this volume.

Radiation Channeling and Scattering

CHARACTERIZATION OF IRRADIATED METALS BY THE ION CHANNELING TECHNIQUE*

S. T. Picraux
Sandia National Laboratories[†]
Albuquerque, New Mexico 87185

The application of ion channeling to the study of irradiation damage in metals is reviewed. In combination with ion backscattering this technique gives depth information about the defects present within the first few micrometers of the surface; in many cases, the depth distribution of the defects present can be obtained, with depth resolutions ~100Å. In addition, for impurity-associated defects the crystallographic site of the impurity can often be determined by channeling location measurements. Quantitative analysis of the defect profile requires that a single type of defect dominates the scattering of particles out of channeled trajectories and this process is characterized by two defect-specific quantities: the direct scattering factor and the dechanneling factor. The channeling technique has been shown to be sensitive to a wide range of defects, including interstitials, dislocations, stacking faults, microtwins and amorphous clusters, and is most powerful when used in combination with the complementary technique of transmission electron microscopy (TEM). The application of ion channeling to the characterization of irradiated metals is illustrated here by four examples: 1) dislocation profiling in ion-implanted aluminum; 2) amorphous layer formation in Ti-implanted iron; 3) blister formation in He-irradiated niobium; and 4) deuterium trapping at defects in D-implanted iron.

*This work supported by the U.S. Department of Energy, DOE, under contract DE-AC04-76-DP000789.

[†]U.S. Department of Energy facility.

Introduction

Ion channeling (1,2) has been used to study lattice disorder in metals, semiconductors and insulators, and to study the site location of solute atoms in association with defects. The channeling technique is usually used in combination with ion backscattering, which often allows disorder profiles to be determined with depth resolutions of $\geq$ 100Å to probing depths of $\lesssim$ 10 μm (3). Thus the technique is ideally suited to probe near surface disorder resulting from ion implantation, surface layer growth, or other surface treatments. While there have been many studies of ion implantation disorder in semiconductors (4), little quantitative work has been carried out on irradiated metals by the channeling technique.

Typicaly, ion channeling experiments utilize a Van de Graaff accelerator to produce monoenergetic helium or proton beams of energies 0.3 to 3.0 MeV. The beam is collimated to within ~ 0.05^0, fwhm, and the crystal is oriented with a goniometer to a similar angular resolution. For greatest versatility the Rutherford backscattering of the incident beam of particles is detected and energy analyzed by a gold surface barrier detector. In some cases ion-induced x-rays or nuclear reaction products are also monitored to enhance elemental sensitivites. Detailed descriptions of the experimental configurations are given elsewhere (3).

The ion channeling technique derives its sensitivity from the enhanced scattering of channeled particles into nonchanneled trajectories that results when atoms are displaced from perfect host-atom sites. The influence of defects on the channeled spectrum depends on the defect configuration and contrasts markedly, for example, between dislocations, amorphous clusters and interstitials in specific sites. The channeling technique is highly complementary to transmission electron microscopy (TEM) studies of defects and many detailed disorder studies today by ion channeling also include TEM analysis. While both techniques are structural probes which detect deviations from the perfect periodicity of the crystal, there are important differences. TEM is best suited to identify the defects present and determine their lateral distribution, but it is primarily sensitive to larger defects, such as clusters of point defects, and to extended defects, such as dislocations. Ion channeling is best suited to determine the number density and depth profile of the defects, and maintains its sensitivity down to a defect size corresponding to single atom displacements $\geqslant$0.1Å. When solute atoms are involved, as in solute-defect complexes, it is sometimes possible to determine the solute lattice site by channeling, and thereby further characterize the defect configuration (2,5).

In this review, the principles of the ion channeling technique used to detect and quantitatively determine the depth profile of defects in single crystals will be outlined. Then selected examples will be given from the area of irradiation damage studies in metals to illustrate the range of applications of the technique to this area. Finally, the relative advantages and limitations of the technique will be discussed.

Disorder Analysis by Channeling

Perfect Crystal

For a perfect crystal, alignment of an energetic beam of ions with a major crystal direction results in most particles undergoing special "channeled" trajectories as they oscillate between the rows or planes of atoms (1). These particles are reflected off the atom rows or planes much like a stone skipping off water, with a minimum impact parameter of

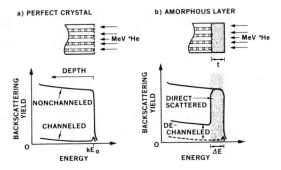

Figure 1 – Illustration of backscattering–channeling spectra for a) a perfect crystal, and b) a crystal covered by an amorphous layer.

$\approx 0.1\text{Å}$ and angles $\leqslant 1^0$. Thus, the channeled particles do not come close enough to nuclei to undergo Rutherford backscattering, and only dechanneled particles are detected in a backscattering measurement. Some particles are dechanneled even in a crystal without defects, due to scattering by the exposed rows or planes as the particles enter the crystal, and also due to multiple scattering of particles out of channeling directions by the electrons and by the vibrating host atoms, distorted from equilibrium lattice positions by their thermal vibration. Typically, ~95 to 98% of the particles are channeled near the surface for major axial directions and ~60 to 90% for major planar directions.

Since the particles steadily lose energy due to electronic excitations as they penetrate the solid, the measured backscattering signal as a function of energy can be converted from an energy scale to a depth scale (3). A schematic diagram of the backscattering spectra for a perfect crystal is shown in Fig. 1a. For a nonchanneling direction, the scattering yield is proportional to the host atom concentration. The yield for the beam incident along the channeling direction gives a measure of the dechanneled fraction which increases as a function of depth. For channeling analyses the scattering yield along the channeling direction is usually normalized at each depth by the nonchanneled scattering yield.

Crystal with Defects

Distortions in the crystal structure, resulting from defects, scatter channeled particles into nonchanneling trajectories. This process is referred to as dechanneling. In addition, when local displacements are greater than ~0.1Å the channeled particles may be directly backscattered to the detector, and this process is referred to as direct-scattering. These two processes are illustrated schematically in Fig. 2 and are the basis for the detection of disorder by the ion channeling technique.

As seen in Fig. 2, certain types of defects such as isolated interstitials contribute to both dechanneling and direct scattering processes. In contrast, other defects, such as the edge dislocation, involve primarily a large region of small local distortions in the atom positions, which are too small for direct scattering but give appreciable dechanneling. These two defect-specific quantities, direct scattering and dechanneling, are thus the required input for the analysis of radiation damage by the channeling technique.

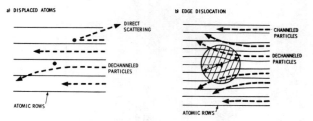

Fig. 2 - Schematic of direct scattering and dechanneling of channeled
particles by a) displaced atoms, and b) edge dislocation.

Direct Scattering by Defects

Direct scattering is illustrated schematically by the channeled spectrum shown in Fig. 1. First, for a perfect crystal oriented in a nonchanneling direction, all the atoms at a given depth in the crystal contribute equally to the backscattering yield. For monoatomic material, a
uniform change in the scattering yield versus energy is observed, with the
surface corresponding to the high energy edge of the spectrum and increasing depths corresponding to decreasing energies. When the beam is aligned
along the channeling direction, direct scattering occurs for atoms at the
surface of the crystal. The backscattering at lower energies corresponds to
that fraction of the particles which are dechanneled even in a perfect
crystal.

If the surface of the crystal is covered by an amorphous layer (Fig.1b),
then in the high energy region the channeled spectrum is identical to that
for a nonchanneling orientation. The energy width of this region corresponds
to the amorphous layer thickness. Direct scattering is allowed for all
atoms in this region. After passing through the amorphous layer to the
underlying perfect crystal, the channeled yield is higher than in the perfect crystal because in penetrating the amorphous layer some of the particles
have been scattered to forward angles too large to remain channeled. This
is the dechanneling contribution from the displaced atoms in the amorphous
layer. Some dechanneling always accompanies the direct scattering process.

The direct scattering contribution can be characterized through a
parameter f, which we define as the fractional contribution to direct backscattering in terms of the effective number of atomic scattering centers per
defect. For example, for randomly distributed interstitial atoms $f = 1$, or
for a dispersion of amorphous clusters containing ten atoms per cluster,
$f = 10$.

Dechanneling by Defects

The influence of dechanneling on the scattering spectrum in the absence
of direct scattering is illustrated in Fig. 3 for a network of edge dislocations. In Fig. 3a, the TEM micrograph shows the regular array of misfit
dislocations in Si resulting from a phosphorous diffusion at high concentrations (6,7). The network occurs at a single depth below the surface
(~4500Å)and can be readily seen in the channeling spectra of Fig. 3b
for both the <111> axial and {110} planar directions by the sudden
increase in yield at that depth (7). In the absence of direct scattering the
height of the channeled yield is directly proportional to the fraction of the
beam which has been dechanneled. Thus, the sudden step indicates a region of
rapid dechanneling due to the dislocation network.

The dechanneling process can be characterized by a dechanneling factor, σ_D, which is characteristic of the particular defect present. The physics of the dechanneling process is then contained within this dechanneling factor and the incremental probability for dechanneling, dP_D, over an interval of depth dz is given by

$$\frac{dP_D}{dz} = \sigma_D \, n_D(z) \quad , \tag{1}$$

where n_D is the density of defects at depth z. For many defects, theoretical estimates of σ_D are available (8); alternatively, σ_D can be measured in a calibration sample for which the total density of defects is known. For example, for the dislocation dechanneling case of Fig. 3, the dechanneling factor can be expressed as a dechanneling width per unit length of dislocation, $\sigma_D \approx \lambda$. It is observed for this orientation (Fig. 2b) that $\lambda \sim 50\text{Å}$ for the axial direction and $\lambda \sim 350\text{Å}$ for the planar direction In general σ_D depends on the channeling parameters and, for example, $\lambda \propto (E/Z_1Z_2)^{1/2}$, where E is the energy of channeled particles of atomic number Z_1 incident on a target of atomic number Z_2 (8).

Defect Depth Profiles

When the measured yields for channeling are normalized by the corresponding nonchanneled yield they are referred to by χ, the fractional yield along the channeling direction. For disorder profile analysis we treat the beam in the crystal as composed of two components, the channeled and the dechanneled fractions. As illustrated in Fig. 4, for the virgin (undamaged) crystal the dechanneled fraction is given by χ_V and for a crystal containing disorder it is given by χ_R. Since the sum of the dechanneled and channeled fractions of the beam is unity, the channeled fraction in the disordered crystal is $1 - \chi_R(z)$ and the dechanneling at depth z is simply given by the total fraction of the beam which is channeled times the probability for dechanneling,

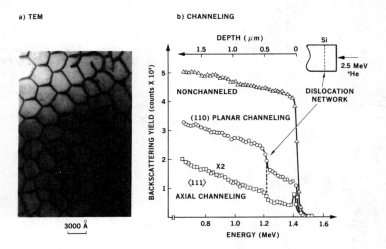

Fig. 3 – Analysis of P-diffused silicon by a) TEM, and b) 2.5 MeV [4]He channeling along {110} planar and ⟨111⟩ axial directions. Misfit dislocation network is observed at a depth of 4500Å (4).

287

$$\frac{d\left[1 - \chi_R(z)\right]}{dz} = -\left(1 - \chi_R(z)\right) \ P_T(z) \quad , \tag{2}$$

where the probabilities for dechanneling in a disordered and a perfect crystal are assumed to be additive, $P_T \simeq P_D + P_V$. This together with the fact that the dechanneled fraction is initially the same in the perfect and defected crystal at $z = 0$ gives

$$\chi_R(z) = \chi_V(z) + \left(1 - \chi_V(z)\right) \ \left[1 - \exp\left(-\int_O^z \sigma_D \ n_D(z')dz'\right)\right] \tag{3}$$

for the dechanneled fraction in the crystal with defects.

The normalized yield for channeling in the crystal with defects (Fig.4) is χ_D and is due to a scattering contribution from the dechanneled fraction of the beam by all the atoms plus the contribution from the channeled fraction of the beam by those defects which contribute to direct scattering. Then

$$\chi_D(z) = \chi_R(z) + (1 - \chi_R(z)) \ f \ \frac{n_D}{n} \ , \tag{4}$$

where for the channeled fraction of the beam, $1 - \chi_R$, the factor f gives a fractional atomic scattering per defect for a density n_D of defects in a crystal of total atomic density n. The first term on the right-hand side of Eq. 4 represents the dechanneling contribution and the second term the direct scattering contribution in which channeled particles undergo close encounter collisions at defects and are directly scattered into the detector without first becoming dechanneled.

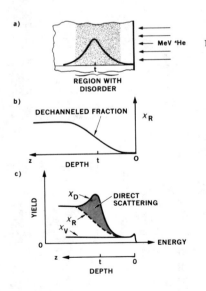

Fig. 4 – Schematic of MeV ^{4}He channeling-backscattering disorder analysis illustrating defect dechanneling and direct scattering contributions to the scattering yield: a) region of disorder below the surface with maximum disorder at depth t ; b) fraction , χ_R, of beam which has been dechanneled in passing through the disorder c) ^{4}He scattering yield vs energy (depth) showing the yield observed for a virgin (χ_V) and the disordered (χ_D) crystal with the dechanneled fraction, χ_R, shown by the dashed line.

The defect scattering factor, f, accounts for any difference between the number of defects and the effective number of scattering centers per defect. This also accounts for the fact that the channeled particle flux density is not uniform across the channel. For example, for interstitial defects which consist of single randomly displaced atoms, $f = 1$, whereas for extended defects such as dislocations which consist primarily of many small distortions of the lattice atoms, the relative contribution of the second term in Eq. 4 is negligible and $fn_D/n \approx 0$, so that for purposes of analysis we may use $f = 0$. At present, detailed depth profile analyses have generally been carried out only for the cases of $f = 0$ and 1. For the case of $f = 1$, the analysis requires an iterative procedure starting at the surface and moving in succeeding increments of depths to determine the dechanneled fraction, χ_R, in the next layer which allows the direct channeling contribution and defect density, n_D, to be determined at that layer (2,4). In cases where the direct scattering is a large contribution relative to the dechanneling in the measured spectrum, the disorder depth profile is essentially directly observed riding on dechanneling background; this is shown by the shaded area in Fig. 4c. In contrast, when the dechanneling contribution is large relative to the direct scattering ($f \approx 0$) then several important simplifications occur and this is referred to as the case of "dechanneling only."

Limiting Case of Dechanneling Only

For many of the defects commonly found in irradiated metals, such as dislocations, stacking faults, bubbles, voids, and other highly strained regions, the relative contribution to direct scattering may be small so that we may use $f \approx 0$. Under this condition

$$\chi_D(z) \approx \chi_R(z) \quad , \tag{5}$$

and the defect density is directly given in terms of the measured normalized yields, χ, for the disordered (D) and virgin (V) crystal, as

$$n_D(z) = \frac{1}{\sigma_D} \frac{\partial}{\partial z}\left[\ln\left\{\frac{1 - \chi_V(z)}{1 - \chi_D(z)}\right\}\right] \quad . \tag{6}$$

Also, the total disorder, N_D, per unit area between the surface and depth $z = t$ is

$$N_D = \frac{1}{\sigma_D} \ln\left[\frac{1 - \chi_V(t)}{1 - \chi_D(t)}\right] \quad . \tag{7}$$

In a limit of small defect concentrations, we obtain for the defect density the approximation

$$n_D(z) \approx \frac{1}{\sigma_D} \frac{d\chi_D}{dz} \quad . \tag{8}$$

In this case the defect concentration is directly given by the slope of the normalized channeling spectrum divided by the dechanneling cross section for that defect.

Impurity-Associated Defects

When impurities or alloying elements are associated with the defects, direct detection of the impurity provides an additional probe of the defect. Heavier mass elements can be sensitively distinguished from the host atoms due to the difference in the scattered energy of the analysis beam. Often

289

intermediate or low mass elements can be detected through their emitted x-rays or through nuclear reactions induced by the incident beam. There are two important advantages of detecting the defect-associated impurity. First, a greater fraction of impurity atoms may be associated with the defects. Therefore, sensitivity to the defects present may be enhanced. For example, if each defect has one impurity atom associated with it and the background concentration of solutionized impurity atoms is small, then the impurity concentration profile directly gives the profile of defect centers. Second, in cases of well-defined impurity atom location, the ion channeling technique of lattice location can be used to determine the crystallographic site of the impurity within the host lattice (5). This gives specific information on the structural nature of the defect center. In certain cases such information cannot presently be obtained by any other technique. Although discussion of the channeling location technique is beyond the scope of this paper, one example of its application will be given for hydrogen-associated defects in the following section. In addition, impurity-associated defect studies by channeling are discussed in the accompanying paper in this proceedings by M. L. Swanson.

Applications to the Study of Irradiated Metals

Analysis of the disorder by ion channeling measurements was shown in the previous section to depend on the nature of the defect through two parameters: the defect dechanneling factor, σ_D, and the direct scattering factor, f. Transmission electron microscopy (TEM) provides an independent determination of the nature of the defects present in many cases and thus is complementary to ion channeling studies. Also, as demonstrated in Figs. 2 and 3, qualitative inspection of the channeling spectra for the presence or absence of a large direct scattering peak often allows one to determine if f is a significant factor for the defects being studied. Finally, quantitative measurements of the dechanneling at a given depth as a function of incident beam energy, E, often provides for a self-consistent check on other interpretations of the defects giving rise to the observed dechanneling. For example σ_D for dislocation lines increases as $E^{1/2}$, for voids is approximately independent of energy and for randomly distributed interstitials or amorphous zones decreases as $\sim E^{-1}$ (2, 4, 8). Knowledge of these energy dependencies can be used to enhance sensitivity to the defects being studied.

Given preliminary measurements to sufficiently characterize the system, the total disorder and its depth profile can be obtained by procedures which have been outlined. High concentrations of disorder are usually required (e.g., $\sim 10^9 - 10^{10}$ dislocations/cm^2, or ~ 1-10% of the host atoms off-site for randomly displaced interstitials) and good depth resolutions (~ 100Å) can be obtained. Double-alignment channeling techniques can often enhance the sensitivity to defects by a factor $\geqslant 10$,(9), but this method is not so convenient in practice. Since channeling measurements are essentially non-destructive, it is often convenient to study the evolution of the disorder as a function of irradiation conditions, material modification treatments, or post-irradiation thermal treatments. Such measurements can be highly complementary to TEM analysis since while a survey of the larger defects can be readily obtained by TEM, it can be quite tedious to obtain quantitative depth information at high defect densities by TEM.

In what follows we give four examples of the application of ion channeling to disorder studies in irradiated metals. All the cases discussed involve charged-particle irradiation, generally referred to as ion implantation. The technique is particularly well suited, though not restricted, to this area of study.

290

Ion Implantation Damage in Aluminum

Ion implantation modification of metals is of interest both for producing favorable properties such as corrosion resistance, hardness, wear resistance and high temperature superconductivity, and also for investigating new metastable materials. The many dpa (displacements per atom) typically produced as the implanted atoms come to rest influence material properties and thus techniques to monitor the resulting radiation damage can help to clarify the origin of the observed material property effects. In addition, the degree to which the damage remains localized in the region of primary energy deposition by the ions is not well understood for the metals. In some metals such as gold (9) and copper (10), the disorder has been reported to extend to depths much greater than the ion range; suggested origins of this effect have included defect migration and the high stresses generated within the implant layer. However, there have been few measurements of the disorder depth profile relative to the ion profiles in metals.

One recent example of such depth profile studies is implanted aluminum (11,12). In aluminum the main defects observed after implantation to $<3 \times 10^{16}$ ions/cm^2 at room temperature are dislocations. An example of a TEM micrograph for Ni-implanted aluminum is shown in Fig. 5a and b for a low, 2×10^{15} Ni/cm^2, and high, 1×10^{16} Ni/cm^2, implantation fluence, respectively.

The network of dislocations in Fig. 5b is of mixed type (i.e., both edged and screw nature). This results from clustering of the mobile vacanices and interstitials produced by the incoming nickel ions to form dislocation loops, (Fig. 5a), and these subsequently grow into tangled dislocation networks.

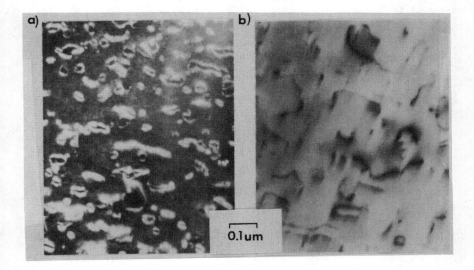

Fig. 5 – TEM micrographis of aluminum implanted to a) 2×10^{15} Ni/cm^2, and b) 1×10^{16} Ni/cm^2 (11)

The {111} planar channeling spectrum corresponding to a 7×10^{15} Ni/cm^2 implant in aluminum is shown in Fig. 6. Also seen is the backscattering spectrum for a nonchanneled direction which provides the nickel depth profile shown in Fig. 7. Since for dislocations $f \approx 0$ and to first order the density is proportional to the slope of the channeled spectrum (Eq. 8), the region of disorder in Fig. 6 corresponds to the region of rapid increase in the {111} yield (0 - 2000Å). The dechanneling cross section per unit length of line dislocation for planar channeling is given approximately by (13)

$$\sigma_D \approx \left(\frac{Eb}{\alpha Z_1 Z_2 e^2 N_P} \right)^{1/2} , \tag{9}$$

where E is the beam energy, b is the magnitude of the Burger's vector, Z_1 and Z_2 the atomic numbers of the projectile and target atoms, N_P the atom density of the planes (equals crystal density times planar spacing) and for randomly oriented dislocations of mixed type the constant $\alpha \approx 8.6$. In the implanted aluminum, the dechanneling was observed to vary with incident beam energy E as given by Eq. 9. The depth profile of the disorder is then given in Fig. 7 using Eqs. 6 and 9.

The disorder for Ni-implanted aluminum (Fig. 7) extends over the same depth region as the nickel atoms. The nickel profile is spread to somewhat deeper depths than predicted by random stopping theory (14), presumably because of some channeling of the implanted nickel. The corresponding spreading is observed in the disorder profile. Also comparison with theory by Brice (14) for the damage energy shows that there is a reduction in the observed disorder within the first ~ 500 to 1000Å, possibly due to the surface acting as a competing sink for defects. Reduction in the enhancements over similar depths have been observed in enhanced diffusion studies of ion irradiated aluminum (15).

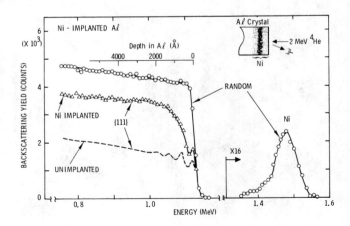

Fig. 6 - Analysis of aluminum implanted at room temperature with 7×10^{15} Ni/cm^2 at 150 keV by 2 MeV ^{4}He channeling (11).

For Zn-implanted aluminum the disorder and zinc profiles are again similar but the total amount of disorder measured by channeling is reduced. Zinc is soluble in aluminum for these implanted concentrations and both Zn and Aℓ-implanted aluminum are observed to give similar disorder as measured by TEM and by channeling analysis (12). Thus, for the zinc implants at room temperature the observed disorder is characteristic of the intrinsic radiation damage effects found for self implantation into aluminum and not strongly modified by the presence of the zinc. In contrast, corresponding fluences of nickel result in solute concentrations greatly exceeding the solid solubility (11). The resulting disorder is observed to be appreciably higher than for the zinc in aluminum as measured by channeling, although similar dislocation densities are observed by TEM. The implantation disorder is localized in the region of the implanted nickel and has an energy dependence characteristic of heavily strained regions where dechanneling predominates. This suggests that the implanted nickel atoms serve to stabilize additional disorder, possibly small Ni-Aℓ clusters not yet observed by the TEM measurements.

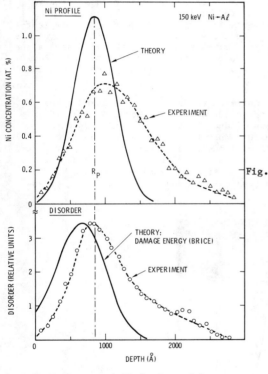

Fig. 7 – Results of analysis of backscattering–channeling spectra of Fig. 6 for disorder and nickel profile in aluminum. Also given are the theoretical profiles by Brice for nickel atoms and energy into atomic displacements (14).

Amorphous Layers on Ti-Implanted Iron

High fluence ion implantation has been used to form metastable amorphous alloys on the surfaces of metals (16). Studies to determine which systems become amorphous are of fundamental interest because the ion damage cascade can be viewed as an ultra-fast quench ($\sim 10^{15}$ °K/s) and essentially any element in the periodic table can be implanted. Also there is interest in the surface properties which can be achieved due to the basic characteristics of amorphous alloys, e.g., improved corrosion resistance, harder surfaces and wear resistance.

A recent study of amorphous layer formation on Ti-implanted iron illustrates the value of combined ion channeling and TEM analysis for understanding irradiation disorder effects (17). It was demonstrated that in this system the amorphous phase was not due to the usual binary combination of the implanted and target species. Instead, the amorphous layer contained carbon in addition to titanium and iron.

After high fluence titanium implantaticn the surface layer was shown to be amorphous by TEM diffraction analysis, as shown in Fig. 8. While at the highest fluences the layer was uniformly corverted into amorphous material, lower fluence implantations left isolated crystallites of α-iron as shown in a dark field micrograph of Fig. 9a. By channeling analysis (Fig. 9b) it was found that the layer was initally much thinner than the implanted Ti range and grew inward from the surface with increasing titanium implantation fluence. This suggested that another, surface-limited step such as impurity introduction was responsible for the formation of the amorphous layer. Ion-induced nuclear reaction analysis confirmed that carbon (and not oxygen) was incorporated into the implanted layer, thus giving rise to a ternary amorphous alloy. The carbon was believed tc be introduced from vacuum hydrocarbons deposited on the surface which are broken down by the implanting beam and subsequently diffuse into the sample. The ternary nature of the amorphous phase was directly confirmed by showing that carbon implants alone did not convert the layer into an amorphous phase but that combined titanium and carbon implants could form a much deeper amorphous layer corresponding to the full ion ranges (17).

The channeled spectra can be analyzed to determine the quantity of amorphous material vs. depth and selected results are shown in Fig. 10. Detailed analysis would utilize Eqs. 3 and 4 with f = 1, where n_D/n is the fractional number of atoms which do not have long range order and therefore correspond to the quantity of amorphous material present at that depth. For small amounts of disorder, as for all the thinner layers of Fig. 9b, each channeled particle is scattered on average only once by displaced atoms (single scattering approximation) and the dechanneling cross section is

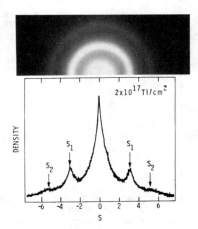

Fig. 8 – Electron diffraction pattern and associated densitometer trace from thin amorphous layer in iron after room temperature implantation of 2×10^{17} Ti/cm^2 (17).

$$\sigma_D = \frac{\pi z_1^2 z_2^2 e^4 d}{E^2 \psi_{1/2}^2} \quad , \tag{10}$$

where d is the atom spacing along the crystal rows and $\psi_{1/2}$ is the critical angle for channeling. At higher disorder levels, e.g., the 2×10^{17} Ti/cm^2 implant of Fig. 9b, numerical tabulations which account for the transition into a multiple scattering regime are required (18). In many practical applications, a satisfactory first estimate of the disorder profile is obtained by a straight line background subtraction for the channeled fraction of the beam (e.g., see dashed line in Fig. 4c) and that has been done in the analysis for Fig. 10. The analysis is necessarily approximate here because of the presence of some yield from the titanium backscattering which overlaps the iron yield.

From Fig. 10 it is seen that the surface layer is not fully converted to amorphous material at the lower implant fluences. This is consistent with the crystallites which were observed by TEM to be contained within the amorphous layer (Fig. 9a). In the upper part of Fig. 10 one of the measured titanium implant profiles and the calculated carbon profile (based on the measured total quantity of carbon) can be compared to the amorphous layer profile. The amorphous layer is seen to be well described by a process by which carbon is introduced into the sample and diffuses inward to form a ternary Fe-Ti-C alloy.

The unexpected formation of this amorphous ternary phase on the implanted surface of iron is quite interesting because of its possible relation to favorable changes in surface properties. For example, this layer was found to be very resistant to aggressive electropolish environments suggesting

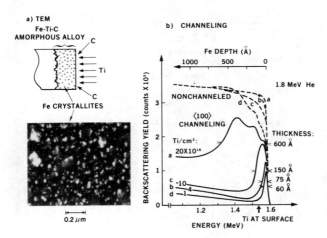

Fig. 9 - a) TEM dark field micrograph of iron crystallites in amorphous layer using (110) iron reflection after 4×10^{16} Ti/cm^2 implant.

b) Channeling analysis with 1.8 MeV ^{4}He along the <100> axis in iron as a function of titanium implant fluence. All implant energies are 180 keV except highest fluence where multiple energy implants were used to maximize the titanium concentration (17).

295

it may have good corrosion resistance (17). Also, titanium implantation into bearing steel has been shown to reduce the wear rates by a factor ~10 and carbon levels similar to that in our amorphous alloys were observed by sputtering Auger analysis to extend inward from the surface (19). Both of these favorable material properties have been attributed to amorphous alloys in bulk materials. Thus it is of interest to monitor transactions to the amorphous phase by the ion channeling technique during implantation in metals.

Helium Irradiation-Induced Blistering of Niobium

Disorder analysis by the channeling technique now appears sufficiently advanced to be of value for studies of the high damage levels in metals anticipated in fusion reactor environments. Of particular interest is the formation of voids and gas bubbles, and the kinetics of their formation and evolution as a function of fundamental parameters such as flux, temperature, and alloy solute concentrations. Many studies have relied on SEM or TEM analysis. In cases where single crystals are available, ion channeling should provide an ideal complement because of its non-destructive nature.

Dechanneling by gas bubbles has been demonstrated in bulk metals by transmission channeling (20) and channeling has been used to look at the evolution of a He-implanted surface as it proceeds through the stages of helium bubble formation and finally blistering of the surface. We briefly describe some early studies in the latter area (21). In Fig. 11b are shown

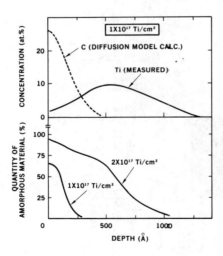

Fig. 10 – Analysis of backscattering-channeling spectra for disorder depth profile. The titanium profile was measured by 6 MeV backscattering and the total carbon determined by (d,p) nuclear reaction analysis (17).

the channeling spectra for 4 keV He-implanted niobium as a function of implantation fluence. In Fig. 11a the corresponding scanning electron micrograph (SEM) of the niobium surface after the highest fluence (4×10^{17} He/cm^2) shows the many spherical-like domes of material which have formed due to blistering. In the channeling mesurements, analysis was made with the detected backscattering also along a channeling direction (referred to as double alignment channeling) to further enhance the sensitivity to this phenomena. For 5×10^{16} He/cm^2 the channeling spectrum exhibits a region of rapid dechanneling between the surface and 500Å, corresponding to the region in which the helium comes to rest. This corresponds to a high density of defects of the type which would primarily give rise to dechanneling, such as may result for dislocations and for helium gas bubbles. At higher fluences blistering begins to occur, the material is greatly distorted, and the thin layer corresponding to the blister thickness loses its crystal orientation relative to the beam direction. At the highest fluence a loss of channeling occurs in this highly distorted material over a depth reported to correspond to the thickness of the blister lids (21).

In conjunction with such measurements it is also possible to determine the lateral stress in the implanted layer, through in situ mesurments of the bending of a cantilever beam formed by the sample (22). This is illustrated in Fig. 12 and provides a direct measure of the total stress in the He-implanted region integrated over the depth dimension. These measurements

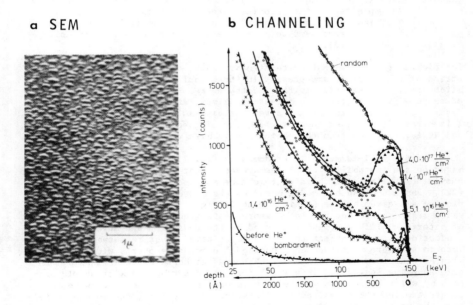

Fig. 11 - Niobium implanted with 4 keV He as seen by a) scanning electron microscopy after 4×10^{17} He/cm^2, and b) channeling analysis as a function of fluence for <100>/<111> double alignment channeling of 150 keV protons (21).

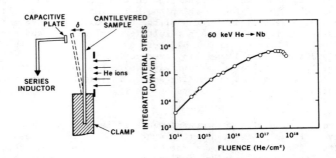

Fig. 12 - Lateral stress mesurement a) schematic, and b) results for 60 keV
He implantation into niobium, where the stress in the plane of
the beam is integrated over the depth of the implanted region
in this measurement to give lateral integrated stress in units
dynes/cm (22).

for niobium (22) show that quite high stresses, of the order of the yield
stress of the material are present in these implanted layers prior to
blistering. For example, in Fig. 12 the 60 keV He in niobium has a projected range ~2000Å and thus the measured maximum lateral stress of 6 x 10^5
dyne/cm corresponds to average stress values ~10^{10} dynes/cm^2 (~10^9 Pa).
This stress is believed to arise from the microscopic helium in the lattice,
and at later stages from the helium gas bubbles which nucleate at defects.

For the well-formed blister domes of Fig. 11a a simple mechanical model
based on the instability of plates suggests that an approximate relationship
should hold between the blister thickness and most probable bister diameter
(22). Results of such a model based on the measured stress valued for
niobium are given by the straight line in Fig. 13. Although the exact details of the agreement should be approximate since the mechanical properties
of materials will be somewhat altered by the radiation and any impurity introduction, it is apparent that this description gives reasonable agreement
both in absolute magnitude and functional dependence over a wide range of
blister diameters as determined by SEM and thickness as determined by SEM
or channeling analysis. This model does not describe the microscopic process by which helium gas bubbles are formed and subsequently grow to the
point at which the catastrophic blistering event occurs, and some have felt
that the vertical force due to helium gas pressure at failure is a factor in
the final lifting of the blister lids. However, while the details of the
exact process by which the blister formation is initiated are difficult
to describe in detail, the resulting mechanical deformation of the type
shown in Fig. 11a is consistent with the relationships that would be anticipated from macroscopic considerations.

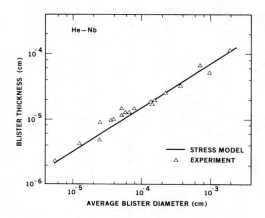

Fig. 13 – Measured and predicted blister thickness vs. average blister
diameter for helium implants at room temperature in niobium (22).

Hydrogen-Implanted Iron

In addition to determing the amount and depth profile of the disorder
present by channeling disorder analysis, it is also possible to determine
the lattice location of impurities associated with defects by channeling
location analysis. Hydrogen is an important element in Fe-based systems
and there is considerable interest in understanding its trapping and transport, since these influence hydrogen accumulation and embrittlement in
iron alloys. However, little is known about the fundamental interactions
between hydrogen and defects in metals. Thus hydrogen trapping is an
important area for study. We briefly illustrate channeling location
analysis by recent results for deuterium trapping in iron (23).

In these studies the hydrogen isotope deuterium was implanted into
iron and trapping occurred as a result of the lattice defects induced by
the deuterium implantation. Thermal release calculations indicated that an
observed deuterium release stage at 260 K corresponded to a single trap of
binding energy 0.5 eV. Thus, a thermal treatment was chosen to fully populate this 0.5 eV trap. The channeling angular distributions were then
measured by detecting the backscattering signal from the Fe lattice and the
nuclear reaction signal from the deuterium atoms simultaneously as the
crystal was tilted about the <100> axis. An example of the angular distribution for the deuterium and iron signals is shown in Fig. 14. The very
sharp signature from the deuterium atoms is quite distinct from the normal
reduction in yield from the Fe along the channeling direction. This indicates that the deuterium is not in a substitutional site but occupies a
well-defined interstitial site. The location determined from angular distributions along this and other crystal directions is shown on the left-hand
side of Fig. 14. The deuterium sits within an octahedral cavity of Fe atoms
in a site displaced ≈0.4 A from the octahedral interstitial site in the

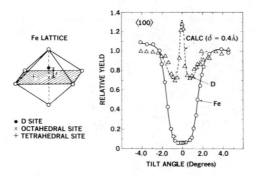

Fe LATTICE

• D SITE
× OCTAHEDRAL SITE
+ TETRAHEDRAL SITE

⟨100⟩

CALC (δ = 0.4Å)

D

Fe

RELATIVE YIELD

TILT ANGLE (Degrees)

Fig. 14 – Schematic of site positions and ⟨100⟩ axial channeling angular
distribution for deuterium implanted in iron at 15 keV to
$1 \times 10^{16}/cm^2$ and annealed to 200 K (23).

direction of a nearest neighbor Fe site. After annealing to higher temperatures where the deuterium is released from this 0.5 eV trap the channeling results indicate deuterium occupies a new site of lower symmetry.

It has been hypothesized that the Fe lattice site is vacant directly
above the position of the deuterium shown in Fig. 14, and that this corresponds to the structure of the hydrogen–vacancy center in iron (23). The
coupling between the hydrogen and the defect cannot be directly monitored by
the channeling technique but only inferred from location and migration measurements, and from other knowledge about defects in the material being
studied. However, studies such as these in iron and other metals (24) have
provided a basis for further experimental and theoretical investigations of
the structural and electronic nature of fundamental hydrogen–defect centers
in metals (25).

Summary of Technique Advantages and Limitations

Several of the most important advantages and limitations of the ion
channeling technique for disorder analysis are summarized in Table 1. The
channeling technique is sensitive to a wide variety of defects ranging from
localized defects such as interstitials to extended defects such as dislocations. The channeling technique is particularly useful for cases of
high densities of defects where complementary techniques such as TEM may be
difficult to apply quantitatively, for example due to a lack of sensitivity
for point defects or to interference by strain contrast and overlapping
images for extended defects. Under conditions where a single type of defect
dominates, the channeling technique can be quite quantitative. In addition,
the technique is non-destructive, since the depth distribution of defects
can be determined without layer removal techniques and usually the damage
introduced by the analysis beam is small relative to the disorder being
analyzed. In conjunction with channeling studies, it is often valuable to
perform selected TEM analyses to provide additional information on the defects present and to give further self-consistency to the interpretation of

the channeling results. In addition, by the channeling lattice location technique, information about the structure of solute-defect centers can be obtained.

TABLE I - Advantages and Limitations of Channeling Technique for Disorder Analysis

A. ADVANTAGES
   1. Nondestructive
   2. Quantitative
   3. Depth profiles (resolution ~100Å)
   4. High damage levels (TEM analysis difficult)
   5. Lattice site information (especially impurity-associated defects)
   6. Complements TEM analysis

B. LIMITATIONS
   1. Single crystals required
   2. Relatively low sensitivity
   3. Poor lateral resolution (typically ~1 mm)
   4. Analysis restricted to near surface ($\lesssim$10 μm)
   5. Quantitative analysis requires single class of defects dominate

Some of the major limitations of the channeling technique are that single crystals are required and, as typically used, lateral resolution is poor (~1 mm). In addition, the technique probes relatively shallow depths ($\lesssim$ 10 μm). Channeling analysis is not well suited for categoguing the variety of host lattice defects present and can be most profitably applied when the dechanneling cross section and number density of defects is such that dechanneling by given class of defects dominates over other defects present. Under such conditions, quantitative studies of the disorder density can be carried out, allowing the influence of irradiation parameters on materials to be studied in detail.

Conclusions

While the channeling technique has recently been demonstrated to hold promise for application to radiation damage studies in metals, only limited work of a quantitative nature has been carried out in this area. Several examples have been presented here to illustrate the type of information and range of possibilities available. The technique is particularly well suited to the relatively high damage levels found for ion beam irradiation of materials, including ion beam simulations of neutron damage, and to conditions which might be anticipated in fusion reactor environments. The nondestructive nature of the technique makes it well suited for investigation of mechanisms and kinetics associated with primary and secondary defect production and evolution. Also, in the case where solutes are associated with the defects, ion beam analysis generally allows the solute depth distribution within the matrix to be monitored as well. For example, this can be related to the damage depth distributions. In favorable cases the crystallographic site of the solute atoms can also be determined from channeling angular scans. Finally, the channeling technique is most valuable when used in conjunction with other defect-specific techniques such as transmission electron microscopy.

References

1. D. S. Gemmell, "Channeling and Related Effects in the Motion of Charged Particles Through Crystals," Reviews of Modern Physics, 46 (1974) pp 129-227.

2. L. C. Feldman, J. W. Mayer and S. T. Picraux, Applications of Channeling to Solid State Science; Academic Press, Inc., New York, NY 1982.

3. W. K. Chu, J. W. Mayer, and M-A. Nicolet, Backscattering Spectrometry; Academic Press, Inc., New York, NY 1978.

4. E. H. Eisen, "Application to Radiation Damage" in Channeling, Ed. by D. V. Morgan, pp 415-434; J. Wiley, New York, NY 1973.

5. S. T. Picraux, "Lattice Location of Impurities in Metals and Semiconductors" in New Uses of Ion Accelerators, Ed. by J. F. Ziegler, pp 229-281; Plenum Press, New York, NY 1975.

6. W. F. Tseng, J. Gyulai, T. Koji, S. S. Lau, J. Roth and J. W. Mayer, "Investigation of Dislocations by Backscattering Spectrometry and Transmission Electron Microscopy", Nuclear Instruments and Methods, 119 (1978) pp 615-617.

7. Private communication, S. T. Picraux, J. A. Knapp and E. Rimini, Sandia National Laboratories, Feb., 1980.

8. Y. Quere, "Dechanneling of Fast Particles by Lattice Defects," Journal of Nuclear Materials, 53 (1974) pp 262-266.

9. P. P. Pronko, "Direct Observation of Self-Interestitial-Type Defects in Metals Through Combined Single-Double Alignment Channeling Backscattering," Physical Review, B16 (1977) pp 4753-4755.

10. D. K. Sood and G. Dearnaley, "Radiation Damage in Copper Single Crystals," Journal of Vacuum Science and Technology, 12 (1975) pp 445-449.

11. S. T. Picraux, D. M. Follstaedt, P. Baeri, S. U. Campisano, G. Foti, and E. Rimini, "Depth Profile Studies of Extended Defects Induced by Ion Implantation in Si and Al," Radiation Effects, 49 (1980) pp 75-80.

12. S. T. Picraux, E. Rimini, G. Foti and S. U. Campisano, "Dechanneling by Dislocations in Ion-Implanted Al, "Physical Review, B18 (1978) pp 2078-2096.

13. J. Mory and Y. Quere, "Dechanneling by Stacking Faults and Dislocations," Radiation Effects, 13 (1972) pp 57-66.

14. D. K. Brice, Ion Implantation Range and Energy Deposition Distributions," Vol. 1; IFI/Plenum, New York, N.Y., 1975.

15. S. M. Myers and S. T. Picraux, "Enhanced Diffusion of Zn in Al Under High-Flux Heavy-Ion Irradiation," Applied Physics Letters, 46 (1975) pp 4774-4776.

16. W. A. Grant, "Ion-Implantation and Irradiation Studies Using Amorphous Metals" Nuclear Instruments and Methods, 182/183 (1981) pp 809-826.

17. D. M. Follstaedt, J. A. Knapp and S. T. Picraux, "Carbon-induced Amorphous Surface Layers in Ti-implanted Fe," Applied Physics Letters, 37 (1980) pp 330-333.

18. P. Sigmund and K. B. Winterbon, "Small-Angle Multiple Scattering of Ions in the Screened Coulomb Region," Nuclear Instruments and Methods, 119 (1974) pp 541-557.

19. C. A. Carosella, I. L. Singer, R. C. Bowers and C. R. Gossett, "Friction and Wear Reduction of Bearing Steel via Ion Implantation", in Ion Implantation Metallurgy, Ed. by C. M. Preece and J. K. Hirvonen, pp 103-115; The Metallurgical Society of AIME, Warrendale, PA 1980.

20. D. Ronikier-Polonsky, G. Desarmot, N. Housseau and Y. Quere, "Dechanneling by Gas Bubbles in a Solid," Radiation Effects, 27 (1975) pp 81-88.

21. J. Roth, R. Behrisch and B. M. U. Scherzer, "Blistering of Niobium Due to Low Energy Helium Ion Bombardment Investigated by Rutherford Back-scattering" in Applications of Ion Beams to Metals, Ed. by S. T. Picraux, E. P. EerNisse and F. L. Vook, pp 573-584; Plenum Press, New York, N.Y. 1974.

22. E. P. EerNisse and S. T. Picraux, "Role of Integrated Lateral Stress in Surface Deformation of He-Implanted Surfaces," Journal of Applied Physics, 50 (1977) pp 9-17.

23. S. M. Myers, S. T. Picraux and R. E. Stoltz, "Defect Trapping of Ion-Implanted Deuterium in Fe," Journal of Applied Physics, 50 (1979) pp 5710-5719.

24. See, for example, A. C. Chami, J. P. Bugeat and E. Ligeon, "Solid Solutions of the Hydrogen-Magnesium System Produced by Implantation," Radiation Effects, 37 (1978) pp 73-81.

25. S. T. Picraux, "Defect Trapping of Gas Atoms in Metals", Nuclear Instruments and Methods, 182/183 (1981) pp 413-437.

A COMPARISON OF ION CHANNELING WITH OTHER SELECTIVE
METHODS FOR STUDYING INTERSTITIAL TRAPPING IN METALS

M. L. Swanson

Atomic Energy of Canada Research Company Limited
Chalk River Nuclear Laboratories
Chalk River, Ontario, Canada K0J 1J0

The trapping of vacancies and self-interstitial atoms (SIA) by solute atoms in metals has been studied by ion channeling and by several other selective methods in recent years. These techniques probe the structure of defect clusters containing only a few atoms, and thus provide the basis for understanding the growth of larger microstructures, which are important for alloy development. Ion channeling is a relatively direct method of 'seeing' the position of solute atoms in a crystal lattice. It is thus especially useful for observing the trapping of SIA by small solute atoms, which results in the formation of mixed dumbbells, a configuration in which a solute atom and a host atom straddle a lattice site. Ion channeling has identified mixed dumbbells with $<100>$ configuration in f.c.c. metals, and with $\sim<40\bar{4}3>$ configuration in h.c.p. metals. Since the absolute concentration of mixed dumbbells can be measured to an accuracy of about 10% by channeling, the creation and annihilation of mixed dumbbells can be monitored under a variety of conditions to obtain the following defect data: (1) the production rate of SIA by irradiation, (2) the mobility of SIA, (3) the trapping efficiencies of solutes for SIA, (4) the trapping configurations, (5) the thermal stability of different trapping configurations, (6) the annihilation of mixed dumbbells by vacancies, and (7) the trapping of vacancies by other solutes.   The channeling data are shown to agree in most cases with Mössbauer, perturbed angular correlation, diffuse X-ray scattering, internal friction, positron annihilation and electrical resistivity results.

1. Introduction

In recent years, several selective methods have been developed for the study of point defects in metals (1-30). By the use of these techniques, the concentrations of particular defects or classes of defects can be monitored during an irradiation or annealing experiment. Thus the processes of defect formation, migration and interaction can be investigated in much greater detail than was previously possible by the study of bulk properties such as electrical resistivity (31-35).

Of these selective techniques, positron annihilation (1-5) is very sensitive to vacancy-type defects since holes in a metal lattice, being negatively charged, trap positrons strongly. The Mössbauer effect (6-10) is especially useful for looking at self-interstitial atoms (SIA) which are trapped at ^{57}Fe atoms, since the resultant compression of the Mössbauer atom leads to an isomer-shifted line with some quadrupole splitting. The channeling technique (11-15) is useful for investigating interstitial-type defects in which a solute atom is displaced an appreciable distance (0.04 nm) into a crystallographic channel (e.g. the mixed dumbbell, consisting of a solute atom and a host atom straddling a lattice site). Diffuse X-ray scattering (16-20) has been used to study both intrinsic defects, in particular SIA, and clusters of such defects. Mechanical relaxation effects (21-25) are very sensitive to small concentrations of reorientable defects but lack some selectivity. All of these techniques measure only the average properties of many defects, although they are selective in the type of defect. In contrast, field ion microscopy (26-30) is a method by which individual defects such as vacancies, interstitials or solute atoms can be identified.

It is the purpose of the present review to compare the results obtained by the channeling technique with results from other selective methods, as well as with electrical resistivity data. In particular, the trapping of SIA by solute atoms to form the mixed dumbbell configuration will be considered. From channeling data, the absolute concentration of mixed dumbbells can be measured to an accuracy of 10%. In dilute solid solutions containing ~0.01-0.1% solute atoms, and at low defect concentrations, almost all freely mobile SIA are trapped by the solute atoms. Thus by measuring the concentration of mixed dumbbells as a function of irradiating particle, energy, temperature and fluence, and by following the annealing processes after irradiation, the following information can be obtained.

1. The production rate of SIA can be measured as a function of irradiating particle energy. (Threshold measurements can be made which are independent of the direct displacement of light impurity atoms.)

2. The kinetics of self-interstitial migration can be determined by observing the annealing temperature at which mixed dumbbells are formed.

3. The trapping efficiency e_t of a solute atom relative to the trapping efficiency e_v of a vacancy for self-interstitials can be determined by measuring the rate of increase of mixed dumbbell concentration C_{di} with increasing particle fluence in the temperature region where SIA are mobile.

4. The trapping configurations of defect-solute atom complexes can be found from detailed channeling measurements for different crystallographic axes and planes.

5. The thermal stability (i.e. binding energy) of different trapping configurations can be determined by monitoring the mixed dumbbell concentration during annealing.

6. The annihilation of mixed dumbbells with vacancies can be studied, yielding the migration energy of the mobile defect (either vacancies or mixed dumbbells).

7. The trapping of vacancies (or of mixed dumbbells) by other solutes can be studied by measuring the effect of these solutes on vacancy-mixed dumbbell annihilation. In addition, the clustering of vacancies at solute atoms can be investigated directly by measuring the displacement of the solute atoms.

2. Survey of Experimental Methods

In this survey, the sensitivity, scope and limitations of the various selective methods of studying point defects will be emphasized. Further details of the techniques can be found in the references.

2.1 Channeling

2.1.1 Lattice Location

The channeling of medium energy ions (e.g. 1 MeV He$^+$ ions) in crystals can be used to determine the position of solute atoms in a crystal lattice, and thus to investigate the displacement of solute atoms from lattice sites when they trap point defects (11-15). The technique is based on the fact that ions which are directed near close packed rows of atoms in a crystal are steered into the open spaces (channels) between these rows by a series of low angle screened Coulomb collisions (Fig. 1). Consequently, the channeled ions do not penetrate closer than the screening distance to the atomic nuclei, and the probability of large angle Rutherford collisions, nuclear reactions or inner shell X-ray excitation is greatly reduced compared with the probability of such events from a non-channeled (random) beam of ions. For incident angles ψ close to the channeling direction, a strong attenuation in yield of ions backscattered from host atoms is thus observed (Fig. 1).

The lattice position of solute atoms is found by measuring the normalized yield χ_i of particles or X-rays emanating from interaction of channeled ions with solute atoms for different channels. χ_i is defined as the ratio of the yield from an aligned beam to the yield from a randomly directed beam of ions for a given depth increment (i.e. for a given energy increment of detected particles). When solute atoms are on normal substitutional lattice sites, the yield χ_i is the same as the normalized yield χ_h from host atoms, which is of the order of 0.01-0.05 for open channels in relatively perfect crystals. However, when solute atoms are displaced from lattice sites, their projection into different <lmn> channels results in large increases in $\chi_i^{<lmn>}$ for these channels (Fig. 1). Thus a simple and direct determination of solute atom

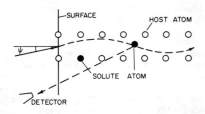

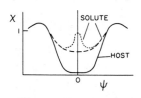

Fig. 1 - Schematic view of the channeling of ions directed at an angle ψ to a close packed row of atoms in a crystal. The normalized yield χ of ions which are backscattered from host atoms shows a strong dip at $\psi = 0$. If 50% of solute atoms are displaced into the channel, the normalized yield of ions backscattered from the solute atoms is approximately half the random yield ($\chi \simeq 0.5$ at $\psi = 0$). If the displaced solute atoms are located near the center of the channel, a peak in yield may occur (dotted line).

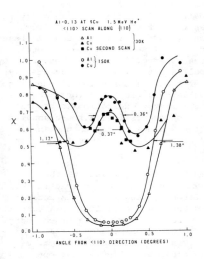

Fig. 2 - Angular dependence of the normalized backscattering yields of 1.5 MeV He$^+$ ions from Cu and Al atoms at 30 and 150 K in an Al-0.13 at.% Cu crystal, after irradiation at 70 K with 9×10^{15} 1.5 MeV He$^+$ ions/cm^2. The angular scans were along a {110} plane. The yields were measured from a depth interval of 100-200 nm (13).

308

positions is obtained by observation of χ_i and χ_h for only two or three different channels. For example, if solute atoms lie in the body-centered position for a f.c.c. lattice, $\chi_i^{<111>} = \chi_h^{<111>}$ since this position is completely shadowed by <111> strings of atoms. On the other hand, $\chi_i^{<110>} > 1$, because the solute atoms lie in the center of <110> channels.

Although the position of solute atoms is thus easily "seen" by channeling, an exact determination of lattice positions (to ~ 0.01 nm) requires knowledge of the ion flux distribution within a given channel. Because of the steering action involved in channeling, the concentration of channeled ions is not uniform in a channel, but is peaked near the center. This "flux peaking" effect can be used to good advantage in lattice location studies, because a peak in yield from solute atoms (Fig. 1) indicates that the solute atoms are displaced to positions near the center of the channel. This effect is shown clearly in Fig. 2, where Cu solute atoms in an Al-0.13 at.% Cu crystal have been displaced into <110> channels by formation of <100> Al-Cu mixed dumbbells (13). The magnitude of the flux peaking can be calculated either analytically (36), or with a Monte Carlo computer simulation method (37). Agreement is generally very good between the two approaches (36). If the yields are measured from very shallow depths, the Monte Carlo method is superior, because the variation of ion flux with depth can be calculated.

From channeling results, the apparent displaced fraction $f_{di}^{<lmn>}$ of solute atoms into an <lmn> channel is determined by the relationship (38)

$$f_{di}^{<lmn>} = (\chi_i^{<lmn>} - \chi_h^{<lmn>})/(1 - \chi_h^{<lmn>}) \qquad (1)$$

This quantity represents the true displaced fraction of solute atoms for the given channel only if the ion flux is uniform or if the solutes are distributed at random. Generally, neither of these criteria is fulfilled. The true displaced fraction is $f_{di}^{<lmn>}/F_i^{<lmn>}$, where $F_i^{<lmn>}$ is the normalized ion flux at the position of the displaced solute atoms. A useful quantity for irradiation damage experiments is the concentration of displaced solute atoms,

$$C_{di} = f_{di}^{<lmn>} C_i/(F_i^{<lmn>} g^{<lmn>}) \qquad (2)$$

where C_i is the solute concentration and $g^{<lmn>}$ is the fraction of displaced solute atoms which project into the given channel.

When a unique position is specified for the displaced solute atoms, C_{di} can be determined from eq. (2) to an absolute accuracy of about 10% with no adjustable parameters. The main sources of error arise from a statistical error of about 5% in measured yields which determine $f_{di}^{<lmn>}$, and an error of 5-10% in the calculated value of the ion flux $F_i^{<lmn>}$. In addition, possible systematic errors in $f_{di}^{<lmn>}$ arise from errors in depth scales. If solute atoms were clustered, the value of C_i would also be subject to error.

In cases where solute atoms are displaced from lattice sites by trapping SIA to form mixed dumbbells, C_{di} is a direct measurement of the concentration of mixed dumbbells, as discussed later.

2.1.2 Dechanneling

As channeled ions penetrate into a crystal, they can be deflected out of channels (dechanneled) by multiple scattering from displaced atoms and from electrons (11-13,39). At shallow depths, the dechanneling from displaced atoms dominates. Thus the rate of increase in normalized yield χ with depth z is a measure of the sum of atomic displacements including those arising from thermal vibrations. The increase in this rate $\Delta(d\chi/dz)$, caused by irradiation, is a measure of the radiation damage. When both $\Delta(d\chi/dz)$ and the radiation-induced increase Δc_{di} of displaced solute atoms decrease during annealing, it can be concluded that defect annihilation is the cause of the recovery stage. If an anneal alters Δc_{di} but not $\Delta(d\chi/dz)$, it can be concluded that a change in trapping configuration, rather than defect annihilation, has occurred.

2.1.3 Defect Studies Using Channeling

Experimentally, channeling is a convenient tool for radiation damage studies, as data can be accumulated rapidly in situ. A sample is mounted on a goniometer in an accelerator chamber. A self-contained He refrigeration unit is attached to the sample holder by means of a Cu braid (40). The single crystal sample can be irradiated in situ at any desired temperature from 20-300 K (or higher if the Cu braid is not attached), by using a resistance heater. A shield at ~ 20 K surrounds the sample to avoid contamination. Usually $^4He^+$ ions are used for both irradiation and for analysis (11-15). In a Van de Graaff accelerator, heavier ions can also be used for damaging the crystal, as changing to $^4He^+$ for analysis is quick. The power introduced into the sample to create a defect concentration of about 10^{-3} in about 10^3 s is only $\sim 10^{-2}$ W, which is easily removed by conduction along the Cu braid.

If electron irradiations are desired, a second accelerator is preferred. The most convenient arrangement would allow electron or ion beams to enter a fixed sample chamber, rather than transporting the sample chamber.

2.1.4 Limitations

The channeling method requires good single crystals, as the sensitivity depends on a low χ_h. This means that both the bulk imperfections, such as mosaic spread, and surface deformation must be minimized. Low temperatures are also helpful. In practice, any annealed single crystal which shows sharp back reflection Laue X-ray patterns is suitable. Electropolished or chemically polished surfaces are satisfactory.

The radiation damage introduced by the analyzing beam is an important limitation of the channeling method. Normally, a complete angular scan through a low index axial channel will introduce 10^{-3} - 10^{-4} atomic fraction of defects, depending on detector geometry and solute mass and concentration. However, measurements in channeling directions produce damage which is proportional to the normalized yield, χ_h, so that if poor channeling directions are avoided, the damage from the analyzing beam is minimized. Nevertheless, from these considerations it is seen that the channeling method is basically a defect-creating technique, so that very low defect concentrations cannot be investigated.

A variety of solute atom displacements or trapping configurations cannot easily be separated by the channeling method. Thus the selectivity is restricted to simple cases. On the other hand, only the displacement of specific solute atoms is observed, and several solutes could be investigated simultaneously by using different methods, such as backscattering, nuclear reactions (41) and characteristic X-rays (42).

2.1.5 Scope, Advantages and Sensitivity

The trapping of SIA by any solute atom can be studied. The preferred method, backscattering (because of the ease of obtaining an energy versus depth relationship), can be used only with solutes appreciably heavier than the host atoms. Nuclear reactions are suitable for light elements (41), such as ^1H, ^2H, He, C, N, O, Be. Characteristic X-rays can be used for intermediate mass solutes in heavier host crystals (42). The depth measurement is somewhat more difficult using X-rays, but various methods can be used to overcome this problem.

The depth of analysis for channeling studies is typically 50-500 nm, so that near-surface problems do not arise. The damaged layer can easily be removed so that experiments can be repeated using the same sample.

The displacement of solute atoms from lattice sites is observed quite directly and unambiguously. The dependence on calculations and assumed models is minimal. The absolute concentration of defects can be measured to an accuracy of $\sim 10\%$ in favourable cases; e.g. the Al-Ag mixed dumbbell. This is important for comparison with theories of defect production and trapping.

The accumulation of data by present pulse height analysis techniques is rapid and reliable, so that a complete irradiation, lattice location and annealing experiment can be completed in a few hours.

When the trapping of SIA at solute atoms to form mixed dumbbells is studied, one requires that an appreciable fraction of the solute atoms ($\gtrsim 0.1$) have trapped SIA. This requirement is satisfied for solute atom concentrations of 10^{-4} - 5×10^{-3} atomic fraction. The lower value is the practical limit of sensitivity with present particle counting techniques, although in principle this limit could be reduced to 10^{-5} or less. The practical considerations are detector size and damage from the analyzing beam. The upper value of 5×10^{-3} is a function of the concentration of point defects which can be introduced into metals by irradiation.

2.2 Positron Annihilation

The lifetime of positrons is increased when they are trapped at dilated (negatively charged) regions of a crystal, such as vacancies, voids or dislocations. At the same time, the angular correlation and energies of gamma rays emitted during positron annihilation are affected by the altered momentum distribution of electrons near the defects. Thus measurements of positron lifetimes, Doppler broadening and angular correlation have proven to be a valuable method of investigating vacancy-type defects in metals (1-5).

Positron annihilation is especially suitable for measuring the equilibrium concentration of vacancies in metals, because the high sensitivity (about 10^{-6} atomic fraction), as compared with other equilibrium methods, largely eliminates divacancy contributions. In radiation damage studies, the mobility and clustering of vacancies can be determined, although the method is not sensitive at present to details of cluster size or symmetry. Positron annihilation can be used in a rather wide variety of metals.

The trapping of positrons at vacancies saturates at relatively low vacancy concentrations (about 10^{-5}), so that the range of defect concentrations which can be studied is rather small. Polycrystalline samples of various sizes can be used. The depth investigated is usually a few micrometers, although surfaces can also be studied by using thermalized positrons. Complications in interpreting results may arise from surface effects, and from large uncertainties in the temperature dependence of trapping rates. Also, the differences in the trapping and annihilation rates of positrons at various defects, such as isolated vacancies and vacancy-solute pairs, are not known.

2.3 Hyperfine Interactions (HFI): Mössbauer Effect and Perturbed Angular Correlation

The Mössbauer effect (6-10) is the recoil-free resonant absorption of gamma rays which are emitted by nuclear decay. It is made possible in a solid because the recoil is absorbed by the lattice during gamma ray emission. Because of the very narrow line width of the gamma ray ($\sim 10^{-8}$ eV) as compared with the gamma ray energy ($\sim 10^4$ eV), the Mössbauer effect is a very sensitive method of determining slight deviations in the electronic environment of a nucleus. Thus when a Mössbauer solute atom traps a defect (especially a SIA), an isomer shift in gamma ray energy is caused by the compression of the solute atom. In addition, some electric quadrupole splitting is expected because of the interaction of the nuclear quadrupole moment with the gradient of the electric field produced by the adjacent defect. Since the intensity of the Mössbauer line, or the recoil-free fraction of gammas, varies exponentially with the mean square amplitude of atomic vibration, information about vibrational frequencies of the trapping configuration can be obtained.

There are three ways of obtaining a nucleus in the excited state, from which Mössbauer gamma rays are emitted: radioactive decay, Coulomb excitation and nuclear reaction. The latter two methods provide a means of creating nuclei which do not occur naturally, while at the same time creating defects by the atomic recoils preceding the gamma emission.

In contrast to the Mössbauer effect, which is more sensitive to interstitial-type defects, perturbed angular correlation (PAC) measurements appear to be more sensitive to vacancy-type defects (7). In PAC measurements, the coincidence counting rate of gammas emitted during the cascade decay of a nucleus is measured for fixed detector positions. The frequency of oscillation of this counting rate (nuclear precession frequency) is a direct measure of the strength of the electric field gradient which interacts with the nuclear quadrupole moment at the decaying nucleus.

In both of these HFI methods, as in the channeling technique, only the trapping of defects at the emitting nucleus is detected. Although a wide range of nuclei could be used, most of the results to date have been obtained using ^{57}Co (Fe) and ^{111}In (Cd). The methods are extremely sensitive to small defect concentrations, down to 10^{-8} atomic fraction. Different defects can be distinguished in a single sample. Information on defect symmetry can be obtained, although the extraction of specific configurations is difficult. The magnitude of solute atom relaxations cannot be found. Rather long counting times are required to accumulate Mössbauer data.

2.4 Diffuse X-ray Scattering

The diffuse scattering of X-rays near the Bragg reflections (Huang scattering) is a measure of the long range displacement field of defects, and thus can be used to determine the symmetry and absolute concentration of defects (16-20). Huang scattering is much more sensitive to interstitials than to vacancies because of the larger strain fields surrounding interstitials. The clustering of SIA can be studied by Huang scattering, since as a first approximation the scattering intensity is linearly proportional to the number of SIA per cluster. The diffuse scattering between the Bragg peaks is a measure of the atomic displacements close to defects, and thus is an even more sensitive measure of defect symmetry. Information about vacancy clustering can also be obtained from small angle scattering.

Diffuse X-ray scattering measurements require very high intensity X-ray sources, which have only recently become available. Good single crystals are required. The experiments are relatively difficult, and require on-line computer analysis of data. The scattering data is compared with model calculations of displacements around defects. A relatively high defect concentration (0.5×10^{-3}) is needed for these experiments, so that the interpretation of results for isolated defects could be complicated by defect clustering.

2.5 Mechanical Relaxation

Mechanical relaxation is commonly studied by measuring internal friction or the elastic aftereffect (21-25). In the former method, the rate of decay of the oscillation amplitude of a vibrating sample, usually a torsion pendulum, is measured. The damping Q^{-1} of the oscillation is caused by the absorption of energy by defects which reorient or oscillate with a time constant τ which is related to the activation energy for reorientation E^R and the temperature T by $\tau = \tau_o \exp(E^R/kT)$. If the defect migrates with the same activation energy as it reorients, the elastic aftereffect, rather than internal friction, is used because fewer defect jumps occur during the aftereffect measurement.

Mechanical relaxation is an extremely sensitive measurement, as only about 10^{-14} defects are required to cause an appreciable change in dislocation damping. In fact, it is this sensitivity which sometimes makes interpretation of results difficult. In order to study point defects, the dislocation damping effects must be largely eliminated, either by reducing the dislocation concentration or by pinning the dislocations firmly by irradiation. The orientation of defects can be determined by mechanical relaxation, and several different defects can be distinguished by their different time constants. Thin-walled single crystal tubes have been used as torsion pendulums for defect symmetry measurements. For point defect studies, the defect concentrations used are about 10^{-4}.

2.6 Field Ion Microscopy

Field ion microscopy enables atoms and defects to be observed directly (26-30). With this technique, a very high potential ($\sim 10^4$ V)

313

is applied between a finely pointed sample tip (which is positively charged) and a fluorescent screen, both of which are located in a vacuum chamber containing a small amount of He gas. When a He atom approaches the sample surface, it becomes ionized by the high field and is accelerated in a straight line to the screen. Thus an enormously enlarged image of surface ions is projected onto the screen. Interstitial atoms are usually seen as extra bright spots, and vacancies as dark spots on the screen. However, contrast effects can give misleading results (27). Field evaporation of surface layers produces a clean surface and also enables interior defects to be revealed, but only under the surface conditions of the experiment. Individual solute atoms can be identified by field evaporation followed by time of flight measurements.

The great advantage of field ion microscopy is that individual defects can be seen directly. SIA and vacancies can be counted, so that the structure of defect cascades and the mobility of SIA can be measured. There is some possibility that the large surface distortion causes a perturbation of the defect structure, and consideration must be given to surface artifacts created by the field evaporation. The method has until recently been restricted to certain high melting point metals such as W and Pt. Defect symmetries have not as yet been unambiguously determined.

2.7 Electrical Resistivity

Any defects in a crystal produce an increase in electrical resistivity. Thus the selectivity of resistivity measurements is poor. However, the measurements are easy, the sample requirements are not stringent, and the sensitivity is very high (about 10^{-10} atomic fraction of Frenkel pairs). For these reasons, electrical resistivity measurements have provided the bulk of data about irradiation damage (31-35).

Resistivity cannot be used to discriminate between different types of defects, although some distinction between defect types may be achieved by using magnetoresistance measurements. The absolute concentration of defects is not determined, but can be related to scattering theories. The contributions to resistivity of different defect species may not be additive, thus leading to errors on account of this deviation from Matthiessen's rule. That is, the influence of dislocation or phonon scattering may alter the resistivity of point defects.

3. Comparison of Results

3.1 Production of SIA

Under conditions where SIA are mobile, where vacancies are immobile, and where the concentration of these defects is much less than the solute concentration, almost all freely mobile SIA are trapped by the solute atoms. For small solutes, the trapping configuration is the mixed dumbbell. Thus, by measuring the concentration of mixed dumbbells using channeling, the absolute production rate of mobile SIA can be determined. Most studies of SIA production rates have been done using electrical resistivity or electron microscopy (35,42a). Few detailed channeling data are available as yet. However, by channeling, one could study the threshold energy E_d for defect creation as a function of particle mass and energy, and of target orientation (42a). Sub-threshold displacements could be investigated. Of particular interest would be the nature of collision cascades; that is, the number of free SIA created as a function of energy deposited in cascades. Another interesting subject is the probability that correlated recombination of SIA with vacancies is interrupted by trapping at solute atoms. Measurement of the mixed dumbbell production rate as a function of solute concentration would give data for comparison with the theory of these kinetic processes (42b).

3.2 Mobility of Self-Interstitials

In some of the earliest radiation damage experiments in Cu it had been concluded that self-interstitials became freely mobile near 50 K (31). The evidence was obtained from the kinetics of the annealing processes as observed by electrical resistivity measurements. A second order recovery process (that is, a process in which the temperature of maximum recovery rate decreased with increasing defect concentration) was observed near 50 K (stage I_E). All lower temperature recovery processes were independent of defect concentration and thus were attributed to close Frenkel pair recombination and to correlated vacancy-SIA recombination. Moreover, after electron irradiation the recovery of electrical resistivity was almost complete at the end of this stage I annealing, so that most of the interstitials had recombined with vacancies. This general result has been supported by numerous later experiments (32-35).

The effect of doping a metal with solute atoms was to suppress the stage I recovery, because the self-interstitials were trapped by the solute atoms. With increasing concentration of solute atoms in the range 10^{-4} - 10^{-2} atomic fraction, almost complete suppression of stage I recovery, including the close pair stages, was observed (33-35).

The newer, selective methods of measurement have verified these earlier measurements. Mössbauer (8-10), internal friction (25) and channeling (12-15) results all show that defects produced by trapping of self-interstitials at solute atoms appear only when the annealing temperature is raised to the end of stage I; that is, where self-interstitials become mobile. This is shown in the case of Al in Fig. 3. Here, the mobility of SIA from 35-50 K is demonstrated by the strong recovery in electrical resistivity for deuteron-irradiated pure Al, by the channeling detection of Al-Ag mixed dumbbells in He$^+$ irradiated Al-0.08% Ag, by the increase in the ^{57}Fe Mössbauer defect line in neutron-irradiated Al-0.002% Co, and by the appearance of several internal friction peaks in electron-irradiated Al-0.04% Fe. In addition, diffuse X-ray scattering experiments (16-20) on Al have shown that interstitial clustering occurs above 45 K.

For Zr, electrical resistivity data (43) indicated that self-interstitials became mobile at 100-120 K. Channeling results support this conclusion (15). As shown in Fig. 4, Au solute atoms in an irradiated Zr-0.2% Au crystal were displaced from lattice sites during annealing in this temperature range.

In the case of Mg, the mobility of SIA was not known from electrical resistivity results (44-45). The channeling data (46) showed conclusively that SIA were mobile below 30 K in Mg, as a high concentration of Mg-Ag mixed dumbbells were created during irradiation at that temperature in a Mg-0.18 at.% Ag alloy. To date, lower temperature irradiations for channeling measurements have not been performed.

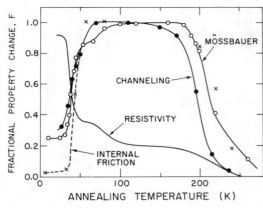

Fig. 3 - A comparison of annealing data in irradiated Al and dilute Al alloys (600 s pulse anneals). For the different experiments, the ordinate F is defined as follows:

(a) Resistivity: fraction of the electrical resistivity increment remaining after annealing 20 MeV deuteron-irradiated pure Al (42c).

(b) Mössbauer effect: fraction of ^{57}Co solute atoms which had trapped self-interstitials in a neutron-irradiated Al-0.0016% Co alloy (fluence 0.7 x 10^{18} cm^{-2} at 4.6 K), as measured by the area of the Mössbauer defect line (10). The results were normalized against the fraction 0.38 at 160 K.

(c) Channeling: fraction of Ag atoms which had formed mixed dumbbells in 1 MeV He$^+$ irradiated Al-0.082% Ag as measured by channeling. The results were normalized against the fraction 0.39 at 110 K, and the data below 70 K were taken from a separate experiment.

(d) Internal friction: fraction of internal friction peaks 1 and 4 which remained after annealing 3 MeV electron-irradiated Al-0.04% Fe (25).

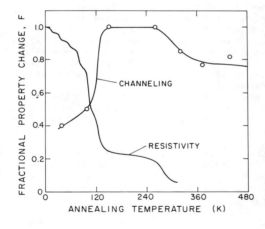

Fig. 4 - A comparison of annealing data for irradiated Zr. In the different experiments, the ordinate F is defined as follows:

(a) Resistivity: fraction of the electrical resistivity increment which remained after annealing electron-irradiated pure Zr (43). (Electron energy 1.5 MeV, maximum change in resistivity 5.5 x 10^{-8} Ω cm, annealing pulses 300 s).

(b) Channeling: fraction of Au atoms which were displaced into [0001] channels in He$^+$-irradiated Zr-0.2% Au (15). The results were normalized against the fraction at 260 K (2 MeV He$^+$ fluence 3 x 10^{15} cm^{-2}, annealing pulses 600 s).

In some instances, such as Au, where it is not known how mobile SIA are (35), the selective methods of measuring the trapping of SIA could be used to determine the migration energies of SIA. Evidence for clustering of SIA at 5 K has been obtained from diffuse X-ray scattering in Mg, Nb, Cd and Au, suggesting that SIA are mobile at that temperature.

By field ion microscopy, the long range migration of SIA to the surface of a W crystal could be observed directly (28). This migration occurred at ∿38 K. In the case of Pt, SIA became mobile at 15 K in approximate agreement with resistivity data (35).

3.3 Trapping Efficiencies

Electrical resistivity measurements in the temperature range where SIA are freely mobile but where vacancies are immobile (50-180 K for Al or Cu) have been used to determine the trapping efficiency e_t of solute atoms for SIA (47-49). Usually the ratio e_v/e_t is determined, where e_v is the trapping efficiency of vacancies for SIA, from the experimental rate of increase of reciprocal damage rate $d\phi/d\rho$, using the equation

$$\frac{d\phi}{d\rho} = \frac{1}{\rho_F{}^t f \sigma_d} \left(1 + \frac{\Delta\rho}{\rho_F{}^t C_i} \frac{e_v}{e_t}\right) \tag{3}$$

Here ϕ is the irradiation fluence, ρ is the electrical resistivity, f is the fraction of interstitials escaping close pair and correlated recombination, σ_d is the displacement cross section for the production of interstitials as measured by the damage rate at 4 K, $\rho_F{}^t$ is the resistivity increment per unit concentration of Frenkel defects (vacancies and trapped interstitials), C_i is the concentration of solute atoms and $\Delta\rho$ is the irradiation-induced increase in residual resistivity.

This equation represents the chemical rate theory approximation and assumes unsaturable traps if e_t is independent of ϕ. In addition, $\rho_F{}^t$ is assumed to be independent of the environment of the interstitial; i.e. of the cluster size. Eq. (3) thus applies only in the rare case where single trapping predominates. From these measurements, which determine the total trapping of SIA in all configurations, it was found that e_t/e_v decreased with increasing temperature as T^{-2} (Fig. 5), which is a much stronger temperature dependence than expected from a simple elastic interaction between SIA and solute atoms (47-49). One possible explanation was that a series of shallow trapping configurations existed, which gradually emptied as the temperature was raised. Only the deepest configuration, the mixed dumbbell, would remain at temperatures just below stage III recovery.

Channeling measurements provide a selective method of determining the probability that SIA are trapped in the mixed dumbbell configuration, since this configuration produces a large, easily measurable displacement of solute atoms from lattice sites. In shallower trapping configurations, the SIA retain their normal form and the solute atoms are displaced only a small amount from lattice sites. Thus, as outlined in Section 2, the concentration C_{di} of displaced atoms, as measured by channeling, is an absolute and direct measure of the concentration of mixed dumbbells. The rate of production of mixed dumbbells is shown in Fig. 6 for irradiations of Al-0.1 at.% Ag with He^+ ions at temperatures of 25, 70 and 150 K (14).

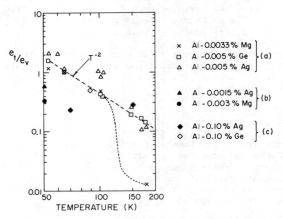

Fig. 5 - Experimental ratios of the respective efficiencies e_t and e_v with which solute atoms and vacancies trap SIA.   The values   (a) were obtained from electrical resistivity damage rates (48), the values (b) from suppression of recovery stage I_E (51), and the values (c) from channeling results for mixed dumbbell production rates.

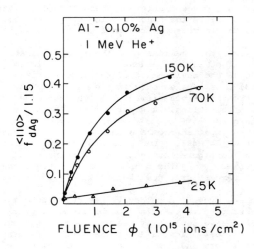

Fig. 6 - The fraction of Ag atoms which form mixed dumbbells $f_{dAg}^{<110>}$ / $F_i^{<110>}$ in an Al-0.10% Ag crystal as a function of the irradiation fluence of 1 MeV He$^+$ ions at 25, 70 and 150 K (14).   The measurements were taken at 25 K in the first case and 35 K in the others. $f_{dAg}^{<110>}$ is the apparent fraction of displaced Ag atoms in a <110> channel (see Equation (1)), and the calculated ion flux $F_i^{<110>}$ at the displaced Ag atoms is 1.15.

318

Three important pieces of information can be obtained from this figure:

(1) the fraction of interstitials which are trapped spontaneously during irradiation;
(2) the probability that the trapping configuration is the mixed dumbbell, rather than shallow configurations; and
(3) the trapping efficiency of Ag atoms for SIA in aluminum.

From Fig. 6, the initial mixed dumbbell production rate at 25 K was only 4% of that at 70 K. Thus the spontaneous trapping of SIA at Ag atoms to form mixed dumbells was quite small. Spontaneous trapping can arise from dynamic trapping of SIA by the termination of collision chains, or from creation of interstitials within the capture volume of a solute atom. Measurement of this quantity as a function of irradiating particle and energy, and as a function of solute concentration, could give valuable information about the dynamic nature of radiation damage (size of collision cascades and the length of collision chains). Such detailed experiments have not yet been reported. From a simplistic point of view, the present results indicate that the spontaneous capture volume of a Ag solute atom is 40 atomic volumes. This volume would include the volume around a solute atom within which the SIA spontaneously forms the mixed dumbbell, plus the number of atomic volumes sampled by the average collision chain.

Since initially all free SIA are trapped at the Ag solute atoms, the initial production rate of mixed dumbbells (from Fig. 6), divided by the production rate of free interstitials, gives the fraction P_d of trapping configurations which are mixed dumbbells.

$$P_d = \frac{1}{f\sigma_d} \left(\frac{d\, C_{di}}{d\phi} \right)_{\phi = 0} \tag{4}$$

From the data of Fig. 6 and using $\sigma_d = 3.2 \times 10^{-19}$ cm^2, (from the 2 MeV damage rate data of Merkle and Singer (50), scaled to 1 MeV) and f $\simeq 0.5$, the data for both 70 and 150 K indicate that $P_d \gtrsim 1$, so that shallow trapping is small at both temperatures. This is in agreement with Mössbauer results (10) for Al-0.002 at.% Co. Thus it appears that the previously mentioned interpretation of the electrical resistivity data of electron-irradiated Al alloys must be modified (47-49).

The rate of decrease of slopes of the curves in Fig. 6 gives directly the ratio of trapping efficiency e_v of vacancies to trapping efficiency e_t of solute atoms for SIA (14). The results, using a trapping model involving only single and double SIA trapping at solute atoms, were $(e_v + e_{2t})/e_t = 5.6$ at 70 K and 4.6 at 150 K, where e_{2t} is the trapping efficiency of a mixed dumbbell for a second SIA. It is generally assumed that $e_{2t} \simeq e_t$. In that case, the resultant ratio $(e_v/e_t)_{150\ K} = 3.6$, which is close to the value 4.0 found from electrical resistivity measurements (See Fig. 5). However, the present value $(e_v/e_t)_{70\ K} = 4.6$ is much greater than that from resistivity measurements, 1.0. The difference in results at the lower temperatures could be partly due to the higher solute concentrations used in the channeling work.

From the magnitude of the suppression of the correlated recovery stage I_E by solute atoms, the trapping efficiency e_t can also be found. The result obtained by Rizk et al., using electrical resistivity mesurements, (51) for Al-.0015% Ag was $e_v/e_t = 1.7$, valid at ~ 50 K. This value is intermediate between the electrical resistivity damage rate and channeling values (See Fig. 5).

3.4 Trapping Configurations

3.4.1 Self-interstitial Configurations

The stable form of the self-intersitial in f.c.c. metals has been calculated to be the <100> dumbbell or split interstitial (52-54) and in b.c.c. metals the <110> dumbbell (54,55). In this configuration, 2 host atoms symmetrically straddle a lattice site. Although large uncertainties still exist in such calculations, it has been found for the f.c.c. lattice that the <100> split interstitial is the stable form, independent of the interatomic potential used. The crowdion, which can be considered as a <110> compressional wave in the f.c.c. lattice, has a higher energy. In h.c.p. metals there may be more than one stable form in a given metal, and the configuration may vary from one metal to another, depending on the c/a ratio. Diffuse X-ray scattering results indicate that in Zn the SIA has [0001] configuration (56).

Experimentally, the <100> dumbbell configuration has been found to be the dominant one for self-interstitials in f.c.c. metals. The most direct and convincing evidence has been the diffuse X-ray scattering results (16-19) in pure Al and Cu. A comparison of the experimental scattering results with the theoretical values for a <100> type distortion is shown in Fig. 7. It is seen that excellent agreement exists only for the <100> configuration.

Measurements of elastic aftereffect in electron irradiated Al have also shown that the <100> self-interstitial dominates (23). However, no such relaxation process was observed (24) for Cu. Various reasons for this lack of success in Cu have been proposed, mainly relating to the fact that the relaxation is extremely difficult to observe experimentally, because the reorientation of the defect also involves a defect jump, so that defect annihilation occurs during the measurements. Thus any reduction in interstitial concentration, or in the dipole force tensor of the <100> interstitial in Cu as compared with Al could reduce the effect below the limit of sensitivity of the apparatus.

Indirect evidence for <100> SIA has also been provided by measurements of elastic constants (57) in neutron irradiated Cu. The strong temperature dependence of the shear moduli in the irradiated samples is consistent with the low frequency vibrational modes (54) calculated for <100> SIA. Similarly, the large decrease is shear moduli (58) caused by electron irradiation of Al at 4 K supports the model of the <100> SIA, which has weak shearing forces.

3.4.2 Channeling Evidence for Mixed Dumbbells

The trapping configurations of SIA at solute atoms are related to the stable SIA configuration. The <100> SIA in f.c.c. metals migrates by a replacement mechanism, as shown in Fig. 8, during which a neighbouring host atom is pushed aside to form half the SIA. If the migrating SIA encounters a substitutional solute atom, the solute atom can become part of the interstitial, forming a mixed dumbbell. In simple terms, since SIA compress the lattice, small solute atoms tend to trap SIA to form mixed dumbbells by reducing the compressional energy. For large solute atoms, the shallow trapping configuration, in which a SIA retains its form, is

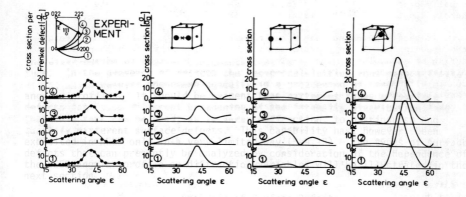

Fig. 7 - Comparison of calculated diffuse X-ray scattering cross section per Frenkel pair with experimental results for Al irradiated with 3 MeV electrons at 4.5 K (56). The insert given in fig. (a) shows the four Ewald circles in reciprocal space along which the scattering intensities were measured. The inserts in (b) to (d) shows the <100> split, octahedral and tetrahedral interstitial configurations for which intensity patterns were calculated along these same Ewald circles.

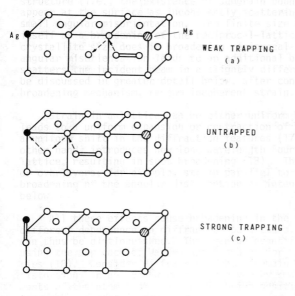

Fig. 8 - Trapping of self-interstitials at solute atoms in Al. (a) One possible weak trapping position of a self-interstitial atom near a Mg solute atom. The thermal release of the self-interstitial from this trap is illustrated by the dotted lines. (b) At a sufficiently large distance from solute atoms, the self-interstitial may be considered untrapped. Further migration to a strong trap (Ag) is indicated by the dotted lines. In reality, the migration will be a three-dimensional random walk.

(c) The strong trapping position, as observed by channeling measurements. The solute atom shares a <100> split interstitial configuration with a host atom.

321

usually preferred (13,54). In this case, the neighbouring solute atom is hardly displaced from its lattice site (Fig. 8).

It should be noted that other metastable configurations of SIA (e.g. the crowdion) could be transformed to the mixed dumbbell configuration during the trapping process. Thus channeling measurements of solute atom displacements do not give direct evidence about SIA configurations. The double alignment technique can in principle provide the required sensitivity to measure SIA configurations (59), but in practice defect clustering would complicate the measurements.

Since the lattice positions of solute atoms can be measured accurately by channeling before and after they have trapped SIA, the type of trapping configuration can be determined. The mixed dumbbell, in which the solute atom is displaced approximately 2/3 of the distance to the body-centered position, is especially easy to detect by channeling because of the resultant peaking effect (Section 2). Although all determinations of mixed dumbbell configurations by channeling have been performed on He^+-irradiated crystals, in which the irradiation damage is somewhat more complex than in electron-irradiated crystals, the results indicate that the configuration is not markedly affected by this additional complexity. A comparison of interstitial trapping caused by electron and neutron irradiation of Al-0.005% Co, as measured by the Mössbauer effect (9,10), indicates that multiple trapping is more difficult to avoid in the neutron irradiation case, probably because of the larger numbers of diinterstitials produced. However, the isomer shift of the trapping configuration was not appreciably different, indicating that the configuration was not altered significantly by the presence of additional trapped SIA. The channeling measurements also indicated that in some cases the displacement of solute atoms in mixed dumbbells was not changed noticeably by very high irradiation fluences (12), where multiple trapping was to be expected.

The channeling results have shown that in Al the solute atoms Cr, Mn, Cu, Zn, Ge and Ag are displaced from lattice sites along <100> directions when they trap SIA. Thus the trapping configuration is the <100> mixed dumbbell (12,13,36). The evidence for this configuration is summarized as follows: (See Fig. 9 for projection of <100> displacements into different channels).

(a) In cases where the solute is displaced $\gtrsim$ 2/3 of the distance to the body-centered position, a pronounced peak in yield from solute atoms is observed for the <110> direction. An example of this is shown in Fig. 2, where Al-Cu mixed dumbbells were created by He^+- irradiation at 70 K. Approximately 50% of the Cu atoms were in mixed dumbbells. The peaking occurs only if the displacement of the solute atoms along <100> directions is greater than 1.2 Å, according to both Monte Carlo computer simulations (37) and to analytical calculations of ion flux profiles in the <110> channels (36). Thus the displacement of the Cu atoms was large, implying either a mixed dumbbbell or a body-centered position for Cu atoms. The body-centered position was excluded for most mixed dumbbells by <111> channeling results, since solute atoms in the body-centered position are completely shadowed for <111> channels. From a comparison of <110>, <100> and <111> channeling results (Table I) (36), the displacement of solute atoms in Al alloys has been determined (Table II). These results were obtained from a comparison of experimental yields of backscattered He ions in those channeling directions with calculated yields, using an analytical method. In the case of Al-Mn dumbbells, the result of this calculation agreed with the displacement found from a Monte Carlo calculation (37), using the magnitude of the <110> peak in Mn yield.

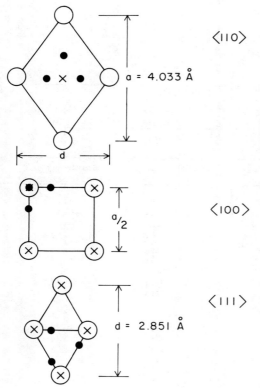

⟨110⟩

a = 4.033 Å

d

⟨100⟩

a/2

⟨111⟩

d = 2.851 Å

Fig. 9 - Projected positions (●) of solute atoms in mixed dumbbells for <110>, <100> and <111>axial channels of Al. The positions are shown for mixed dumbbells in which the solute atoms are displaced 1.3 Å from normal lattice sites (o) along <100> directions. This displacement is two-thirds the distance to the body-centered site (X). The lattice constant is a and the interatomic distance is d.

Table I. Channeling Measurements of the Apparent Displaced Fractions of Solute Atoms $f_{di}^{\langle lmn\rangle}$ in Irradiated Dilute Alloys (see eq. (1))

Host	Solute	$f_{di}^{\langle 110\rangle}$	$f_{di}^{\langle 100\rangle}$	$f_{di}^{\langle 111\rangle}$	$f_{di}^{\{100\}}$	$f_{di}^{\{111\}}$
Al	0.07% Cr	0.31	0.06	0.09	0.02	
	0.09% Mn	$\left\{\begin{array}{l}0.51\\0.64\end{array}\right.$	0.10	0.22	0.12	
	0.13% Cu	0.59	0.09			
	0.09% Zn	0.32			0.12	0.44
	0.08% Ag	$\left\{\begin{array}{l}0.30\\0.37\end{array}\right.$	0.10	0.16	0.14	0.34
	0.10% Ge	0.25			0.11	0.24
Cu	0.25% Be	0.37	0.06	0.14		

Table II. Trapping Configurations of Self-interstitials from Channeling Data

Host Atom	Solute Atom	$\Delta V/V$ [a]	Trapping Configuration [b]	Solute Atom Displacement(nm)
Al	Cr	-0.57	$\langle 100\rangle$ m.d.	0.146
	Mn	-0.47	$\langle 100\rangle$ m.d.	0.143
	Fe	~-0.38	m.d.	
	Cu	-0.38	$\langle 100\rangle$ m.d.	0.148
	Zn	-0.06	$\langle 100\rangle$ m.d.	
	Ag	0.001	$\langle 100\rangle$ m.d.	0.139
	Ga	0.05	~$\langle 100\rangle$ m.d.	
	Ge	0.13	$\langle 100\rangle$ m.d.	0.11
	Sn	0.24	s.t.	
	Mg	0.41	s.t.	
Cu	Be	-0.26	$\langle 100\rangle$ m.d.	0.132
	Ag	0.44	s.t.	
	Au	0.48	s.t.	
	Sb	0.92	s.t.	
Mg	Ag	-0.63	$\langle 40\bar{4}3\rangle$ m.d.	
	Bi	0.23	s.t.	
Zn	Au	-0.11	s.t.	
Zr	Au	-0.27	~$\langle 40\bar{4}3\rangle$ m.d.	
Fe	Te	0.09	s.t.	
	Au	0.44	s.t.	

[a] $\Delta V/V$ is the expansion of the host lattice caused by alloying, expressed in atomic volumes per solute atom

[b] m.d. is the mixed dumbbell, s.t. is shallow trapping

In all of these calculations <100> displacements were assumed. <111> displacements of $\simeq 2.8$ Å for Al-Cu would also be consistent with the axial channeling results (60), but not with the planar results to be discussed.

(b) A comparison of {100} planar channeling yields with <110> axial yields is the most direct method of determining whether the solute atom displacement is in <100> directions (12,38), since 2/3 of <100> type displacements are shadowed in a given {100} plane. Thus the ratio of apparent displaced fractions

$$f_{di}^{<110>}/f_{di}^{\{100\}} = 3 \ F_i^{<110>}/F_i^{\{100\}} \qquad (5)$$

where $F_i^{<lmn>}$ is the normalized ion flux in the <lmn> channel at the position of the displaced solute atoms (See Eq. (2)). As a first approximation $F_i^{<110>}/F_i^{\{100\}} = 1$, so that a ratio of $f_{di}^{<110>}/f_{di}^{\{100\}} \simeq 3$ is a strong indication for <100> displacements. No other displacements can give such large ratios. Experimentally, these ratios were close to 3 for the solutes mentioned above (Table I). In cases where peaking in solute yield was observed for <110> channels (e.g. Al-Cr, Al-Mn), the ratio was greater than 3, because $F_i^{<110>} > F_i^{\{100\}}$.

The magnitude of solute atom displacements in mixed dumbbells (Table II) agrees qualitatively with the theory of Dederichs et al. (54) which states that the displacement of the solute atom increases with decreasing size of the solute atom (as determined from lattice parameter data in alloys). However, no case has been seen where the solute atom is displaced into the body-centered position, as predicted for very small solutes.

In the h.c.p. metals Zr and Mg, doped with $\sim$0.2% Au or Ag, results (15,46) indicated that Zr-Au and Mg-Ag mixed dumbbells were formed by the trapping of SIA (See Fig. 4). The configuration was close to the $<40\overline{4}3>$ axis, which is the direction between next nearest neighbours (analogous to the <100> axis in f.c.c. lattices). This configuration is similar to that calculated (61) for SIA in Zr. No evidence was obtained for a [0001] configuration which was measured (56) for SIA in Zn. It should be noted, however, that the c/a ratio for Zn is 1.86 (which would favour a [0001] configuration), as compared with 1.6 for Mg and Zr.

3.4.3 Shallow Trapping

Large solute atoms generally trap SIA weakly, because both of these defects compress the lattice. In f.c.c. metals, a combination of electrical resistivity (47-49) and channeling (12,13,62,63) results have demonstrated that SIA retain their identity when trapped by large solute atoms (see Fig. 8) and that the SIA are released during stage II recovery (50-180 K for Al). (See Table II). The channeling results show that large solute atoms, such as Sn in Al (12), or Au in Cu (62), are not displaced appreciably from lattice sites during irradiation at temperatures where SIA are mobile. Since electrical resistivity data (47-49) indicate that these solutes do trap SIA in stage I, but release them in stage II, it follows that the trapping configuration is a weak one in which the SIA is located adjacent to or near the solute atom.

325

3.4.4 Comparison with Other Data

Although trapping configurations have been determined mainly by channeling measurements, the results of other techniques generally support the channeling results.

(a) Diffuse X-ray scattering (20). In electron-irradiated Al-0.2% Ge crystals it has been shown that the <100> interstitial persists up to ∿150 K, whereas in pure Al it is replaced by SIA clusters as annealing proceeds beyond 45 K. This result indicates that the <100> SIA configuration is stabilized by trapping at Ge atoms, but the data is not sufficient to identify mixed dumbbells.

(b) Mössbauer effect (8-10). A new isomer shifted ^{57}Fe Mössbauer line with weak quadrupole splitting is produced by irradiation of dilute alloys of Al containing ^{57}Co. This line has been shown by annealing results to be due to the trapping of Al SIA by the ^{57}Co atoms (Fig. 3). The central shift of the line was +0.42 mm s^{-1}, which was mainly an isomer shift. The magnitude of this shift indicates a large increase in electron density of the Mössbauer atoms and therefore a large decrease of ∿40% in atomic volume of the Fe atoms which have trapped SIA. Thus these Fe atoms were squeezed into interstitial positions. The position cannot be the body-centered one because the cubic symmetry of that position would not result in the observed quadrupole splitting. From the temperature dependence of the Debye-Waller factor of the isomer shifted line, both a high jump frequency of the interstitial Fe atoms in a so-called interstitial cage, and low frequency vibrational modes of the interstitial Fe atoms were deduced (8-10, 63a). Although these results were first interpreted in terms of a <100> Al-Fe mixed dumbbell configuration (54, 63a), more recent experiments on single crystals indicate that the cage motion of the interstitial Fe atoms is consistent with <111> rather than <100> symmetry (63b). The positions of the Fe atoms were calculated to be displaced 0.052 nm from the body-centered site in <111> directions. It will be noted that the previously discussed channeling results indicated that Cr, Mn and Cu atoms in <100> mixed dumbbells were displaced almost the same distance, 0.06 nm, from the body-centered position, but in this case along <100> directions. The reason for the different behaviour for Fe atoms is not known.

It is interesting to note that the isomer shift was constant during annealing of irradiated Al-Co crystals up to at least 180 K, under conditions where multiple trapping of SIA at Co atoms had occurred. Thus the trapping configuration was not altered by the trapping of further SIA.

(c) Internal friction. Anelastic relaxation measurements (25) on Al crystals doped with 0.04 at.% Fe have also given evidence concerning interstitial trapping. After electron irradiation of this alloy, several internal friction peaks were seen which were not present for pure Al. Since these peaks all vanished in stage III recovery near 200 K (see Fig. 3) they can be attributed to SIA-solute atom complexes. None of these relaxation peaks had <100> symmetry. It was predicted (54) that the solute atom in a mixed dumbbell would easily move to adjacent solvent atoms in a "cage motion", in which the solute atom was always bound to a solvent atom. For <100> mixed dumbbells, this reorientation of the mixed dumbbell would cause an internal friction peak which should exhibit <100> symmetry. The fact that no relaxation peaks having <100> symmetry were observed for Al-Fe thus supports the Mössbauer data. However, for Al-Zn crystals, a defect having <100> symmetry was observed, in agreement with channeling data for Al-Zn alloys.

326

3.5 Thermal Stability

The channeling data for Al alloys indicate that the mixed dumbbell is generally stable up to stage III recovery ($\sim$ 200 K). In the case of Al-Cu dumbbells, this stability is shown by the detailed angular scans of Fig. 2. Since the magnitude of the peaking effect in Cu yield for a <110> angular scan was almost the same at measuring temperatures of 30 and 150 K, it is clear that the displacement of Cu atoms was unchanged and the configuration was unaltered.

The thermal stability of mixed dumbbells has also been demonstrated in several other Al alloys by measuring the isochronal recovery of $f_{di}^{<110>}$, as shown in Fig. 10(a). (See also Fig. 3.) Any change in trapping configuration would be reflected in altered displacement of solute atoms, which is most sensitively detected in f.c.c. metals by measuring <110> yields. The results of Fig. 10 also indicate that shallow trapping of SIA at Ag or Mn solute atoms is not significant, because the release of SIA from the shallow traps would be accompanied by an increase in the concentration of mixed dumbbells. Some shallow trapping may occur at Fe and Cu atoms.

As shown in Figs. 4 and 10(b), mixed dumbbells in Zr, Cu and Mg crystals were stable up to at least 260, 200 and 140 K respectively.

The large binding energy of mixed dumbbells has also been shown quite conclusively by electrical resistivity measurements of damage rates at various temperatures (47-49). It was shown that small solutes in Al and Cu continued to trap SIA up to stage III recovery, whereas large solutes (Pd, Sb, or Au in Cu; Mg in Al) lost their trapping capability somewhere in stage II (Fig. 5).

It was not known whether mixed dumbbells, in the absence of vacancies, would be stable beyond stage III recovery. This question will be discussed in the next section.

The thermal stability of Al-Cr mixed dumbbells in irradiated Al was investigated in some detail by channeling, especially in the temperature range 50-70 K, where a change in trapping configuration was inferred from electrical resistivity data (64). The channeling data (65) showed that no significant change in mixed dumbbell configuration occurred in that temperture range.

In the case of Al-Zn dumbbells, however, channeling results (38) indicated clearly that changes in the trapping configuration occurred between 70 and 150 K. As shown in Fig. 11, recovery of $f_{dZn}^{<110>}$ occurred in three stages, which corresponded to the recovery stages in electrical resistivity of electron- (66) and neutron-irradiated (67) Al-Zn alloys. The first two stages, at $\sim$90 K and $\sim$130 K, corresponded to changes in the trapping configuration. Evidence for this conclusion was: (a) the dechanneling increased during these recovery stages, indicating that defects were not annihilated; and (b) the displacement of Zn atoms was changed from a predominantly <100> displacement at 70 K to a non-<100> displacement at 150 K. Complete recovery of the dechanneling increment, the displacement of Zn atoms and the electrical resistivity increment occurred at $\sim$200 K, as observed for other Al alloys. These results agree with internal friction experiments, in which a SIA-Zn atom complex with <100> symmetry vanished during annealing near 130 K (67a).

327

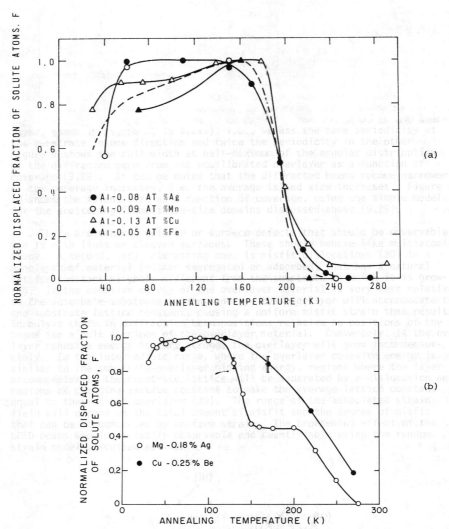

Fig. 10 - The effect of isochronal annealing (600 s pulses) on the displacement of solute atoms in Al, Mg and Cu alloys after irradiation (13,46). All measurements were at 30-40 K. Here F is the normalized displaced fraction of solute atoms remaining after annealing at the indicated temperatures (normalized to the maximum displaced fractions observed during the series of anneals).

(a) <u>Al alloys</u>. The maximum displaced fractions were 0.38, 0.36, 0.46 and 0.29 for alloys containing Ag, Mn, Cu and Fe respectively. The <110> channel was measured for all alloys except Al-0.09% Mn, where the <111> channel was used. The dashed line is the fraction of the irradiation-induced dechanneling increment which remained after annealing the Al-0.13 at.% Cu crystal.

(b) <u>Mg and Cu alloys</u>. The maximum displaced fractions were 0.31 for the Mg alloy (for a <11$\bar{2}$0> channel), and 0.20 for the Cu alloy (for a <110> channel). The irradiation fluences were 1.9 x 10^{15} cm^{-2} of 1 MeV He$^+$ and 10^{16} cm^{-2} of 0.6 MeV D$^+$ respectively.

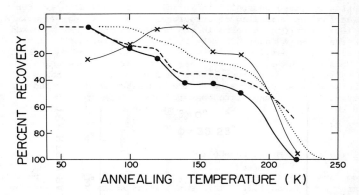

Fig. 11 - Isochronal recovery (600 s pulse anneals) of irradiation-induced
Zn atom displacements (●) $((\Delta f_{dZn}^{<110>})_{max} = 0.38)$ and dechanneling (x) in an
Al-0.09 at.% Zn crystal, as measured at 40 K in a <110> direction.   The
recovery of electrical resistivity changes $\Delta\rho$ for electron-irradiated Al-0.3
at.% Zn (66) (--- $\Delta\rho_0$ = 2 x 10^{-9} Ω cm) and neutron-irradiated Al-0.1 at.%
Zn (67) (....$\Delta\rho_0$ = 2 x 10^{-7} Ω cm) are also shown.

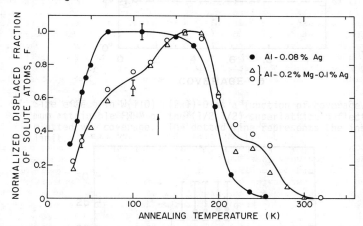

Fig. 12 - The effect of annealing (600 s pulse anneals) on the normalized
displaced fraction F of Ag atoms for irradiated Al-0.08% Ag and Al-0.2%
Mg-0.1% Ag.   F = $\Delta f_{dAg}^{<110>}/(\Delta f_{dAg}^{<110>})_{max}$, where $\Delta f_{dAg}^{<110>}$ is the increase
in apparent displaced fraction of Ag atoms (eq. (1)) caused by He^+
irradiation.   The maximum values of this increase were 0.39 for the
Al-0.08% Ag crystal (see Fig. 3) and 0.18 (symbol Δ) or 0.29 (symbol O)
for the Al-0.2% Mg-0.1% Ag crystal (63).   The He ion fluence was 5 x 10^{15} cm^{-2} in
each case.   The arrow indicates that temperature (129 K) at which a recovery
substage in electrical resistivity occurred for neutron-irradiated Al as a
result of Mg doping (68).

The thermal stability of weak traps can be studied by monitoring the concentration of mixed dumbbells in irradiated ternary alloys. For example, during the annealing of irradiated Al-0.2 at.% Mg-0.1 at.% Ag crystals (63), a strong increase in the concentration of Al-Ag mixed dumbbells was observed near 125 K by channeling measurements (Fig. 12). This increase was clearly due to the release of SIA trapped weakly by Mg atoms. This conclusion is in complete agreement with electrical resistivity results. A recovery stage in resistivity occurred at 129 K for neutron irradiated Al-0.1 at.% Mg, which was attributed to relese of SIA from Mg traps (68). Damage rate measurements on electron-irradiated Al-Mg alloys in the stage II recovery range were also consistent with the detrapping of SIA from Mg atoms near 125 K (47-49) (Fig. 5).

3.6 Annihilation of Mixed Dumbbells

Mixed dumbbells are annihilated during stage III recovery near (200 K) by combining with vacancies. This combination could be achieved by

(a) release of SIA from the solute atoms,
(b) migration of mixed dumbbells to vacancies, or
(c) migration of vacancies to mixed dumbbells.

In the case of Al alloys, the possibility (a) is excluded by the ternary alloy results shown in Fig. 12. If SIA were released from solute atom traps in stage III, complete recovery of Ag atom displacements would occur, whether a binary or ternary alloy were used. However, the results show that complete stage III recovery is delayed by the addition of Mg solute atoms to an Al-0.1 at.% Ag alloy. This delay is attributed to trapping of migrating mixed dumbbells or vacancies at the Mg atoms (63). The Mössbauer results of Vogl et al. also showed, via arguments of annealing kinetics, that mechanism (a) did not occur in Al-Co alloys (9).

There is now considerable evidence from a variety of experimental methods that mechanism (c) is correct: stage III recovery in Al, Cu and other f.c.c. metals is due to vacancy migration. (However, it is sometimes difficult to separate stage III and stage IV recovery in Al.) The Mössbauer data for Al-Co(Fe) showed that no clustering of Co atoms occurred during stage III recovery (see Fig. 3), and thus indicated that mixed dumbbell migration did not occur. From the internal friction data of Fig. 3 it is seen that a variety of complexes involving Fe atoms and SIA disappeared simultaneously during stage III recovery. Thus vacancy migration is inferred. Positron annihilation data in Cu (4) and Ni (69) indicated that vacancy clustering occurred during stage III recovery. Diffuse X-ray scattering results (18) for Cu also showed that vacancy clustering occurred in stage III. Channeling evidence for vacancy migration in stage III is provided by the displacement of large solute atoms such as Au in Cu (62) or Sn in Al (12) during stage III recovery (see the next section). Other vacancy annealing data have been summarized by Balluffi (70).

The significance of the assignment of vacancy migration to stage III recovery is that vacancies must have a relatively low migration energy. That is, for Al, recovery at $\sim$ 200 K (Fig. 3) for high fluence irradiations ($\sim 10^3$ jumps to annihilation) implies an activation energy for migration of $E_{1V}{}^M = 0.5$ eV. For Cu, $E_{1V}{}^M = 0.6$ eV. These values are somewhat lower than those previously derived from diffusion data combined with data for vacancy formation energies (71), and could reflect the influence of nearby defects.

330

An interesting question concerning the annihilation of mixed dumbbells is whether they would persist beyond stage III recovery in the absence of vacancies. In other words, what is the binding or migration energy of mixed dumbbells? This question is crucial in calculating the creep and void swelling of reactor materials. If the mixed dumbbell binding energy is at least as large as the migration energy of vacancies, the resultant stabilization of interstitial defects causes a greatly increased vacancy-interstitial recombination rate during irradiation, and thus reduces creep and void swelling (72,73).

In order to measure the binding energy of mixed dumbbells, it is necessary to eliminate or trap vacancies, which normally migrate in stage III and annihilate the mixed dumbbells. Mixed dumbbells could be produced in isolation by bombarding a crystal with ions of low enough energy that the vacancies would be created at the surface and the SIA would be injected into the bulk of the crystal. An experiment of this type would be difficult to perform by the channeling method because in order to achieve suitable counting statistics vacancies would be introduced by the analyzing beam. Other more sensitive techniques, such as internal friction or the Mössbauer effect, might be suitable for such an experiment.
Alternatively, vacancies could be stabilized by trapping them beyond stage III recovery. The trapping of vacancies by Mg atoms in an Al-0.2 at.% Mg-0.1 at.% Ag alloy, as shown in Fig. 12, delayed the annihilation of Al-Ag mixed dumbbells, indicating that they were stable up to at least 260 K. An attempt has been made to trap vacancies more strongly at implanted ^{3}He ions in an Al-0.1 at.% Ag crystal, with negative results (74).

3.7 Vacancy Trapping

It was shown in Fig. 12 (Section 3.5, 3.6) that the annihilation of Al-Ag mixed dumbbells by migrating vacancies was delayed from 200 K to 260 K by the addition of Mg solute atoms (63). It can be concluded that vacancies are trapped by the Mg atoms, and this difference in temperature corresponds to a vacancy-Mg binding energy of $E_{VMg}{}^B = 0.15$ eV, assuming a migration energy of single vacancies of $E_{1V}{}^M = 0.5$ eV. This binding energy is consistent with quenching data (75). By this procedure, the channeling method can be used to find binding energies of vacancies to a variety of solute atoms.

The multiple trapping of vacancies by large solute atoms has been studied by channeling, positron annihilation, and HFI measurements. Channeling is not a sensitive method of measuring the trapping of single vacancies by solute atoms, since the resultant relaxation of a solute atom is not expected to be more than 0.01 nm (76). However, multiple trapping of vacancies could lead to a large displacement of solute atoms from lattice sites. Such large displacements of Ag, Sb and Au solute atoms in irradiated Cu crystals have been observed during stage III annealing (62), and were attributed to the trapping of several vacancies at each solute atom. Recent channeling results for Al-0.03 at.% Sn crystals (77) have indicated that unique configurations of vacancies around Sn atoms are created during stage III annealing after alpha particle irradiation. The evidence for this conclusion is shown in Fig. 13, where it is seen that the apparent displaced fraction of Sn atoms into <110> channels was increased greatly during annealing near 200 K, whereas the change in the <111> displacement was much smaller. Detailed channeling data indicated that the Sn atoms were located at the center of a tetrahedron of 4 vacancies, at the center of an octahedron of 6 vacancies (the body-centered position), and probably also at the center of a triangle of 3 vacancies

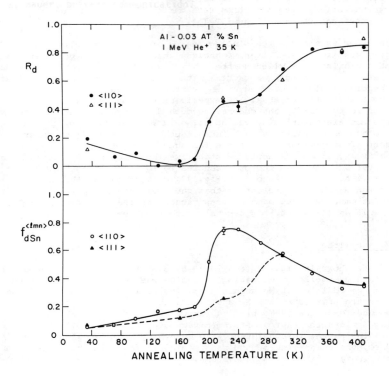

Fig. 13 - Isochronal annealing (600 s pulse anneals) of a water quenched
Al-0.03 at.% Sn crystal after irradiation with 1 MeV He ions at 35 K to a
fluence of 1.0×10^{16} cm^{-2}.   In the lower part of the figure, the effect of
annealing on $f_{dSn}^{<110>}$ and $f_{dSn}^{<111>}$ is shown.   In the upper part, the fractional
recovery R_d of the irradiation-induced dechanneling increment is shown
(77).

in a {111} plane. Subsequent irradiation at 70 K, where SIA but not vacancies are mobile, reduced $f_{dSn}^{<110>}$ because the size of the vacancy clusters was reduced by absorption of SIA. The difference between <110> and <111> channeling behaviour vanished after further annealing to 300 K, indicating a change in the vacancy configuration. The irradiation-induced dechanneling increment recovered only partially during the 200 K annealing stage (Fig. 13), because of the defect clusters remaining.

Perturbed angular correlation (PAC) studies of ^{111}In in Al, Cu and Ni crystals have provided convincing evidence that vacancies can be trapped by the In atoms in a variety of configurations (7, 78-82). In the case of quenched or electron-irradiated Al or Cu (where the ^{111}In was introduced by diffusion), the PAC results indicated that unique electric field gradients were created at the ^{111}In atoms during annealing in stage III (78-80). Since these gradients were the same after quenching and after electron irradiation, it was concluded that vacancy trapping was responsible. Similar hyperfine interaction results for ^{111}In-implanted Ni indicated that two unique configurations of vacancy clusters surrounding In atoms occurred (81,82). Suggested configurations were an In atom surrounded by three vacancies in a (111) plane or by a tetrahedron of four vacancies.

In plastically deformed Ni, the multiple trapping of vacancies by Sb solute atoms during annealing above 350 K was demonstrated by positron annihilation studies (69). Annealing at higher temperatures produced microvoids, as observed by electron microscopy.

4. CONCLUSIONS

From a comparison of channeling and other data, the following conclusions can be made concerning irradiation-induced defects in metals.

(a) SIA are mobile in Al at 45 K, in Cu at 50 K, in Zr at 120 K and in Mg below 30 K.

(b) The ratio of the trapping efficiency e_v of vacancies to the trapping efficiency e_t of Ag solute atoms for SIA in Al was $(e_v/e_t)_{150\ K}$ = 3.6 and $(e_v/e_t)_{70\ K}$ = 4.6. These values refer to mixed dumbbell trapping, which was found to dominate. The value at 150 K agreed with that obtained from electrical resistivity results, but the value at 70 K was about a factor of 5 greater than the resistivity result.

(c) Mixed dumbbells were the dominant configuration resulting from the trapping of SIA by small solute atoms, but SIA retained their identity when trapped by large solute atoms (shallow trapping). The mixed dumbbell had <100> orientation in the f.c.c. metals Al and Cu, and $\sim<40\bar{4}3>$ orientation in the h.c.p. metals Mg and Zr.

(d) Mixed dumbbells were thermally stable up to at least stage III recovery (with the exception of Al-Zn), while SIA were released from shallow traps in stage II.

(e)   Mixed dumbbells were annihilated during stage III recovery by vacancy migration (at $\sim$ 200 K for Al and $\sim$ 240 K for Cu).

(f)   Vacancies were trapped by large solute atoms. In the case of Sn solutes in Al, specific geometric configurations of vacancies were created by this trapping, leading to large Sn atom displacements into <110> channels.

Acknowledgements

The support of the CRNL Solid State Science Branch, and in particular J.A. Davies, I.V. Mitchell and J. Lori is gratefully appreciated. Most of the channeling work was done in collaboration with L.M. Howe, A.F. Quenneville, N. Matsunami and F. Maury.

References

1.   A. Seeger, J. Phys. F.: Metal Phys., 3, (1973) p. 248.
2.   B.T.A. McKee, A.G.D. Jost and I.K. MacKenzie, Can. J. Phys., 50, (1972) p. 415.
3.   B.T.A. McKee, W. Triftshäuser and A.T. Stewart, Phys. Rev. Letters, 28, (1972) p. 358.
4.   M. Eldrup, O.E. Mogensen and J.H. Evans, J. Phys. F.: Metal Phys., 6, (1976) p. 499;
     S. Mantl and W. Triftshäuser, in Fundamental Aspects of Radiation Damage in Metals, M.T. Robinson and F.W. Young, Jr., eds., USERDA Pub. CONF-751006-P2, p. 1122 (1975).
5.   Positron Annihilation (Proc. Fifth Int. Conf.), R.R. Hasiguti and K. Fujiwara, eds., Japan Inst. of Metals, Sendai, 1979.
6.   U. Gonser, in Vacancies and Interstitials in Metals, North-Holland, Amsterdam, 1970, p. 649.
7.   See Hyperfine Interactions, 4 (1978).
8.   G. Vogl, W. Mansel and W. Vogl, J. Phys. F.: Metal Phys., 4, (1974), p. 2321.
9.   W. Mansel and G. Vogl, J. Phys. F.: Metal Phys., 7, (1977) p. 253.
10. W. Mansel, H. Meyer and G. Vogl, Rad. Effects, 35, (1978) p. 69.
11. J.A. Davies, Channeling in Solids, D.V. Morgan, ed., Ch. 11; John Wiley and Sons, New York, 1973.
     S.T. Picraux, New Uses of Low Energy Accelerators, J.F. Ziegler, ed., Plenum Press, New York, 1975, p. 229.
     D.S. Gemmell, Rev. Mod. Phys., 46, (1974) p. 129.
12. M.L. Swanson and F. Maury. Can. J. Phys., 53, (1975) p. 1117.
13. M.L. Swanson, L.M. Howe and A.F. Quenneville, J. Nucl. Mat., 69,70, (1978) p. 372.
14. M.L. Swanson and L.M. Howe, Rad. Effects, 41, (1979) p. 129.
15. M.L. Swanson, L.M. Howe, A.F. Quenneville and J.F. Watters, J. Nucl. Mat., 67, (1977) p. 42.
16. P.H. Dederichs, J. Phys. F.: Metal Phys., 3, (1973) p. 471.
17. P. Ehrhart and W. Schilling, Phys. Rev., B8, (1973) p. 2604.
18. P. Ehrhart and V. Schlagheck, J. Phys. F.: Metal Phys., 4, (1974) p. 1575, 1589.
19. P. Ehrhart, J. Nucl. Mat., 69,70, (1978) p. 200.
20. B.C. Larson and H.G. Haubold, J. Nucl. Mat., 69,70, (1978) p. 758.
21. A.S. Nowick and B.S. Berry, Anelastic Relaxation in Crystalline Solids, Academic Press, New York, 1972.

22. J.L. Snoek, Physica, 6, (1939) p. 161, 591.
23. V. Spirić, L.E. Rehn, K.-H. Robrock and W. Schilling, Phys. Rev., B15, (1977) p. 672,
    K.-H. Robrock, L.E. Rehn, V. Spirić and W. Schilling, Phys. Rev., B15, (1977) p. 680.
24. L.E. Rehn and K.-H. Robrock, J. Phys. F.: Metal Phys., 7, (1977) p. 1107.
25. L.E. Rehn, K.-H. Robrock and H. Jacques, J. Phys. F.: Metal Phys., 8, (1978) p. 1835.
26. E.W. Müller, Adv. Electronics and Electron Phys., 13, (1960) p. 84.
27. D.N. Seidman and K.H. Lie, Acta Met., 20, (1972) p. 1045.
28. D.N. Seidman, J. Phys. F.: Metal Phys., 3, (1973) p. 393.
29. A. Wagner, T.M. Hall and D.N. Seidman, J. Nucl. Mat., 69,70, (1978) p. 413.
30. C.-Y. Wei and D.N. Seidman, Rad. Effects, 32, (1977) p. 229;
    J. Nucl. Mat., 69,70, (1978) p. 693.
31. J.W. Corbett, R.B. Smith and R.M. Walker, Phys. Rev., 114, (1959) p. 1452.
32. J.W. Corbett, Electron Radiation Damage in Semiconductors and Metals, Solid State Phys. Suppl. 7, Academic Press, New York, 1966.
33. T.H. Blewitt, R.R. Coltman, C.E. Klabunde and T.S. Noggle, J. Appl. Phys., 28, (1957) p. 639.
34. A. Sosin, Lattice Defects and Their Interactions, R.R. Hasiçuti, ed., Gordon and Breach, New York, 1967, p. 235.
35. W. Schilling, G. Burger, K. Isebeck and H. Wenzl, Vacancies and Interstitials in Metals, Proc. 1968 Jülich Conf.,North-Holland, Amsterdam, 1970, p. 255.
36. N. Matsunami, M.L. Swanson and L.M. Howe, Can. J. Phys., 56, (1978) p. 1057.
37. J.H. Barrett, in Proc. 4th Conf. on Applications of Small Accelerators (Denton, Texas, 1976), IEEE Publication Number 76CH1175-9NTS (IEEE, New York).
38. M.L. Swanson, L.M. Howe and A.F. Quenneville, Phys. Stat. Sol., (a) 31, (1975) p. 675.
39. J. Lindhard, Kong. Danske Vid. Selsk. mat.-fys. Medd., 34, No. 14 (1965).
40. J. Bøttiger, J.A. Davies, J. Lori and J.L. Whitton, Nucl. Inst. Meth., 109, (1973) p. 579.
41. Ion Beam Handbook for Material Analysis, J.W. Mayer, E. Rimini, eds., Academic Press, New York 1977.
42. J.F. Chemin, I.V. Mitchell and F.W. Saris, J. Appl. Phys., 45, (1974) p. 532.
42a. P. Vajda, Rev. Mod. Phys., 49, (1977) p. 481.
42b. K. Schroeder, Rad. Effects, 5, (1970) p. 255; K. Schroeder and W. Heidrich, Phys. Letters, 43A, (1973) p. 315.
42c. K. Herschbach and J.J. Jackson, Phys. Rev., 153, (1967) p. 694.
43. H.H. Neely, Rad. Effects, 3, (1970) p. 189.
44. T.N. O'Neal and R.L. Chaplin, Phys. Rev., B5, (1972) p. 3810.
45. J. Delaplace, J. Hillairet, J.C. Nicoud, D. Schumacher and G. Vogl, Phys. Stat. Sol., 30, (1968) p. 119.
46. L.M. Howe, M.L. Swanson and A.F. Quenneville, Rad. Effects., 35, (1978) p. 227.
47. A. Kraut, F. Dworschak and H. Wollenberger, Phys. Stat. Sol., (b) 44, (1971) p. 805.
48. H. Wollenberger, J. Nucl. Mat., 69,70. (1978) p. 362.
49. F. Dworschak, Th. Monsau and H. Wollenberger, J. Phys. F.: Metal Phys., 6, (1976) p. 2207.
50. K.L. Merkle and L.R. Singer, Appl. Phys. Lett., 11, (1967) p. 35.

51. R. Rizk, P. Vajda, F. Maury, A. Lucasson and P. Lucasson, J. Appl. Phys., $\underline{47}$, (1976) p. 4740.
52. J.B. Gibson, A.N. Goland, M. Milgram and G.H. Vineyard, Phys. Rev., $\underline{120}$, (1960) p. 1229.
53. R.A. Johnson, J. Phys. F.: Metal Phys., $\underline{3}$, (1973) p. 295.
54. P.H. Dederichs, C. Lehmann, H.R. Schober, A. Scholz and R. Zeller, J. Nucl. Mat., $\underline{69,70}$, (1978) p. 176.
55. R.A Johnson, Phys. Rev., $\underline{134}$, (1964) p. A1329.
56. W. Schilling, J. Nucl. Mat., $\underline{69,70}$, (1978) p. 465.
57. J. Holder, A.V. Granato and L.E. Rehn, Phys. Rev. B, $\underline{10}$, (1974) p. 363.
58. K.-H. Robrock and W. Schilling, J. Phys. F.: Metal Phys., $\underline{6}$, (1976) p. 303.
59. K. Morita, Rad. Effects, $\underline{28}$, (1976) p. 65.
60. N. Matsunami, private communication.
61. D.M. Brudnoy, private communication.
62. M.L. Swanson, L.M. Howe and A.F. Quenneville, Rad. Effects, $\underline{28}$, (1976) p. 205.
63. M.L. Swanson, L.M. Howe and A.F. Quenneville, J. Phys. F.: Metal Phys., $\underline{6}$, (1976) p. 1629.
63a. G. Vogl, W. Mansel and P.H. Dederichs, Phys. Rev. Lett., $\underline{36}$, (1976) p. 1497.
63b. W. Petry, G. Vogl and W. Mansel, Phys. Rev. Lett. $\underline{45}$ (1980) p. 1862.
64. C. Dimitrov, O. Dimitrov and F. Dworschak, J. Phys. F.: Metal Phys., $\underline{8}$, (1978) p. 1031.
65. M.L. Swanson, L.M. Howe, A.F. Quenneville and C. Dimitrov, Can. J. Phys., $\underline{58}$, (1980) p. 1538.
66. C.L. Snead and P.E. Shearin, Phys. Rev., $\underline{140}$, (1965) p. A1781.
67. A. Ceresara, T. Federighi and F. Pieragostini, Phys. Letters, $\underline{6}$, (1963) p. 152.
67a. A.V. Granato, private communication.
68. C. Dimitrov, F. Moreau and O. Dimitrov, J. Phys. F.: Metal Phys., $\underline{5}$, (1975) p. 385.
69. G. Dlubek, O. Brummer, N. Meyendorf, P. Hautojarvi, A. Vehanen and J. Yli-Kauppila, J. Phys. F.: Metal Phys., $\underline{9}$, (1979) p. 1961).
70. R.W. Balluffi, J. Nucl. Mat., $\underline{69,70}$, (1978) p. 240.
71. A. Seeger, in Fundamental Aspects of Radiation Damage in Metals, M.T. Robinson and F.W. Young, Jr., eds., USERDA Pub. CONF.-751006-P1, (1975) p. 493.
72. P.R. Okamoto and H. Wiedersich, J. Nucl. Mat., $\underline{53}$, (1974) p. 336.
73. R.A. Johnson and N.Q. Lam, Phys. Rev., B13, (1976) p. 4364.
74. M.L. Swanson, L.M. Howe and A.F. Quenneville, Proc. 14th Int. Conf. on Atomic Collisions in Solids, Hamilton, 1979. Nucl. Instr. & Meth., $\underline{170}$, (1980) p. 427.
75. J. Takamura, in Lattice Defects in Quenched Metals, R.M.J Cotterill, M. Doyama, J.J. Jackson, M. Meshii, eds., Academic Press, New York, 1965, p. 521.
76. A.C. Damask and G.J. Dienes, Point Defects in Metals, Gordon and Breach, New York, 1963.
77. M.L. Swanson, L.M. Howe and A.F. Quenneville, Phys. Rev. B22, (1980) p. 2213.
78. H. Rinneberg and H. Haas, Hyperfine Inter., $\underline{4}$, (1978), p. 678.
79. H. Rinneberg, W. Semmler and G. Antesberger, Phys. Lett., $\underline{66A}$, (1978) p. 57.
80. Th. Wichert, M. Deicher, O. Echt and E. Recknagel, Phys. Rev. Lett., $\underline{41}$, (1978) p. 1659.
81. C. Hohenemser, A.R. Arends, H. de Waard, H.G. Devare, F. Pleiter and S.A. Drentje, Hyperfine Inter., $\underline{3}$, (1977) p. 297.
82. F. Pleiter, Hyperfine Inter., $\underline{4}$, (1978) p. 710.

A TEM AND CHANNELING STUDY

OF He[+] ION BOMBARDED MOLYBDENUM SINGLE CRYSTALS[*]

J. Greggi[(1)], C. F. Tzeng[(1)], J. R. Townsend[(1)] and W. J. Choyke[(1,2)]
(1) University of Pittsburgh, Pittsburgh, Pennsylvania
(2) Westinghouse Research and Development Center
Pittsburgh, Pennsylvania

Ion beam channeling in association with backscattering techniques can provide an efficient method for detecting and identifying a variety of lattice defects. However, routine application of this technique to defect analysis requires an initial correlation between channeling data and independent determinations of the defect microstructure. In this study, dislocation substructures and helium bubbles have been introduced into molybdenum single crystals by He[+] ion bombardment followed by post bombardment anneals, and channeling studies have been correlated with TEM observations of the microstructure.

Both dislocations and bubbles are found to produce measurable dechanneling rates, but the contribution to dechanneling by bubbles may be obscured in the presence of significant dislocation densities. Furthermore, TEM observations can easily detect microscopic bubbles at concentrations which do not produce significant dechanneling. Channeling measurements, however, are more sensitive than TEM observations of the microstructure to the degree of crystalline disorder created by the initial displacement damage. For planar alignment conditions, the magnitude and energy dependence of the dechanneling probability per bubble is found to be in general agreement with theoretical calculations.

[*]Research sponsored in part by the NSF under Contract No. DMR 78-02598.

337

Introduction

The governed motion of energetic ions between atomic rows or planes is known as channeling. [For a comprehensive review of channeling theory and its application, see Morgan(1)]. This highly directional phenomenon in conjunction with the ion backscattering technique has been used extensively for damage profile and disorder measurements in semiconductors, but has seen only limited application in comparable studies on metals and alloys. For this latter class of materials the backscattered intensity derives essentially from the dechanneling of the ions near some defect which slightly perturbs the atomic regularity of the lattice rather than from the direct backscattering by randomly displaced atoms(2). In fact, Quéré(3) has reviewed the evidence for dechanneling by a number of lattice defects including dislocations, bubbles, G. P. zones, and internal faults and boundaries. Considering the variety of defects which can result in dechanneling, routine application of this technique to metals will be predicated on systematic studies on model systems containing one previously characterized major defect. For example, Merkle et al.(2,4) have investigated the dechanneling produced essentially by small planar defects in gold, and Picraux et al.(5) have studied the dechanneling arising from random dislocation networks in aluminum. In both studies the defects were introduced by ion bombardment and characterized by transmission electron microscopy (TEM).

A range of defect substructures may be introduced into molybdenum by utilizing He^+ ion implantation followed by post implant anneals. These defects include small defect clusters, perfect dislocation loops, random dislocation arrays, and a high density of small helium bubbles(6). Channeling studies employing the backscattering technique are presently being conducted on implanted and annealed crystals, and the backscattered spectra are being correlated with the defect type and density as determined by electron microscopy. This paper presents the preliminary results on dechanneling by bubbles in He^+ implanted and annealed molybdenum single crystals.

Experimental Procedure

The starting material for the channeling studies was high purity zone refined molybdenum single crystals acquired from MRC. These crystals were in the form of rods approximately 12 mm in diameter with a [111] direction aligned along the rod axis. The rods were sectioned into 1 mm thick disks, mechanically lapped with progressively finer abrasives, and finally etched in a solution of 3:1 HCl to H_2O_2. Crystals of suitable quality for channeling experiments were selected on the basis of a low density of subgrains as revealed by the etch, by sharp reflections on back reflection Laue patterns, and by a high ratio of the initial surface peak to the minimum yield as determined by the channeling technique itself.

The crystal disks were subsequently implanted with 2 MeV $^4He^+$ ions from a 2 MV Van de Graaff accelerator equipped with magnetic separation. During implantation the crystals were rocked from 90° to 30° in such a manner as to produce a square wave helium concentration over a depth extending from the projected range R_p (7) to one-half the range, as shown in Fig. 1. One-half of each crystal was masked during the implantation so as to leave a non-implanted reference crystal for the channeling studies. After implantation the crystals were ion milled with 800 eV Kr^+ ions to a depth of approximately 2.2 µm leaving a region of constant helium concentration 0.8 µm in depth for sampling by the channeling ion. The damage profile as calculated by a modified E-DEP-1(8) is also indicated in Fig. 1 for this particular implant geometry. The square wave helium profile results in a broad damage curve which peaks near the front of the helium distribution. However,

338

over the region sampled by the channeling ion after milling ($\approx 2.2 \rightarrow 3.0$ µm) the damage varies by only about a factor of two.

The channeling experiments were performed on a 4.8 meter beam line on the 2 MV accelerator using He$^+$ ions at 1.0, 1.5, and 2.0 MeV, aligned in both the <111> axial and {110} planar directions. The beam was well collimated to produce an angular divergence of 0.03°, and each spectrum was obtained under a constant total fluence as determined by the backscattering from a rotating gold foil which sampled the beam approximately 7% of the time. The aligned spectra were recorded from both the implanted and non-implanted halves of each crystal,

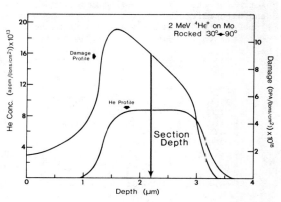

Fig. 1 - Calculated helium and damage profiles for 2 MeV He$^+$ on molybdenum rocked 30° to 90°.

and the random spectra was obtained from the non-implanted half. The energy scale of the backscattered spectra was converted to a depth scale by employing the electronic stopping powers of ^{4}He$^+$ in molybdenum(7) with the assumption that the energy loss of the channeled ions is the same as that for ions impinging along a random direction.

The microstructure for each condition was obtained from single crystal TEM specimens implanted either with the channeling specimens or in groups of eight under identical conditions. After the front surface of the TEM specimens were milled to the required 2.2 µm depth the foils were back thinned with a solution of 7:1 methanol to H_2SO_4 at -75°C. The section depth for both the TEM and channeling specimens after ion milling was measured using a profilometer, and the foil thicknesses of the TEM specimens were measured using standard stereo pair techniques. All subsequent annealing of the implanted crystals was carried out in vacuum at a pressure of $\approx 1.3 \times 10^{-5}$ Pa in a furnace of low thermal mass which permitted rapid heating and cooling of the samples.

Prior to the channeling studies, a survey was made of the microstructures obtained over a wide range of helium concentrations and annealing conditions. Specifically, to produce a high density of bubbles in the annealed condition, helium concentrations of at least 7×10^3 at. ppm are necessary. At present channeling studies have been made on crystals implanted to 7×10^3 at. ppm He in the square wave and annealed at 1200°C for 3 min (hereafter referred to as condition A); further annealed at 1400°C for 30 min (condition B); and implanted to 3.5×10^4 at. ppm He and annealed at 1400°C for 30 min (condition C). Aligned spectra were also obtained from the implanted but unannealed crystals for an initial reference. The initial displacement damage before annealing is approximately 0.6 dpa (displacements per atom) and 3 dpa for the 7×10^3 at. ppm He and 3.5×10^4 at. ppm He implants, respectively.

Experimental Results

A partial summary of the initial TEM survey on helium implanted and annealed crystals is given in Table I, and the 1.5 MeV He$^+$ backscattered spectra for the three annealed conditions, as well as for the implanted and unannealed reference conditions, are shown in Figs. 2-4. On these figures (i) refers to the aligned spectra from the implanted half of the crystal, (n-i) from the non-implanted half, and (r) refers to the random spectrum.

Table I. Summary of TEM Results for Helium in Molybdenum

Annealing Time (min)	Bubble Density N_v (1/cm)^3	Radius $\overline{r}$ (nm)	Mean Squared Radius/Vol $(\overline{r^2}N_v)$ (1/cm)	Dislocation Density (cm/cm^3)
7×10^3 at. ppm He - 1200°C				
(A)* 3	2.7×10^{17}	1.2	4.0×10^3	4×10^{10}
(B)+30 min-1400°C:$\overline{r^2}N_v$ assumed to be 3 $\times 10^3$				2.5×10^{10}
30	1.4×10^{17}	1.5	2.8×10^3	–
300	8.1×10^{16}	1.8	2.4×10^3	–
7×10^3 at. ppm He - 1400°C				
30	8.0×10^{16}	2.1	3.0×10^3	–
300	2.6×10^{16}	3.3	2.8×10^3	
3.5×10^4 at. ppm He - 1400°C				
(C) 30	4.5×10^{17}	1.4	9.2×10^3	6×10^{10}

*Experimental Conditions

The results summarized in Table I indicate that annealing molybdenum containing 7×10^3 at. ppm He for as short a time as 3 min at 1200°C produces a high density ($N_v > 10^{17}$/cm^3) of small bubbles ($\overline{r} \approx 1.2$ nm). Furthermore, with longer annealing times or higher temperatures, the bubble density (N_v) decreases and the average size ($\overline{r}$) increases. However, for the specific case of dechanneling by bubbles, the important defect parameter will be the total projected surface area of bubbles seen by the ion beam as it progresses a unit depth into the crystal. This quantity is given by $\pi \overline{r^2} N_v$ which can be seen from the data summarized in Table I to remain essentially constant for a given helium concentration independent of annealing time or temperature. Quéré(3) has pointed out that this situation should obtain for equilibrium bubbles during coarsening, and experimentally this has been previously verified for helium bubbles in annealed aluminum(9). This result can be used advantageously in the study of dechanneling by bubbles since this component to the spectra should remain essentially constant upon successive anneals until a saturation condition is reached which derives essentially from dechanneling by bubbles alone. Although the relevant bubble statistics for the doubly annealed case (condition B) have not as yet been measured, it is not unreasonable to assume that $\overline{r^2} N_v \approx 3 \times 10^3$/cm for this case. The dislocation densities for the three annealed cases which have been studied by channeling are also indicated in Table I, and are seen to lie in the 10^{10} cm/cm^3 range with condition C exhibiting the highest density at 6×10^{10} cm/cm^3. Between condition A and condition B the dislocation density drops by less than a factor of two from 4×10^{10} cm/cm^3 to 2.5×10^{10} cm/cm^3.

As an initial reference, the backscattered spectra for the implanted but unannealed crystals are shown in Fig. 2 and the corresponding TEM microstructures are shown in Fig. 5. Both microstructures are characterized by a somewhat irregular dislocation array in close association with a high density of small defects exhibiting strain contrast. Although the defect and dislocation densities appear to be comparable in each micrograph, Fig. 5a corresponds to ≈0.6 dpa (7×10^3 at. ppm He) whereas Fig. 5b corresponds to ≈3.0 dpa (3.5×10^4 at. ppm He). The aligned backscattered spectra indicate a considerable increase in the dechanneled yield of the implanted crystals over and above that characteristic of the non-implanted reference crystals. The greater amount of damage deposited in the 3.5×10^4 at. ppm He crystal is now readily apparent from the initially higher dechanneling rate for the aligned spectrum in Fig. 2b than for the aligned spectrum in Fig. 2a (7×10^3 at. ppm He).

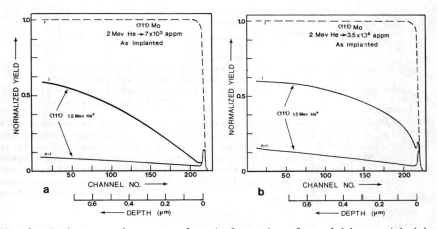

Fig. 2 - Backscattered spectra after implantation; for molybdenum with (a) 7×10^3 at. ppm He and (b) 3.5×10^4 at. ppm He.

A typical microstructure for the annealed crystals is shown in Fig. 6 (condition A). The complex microstructure indicative of the room temperature implants is now replaced by a simple random dislocation network and a high density of small bubbles. Close inspection of the micrograph taken under kinematic conditions to reveal the bubbles (Fig. 6b) shows that a significant fraction of the dislocation line length is decorated by bubbles having a somewhat greater diameter than the general population. The consequence of this microstructural feature on the dechanneling will be made evident in a later section. The backscattered spectra for the lower helium concentration are shown in Fig. 3a for the first anneal (condition A) and Fig. 3b for the second anneal (condition B). These figures now include the spectra from the {110} planar alignment as well as the <111> axial alignment. When compared with Fig. 2a, these spectra indicate that a considerable fraction of the disorder has been removed by the first anneal whereas, after the second anneal (Fig. 3b), the backscattered spectra for the implanted crystals exhibit only a slightly greater yield than those for the non-implanted reference crystals. This result implies that a bubble density of ≈10^{17}/cm^3 is necessary for detection by the channeling technique. For both conditions the separation between the implanted spectra and

341

non-implanted spectra is greater for the {110} planar alignment than for the the <111> axial alignment.

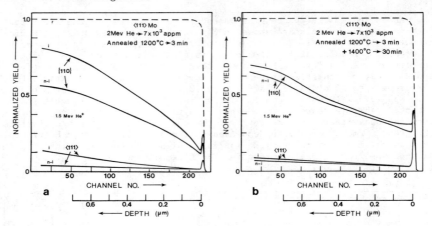

Fig. 3 - Backscattered spectra for 7 x 10^3 at. ppm He annealed at 1200°C for 3 min (a) plus 1400°C for 30 min (b).

The backscattered spectra for condition C are shown in Fig. 4. Again, a considerable reduction in the dechanneled yield after annealing occurs when compared to the implanted and unannealed condition. However, the de-channeled fraction is higher than in the lower dose implants indicative of the higher density of defects present after annealing as summarized in Table I. The break in the slope at approximately 0.5 → 0.6 µm in the <111> axial curve corresponds roughly to the end of range of the He implant for this crystal since the ion milling removed ≈2.4 µm from the surface.

The backscattered yield arising from dechanneling can be related to the defect density and cross section for dechanneling per defect by a first order approach outlined by Merkle et al.(2,4). If (X_i) is the dechanneled fraction (the backscattered yield normalized with respect to the random spectrum) and $(1 - X_i)$ the channeled fraction, then:

$$\frac{d\,X_i}{dz} = (1 - X_i) \sum_j \sigma_j n_j \qquad (1)$$

where (z) is depth, (σ_j) the cross section for dechanneling for a defect of type (j) and density (n_j); and the summation is made over all defect types. If the intrinsic dechanneling as measured in the non-implanted reference crystal is simply additive to that produced by the defects, then Eq. (1) becomes upon integration:

$$\frac{(1 - X_i)}{(1 - X_{n-i})} = \exp\left[-\int \sum_j \sigma_j n_j dz\right] \qquad (2)$$

where (X_{n-i}) is the dechanneled frac-tion at depth z of the non-implanted reference crystal, and

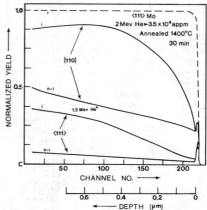

Fig. 4 - Backscattered spectra for 3.5 x 10^4 at. ppm He annealed at 1400°C for 30 min.

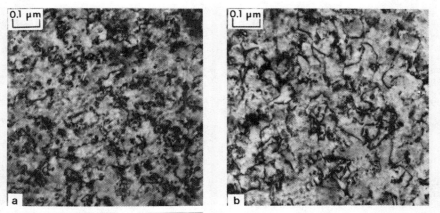

Fig. 5 - Initial microstructure of implanted but unannealed molybdenum with
(a) 7×10^3 at. ppm He and (b) 3.5×10^4 at. ppm He. Two beam
(g = [110]) imaging conditions.

$(1 - X_i)/(1 - X_{n-i})$ is termed the dechanneling parameter. Therefore, at any
point on the channeling spectra:

$$Ln \left[\frac{(1 - X_i)}{(1 - X_{n-1})} \right] = \int \sum_j \sigma_j n_j dz \qquad (3)$$

A plot of Ln of the dechanneling parameter versus depth will consequently
result in a curve whose slope gives the total cross section/unit volume
($- \sum \sigma_j n_j$ in units of 1/length). Such a plot is shown in Fig. 7 for condition C up to a depth corresponding to the limit of the He deposition in

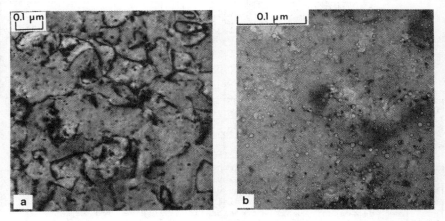

Fig. 6 - Typical microstructure of annealed molybdenum crystals showing (a)
dislocations under g = [110] two beam conditions and (b) bubbles
under kinematic diffracting conditions. Condition A.

the ion milled sample. The resulting linear relationship is predicted by Eq. (3) for a defect density which is independent of depth, and substantiates the approximately constant He concentration and damage deposition with depth indicated in Fig. 1. The experimentally determined total cross sections/unit volume ($\Sigma\ \sigma_j n_j$) arrived at by taking the slopes of the best straight line fit through the data points as indicated in Fig. 7 are summarized in Table II for the three annealed conditions. This data includes the 1.0, 1.5, and 2.0 MeV results from both the <111> axial alignment and the {110} planar alignment. Both the magnitude and energy dependence of this data in relation to the observed microstructures are discussed in the next section.

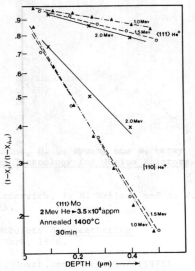

Fig. 7 - Dechanneling parameter versus depth for condition C.

Table II. Experimental Values of Dechanneling Cross Section per Unit Volume

He⁺ Beam Energy (MeV)	Total Cross Section per Unit Volume in Units of (1/µm)					
	7 x 10³ at. ppm He				3.5 x 10⁴ at.ppm He	
	3 min-1200°C		+ 30 min-1400°C		30 min-1400°C	
	<111>	{110}	<111>	{110}	<111>	{110}
1.0	.05	.48	.04(.04)	.21(.24)*	.41	3.8
1.5	.09	.70	.03(.03)	.18(.21)	.53	3.5
2.0	.13	1.65	.04(.04)	.16(.18)	.58	2.0

*Dechanneling probability per bubble.

Background and Discussion

Although the experimental conditions were specifically designed to study dechanneling by bubbles alone, the TEM results indicate that both bubbles and dislocations would contribute significantly to dechanneling in the implanted and annealed crystals. Quéré(10) has theoretically derived the dechanneling cross section for dislocations in the axial alignment. This cross section/unit length of dislocation $\overline{\lambda}$ is given by:

$$\overline{\lambda}_a = \left[b\ da_{TF}\ E/\alpha\ Z_1 Z_2\ e^2 \right]^{1/2} \tag{4}$$

where b is the Burger's vector of the dislocation, d is the atomic spacing along the rows of the channel, a_{TF} is the Thomas-Fermi screening radius, E the energy of the channeled ion, Z_1 and Z_2 the atomic number of ion and

344

target respectively, and e the electronic charge. The cross section/unit length ($\overline{\lambda}$) actually represents the average radius of a cylinder co-axial to the dislocation line which leads to dechanneling when intercepted by a channeled ion. The parameter α depends on dislocation type and is 12.5 for screw dislocations and 4.5 for edges.

For planar dechanneling by dislocations Quéré and Mory(11) give $\overline{\lambda}$ as:

$$\overline{\lambda}_p = \left[bE/8.6 \; Z_1 Z_2 \; e^2 \; N_p \right]^{1/2} \tag{5}$$

where N_p is the atomic density of the planes forming the channel wall. Therefore, the total cross section/unit volume ($\sigma_j n_j$) can be related to ($\overline{\lambda}$) and (ρ) the dislocation density by:(5)

$$(\sigma n)_{\text{dislocations}} = \overline{\lambda} \; \rho / \beta \tag{6}$$

where β, a geometrical term relating the projection of the dislocation array onto a plane perpendicular to the channeled beam, is $(4/\pi)$ for a random dislocation network. The values of these cross sections/volume are tabulated in Table III for the dislocation densities corresponding to conditions A, B, and C as calculated from Eqs. (4), (5) and (6) for ^{4}He$^+$ on Mo. The parameter (α) was taken to be 8.5, the average for edge and screw dislocations.

Table III. Calculated Values of Dechanneling Cross Section
per Unit Volume for Dislocations

He Beam Energy (MeV)	Total Cross Section per Unit Volume in Units of (1/μm)					
	$\rho_\perp$ 4×10^{10} cm/cm^3		$\rho_\perp$ 2.5×10^{10} cm/cm^3		$\rho_\perp$ 6×10^{10} cm/cm^3	
	<111>	{110}	<111>	{110}	<111>	{110}
1.0	.30	1.36	.19	.85	.45	2.0
1.5	.36	1.66	.23	1.04	.56	2.5
2.0	.42	1.92	.26	1.20	.63	2.9

Dechanneling by bubbles has been investigated both experimentally in transmission experiments and theoretically by Ronikier-Polonsky et al.(9). They find that the cross sections for planar dechanneling can be related to the projected surface area of the bubbles by:

$$(\sigma n)_b = \pi \; \overline{r^2} \; N_v \; P \tag{7}$$

where P is the probability that an ion entering a bubble will be dechanneled upon re-entering the crystal at the opposite side. To a first approximation P is independent of bubble size, and its theoretical value of 0.2 to 0.3 for planar dechanneling agrees with the experimental value determined in the transmission experiments(9). Note that the cross section for dechanneling by bubbles is independent of ion energy, whereas the cross section for dechanneling by dislocations follows an $E^{1/2}$ dependence.

345

The experimentally determined cross sections reported in Table II for the lower He dose (conditions A and B) are considerably lower than the calculated values based on the measured dislocation densities alone as tabulated in Table III. This result is not surprising since the measured densities fail to account for the considerable fraction of dislocation core which is actually replaced by the bubbles. However, this fact obscures the relative contributions of the bubbles and the dislocations to the total dechanneling. Also, since there exists a considerable reduction in the observed dechanneling between the first and second anneal, it is uncertain that the predicted saturation level has been reached without performing further annealing studies.

However, condition B should represent the closest situation studied thus far which represents dechanneling by bubbles alone. If the experimental probability per bubble for dechanneling is calculated under this assumption, then $P \approx 0.03 \rightarrow 0.04$ for <111> axial dechanneling and $\approx 0.16 \rightarrow 0.2$ for {110} planar dechanneling. These numbers are indicated in the parentheses in Table II. The value of ≈ 0.2 for the planar dechanneling probability is close to the theoretical value previously cited. The axial values ($0.03 \rightarrow 0.04$) are also consistent with the theoretically anticipated value since they are comparable to the minimum yield for a beam entering a crystal along an axial direction. To the first order, this would correspond to the dechanneling probability, P, for a bubble surface. Furthermore, the energy dependence of the cross sections for this condition is approximately constant (although there is a tendency for the planar results to drop slightly) which is another feature predicted by dechanneling by bubbles.

Since there is a significant drop in the dechanneling cross sections for condition B relative to condition A, a significant portion of the dechanneling for the latter case must result from the dislocations. The energy dependence of the cross sections support this view although this dependence is somewhat greater than the $E^{1/2}$ dependence predicted by Eqs. (4) and (5).

The cross sections for condition C are difficult to correlate with the microstructures at the present time. The <111> axial results show an $E^{1/2}$ dependence consistent with dechanneling by dislocations whereas the planar {110} cross sections show a tendency to drop with increasing energy. Further annealing studies are necessary for this He level before any definitive statements can be made.

Summary and Conclusions

(1) Channeling experiments employing the backscattering technique can detect the presence of bubbles if the specific surface area ($\pi \, r^2 \, N_v$) is approximately $10^4/cm$. This necessitates a high density of bubbles ($N_v \approx 10^{17}/cm^3$) if the average radius $\bar{r} \lesssim 1.5$ nm. These stringent requirements could be relaxed somewhat under planar alignment which is more sensitive to the presence of defects under certain conditions than the axial alignment.

(2) For the case in which most of the observed dechanneling was attributed to bubbles, the probability for dechanneling per bubble (P) is approximately 0.04 for axial alignment and 0.2 for planar alignment. The latter value is in general agreement for predicted planar dechanneling probabilities for bubbles.

(3) In the presence of dislocations (or other defects) which result in strong dechanneling, the presence of bubbles may be undetected by the channeling technique alone.

346

(4) The value of the channeling technique is enhanced if used in conjunction with other techniques such as TEM which indicate the density and spatial arrangement of the defects.

(5) The ion bombardment produces a complex microstructure consisting of a high density of small defects in association with a random dislocation network. Channeling experiments can differentiate the relative degree of disorder introduced by this microstructure more easily than TEM observations.

References

1. Channeling - Theory, Observation and Applications, D. V. Morgan, ed., J. Wiley and Sons, London (1973).

2. K. L. Merkle, P. P. Pronko. D. S. Gemmell, R. C. Mikkelson and J. R. Wrobel, "Dechanneling from 2-MeV He$^+$ Damage in Gold," Physical Review B, 8(3) (1973), pp. 1002-10.

3. Y. Quéré, "Dechanneling of Fast Particles by Lattice Defects," Journal of Nuclear Materials, 53 (1974), pp. 262-7.

4. P. P. Pronko and K. L. Merkle, "Dechanneling from Damage Clusters in Heavy Ion Irradiated Gold," pp. 481-93 in Applications of Ion Beams to Metals, S. T. Picraux, E. P. EerNisse and F. L. Vook, eds., Plenum, New York (1974).

5. S. T. Picraux, E. Rimini, G. Foti and S. U. Campisano, "Dechanneling by Dislocations in Ion-Implanted Al," Physical Review B, 18 (5) (1978), pp. 2078-96.

6. J. Greggi, C. F. Tzeng, W. J. Choyke and J. R. Townsend, unpublished research.

7. J. F. Ziegler, Helium Stopping Powers and Ranges in All Elemental Matter, Vol. 4, Pergamon Press, New York (1977).

8. I. Manning and G. P. Mueller, Computer Physics Communications, 6 (1973).

9. D. Ronikier-Polonsky, G. Desarmot, N. Housseau and Y. Quéré, "Dechanneling by Gas Bubbles in a Solid," Radiation Effects, 27 (1975), pp. 81-8.

10. Y. Quéré, "Dechanneling Cylinder of Dislocations," Physica Status Solidi, 30 (1968), pp. 713-22.

11. J. Mory and Y. Quéré, "Dechanneling by Stacking Faults and Dislocations," Radiation Effects, 13 (1972), pp. 57-66.

atoms in substitution site averages. The enhancement may be further
used in combination with ion backscattering, which often allows crack[?]
profiles to be determined with depth resolu[...]
[...] to identify defects or cracks
[...] surface disorder resulting from ion implantation, surface layer
growth, or other surface treatments. While there have been many studies
of ion implantation disorder in semiconductors (4), little quantitative work
has been carried out on irradiated metals by the channeling technique.

Typically, ion channeling experiments utilize a Van de Graaff accelerator
to produce monoenergetic helium or proton beams of energies 0.5 to 3.0 MeV.
The beam is collimated to within ~ 0.05°, fwhm, and the crystal is
oriented with a goniometer to a similar angular resolution. For greatest
versatility the Rutherford backscattering of the incident beam of particles
is detected and energy analyzed by a gold surface barrier detector. In some
cases ion-induced x-rays or nuclear reaction products are also monitored to
enhance elemental sensitivities. Detailed descriptions of the experimental
configurations are given elsewhere (3).

The ion channeling technique derives its sensitivity from the enhanced
scattering of channeled particles into nonchanneled trajectories that results when atoms are displaced from perfect host-atom sites. The influence
of defects on the channeled spectrum depends on the defect configuration
and contrasts markedly, for example, between dislocations, amorphous clusters
and interstitials in specific sites. The channeling technique is highly
complementary to transmission electron microscopy (TEM) studies of defects
and many detailed disorder studies today by ion channeling also include TEM
analysis. While both techniques are structural probes which detect deviations from the perfect periodicity of the crystal, there are important
differences. TEM is best suited to identify the defects present and determine
their lateral distribution, but it is primarily sensitive to larger defects,
such as clusters of point defects, and to extended defects, such as dislocations. Ion channeling is best suited to determine the number density
and depth profile of the defects, and maintains its sensitivity down to a
defect size corresponding to single atom displacements >0.1Å. When
solute atoms are involved, as in solute-defect complexes, it is sometimes
possible to determine the solute lattice site by channeling, and thereby,
further characterize the defect configuration (2,5).

In this review, the principles of the ion channeling technique used
to detect and quantitatively determine the depth profile of defects in
single crystals will be outlined. Then selected examples will be given
from the area of irradiation damage studies in metals to illustrate the
range of applications of the technique to this area. Finally, the relative
advantages and limitations of the technique will be discussed.

Disorder Analysis by Channeling

Perfect Crystal

For a perfect crystal, alignment of an energetic beam of ions with
a major crystal direction results in most particles undergoing special
"channeled" trajectories as they oscillate between the rows or planes of
atoms (1). These particles are reflected off the atom rows or planes much
like a stone skipping off water, with a minimum impact parameter of

CRYSTAL DEFECT STUDIES USING X-RAY DIFFUSE SCATTERING*

B. C. Larson

Solid State Division, Oak Ridge National Laboratory
Oak Ridge, Tennessee   37830

Microscopic lattice defects such as point (single atom) defects, dislocation loops, and solute precipitates are characterized by local electronic density changes at the defect sites and by distortions of the lattice structure surrounding the defects. The effect of these interruptions of the crystal lattice on the scattering of x-rays is considered in this paper, and examples are presented of the use of the diffuse scattering to study the defects.

X-ray studies of self-interstitials in electron irradiated aluminum and copper are discussed in terms of the identification of the interstitial configuration. Methods for detecting the onset of point defect aggregation into dislocation loops are considered and new techniques for the determination of separate size distributions for vacancy loops and interstitial loops are presented. Direct comparisons of dislocation loop measurements by x-rays with existing electron microscopy studies of dislocation loops indicate agreement for larger size loops, but x-ray measurements report higher concentrations in the smaller loop range. Methods for distinguishing between loops and three-dimensional precipitates are discussed and possibilities for detailed studies of precipitates are considered. A comparison of dislocation loop size distributions obtained from integral diffuse scattering measurements with those from TEM shows a discrepancy in the smaller sizes similar to that described above.

---

*Research sponsored by Union Carbide Corporation under contract W-7405-eng-26 with the U.S. Department of Energy.

Introduction

Detailed use of x-ray diffuse scattering around Bragg reflections for the study of small, lattice defects has increased considerably in the last ten years. Analytical work by Krivoglas (1), Dederichs (2) and Trinkaus (3) was instrumental in developing the present form of the theory for defect clusters, and radiation damage investigations (4-8) have been responsible for much of the experimental activity in this field. Alternate formulations of defect diffuse scattering exist (9) and provide powerful tools for the study of lattice structure and defects; however, in the limited scope of this paper, only the former treatment will be discussed because of its adaptibility to the case of sizeable clusters. No attempt is made to give a comprehensive review of the work in this field. Rather, it is the aim of this paper to indicate the range of information that can be obtained from this scattering and the methods used in this process. A brief review of the mathematical framework will be given, a short discussion of the experimental requirements for diffuse scattering measurements will be presented, and examples of point defect, dislocation loop and solute precipitate studies will be given. The primary purpose of the results shown will be to illustrate the techniques rather than to discuss the physical interpretations. Where meaningful, comparisons with electron microscopy results will be made in order to establish a reference point. Finally, more general comments and conclusions relative to the applicability and limitations of the x-ray method will be discussed.

Theoretical Framework

The theory of diffuse scattering from lattice defects has been discussed in some detail in a number of papers (1-3), and only an outline of the theoretical results required to interpret the scattering results will be given here. The scattering cross-section for randomly distributed defects in a crystal lattice is given by

$$\frac{d\sigma}{d\Omega} (\vec{K}) = (r_e f_h)^2 \left| A (\vec{K}) \right|^2 \tag{1}$$

where the scattering vector $\vec{K} = \vec{h} + \vec{q}$ specifies the measuring position in terms of a reciprocal lattice vector $\vec{h}$ and a position $\vec{q}$ relative to $\vec{h}$. r_e is the classical electron radius, f_h is the atomic scattering factor of the host lattice (including thermal and polarization effects), and $A(\vec{K})$ is the scattering amplitude. The scattering amplitude contains all the physics pertaining to the defect structure and, in general, is composed of the scattering from the defect atoms (sometimes called Laue or direct scattering) and the scattering generated as a result of the displacement of the atoms surrounding the defect site. The form of this is

$$A(\vec{K}) = \sum_j e^{i\vec{K} \cdot \vec{r}_j^d} + \sum_j e^{i\vec{q} \cdot \vec{r}_j} (e^{i\vec{K} \cdot \vec{S}_j} - 1) \tag{2}$$

where $\vec{r}_j^d$ is the position of the defect atom(s) and $\vec{S}_j$ is the displacement of the jth lattice atom from its undistorted position $\vec{r}_j$. The -1 in the second term is inserted to remove the scattering from the periodic lattice (Bragg reflections, tails of Bragg reflections) so that for $\vec{S}_j = 0$, the distortion term vanishes.

350

For single interstitials, $A(\vec{K})$ is given by (4)

$$A(\vec{K})_I = e^{i\vec{K}\cdot\vec{r}_I} + \sum_j e^{i\vec{q}\cdot\vec{r}_j} (e^{i\vec{K}\cdot\vec{S}_j} -1) \tag{3}$$

where $\vec{r}_I$ is the position of the interstitial atom. The first term represents the Laue scattering and the second represents the scattering from the distortion field around the defect. The entirely analogous amplitude for dislocation loops is given by

$$A(\vec{K})_L = \sum_j e^{i\vec{K}\cdot\vec{r}_j^L} + \sum_j e^{i\vec{q}\cdot\vec{r}_j} (e^{i\vec{K}\cdot\vec{S}_j} -1) \tag{4}$$

in which $\vec{r}_j^L$ are the positions of the atoms in the dislocation loop. The sign of $\vec{S}_j$ reverses on going from interstitial to vacancy loops and $\vec{r}_j^L$ changes to account for an intrinsic rather than extrinsic loop. An additional factor of (-1) must be included in the vacancy loop Laue scattering to represent the absence of atoms rather than the presence of additional atoms. The scattering amplitude for coherent precipitates of atoms with scattering factor f_p in a host lattice of scattering factor f_h is given by

$$A(K)_p = \sum_j e^{i\vec{q}\cdot\vec{r}_j^P} (f_p e^{i\vec{K}\cdot\vec{S}_j} -1) + \sum_j e^{i\vec{q}\cdot\vec{r}_j} (e^{i\vec{K}\cdot\vec{S}_j} -1). \tag{5}$$

The first sum runs over the $\vec{r}_j^P$ positions of the precipitate atoms and the second sum over the host atoms outside the precipitate volume. Since there are no additional (or missing) lattice sites involved, the Laue term as in Eqs. (3,4) does not appear as such; however, the first term in Eq. (5) can be considered the Laue scattering here. The factor f_p/f_h occurs because we are replacing host atoms of scattering f_h with ones of scattering f_p. For simplicity here, the change in the host scattering due to the loss of the randomly distributed solute atoms through precipitation has been neglected.

Eqs. (3,4,5) require detailed knowledge of the displacements around the defect sites and require numerical integrations of the lattice sums for evaluation. For detailed studies, this is true in general; however, analytic approximations are available that provide symmetry information and the overall form of the scattering in particular regions (3). When $K\cdot S \ll 1$, the distortion scattering can be expanded so that

$$A^H(\vec{K}) \approx e^{i\vec{q}\cdot\vec{r}}(i\vec{K}\cdot\vec{S}) = i\vec{K}\cdot\vec{S}(\vec{q}) \tag{6}$$

which is the so called Huang diffuse scattering amplitude. Even in cases where $\vec{K}\cdot\vec{S}(\vec{r}) \gtrsim 1$ out to a radius of $r=R_c$ around a clustered defect, Eq. (6) can remain a useful result for $q < 1/R_c$ (i.e. close to the reciprocal lattice point). For the small q region $A^H(\vec{K})$ can be calculated analytically for arbitrary lattice anisotropy as (6)

$$\left|A^H(\vec{K})\right|^2 = \frac{h^2}{q^2} (\gamma_1\pi_1 + \gamma_2\pi_2 + \gamma_3\pi_3) \tag{7}$$

351

where the Υ's are functions of the elastic constants, $\vec{h}$, and $\vec{q}$. The π's contain the defect characteristics in the form of the dipole-force tensor. π_1 is proportional to $(\Delta V)^2$ where ΔV is the volume change due to the defect and π_2 and π_3 are related to the anisotropy (relative to spherical or cubic symmetry) of the distortion field around the defect site.

The next approximation to Eq. (2), obtained by including all the even orders of $\vec{K} \cdot \vec{S}$ leads to

$$\left| A(\vec{K}) \right|^2 \simeq \left| A^H(\vec{K}) \right|^2 - 2i \ A^H(\vec{K}) \ L(\vec{K}) \tag{8}$$

where $L(\vec{K})$ is essentially the Debye-Waller factor associated with the defect. Since $A^H(\vec{K})$ is antisymmetric in q and $L(\vec{K})$ is symmetric in q, Eq. (8) indicates the important result that the scattering for $q < 1/R_c$ contains a term proportional to $(\Delta V)^2 h^2/q^2$ and a term proportional to $(\Delta V)h/q$ so that the symmetric average $((I(q) + I(-q))/2)$ of the intensity parallel and anti-parallel to $\vec{h}$ determines $(\Delta V)^2$ and the sign of the asymmetry determines if the defect contracts (vacancy like) or expands (interstitial like) the lattice. Higher scattering intensities for $q > 0$ imply interstitial like defects where the displacements are outward. A rough approximation for $L(\vec{K})$ can be given for $q < 1/R_c$ as (6)

$$L(\vec{K}) \approx \frac{|h\Delta V|}{10 V_c}^{3/2} \tag{9}$$

where V_c is one atomic volume.

The region of $q > 1/R_c$, for which the above approximations break down, no longer retains the $1/q^2$ falloff but decreases more like $1/q^4$, as a result of the increased interferences. An analytic approximation for $q > 1/R_c$ is given by (6,10)

$$\left| A^A(\vec{K}) \right|^2 \approx \frac{h|\Delta V|}{4\pi V_c q^4} \ \Phi(\vec{q}/q). \tag{10}$$

$\Phi(\vec{q}/q)$ is, in general, a function of angle but has an average value of about 150, which is applicable for both spherical defects and dislocation loops. This amplitude has been given the superscript A (Asymptotic diffuse scattering) and comes mainly from the highly distorted region near the defect site where $\vec{K} \cdot \vec{S} \gtrsim 1$.

Eqs. (8,10) provide the basis for understanding the general features of defect diffuse scattering near Bragg reflections. However, for detailed quantitative information, direct numerical integration of Eqs. (3,4,5) is necessary. These calculations will not be discussed in detail in this paper, although the results of such calculations play an essential role in the results to be discussed later (11,12).

The above discussion applies to the usual differential diffuse scattering measurements in which the measured intensity in symmetric geometry is related to the scattering cross-section of Eq. (1) by

$$I(\vec{K}) = \frac{I_0}{2\mu_0} \sum_i c_i \ \frac{d\sigma_i(\vec{K})}{d\Omega} \ \Delta\Omega \tag{11}$$

352

where I_0 is the incident beam power, μ_0 is the linear absorption coefficient, C_i is the volume concentration of the ith defect and $\Delta\Omega$ is the solid angle subtended at the detector. These measurements correspond to the geometry sketched in Fig. (1a). A useful variation of the scattering geometry is shown in Fig. (1b) in which the scattering from dislocation loops is studied using a wide-open (that is, no collimator defining the diffracted beam)

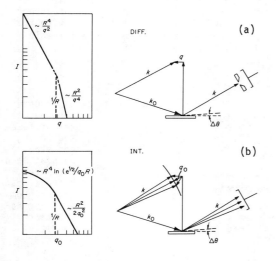

(a)

(b)

Fig. 1- Schematic view of Differential and Integral diffuse scattering geometries and results.

detector (13). Under the approximation of isotropic scattering (which is reasonable for loops in cubic materials), the intensity averaged at $\pm q_0$ $(I^s(q_0))$ for the Integral Diffuse Scattering is given by (6)

$$I^s(q_0) = \frac{I_0(r_e f)^2}{2\mu_0} \left(\frac{h}{k}\right)^2 2\pi\tau \sum_i C_i \left(\frac{b\pi R_i^2}{V_c}\right)^2 \begin{bmatrix} \ln(e^{1/2}\alpha/q_0 R_i) \\[2mm] \frac{\alpha^2}{2q_0^2 R_i^2} \end{bmatrix} \tag{12}$$

where $k = 2\pi/\text{wavelength}$, $q_0 = h\Delta\theta \cos\theta_B$, $\alpha = 1$, $\Delta\theta$ is the angular missetting from the Bragg angle θ_B, and

$$2\pi\tau = \frac{8\pi}{15} + \frac{\pi}{15} \frac{(3\nu^2 + 6\nu - 1)}{(1-\nu)^2} \cos\theta_B \quad . \tag{13}$$

The upper relation in Eq. (12) is for $q_0 \leq \alpha/R_i$ and the lower for $q_0 \geq \alpha/R_i$ where R_i is the radius of the ith dislocation loop. Eq. (12) can be used to determine the loop size distribution through fitting the experimental measurements with C_i as parameters. Eq (11) can be used with differential diffuse scattering measurements to determine the size and concentrations of defects in the same manner.

353

Experimental Considerations

The study of defects ranging from single interstitial atoms to larger clusters and precipitates requires the use of a number of experimental configurations. One of the most useful and versatile configurations has been that of a focussed monochromating system using a curved monochromator crystal together with x-ray generators varying from 2 KW fixed anode systems to 15 KW rotating anode supplies. Several variations of the focussing geometry, such as shown in Fig. (2a), are used (4,7). The important aspect being the selection of only $K\alpha_1$ radiation and the capability of varying the angular divergence of the incident beam (and hence the measurement resolution). This is done by limiting the width of the beam striking

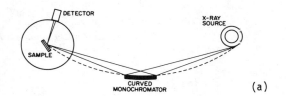

(a)

X—RAY DIFFUSE SCATTERING SYSTEM

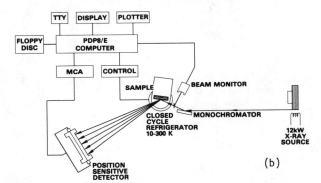

(b)

Fig. 2- a) Focussed monochromator diffuse scattering geometry.
b) Flat monochromator, diffuse scattering geometry with a position sensitive detector.

the curved monochromator crystal. For beam divergences 0.1 - 1°, incident beam powers of 10^7 - 10^9 photons/sec (depending on the vertical resolution) can be achieved and are adequate for Huang diffuse scattering from single interstitials. In larger angle, point defect studies, beam powers >5 x 10^{10} photons/sec have been achieved using total reflecting mirrors in conjunction with an 100 KW x-ray source (5). Because of the low scattering rates between the Bragg reflections, a large multi-detector (~100) array was also necessary.

In the study of larger defect clusters, higher resolution is required and another technique employing a flat monochromator, 1 meter incident and diffracted beam distances and a position sensitive detector system has been used as shown in Fig. (2b). The relatively low beam power of ~10^6 photons/sec realized in this scheme is offset by the position sensitive detector. Integral diffuse scattering measurements are considerably less demanding and can be carried out with 10^5 - 10^6 photons/sec because of the large solid angle subtended by the detector. A 2 KW fixed anode generator and flat monochromators suffice for this geometry.

The sample requirements for diffuse scattering measurements discussed in this paper are first that they be single crystals. This can be a rather severe requirement for alloyed and compound materials, but has not proved to be too serious for pure materials. Sample sizes of 0.5 cm² or greater are normally used, although smaller sizes are in principal possible in most cases. For detailed studies, the mosaic spread and subgrain misorientations must be < 1/hR (where R is the defect radius) to ensure that q is well defined substantially into the Huang scattering region. For 10 Å radius clusters, mosaic widths ~0.2° would be possible, but for larger clusters, rather high quality crystals are required.

The subtraction of non-defect diffuse scattering must be carried out before analysis (4-7). This requires either identical crystals (one with and one without defects) or that measurements be made before and after the defects are introduced. Bragg reflections, Bragg tails, Compton scattering, and thermal diffuse scattering fall into this catagory. The thermal scattering can be minimized by measuring at temperatures ~10 K.

Measurement Results

Single Interstitials

Point defects can be characterized by x-ray diffuse scattering, through the symmetry of the distortion field surrounding the defect site and by the atomic configuration at the defect site. The symmetry of the distortion field is obtained from Huang diffuse scattering measurements close to reciprocal lattice points (small q), and the atomic configuration is determined from larger angle scattering between the Bragg reflections. As indicated in Eq. (7), Huang scattering measurements in appropriate directions lead to a determination (3) of the dipole-force tensor of interstitials and therefore differentiate between the several possible interstitial configurations. Without dealing in detail with the individual parameters in Eq. (7), measurements with $\vec{q}$ parallel and perpendicular to $\vec{h}$ at the (hoo), (hho) and (hhh) reflections provide all the obtainable information on the symmetry of the distortion field. As shown in Table I (4), cases of well defined symmetry can be determined by the identification of positions of zero Huang scattering.

Table I. Planes (P), Surfaces (S) or Lines (L) of
Zero Huang Scattering for Characteristic
Defect Symmetries.

Defect Symmetry	Reflection		
	(hoo)	(hho)	(hhh)
Cubic	P⊥ [100]	P⊥ [110]	S⊥[111]
Tetragonal <100>	P⊥ [100]	L‖ [011]	-
Trigonal <111>	-	L‖ [110]	-
Orthorhombic <110>	-	-	-

355

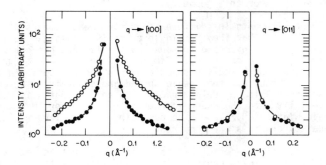

Fig. 3- Huang diffuse scattering at the 400 reflection from interstitials
in electron irradiated (open circles) and unirradiated (filled
circles) aluminum.

An application of the results of Table I is shown in Fig. (3). Interstitial
scattering from low temperature electron irradiated aluminum was found (4)
to be high along the [100] direction at the 400 reflection, but essentially
zero along the [011] direction at this reflection. Of the cases listed in
Table I, cubic and tetragonal symmetry would be allowed, and other
measurements indicated that all configurations other than the <100>-split
interstitial, the octahedral, and the tetrahedral could be ruled out.

While lattice statics calculations have shown that each of these
interstitial configurations have essentially cubic distortion fields (5),
these configurations have significantly different atomic arrangements.
Therefore the scattering at large angles between the Bragg reflections (which
is quite sensitive to the Laue scattering) was able to show the <100>-split
interstitial to be the actual configuration. The same configuration was
found for interstitials in copper (5) and Fig. (4) indicates the large angle
scattering results (smoothed to emphasize the general features) supporting
the <100>-split interstitial in copper. The calculated intensity contours

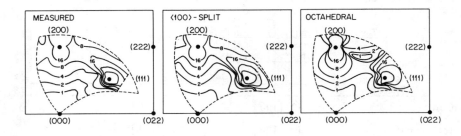

Fig. 4- Large angle scattering from interstitials in electron irradiated
copper, with calculated scattering for the <100>-split and octahedral
interstitial configurations.

were obtained using Eq. (3). The important result then, is that through
scattering measurements both near-to and far-from Bragg reflections, single
atom defects in the concentration range of 10^{-4} - 10^{-3} (atomic fraction) can
be studied with diffuse scattering. These techniques apply to cases of gas
atom interstitials and single sublattice interstitials as well as self
interstitials, of course (7).

Clustering of Point Defects

The clustering of point defects into 3-dimensional aggregates or dislocation loops can be observed in the symmetric averaged scattering
$[I(\vec{q})+I(-\vec{q})]/2$, through the presence of a $1/q^4$ falloff of the diffuse
intensity at large q (as suggested by Eq. (10)) in addition to the $1/q^2$
decrease of the intensity in the Huang region. This effect is shown clearly
in Fig. (5) where 300 K thermal annealing of low temperature electron
irradiated copper (14) is shown to result in increased intensity in the Huang
region and a change to a $1/q^4$ dependence for $q/h > 0.005$. The increased
intensity results from the coherent scattering of point defects that have
aggregated, compared to that for individual point defects scattering
incoherently. According to the discussion of Eq. (10), the position (q_c) of
the transition to the $1/q^4$ dependence provides an estimate of the cluster
size (i.e. $R_c \approx 1/q_c$); however, when distributions of sizes are present, this
estimate can be rather misleading because of the weighting of the scattering
by R^4 and R^2 in the Huang and asymptotic regions, respectively. As will be
discussed below, direct fitting of detailed diffuse scattering calculations
is necessary for accurate size determinations.

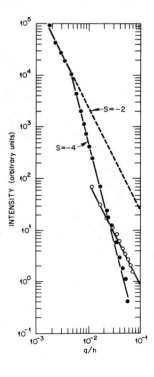

Fig. 5- Huang diffuse scattering from
single interstitials (lower
curve) and dislocation loops
(upper curve) in electron
irradiated copper at the 220
reflection with q along [110].
S denotes the slope of the
lines.

Diffuse Scattering Calculations for Dislocation Loops

Although the asymmetry in the Huang scattering intensity (Eq. (8)) can be used to differentiate between the presence of vacancy or interstitial loops, the R^4 weighting of the intensities in the Huang region strongly biases the interpretation toward the larger loops. This makes it somewhat difficult to study small loops and clusters in the presence of large clusters. In the asymptotic region the weighting for loops is R^2, which gives equal scattering weight to loops with equal total areas. Therefore the intensity in this region is proportional to the number of point defects condensed into loops. This fact makes the asymptotic scattering more appealing for measurements, and detailed calculations for vacancy and interstitial loops on {111} planes in fcc materials have shown (12,15) the 222 reflection to be particularly favorable for distinguishing between vacancy and interstitial loops. This is indicated in Fig. (6) where contour plots of the scattering from 20 Å Frank loops of both vacancy and interstitial type are shown in a form multiplied by q^4 to remove the average $1/q^4$ falloff. The important result is the peak in the scattering at $q > o$ for interstitial loops and at $q < o$ for the vacancy case. This peaked intensity has been identified as Bragg-like scattering from the strained region near the loops, and because the magnitude of the strain in this region is $\varepsilon \approx -b/4R$, this peak can be expected at (12)

$$q = -\varepsilon h \approx \frac{bh}{4R} .$$

(14)

Not only is the vacancy-interstitial nature of the loop specified, but also the size can be determined through comparisons with numerical scattering calculations, since ε is a function of R.

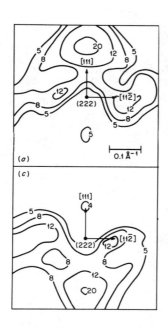

Fig. 6- Calculated diffuse scattering intensity (multiplied by q^4) for Frank loops of interstitial (upper) and vacancy (lower) loops at the 222 reflection in an isotropic material.

Vacancy and Interstitial Loop Size Distributions

The above characteristics were observed in earlier calculations (12) and were seen experimentally (16) on 4 K electron irradiated copper that had been annealed to 200 K. The presence of interstitial type loops in copper (at 200 K) resulted in a highly asymmetric scattering similar to that indicated in Fig. (6a). Calculations for anisotropic crystals have been carried out using anisotropic displacement field data as generated by Ohr (17). Similar results were obtained. These anisotropic calculations have been applied to the analysis of diffuse scattering measurements from copper irradiated with 60-MeV Ni ions at 300 K. Scattering measurements for $\vec{q}$ along [111] near the 222 reflection are shown in Fig. (7) and the analysis of this scattering is shown in Fig. (8). These results, shown here for the first time, demonstrate the size distribution and concentration determination possibilities using this technique. The details of the measurement and analysis will be published elsewhere. The observation of nearly equal intensities for $q > o$ and $q < o$ in Fig. (7) indicates, qualitatively, the presence of both types of loops in the sample, while the differences in the shapes and positions of the maxima imply the vacancy loops are smaller than the interstitials (recall that the vacancy loop scattering is predominantly at $q < o$). Fitting of these data with numerical calculations for vacancy and interstitial loops of 10, 20, 30, 40 and 60 Å sizes and treating the concentration of each loop size and type as parameters, lead to the size distributions in Fig. (8). Although the fitting was done with five specific sizes, the results have been (artifically) spread over 10 Å ranges and reduced to units of loops/cm^3/Å (the 60 Å result is spread over 30 Å) so that the area under Fig. (8) yields the total loop concentrations.

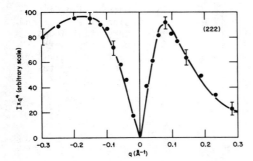

Fig. 7- Measured diffuse scattering from ion irradiated copper at the 222 reflection.

The vacancy loops can be seen to have smaller sizes than the interstitial loops, but the significantly larger numbers of vacancy loops offset the size difference such that the number balance of point defects in the combined distribution turns out to be 53.5% vacancies and 46.5% interstitials. These values may be considered equal within the estimated uncertainties (±7%) of the analysis and therefore suggest that all the vacancies are condensed into dislocation loops. Existing electron microscopy results on copper, irradiated under various conditions have been inconclusive on this point, reporting values ranging from a factor of two to less than half as many vacancies in loops as interstitials in loops (18,19). Fig. (9) is a semi-log plot of the combined size distribution together with published (20) TEM results superposed for comparison. Both results refer to the damage peak region near the end of range of the ions (20). The agreement in the larger sizes is quite

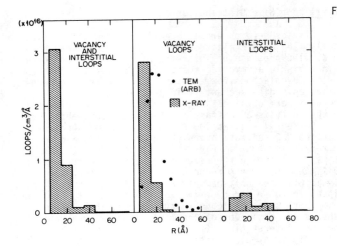

Fig. 8- Size distribution for loops in ion irradiated copper obtained from the x-ray diffuse scattering in Fig. (7). The TEM results are for vacancy loops in 30 KeV Cu irradiated copper (21). The TEM vertical scale is arbitrary.

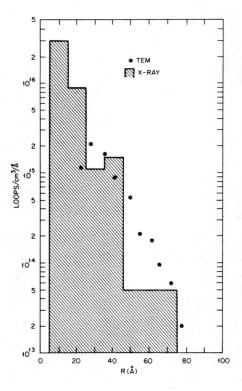

Fig. 9- Combined size distribution of vacancy and interstitial loops from Fig. 8, with TEM results on a similarly irradiated crystal (20).

360

reasonable, but the correspondence is not very good for small sizes, which are predominantly composed of vacancy loops in the x-ray results. A direct comparison of the vacancy loop distribution determined by x-rays with vacancy loops imaged by TEM in low energy (30 KeV) copper ion irradiated copper (21) at 293 K, is made in Fig. (8). These microscope results are scaled arbitrarily, so only the size of the loops and the shape of the distribution are of significance. Although the microscope result indicates slightly larger sizes than does the diffuse scattering, both methods show rather narrow distributions. Considering the differences in irradiation conditions, the close similarity in the results seems encouraging. However, this is the first direct comparison of the size distributions of the individual dislocation loop types determined by x-rays with those from TEM, so firm conclusions about the above results may be somewhat premature.

Loops and Precipitates

Huang diffuse scattering can be used to differentiate between planar loops and three dimensional precipitate clusters in a manner similar to that applied in point defect symmetry determinations (see Table I). An example of this is shown in Fig. (10) where the Huang diffuse scattering from dislocation loops in neutron irradiated copper and cobalt precipitates in Cu(1%)Co are shown (22). Characteristic differences can be seen between the scattering in the two cases. The scattering is rather isotropic from the dislocation loops while the precipitate scattering perpendicular to the [100] direction tends to vanish. In addition, the asymmetry in the scattering, along the [100] direction, has the opposite sense for the cobalt precipitates, compared to that for the loops. From Table I, it can be seen that these measurements would rule out cubic or tetragonal defects in the neutron irradiation case and rule out trigonal or orthorhombic defect symmetry in the precipitate case. Measurements at the 111 reflection completed the characterization as cubic or spherical symmetry for the precipitates and {111} loops in the case of the neutron irradiated copper. Both of these results

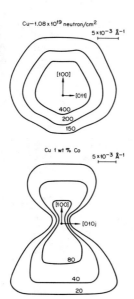

Fig. 10- Huang scattering from dislocation loops (upper) in neutron irradiated copper and precipitates (lower) in 570°C aged Cu(1%)Co.

361

are in agreement with known electron microscopy results (22,23) for these systems. The higher intensity for q>o in the loop case and higher intensity for q<o in the cobalt scattering can be related to the second term in Eq. (8) and indicates the cobalt precipitates cause an inward displacement of the lattice surrounding them, while the asymmetry in the loop scattering indicates an outward displacement. An outward displacement implies interstitial loops; however, since vacancy loops are known to be present as well, the larger sizes of the interstitial loops in connection with the R^4 scaling can be assumed to cause them to dominate the scattering. This possibility was discussed in the calculational paragraph above.

Diffuse Scattering from Coherent Precipitates

Although a considerable amount of small angle scattering (forward direction) has been directed toward the study of precipitates, x-ray scattering near Bragg reflections has received relatively little attention in this regard. The characteristic features associated with scattering from precipitates can be seen in the calculations shown in Fig. (11). These calculations were based on Eq. (5), using $f_h = f_p$ for simplicity, and choosing the strain inside the precipitate to be -0.014, the radius to be 70 Å and assuming an isotropic crystal with the lattice parameter of copper. $\vec{q}$ is along the (400) reciprocal lattice vector. The important features in the scattering are the $1/q^2$ Huang scattering region, the asymmetry (corresponding to inward displacements) caused by the strain ϵ = -0.014, the oscillating intensity for q<o with a $1/q^4$ falloff of the envelope, and the smoothly varying intensity for q>o with a shoulder at $q \sim 0.1$ Å^{-1}. The oscillations

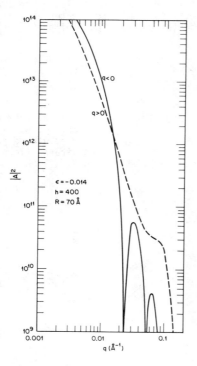

Fig. 11- Calculated diffuse scattering from spherical precipitates in an isotropic crystal.

are a result of interferences (10) in the scattering from the highly strained region outside the precipitate, and the shoulder at $q \sim 0.1$ A^{-1} can be shown to result from the Laue scattering from the precipitate particle. Experimental identification of this shoulder position can then be used to measure the strain inside the particle as

$$\varepsilon = -q/h \tag{14}$$

Measurements on Cu(1%)Co have shown (24) the existence of a shoulder at $q \sim 0.1$ A^{-1} which would appear to be consistent with a TEM measurement of $\varepsilon = -0.014$ using Moire fringe measurements (25). However, Fig. 11 does not consider the anisotropy of the copper lattice so this comparison cannot be regarded as complete. The possibility of measuring the misfit strain inside the particle is demonstrated, though, and further work should allow size and concentration determinations using anisotropic calculations.

Integral Diffuse Scattering

Although differential diffuse scattering measurements (which are characterized by a single scattering vector, $\vec{K}$) provide the most detailed information about defect structures and are the most amenable to numerical calculations for analyses, integral diffuse scattering (6,13,26,27) as illustrated in Fig. 1b) can be useful also. Because of the integration of the intensity over a large solid angle (limited in effect by the range of the scattering for clustered defects) accurate measurements can be made with modest equipment. However, the approximations associated with Eq. (12) and the symmetric averaging $(I(\Delta\theta)+I(-\Delta\theta))/2$ result in the loss of vacancy-interstitial type determinations, and the detailed calculations using Eq. (2) enter only indirectly through integration. Nevertheless, as shown in Fig. (12a, b) for neutron irradiated copper (6), results can be obtained that yield

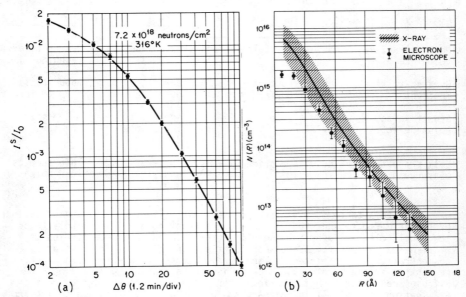

Fig. 12- a) Integral diffuse scattering at the 111 reflection from neutron irradiated copper b) Size distribution of dislocation loops obtained from (a) with TEM results on a similarly irradiated sample.

dislocation loop concentrations and size distributions. Fig. (12a) contains integral diffuse measurements on copper irradiated with 7.2×10^{18} n/cm^2 at 316°K. The logarithmic q_0 dependence at small q_0 can be seen, as can the $1/q_0^2$ dependence for larger q_0. The results of fitting the data in Fig. (12a) yielded the size distribution in Fig. (12b) (solid line). These results compare favorably with the electron microscopy results on a similarly irradiated sample shown as the solid circles. It is interesting to notice that in this case, as well as in Fig. (9), the largest discrepancy occurs at the small sizes. This method has been employed in studies of dislocation loops in electron, neutron, and ion irradiated metals, and in cases where comparisons with electron microscopy could be carried out (12), similar results to those in Fig. (12b) were obtained. Direct comparisons of integral diffuse scattering and differential diffuse scattering results are in progress. As suggested by Fig. (9); however, the high concentrations of small loops seem to be present in both the integral and differential methods of investigations.

Conclusion

We have seen that diffuse scattering can be used effectively in investigations of defects as small as single interstitials up to defect clusters that could, in principle, range to several hundred Angstroms. Larger clusters are not precluded (27), but will in general require special considerations on measuring resolution and crystal perfection. In order to make a rough overall evaluation of the diffuse scattering technique compared to the electron microscope technique for studying these defects, positive (+) and negative (-) aspects of both techniques are listed in Table II.

Table II. Positive (+) and Negative (-) Aspects of Diffuse Scattering and Electron Microscopy techniques.

X-Ray Diffuse Scattering		Electron Microscopy	
+	-	+	-
Non-destructive technique	Single crystals required	Polycrystalline or single crystal	Destructive technique
Good statistics inherent	higher concentrations required	Lower concentrations possible	Good statistics tedious
Sensitive to point defects	>500 A sizes difficult	Larger sizes possible (sample limited)	>10 A sizes required
Bulk property measurement	Results require analysis	Direct observation (usually)	Thin samples (non-bulk)

From this table it would appear that initial survey work, investigations of larger defects, and studies where concentration levels are likely to be low, would clearly favor the use of electron microscopy. For studies of small defects, investigations where subtle changes in defect sizes or concentrations are of interest, and when measurements are required after sequential sample treatments, such as isochronal annealing or ageing, the diffuse scattering technique offers significant advantages. An overall examination of Table II indicates that in many ways, diffuse scattering and electron microscopy are complementary tools. Although the study of point defects and very small clusters by diffuse scattering requires a significant experimental effort, this size region is inaccessible to electron microscopy so that such an effort is justified. The availability of synchrotron radiation sources and the construction of sophisticated scattering equipment at synchrotron sites will, however, significantly reduce the barriers for presently difficult diffuse scattering measurements and make these experiments within the capabilities of a much larger community.

Because of the requirement of single crystal samples with little or no deformation tolerable, diffuse scattering measurements are not well suited for studying uncharacterized samples such as obtained from structural components. Rather, the power of diffuse scattering is realized by devising experiments, (on controlled and well characterized samples) that will provide information of general application. This is, of course, the philosophy of fundamental defect studies in any case.

The technique of obtaining the size distribution of vacancy loops separate from interstitial loops by diffuse scattering would appear to be a particularly useful contribution. Although electron microscopy has this capability, it is a very tedious process that tends to inhibit its routine use for small loops. The development of theoretical diffuse scattering calculations, for analysis of scattering measurements, has played an important role in the loop studies, and will continue to play a critical role in the further development of defect studies by diffuse scattering. As the capabilities of the diffuse scattering technique improve, it will be increasingly important to make use of coordinated investigations of defects and defect structures, by both diffuse scattering and TEM, in order that the most desirable attributes of both techniques may be exploited.

Acknowledgement

The author would like to thank J. F. Barhorst for help in collecting and analyzing some of the data presented in this paper.

References

1. M. A. Krivoglaz, Theory of X-Ray and Thermal Neutron Scattering by Real Crystals, Plenum Press, New York, 1969.

2. P. H. Dederichs, "The Theory of Diffuse X-ray Scattering and its Application to the Study of Point Defects and their Clusters," J. Phys. F: Metal Phys. 3, (1973) pp. 471-496.

3. H. Trinkaus, "On the Determination of the Double-Force Tensor of Point Defects in Cubic Crystals by Diffuse X-ray Scattering," Phys. Stat. Sol. (b), 51, (1972) pp. 307-319.

4. P. Ehrhart and W. Schilling, "Investigation of Interstitials in Electron Irradiated Aluminum by Diffuse X-ray Scattering Experiments," Phys. Rev. B8, (1974) pp. 2604-2621.

5. H. -G. Haubold, "Study of Irradiation Induced Point Defects by Diffuse Scattering," pp. 268-283 in Fundamental Aspects of Radiation Damage, M. T. Robinson and F. W. Young, Jr., ed.; ERDA CONF-75 1006 - P1, Oak Ridge, TN, 1976.

6. B. C. Larson, "X-ray Studies of Defect Clusters in Copper," J. Appl. Cryst. 8, (1975) pp. 150-160.

7. H. Peisl, "X-ray Scattering from the Displacement Field of Point Defects and Defect Clusters," J. Appl. Cryst. $\underline{8}$, (1975) pp. 143-149.

8. B. von Guerard and J. Peisl, "Diffuse X-ray Study from Copper after Neutron Irradiation at 4.6 K," pp. 287-294 in (same as Ref. 5).

9. M. Moringa and J. B. Cohen, "The Defect Structure of VO_x.II. Local Ionic Arrangements in the Disordered Phase," Acta. Cryst. $\underline{A35}$, (1979) pp. 975-989.

10. H. Trinkaus, "Der Reflexferne Teil der Diffusen Streuung von Roentgen-strahlen," Z. Angew. Phys. $\underline{31}$, (1971) pp. 229-235.

11. H. -G. Haubold and D. Martinsen, "Structure Determination of Self-Interstitials and Investigation of Vacancy Clustering in Copper by Diffuse X-ray Scattering," J. Nucl. Mat. $\underline{69,70}$ (1978) pp. 644-649.

12. B. C. Larson, "X-ray Studies of Irradiation Induced Dislocations in Metals," pp. 820-838, in (same as Ref. 5). P. Ehrhart, B. C. Larson and H. Trinkaus, "Diffuse Scattering from Dislocation Loops," to be published. B. C. Larson and W. Schmatz, "Huang Diffuse Scattering from Dislocation Loops," Phys. Stat. Sol. (b) $\underline{99}$, (1980) pp. 267-275.

13. B. C. Larson and F. W. Young, Jr., "A Comparison of Diffuse Scattering by Defects Measured in Anomalous Transmission and Near Bragg Reflections," Zeit. Naturforsch. $\underline{28a}$ (1973) pp. 626-632.

14. P. Ehrhart, H. -G. Haubold, and W. Schilling, "Investigation of Point Defects and their Agglomerates in Irradiated Metals by Diffuse X-ray Scattering," Advances in Solid State Physics, $\underline{14}$ (1974) pp. 87-110.

15. P. Ehrhart and B. C. Larson, "Calculation of Diffuse Scattering from Dislocation Loops," Annual Progress Report, Solid State Division, Oak Ridge National Lab., Oak Ridge, TN, (1977) pp. 84-85.

16. P. Ehrhart and V. Schlagheck, "Investigation of Interstitial Clusters in Copper by Measurement of the Huang Diffuse Scattering," pp. 839-845 in (same as Ref. 5).

17. S. M. Ohr, "Displacement Field of a Dislocation Loop in Anisotropic Cubic Crystals," Phys. Stat Sol. (b) $\underline{64}$, (1974) pp. 317-323.

18. M. Rühle, F. Häusermann, and M. Rapp, "Transmission Electron Microscopy of Point Defect Clusters in Neutron Irradiated Metals," Phys. Stat. Sol. $\underline{39}$ (1970) pp. 609-620.

19. J. Narayan and S. M. Ohr, "The Nature of High Energy Neutron Damage in Copper and Gold," J. Nucl. Mat. $\underline{85,86}$ (1979) 515-519.

20. J. B. Roberto and J. Narayan, "Ni Ion Damage in Cu and Nb," pp. 120-126 in (same as Ref. 5).

21. C. A. English, B. L. Eyre, J. Summers, and H. Wadley, "The Behavior of Collision Cascades at Elevated Temperatures in Copper and Molybdenum Crystals," pp. 910-917 in (same as Ref. 5).

22. B. C. Larson and W. Schmatz, "Huang Diffuse Scattering from Dislocation Loops and Cobalt Precipitates in Copper," Phys. Rev. $\underline{B10}$ (1974) pp. 2307-2314.

23. H. P. Degischer, "Diffraction Contrast from Coherent Precipitates in Elastically Anisotropic Materials," Phil. Mag., 26 (1972) pp. 1137-1152.

24. W. Schmatz and B. C. Larson, unpublished.

25. C. G. Richards and W. M. Stobbs, "Determination of in situ Misfits of Coherent Particles," J. Appl. Cryst. 8 (1975) pp. 226-228.

26. B. C. Larson and F. W. Young, Jr., "Effect of Temperature on Irradiation Induced Dislocations in Copper," J. Appl. Phys. 48 (1977) pp. 880-886.

27. J. R. Patel, "X-ray Diffuse Scattering from Silicon Containing Oxygen Clusters," J. Appl. Cryst. 8 (1975) pp. 186-190.

X-RAY SCATTERING INVESTIGATION OF MICROALLOYING

AND DEFECT STRUCTURE IN ION IMPLANTED COPPER

S. Spooner

Fracture and Fatigue Research Laboratory
Georgia Institute of Technology
Atlanta, Georgia 30332

The double-crystal method for x-ray scattering analysis of radiation described by B. C. Larson (1) has been applied to the investigation of aluminum implanted copper. The interpretation of x-ray observations is based on effects of lattice strain in the surface microalloy and the presence of dislocation loops which originate from implantation damage. The copper crystal with a dislocation less than 10^5 cm/cm^3 was implanted with aluminum to a dose of 2×10^{16} ions/cm with energies up to 200 keV. The response of the implanted crystal to annealing at 500 C and 600 C was determined. The quantitative use of the x-ray technique to assess implantation effects and the limitations of the technique are discussed.

This research was sponsored by the Office of Naval Research under Contract N00014-78-C-0270.

Introduction

X-ray diffraction is an effective method for analyzing radiation damage particularly for quantitative measurement of lattice strain effects associated with defect clusters (1). In recent years there have been a variety of x-ray diffraction investigations of ion implantation damage produced in single crystals based on double-crystal measurements. Komenou et al. (2) observed x-ray scattering Pendellosung interference in rocking curves from Ne^{+}-implanted garnet films which Speriousu (3) interpreted according to a kinematic diffraction theory incorporating strain and damage distributions as a function of depth. Afanasev et al. (4) have used dynamical theory for calculating the scattering from a silicon crystal with disturbed layers. Yamagishi and Nittono (5) studied Ar^{+} ion-implanted copper whiskers with both x-ray topography and a triple-crystal diffraction method to assess lattice strain response with dose and annealing. In the foregoing studies (2-5) no absolute intensity measurements were made so that analysis of structural changes depended mostly upon scattering distribution shape. In the present study, absolute reflectivity measurements are used to study the effects of Al^{+}-ion damage in copper due to low energy (200 keV) and high dose (2 x 10^{10} ions/cm) using a double-crystal diffraction method. Both surface alloying and implantation damage are under consideration for their important influence on fatigue crack initiation (6). Because radiation damage production of point defect clusters enters our work in a fundamental way, this paper offers an example of the utility of x-ray scattering techniques in radiation damage research.

The principle challenge in this x-ray study was to find an effective x-ray method for investigating the damage and surface alloying effect in an implanted layer which is much thinner than the sampling depth of x-rays. In addition, there was the consideration of which theoretical analysis of scattering intensity would be most appropriate to describe the combined damage and surface alloying scattering effects. This question was approached from two perspectives; (a) use of dynamical theory of diffraction for the analysis of lattice strain due to surface alloying (7,8) and (b) use of kinematic theory for the description of scattering from defect clusters (1). It is shown that the scattering data are dominated by implantation damage defect clusters and that the kinematic theory is most appropriate for the description of scattering in the case at hand. Furthermore, it is shown that a quantitative evaluation of implantation damage can be obtained from the absolute reflectivity measurements made in the double-crystal method.

X-Ray Scattering Models

The structure the implanted region is modeled by placing of point defect clusters within a surface layer which has a lattice parameter that is expanded by implantation alloying. As yet, no single formulation for scattering intensity gives a calculation of the scattering from the combined defect cluster and lattice distortion effects. Instead, we make a calculation for the case of scattering from a defect-free surface alloy on one hand and a calculation for the scattering from defect clusters in a unalloyed matrix on the other hand. The measured x-ray scattering effects are then used to determine the manner in which the two calculations might be applied to represent the scattering from the implanted layer.

For a surface alloy layer free of defects, the dynamical theory of x-ray scattering can be used to calculate the reflectivity of x-rays as a function of crystal rotation in a double-crystal rocking curve. In a two-crystal arrangement, the first crystal which is not implanted is set to maximum reflectivity. The second crystal is rotated about an axis perpendicular to the scattering plane (defined by the incident and reflected x-ray beams.) The resulting reflectivity curve is the convolution of the reflection characteristic of the first crystal with the reflectivity of the second crystal. Larson (7,8) has adapted, for this surface alloy problem, a method of calculation used by Klar and Rustichelli (9) for neutron scattering from elastically bent crystals. The reflectivity from a crystal is obtained by the computation of the real and imaginary components of the complex scattering amplitude of the reflected radiation. Two coupled differential equations – one for real and one for imaginary components – are integrated numerically. The integration is dependent upon initial values of the amplitude components and the variation in the Bragg angle for the crystalline sublayers due to the elastic lattice distortion arising from bending or composition change. Full algebraic development of the theory can be found in papers by Larson and Barhorst (8) and Klar and Rustichelli (9). The equations requiring integration express the derivatives of the real (X_1) and imaginary (X_2) scattering amplitude components with respect to a variable A which is proportional to depth measured relative to the external surface:

$$\frac{dX_1}{dA} = k(X_1^2 - X_2^2 + 1) + 2X_2(X_1 - y) - 2gX_1 \tag{1}$$

$$\frac{dX_2}{dA} = -(X_1^2 - X_2^2 + 1) + 2X_1(kX_2 + y) - 2gX_2 \tag{2}$$

where k and g are constants which depend on x-ray absorption and the parameter y contains the misset angle, $\Delta\theta$, for the rocking curve as follows:

$$y = C_1 \Delta\theta - C_2 \tag{3}$$

where C_1 and C_2 are constants dependent on x-ray scattering parameters tha are fixed for the Bragg diffraction peak under examination. For the case where the lattice parameter varies with A it is shown (8) that

$$y = C_1(\Delta\theta + \epsilon(A)\tan\theta_B) - C_2 \tag{4}$$

where the variation of the lattice parameter with depth is contained in the strain function $\epsilon(A)$. In the case at hand, $\epsilon(A)$ is determined by the composition of the surface alloy as a function of implantation depth.

The method by which the change in relectivity due to surface alloying is calculated does not require integration over the entire crystal thickness. Instead, one uses the well known results (10,11) for the reflectivity from a perfect crystal as a starting point. The real and imaginary components of the scattering amplitude at a set rocking angle are used as initial values for the integration beginning at a depth below the implanted ions. For the integration back to the surface the effects of surface alloying, $\epsilon(A)$, are allowed to affect the computation of scattering amplitude. A set of these calculations is done for a range of rocking angles where the reflectivity is calculated from,

$$R(\Delta\theta) = X_1^2 + X_2^2 \tag{5}$$

where the amplitude components, X_1 and X_2 are evaluated at the reflecting crystal surface. Note that the result is an'absolute reflectivity value.

Figure 1 shows the calculated results we have obtained at the reflecting in which 2 atomic percent of aluminum is implanted in copper to a depth of approximately 1000 A. The lattice parameter expansion used in the calculation was taken from the data given on linear lattice strain by King (12) equal to +0.0626 per atomic percent of aluminum in copper. A sharp subsidiary peak of 1.4 percent reflectivity is seen at a Bragg angle displaced to a lower angle than the substrate Bragg angle corresponding to the expanded lattice parameter. The small peak width is approximately 2 minutes of arc. The reflectivity is the order of the ratio of implanted layer thickness to the x-ray penetration thickness, $1/2 \mu_o$, where μ_o is the linear absorption parameter.

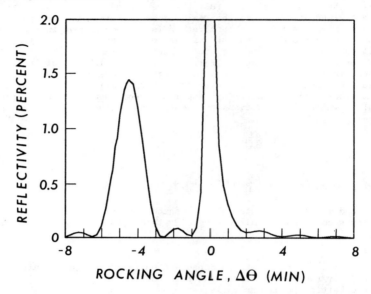

Fig. 1   Calculated reflectivity from a surface implanted to 2 atomic percent of aluminum in copper to a depth of approximately 1000 A The subsidiary peak appears at an angle appropriate for the lattice parameter of this composition.

Consider now the calculation of the scattering from defect clusters in a crystal of uniform lattice parameter. In this case, kinematic diffraction theory is used to calculate the scattering intensity from an isolated defect cluster. The scattering resulting from a collection of defects is the sum of the intensities. This implies that no scattering interference occurs between scattering amplitudes coming from each defect. Larson (1) summarizes the calculation of the scattering intnsity from defect clusters. The experimental geometry used in our experiments is shown in Figure 2 where the scattered x-rays are recieved by a large detector. Each of the scattering vectors is associated with a scattering space vector, q, going from the Bragg spot (at the top) to the surface of the Ewald scattering sphere. In such an experiment, the intensity is averaged over the scattering space vectors, q. q_o is the shortest vector between the Bragg position and the Ewald sphere at a given crystal setting. The measured intensity is called the integral diffuse scattering. The intensity is measured as a function of rocking angle of the crystal in the same geometry used for measurement of dynamical difraction effects described above.

The diffuse scattering from dislocation loops measured close to the Bragg peak is attributed to long range strain fields around the loop and is called Huang scattering. Scattering measured farther away from the Bragg

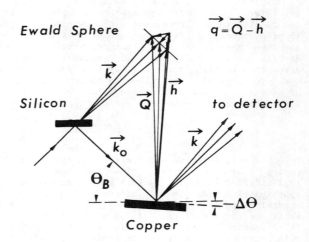

Fig. 2 Scattering geometry for the double crystal method used in this experiment. Upon rocking the crystal the Ewald scattering sphere is swept through the Bragg point. At a fixed crystal setting the diffuse scattering is integrated over a portion of the scattering sphere near the Bragg point.

peak is attributed to short range strain fields and is termed Stokes–Wilson scattering. The diffuse scattering is distributed about the Bragg position in a way dependent on the precise strain field distribution (1,13). The calculation of integral diffuse scattering requires an averaging of the diffuse scattering over the portion of the Ewald scattering sphere which is close to the Bragg position (14). For the scattering from loops of radius R, the Huang scattering smoothly joins the Stokes-Wilson scattering at a scattering parameter $q_o = q_L = \alpha/R$ where $q_o = h\Delta\theta\cos\theta_B$ with d_{hk} spacing, h $=2\pi/d_{hkl}$, θ_B the Bragg angle for reflection from the hkl planes, $\Delta\theta$, the misset angle of the rocking curve. A symmetric diffuse scattering cross section is defined

$$\sigma_h^s(q_o) = 1/2(\sigma_h^s(-q_o) + \sigma_h^s(q_o)) \qquad (6)$$

which is obtained by the average of intensities measure symmetrically above and below the Bragg position ($q_o = 0$). The symmetric diffuse cross sections for Huang and Stokes-Wilson scattering are given by,

(Huang) $$\sigma_h^s(q_o) = (r_e^2 f_h^2 e^{-2M}(h/k)^2 2\pi\tau (b\pi R^2/V_c)^2 \ln(e^{1/2}q_L/q_o) \qquad (7)$$

for $q_o < q_L$, and,

(Stokes-Wilson) $$\sigma_h^s(q_o) = (r_e^2 f_h^2 e^{-2M}(h/k)^2 2\pi\tau (b\pi R^2/V_c)^2 q_L^2/2q_o^2 \qquad (8)$$

for $q_o > q_L$, r_e is the Thompson electron radius (2.82 x 10^{-13} cm), f is the scattering factor, e^{-M} is the Debye-Waller factor, $k = 2\pi/\lambda$, λ= wavelength, is a constant of order 1 which depends on averaging of loop orientations, b= Burgers vector, V_c= atomic volume, The scattering intensity relative to the incident intensity is given by,

$$\frac{I^s(q_o)}{I_o} = \frac{C(R)}{2\mu_o V_c} \sigma_h^s(q_o) \qquad (9)$$

where $C(R)/V_c$ is the density of loops of radius R. From Eqns. (7),(8) and (9) one can obtain loop size and density. Note that ($b\pi R^2/V_c$) equals the number of point defects in the defect cluster.

In summary of the two calculations, the dynamical theory predicts a subsidiary peak which appears at an angle determined by the lattice strain due to alloying. The kinematic theory predicts a diffuse scattering which is proportional to the number and size of loops. Both calculations give the absolute relectivity with no adjustable parameters other than those describing the structure. The dynamical theory calculation depends on the assumption that the surface alloy is crystallographically coherent with the unalloyed crystal. The kinematic theory is likely to be limited in the case of very high defect cluster densities where nonrandom loop distributions may lead to interference between diffuse scattering amplitudes.

Experimental

The calculated strain scattering effects must be measured at small angles near the Bragg diffraction peak of the unaffected crystal. The implant affected region is less than 1 micron and the penetration depth is approxmately $1/2 = 11$ microns. It is required that the bulk of the crystal be perfect (mosaic spread less than 1 minute) in order that the small scattering effects can be measured near the Bragg peak. Furthermore, it is required to subtract a significant background due to the tails of the bulk crystal Bragg peak in order to determine the diffuse scattering intensity due to surface alloying and defect clusters. A convenient approach to this measurement is to translate the crystal between an implanted and implantation-free area on the same crystal. Crystals used in these studies were provided by F. W. Young of Oak Ridge National Laboratory. The crystals were grown by the Bridgeman technique, cut to orientation, then annealed at a few degrees below the melting point for two weeks. The crystal pieces were hardened by neutron irradiation and then further cut and shaped by chemical cutting methods (15). The dislocation density measured by etch pit techniques was less than 10^3 cm^{-2} after shaping procedures were completed.

The two-crystal arrangement consisted of a silicon crystal fixed to diffract the Cu K$_\alpha$ radiation onto the implanted copper crystal. The (333) d-spacing (1.0451 A) of silicon happens to match the (222) d-spacing (1.0436 A) of copper very well so that the system is well focussed to give a narrow rocking curve width. The copper crystal is initially aligned to give a sharp maximum in the rocking curve by adjusting the (111) normal about an axis in the scattering plane. When properly adjusted, the full width at half-maximum (FWHM) of the copper rocking curve is 12.5 arc-seconds. The crystal is mounted on a goniostat which can be translated in the plane of the crystal surface so that rocking curves can be made from the implanted area and masked implantation-free areas. In a typical run, the copper crystal is rocked about an axis perpendicular to the scattering plane at a rate of 5 to 20 arc-seconds per minute while x-ray intensities are recorded continuously at 10 second intervals. The x-ray detector has an active receiving area of 5 cm^2 at a distance of 8 cm so that the subtended solid angle (0.08 steradians) integrates the scattering over a large portion of the Ewald sphere in the vicinity of the 222 Bragg peak of copper.

The implantation of aluminum into copper was chosen for these experiments because the ion penetration was favorable and the microalloy concentration was well below the solubility limit of the aluminum in copper. The details of implantation are given elsewhere (19). The implanted layer was 1200 A thick (16) with a composition of 1.8 atomic percent. The distribution of damage over the alloy thickness was estimated on the basis of calculations by Fritzsche (17) and Winterbon (18). The alloy distribution (solid line) and the damage profile (dashed line) are shown in Figure 3.

374

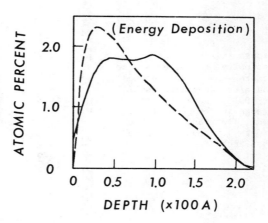

Fig. 3
Distribution of implanted Al$^+$ ions (solid) and the energy deposition (dashed) for the implantation of 2×10^{16} ion/cm^2 with energies up to 200 keV. Note that damage is concentrated toward the surface and that the damage energy is on a relative scale.

Annealing of the specimens was performed as a means to differentiate the sources of scattering in the implanted layers. The crystals were placed in a vacuum of 10^{-8} Torr at 500 C, 600 C and 900 C for 30 minutes. Annealing at 900 C restored the original structure as seen in the rocking curves.

Fig. 4
Rocking curves are shown for the implanted (upper) and implantation-free (lower) crystal. The scattering is expressed as a fraction of the incident beam intensity. Note the larger scattering at low angles.

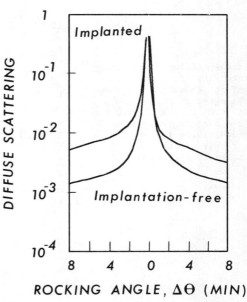

Fig. 5
Excess diffuse scattering intensity for the sample before annealing (dashed) and after annealing (solid) at 500 C. Note that little change in the general level and distribution of the excess intensity occurs upon annealing.

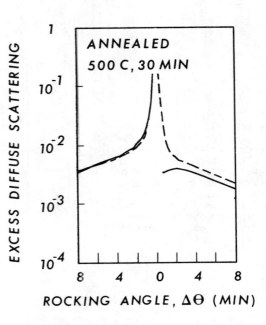

Fig. 6
Excess diffuse scattering intensity for the sample before annealing (dashed) and after annealing (solid) at 600 C. The level and the distribution of the excess intensity changes as a result of the annealing at this temperature.

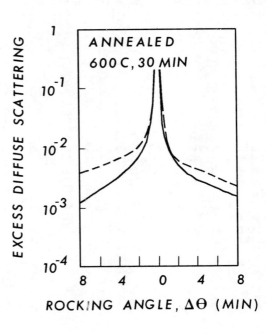

376

Results and Discussion

The rocking curves for implantation-free copper and for aluminum implanted copper were measured on the same crystal. These curves are shown in Figure 4. The diffuse scattering from the implanted crystal is more intense on the low angle side of the Bragg peak position. The excess diffuse scattering is calculated by subtraction of the implantation-free rocking curve intensity from the corresponding intensity in the implanted crystal. The excess diffuse scattering for the implanted crystal is shown in Figures 5 and 6 as a dashed line. The effect of 30 minute anneals on the excess diffuse intensity is show in Figure 5 for annealing at 500 C and in Figure 6 for annealing at 600 C. No large change due to annealing occurs at 500 C while for annealing at 600 C, there is a reduction of scattering and scattering becomes more symmetric with respect to the Bragg peak position.

The observation of a higher diffuse scattering at low rocking curve angles can be attributed to the fact that implanted aluminum expands the copper lattice so that Bragg scattering from the implanted region occurs at a lower angle than that for the implantation-free material. The composition of the implanted layer was estimated to be 1.8 atomic percent. The resulting Bragg position would be displaced to lower angle by 4.2 minutes for the 222 reflection from the copper alloy layer.

The diffuse scattering seen on both sides of the main Bragg position can be compared to calculations of the scattering from dislocation loops. In Figure 7 the excess diffuse scattering is plotted versus the log of the rocking angle according to Eqn. (7) for Huang loop scattering. The rocking angle was measured relative to the supposed Bragg position for the alloy.

Fig. 7
The excess diffuse scattering from the implanted crystal is plotted versus $\ln(\Delta\theta)$ for the intensity above and below the Bragg position assumed to apply for the implanted region of the crystal.

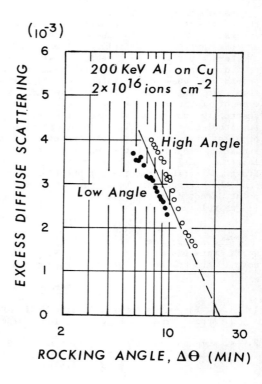

Although there is a displacement between the two sets of points, the average of the high angle and low angle intensity is close to a straight line which yields an estimated loop radius of 25 A.

An estimate of the density of loops can be made by comparing measured reflectivity with Eqn. (9). We use a loop radius of 25 A and a reflectivity of 1 percent at $\Delta\theta$= 2 minutes. Substitution of appropriate constants into Eqn. (9) for a 25 A loop size gives

$$\frac{I^S(q_o)}{I_o}=6.1\text{x}10^{-21}\frac{C}{V_c}\ln(\frac{44}{\Delta\theta(\text{min})})\tag{10}$$

from which a value of C/V_c is 5.3 x 10^{17} loops/cc. (The loops are concentrated by a factor of 40^C in the implanted layer since the above calculation assumes the loops to be uniformly distributed).

The failure to observe a sharp Bragg peak associated with the implanted aluminum and the general agreement with scattering levels calculated for loop scattering point to the conclusion that the kinematic theory for diffraction from an implanted crystal containing loops is appropriate. The annealing at 600 C produces symmetrical scattering which suggests that most of the aluminum is removed from the region where loops persist. Thereby the loop scattering now originates in essentially pure copper. The role of aluminum is seen as simply expanding the lattice in a region where loops persist which, by virtue of severe damage, is no longer strictly coherent with the implantation-free crystal.

Conclusions

Analysis of x-ray diffraction in aluminum-ion implanted copper suggests that defect cluster scattering dominates the observed rocking curve intensity. Alloying in the implanted layer contributes through a shifting of the diffuse scattering to lower angles due to the fact that the defect clusters are formed in a region of aluminum-expanded lattice. The formation of a distinct peak predicted by dynamical diffraction theory does not occur, probably because of the intense defect scattering and the widths of the peak from the thin layer. Problems in the analysis of scattering remain in the area of formulating a model of combined alloying and defect cluster scattering as well as description of very high defect cluster scattering. Nevertheless the simplistic interpretation of x-ray scattering observation provides useful insights into the type and quantity of damage as well as the annealing response of the implanted structure. Measurements carried out to larger q_o will be useful in further definition of the defect structure since Bragg scattering from the implantation-free and implanted layer are avoided and the kinematical theory can be assumed. Size distributions an and total point defect densities are more directly measurable at the larger q_o values (1) as well.

Acknowledgements

The author thanks Dr. B. C. Larson and Mr. Jim Barhorst of the Solid State Division of Oak Ridge National Laboratory for their considerable help in the collection of the data and many useful discussions.

References

1. B. C. Larson, "X-ray Studies of Defect Clusters in Copper," J. Appl. Cryst. , 8 ,pp. 150-160 (1975).

2. K. Komenou, I. Hirai, K. Asama and M. Sakai, "Crystalline and Magnetic Properties of an Ion-Implanted Layer in Bubble Garnet Films," J. Appl. Phys. , 49 ,pp. 5816-5822 (1978).

3. V. S. Speriousu, H. L. Glass and T. Kobayashi, "X-ray Determination of Strain and Damage Distributions in Ion-Implanted Layers," Appl. Phys. Lett. , 34 , pp. 539-542 (1979).

4. A. M. Afanasev, M. V. Kovalchuck, E. K. Kovev and V. G. Kohn, "X-ray Diffraction in a Perfect Crystal with Disturbed Surface Layer," Phys. Stat. Sol. (a) 42 , pp. 415-422 (1977)

5. H. Yamagishi and O. Nittono, "X-ray Study on Lattice Defects in Ar^+ Ion Implanted Copper Whiskers," Nip. Kinz. Gakk. , 43 , pp. 689-695 (1979).

6. A. Kujore, S. B. Chakrabortty and E. A. Starke, "The Effect of Ion Implantation on the Fatigue Properties of Polycrystalline Copper," Nucl. Instr. Meth., 182/183, pp. 949-958 (1981).

7. B. C. Larson, C. W. White and B. R. Appleton, "Unidirectional Contraction in Boron-Implanted Laser-Annealed Silicon," Appl. Phys. Lett., 32, pp. 801-803 (1978).

8. B. C. Larson and J. F. Barhorst, "X-ray Study of Lattice Strain in Boron Implanted Laser Annealed Silicon," J. Appl. Phys., 51, pp.3181-5 (1980).

9. B. Klar and F. Rustichelli, "Dynamical Neutron Diffraction by Ideally Curved Crystals," Nuovo Cimento, 13B, pp. 249-270 (1973).

10. B. E. Warren, X-ray Diffraction, Chapter 14, pp. 315-354, Addison-Wesley Press, Reading, Mass. (1969).

11. W. H. Zachariasen, Theory of X-ray Diffraction in Crystals, Chapter 3, pp. 83-155, Dover Publications, New York (1967).

12. H. W. King, "Quantitative Size-Factors for Metallic Solid Solutions," J. Mat. Sci, 1, pp. 79-90 (1966).

13. B. C. Larson and W. Schmatz, "Huang-Diffuse Scattering from Dislocation Loops and Cobalt Precipitates in Copper," Phys. Rev. , B10, pp. 2307-2314 (1974).

14. B. C. Larson and F. W. Young, Jr. , "A Comparison of Diffuse Scattering by Defects and Measured in Anomalous Transmission and Near Bragg Reflections," Z. Naturforsch. , 28a, pp. 626-632 (1973).

15. F. W. Young, Jr. , "Etch Pit Studies of Dislocations in Copper Crystals Deformed by Bending. I. Annealed Crystals. II. Irradiated Crystals," J. Appl. Phys. , 33, pp. 3553-3564 (1962).

16. J. Keinonen, M. Hautala, M. Luomajari, A. Antilla and M. Bister, "Ranges of $^{27}Al^+$ Ions in Nine Metals Measured by (p,γ) Resonance Broadening," Rad. Eff. , 39, pp. 189-193 (1978).

17. C. R. Fritzche, "A Simple Method for the Calculation of Energy Deposition Profiles from Range Data of Implanted Ions," Appl. Phys. Lett. , 12, pp. 347-353 (1977).

18. K. B. Winterbon, Ion Implantation Range and Energy Deposition Distributions, Vol. 2, Low Incident Ion Energies, Plenum Press, New York (1975)

19. S. Spooner and K. Legg, "X-ray Diffraction Characterization of Aluminum Ion-Implanted Copper Crystals," Ion Implantation Metallurgy, C. M. Preece and J. K. Hirvonen, eds. TMS AIME, pp. 162-170 (1980).

THE USE OF SMALL-ANGLE X-RAY AND NEUTRON SCATTERING FOR CHARACTERIZING

VOIDS IN NEUTRON-IRRADIATED METALS AND ALLOYS*

Robert W. Hendricks
Technology for Energy Corporation
Knoxville, Tennessee 37922

Small-angle x-ray and neutron scattering are powerful analytical tools
for investigating long-range fluctuations in electron (x-rays) or magnetic
moment (neutrons) densities in materials. In recent years they have yielded
valuable information about voids, void size distributions, and swelling in
aluminum, aluminum alloys, copper, molybdenum, nickel, nickel-aluminum, niobium and niobium alloys, stainless steels, graphite and silicon carbide. In
the case of aluminum information concerning the shape of the voids and the
ratio of specific surface energies was obtained. The technique of small-
angle scattering and its application to the study of voids is reviewed in
the paper. Emphasis is placed on the conditions which limit the applicability of the technique, on the interpretation of the data, and on a comparison
of the results obtained with companion techniques such as transmission
electron microscopy and bulk density.

*This research was performed while the author was a Senior Staff Member of
the Metals and Ceramics Division, Oak Ridge National Laboratory, Oak Ridge,
Tennessee 37830, and was sponsored by the Division of Materials Sciences,
U.S. Department of Energy, under contract W-7405-eng-26 with the Union
Carbide Corporation.

Introduction to Small-Angle Scattering

As with all scattering techniques, the amplitude of x-ray or neutron scattering is related to the scattering length density (electron density for x-rays; coherent nuclear scattering length density or magnetic moment density for neutrons) by a Fourier transform. Thus, the characteristic angle at which radiation is scattered and the size cf the scattering object are inversely related. Typically, x-ray wavelengths are in the range 0.5-2.0 Å while neutron wavelengths are 1-15 Å. The scattering objects of interest in small-angle x-ray scattering (SAXS) or small-angle neutron scattering (SANS) are typically 10-10 000 Å. Because the object dimensions are large compared to the wavelength of the incident beam, the resultant scattering occurs in the very small angular region of 0.005-5°. In order that measurements can be made at such small angles, special collimation techniques have been developed during the past 40 years. It is these specialized collimation techniques which have separated the field of small-angle scattering from the more classical fields of high-angle x-ray and neutron diffraction.

What information can be obtained from small-angle scattering? The scattering intensity for a single particle is given by

$$I(\underline{k}) = \sum_m \sum_n f_m f_n \, e^{i\underline{k}\cdot\underline{r}_{mn}} , \qquad (1)$$

where

f_m = the scattering amplitude of the mth atom,
$\underline{r}_{mn}$ = the vector separation of the m and n^{th} atoms, and
$\underline{k}$ = the diffraction vector.

This expression, given generally for x-ray scattering, is appropriate to neutron scattering when there are no inelastic neutron scattering effects. The magnitude of the diffraction vector is given by

$$|\underline{k}| = \frac{4\pi}{\lambda} \sin\theta , \qquad (2)$$

where

λ = radiation wavelength, and
2θ = the scattering angle.

In the case of N randomly oriented, independently scattering, centrosymmetric particles it can be shown that at small scattering angles (1)

$$I(k) = NV^2 \, \Delta\rho^2 \int_0^\infty \bar\gamma(r) \, \frac{\sin kr}{kr} \, 4\pi r^2 dr , \qquad (3)$$

where

$\gamma(r)$ = characteristic function of the particle,
$\Delta\rho$ = difference in scattering length of the particle and its matrix,
and
V = volume of the particle.

The characteristic function has been computed for a variety of randomly oriented particles of various shapes (1).

Although the exact shape of the scattering curve depends on the shape of the scattering particles, there are several general results which are independent of the shape (1).

I. At $k = 0$

$$I(0) = NV^2 \Delta\rho^2 .$$ (4)

II. At small values of k, $\dfrac{\sin kr}{kr}$ can be expanded as a series which leads to (1)

$$I(k) = I(0)e^{-k^2R^2/3} ,$$ (5)

where

R_g is the radius of gyration of the particle. Equation (5) is known as Guinier's law.

III. At large values of k, an approximate expression for the intensity can be obtained by expanding the characteristic function $\gamma(r)$ about small values of r (1). The leading term of the result shows

$$I(k) = \frac{2\pi \Delta\rho^2 S}{k^4} ,$$ (6)

where S is the total interphase surface area. Equation (6) is known as Porod's law.

IV. A Fourier inversion of Eq. (3), evaluated at $r = 0$, yields

$$Q_0 \equiv 4\pi \int_0^\infty k^2 I(k)dk = 8\pi^3 \Delta\rho^2 NV .$$ (7)

The left-hand side of Eq. (7) is defined as the invariant and is denoted by Q_0. The invariant is quickly recognized to be the total scattered intensity integrated over all space.

V. Finally, if the particles are spherical in shape, the characteristic function is easily evaluated, and we find

$$I(k) = NV^2 \Delta\rho^2 \frac{9\pi}{2} \frac{J_{3/2}^2 (kR)}{(kR)^3} ,$$ (8)

where $J_{3/2}(x)$ is a Bessel function, and R is the sphere radius.

If there is a distribution of sphere sizes, then

$$I(k) = \int_0^\infty N(R)I(k,R)dR / \int_0^\infty N(R)dR ,$$ (9)

where $I(k,R)$ is given by Eq. (8) and $N(R)$ is number of particles with radii between R and $R + dR$. It is important to note that $I(k,R)$ is weighted by the square of the volume of the particle. Thus, the forward scattering from a 100 Å diameter particle (void) is 10^6 greater than that from a 10 Å diameter particle. A variety of techniques have been developed to unfold the integral equation of Eq. (9) to recover $N(R)$ under the assumption that the individual particles are spheres and obey the scattering equation of Eq. (8) (2).

Each of these results has been used in the study of voids in irradiated metals and alloys, as will be described in a later section of this review.

Small-Angle Scattering Instruments

In order to measure scattering at the small angles described in the previous section, the incident beam must be collimated very finely. In the early days, such collimation reduced the power of the incident beam such that the diffusely scattered radiation could not be observed over the noise of the detector (i.e., fog with film, and cosmic and electronic noise for photon counting detectors such as proportional or scintillation detectors). The early compromise for SAXS was to collimate the beam very well in one direction, but only very poorly in the orthogonal direction. This led to numerous camera geometries such as the Kratky, Guinier, or Beeman four-slit, all of which are in use today. Such cameras have been very practical for the study of macromolecules in solution, but have serious limitations for metallurgical work. First, the beam size is fairly large (approximately 0.060 by 30 mm), thus requiring large, thin samples. More importantly, the large angular divergence of the incident beam in the long direction (typically $\pm$ 20°) led to serious problems with a form of multiple scattering known as double-Bragg scattering. Only in those cases involving large, thin metal foils with very high scattering power (e.g., GP zones in Al-Zn) could such instruments be used effectively. With considerable effort, Epperson et al. (3) irradiated large aluminum single crystals and studied the small-angle scattering from the radiation induced voids using a Kratky camera. To solve the problem of thinning such large samples to the optimum thickness for the transmission of x-rays [t_{opt} = $1/\mu$, where μ is the linear absorption coefficient] AgK$_\alpha$ (λ = 0.5 Å) was used. These results will be discussed in the next section.

Because neutron beams and sample sizes are typically measured in centimeters rather than millimeters, long-slit neutron machines were never practical. It was not until the advent of position-sensitive detectors (PSDs) that point-geometry small-angle collimation systems became possible. With position-sensitive detectors the electronic and cosmic noise is distributed over numerous counting channels; 10^2 for a linear detector, 10^3–10^5 for an area detector. Thus, the background was decreased from 0.1 to 1 cps for a non-position-sensitive counter to 10^{-3}–10^{-2} cps/channel for a PSD. This enormous increase in sensitivity thus made it possible to work with the significantly decreased incident beam intensity available in point collimation. The first SANS facility, with an array of six linear PSDs, was constructed in Jülich by Schelten (4), and was followed by the big D-11 facility developed at the Institute Laue-Langevin in Grenoble by Ibel (5). The enormous success of these machines spurred the development of many more SANS facilities throughout the world (6-13). In the United States, facilities at the University of Missouri (6), the National Bureau of Standards (14), and Oak Ridge National Laboratory (11,12) are now operational and have appropriate geometries for metallurgical research. Of these, the NSF-funded user-dedicated 30-m facility at ORNL (11) promises to be one of the world's premier machines.

Following the development of successful point collimation SANS facilities using position-sensitive detectors, Schelten and Hendricks (15) constructed the first pinhole SAXS facility by fitting a Keissig film camera with an extended flight path and a linear PSD. They were able to show that the old problem of double-Bragg scattering, especially with polycrystalline foils, was caused principally by the wide divergence of the incident beam in the classical long-slit machines, and that with point collimation the problem was essentially eliminated even for 1.54 Å x-rays. With this instrument as a highly successful prototype, Hendricks (16) developed the first x-ray facility which was equipped with an area detector. This machine utilizes a 6-kW rotating anode x-ray source, graphite monochromatization of the incident radiation, pinhole collimators separated by distances of up to 5 m, sample-to-detector distances of up to 5 m, and a two-dimensional position-sensitive proportional counter. It was designed especially for use with small (2.5 mm diam) metallurgical samples.

The field of small-angle scattering has literally exploded as a result of the application of position-sensitive detectors. The recent developments in instrumentation, theory and data analysis have been reviewed by Schelten and Hendricks (2), while Gerold and Kostorz (17) have reviewed the application of the technique to materials science. The special issue of the Journal of Applied Crystallography which contains these reviews is the proceedings of an international conference on the subject; the interested reader will find numerous papers which exemplify the applicability of small-angle scattering to a wide range of materials.

SAS Studies of Voids in Neutron-Irradiated Metals and Alloys

Small-angle x-ray and neutron scattering have been used to study voids in a wide variety of pure metals, alloys, semiconductors and insulators. Some of the published work is given in Table I. It is clear that the technique has been used extensively. However, it is not the purpose of this paper to review each of these investigations in detail. Rather, selected results from the literature will be used to illustrate the various kinds of information which can be obtained about voids and to emphasize the strengths and limitations of the technique for obtaining such information. The reader should keep firmly in mind throughout this discussion that it may not necessarily be possible to obtain all of the various types of information from any given sample or material.

Table I. Materials Investigated by Small-Angle Scattering

Type of Radiation		SAXS	References	SANS	References
Neutron Irradiated		Al	3,15,18,19	Al	20,21,22
		Mo	23,24	Cu	25,26
		Ni	16,27	GaAs	28
		SiC	29	Nb(Zr)	30
		SiO_2	31	NiAl	32
		Graphite	33,34	Stainless steel	8,30,35
Heavy Ion Irradiated		Nb	36		
		Ni	37		

It is seen from Eq. (3) and the subsequent results that the scattered intensity depends linearly on the number of defects (N), the square of the scattering length density difference between the defect and the matrix ($\Delta\rho^2$) and the square of the defect volume V^2. Thus, any defect which has a scattering length density different from that of the matrix will scatter x-rays or neutrons at small angles. If multiple defects are present in the sample simultaneously, the observed scattering curve will be the weighted average from all the defects and the interpretation of the scattering curve can become quite complex. For the purposes of this discussion, the major defects which may present significant small-angle scattering are voids, small precipitate particles, and high densities of small dislocation loops.

It is thus implicitly assumed in the interpretation of most small-angle scattering data that only one type of defect is present in the sample. Such assumptions have proven to be generally valid in most of the research cited in Table I.

A typical small-angle x-ray scattering curve from voids in a sample of neutron-irradiated aluminum is shown in Figure 1. This pattern was obtained from a disc 3 mm in diameter and 74 μm thick in about 60 min measuring time on the pinhole camera described by Schelten and Hendricks (15). Similar data can now be obtained on the ORNL 10-m facility (16) in 1—2 min and, with an appropriately larger sample (2 cm diam by 1/2—2 cm thick) in a similarly short time on most major SANS facilities. From an analysis of the data of Figure 1, one can obtain the void radius of gyration [Figure 1(b) and Eq. (5)], and the void specific surface [Eq. (6)]. These results are given in Table II and are compared there with results obtained on a similar sample by SANS (ref. 21). The agreement of the results is excellent.

It is important to note that although the sample required for the SANS experiment is significantly larger than that required for SAXS, the SANS experiment is nondestructive and so the samples can be used for other tests (i.e., mechanical properties).

The radius of gyration can often be measured with great accuracy (1%). This is illustrated especially well by data obtained by Hendricks et al. (21) on a series of aluminum single crystals which were irradiated together under seemingly identical conditions along the axis of the RB-7 position of the beryllium reflector of the Oak Ridge High Flux Isotope Reactor (HFIR). The results are shown in Figure 2. The increasing trend in R_g was interpreted to result from an estimated $10°C$ increase in the reactor cooling water as it flowed from the top of the core to the bottom. Data of such accuracy result because SAS measures a bulk average over a very large number of voids (~10^{14}), an averaging process which obviously cannot be obtained by other techniques.

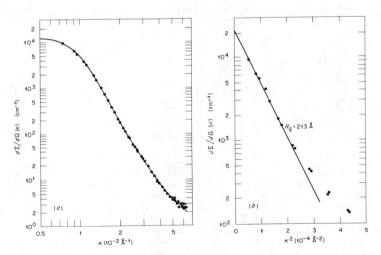

Figure 1. Small angle x-ray scattering from voids in a neutron-irradiated aluminum single crystal (fluence = 2.0×10^{22} n cm^{-2}, E > 0.18 MeV). (a) Scattering function recorded in a single run; (b) Guinier plot of innermost points (from ref. 15).

Table II. Comparison of the Small-Angle X-Ray Scattering Results for
Specimen Al-6 with the Neutron Small-Angle Scattering
Results for Specimen Al-5

Property	X-Rays	Neutrons
$d\Sigma(0)/d\Omega$, forward scattering[*]	(2.7 ± 0.2) $\times 10^4 \times (NZr_T)^2$	(2.60 ± 0.05) $\times 10^4 (Nb_{coh})^2$
R_g, radius of gyration (Å)	213 ± 10	215 ± 4
$\Delta V/V$, specific void volume swelling, %	0.72 ± 0.08	0.78 ± 0.04
S/V, specific void surface ($m^2 \, cm^{-3}$)	1.4 ± 0.3	1.3 ± 0.1

[*]N = number density of Al atoms = 6.06×10^{22} cm^{-3}; Z = number of electrons
per Al atom = 13; r_T = scattering amplitude of a classical electron
= 2.82×10^{-13}; and b_{coh} = nuclear coherent scattering amplitude of Al
= 3.441×10^{-13} cm.

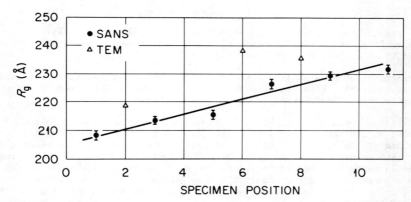

Figure 2. Radius of gyration R_g versus reactor core position (as identified by sample position — Position 1 is at the top, 11 at the bottom)
(see ref. 21).

The swelling and specific surface of voids in the same series of
neutron-irradiated aluminum single crystals are shown in Figure 3 and are compared with similar results from immersion density and transmission electron
microscopy (TEM) measurements. It is seen that the SANS, TEM, and density
measurements are in quite good agreement for the swelling; but that SANS and
TEM do not agree well for the specific surface. This illustrates a particular strength of the small-angle scattering technique. In the derivation
of Eq. (6), from which the results of Figure 3(b) are computed, there are no
assumptions concerning the shape or size distribution of the voids. However, in obtaining a specific surface from TEM micrographs the image of a
non-spherical (facetted) three-dimensional object is projected on a plane

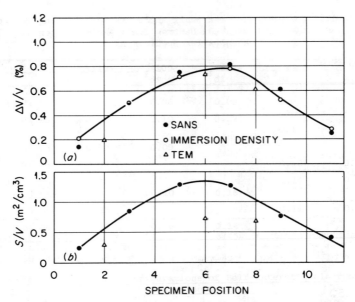

Figure 3(a). Swelling ($\Delta V/V$) versus sample position, and (b) total void surface area per unit volume of sample versus sample position as determined by SANS, TEM, and immersion density (from ref. 21).

(micrograph) whose size is characterized by a circle of radius R from which S/V is computed. Clearly, such procedures can lead to large errors, as is illustrated in Figure 3(b).

If the voids are assumed to be spherical (an assumption to be considered in the next section) then data such as shown in Figure 1 can be unfolded via Eq. (9) with the help of some very sophisticated computer codes to obtain an estimate of the void size distribution. A typical result for voids in neutron-irradiated aluminum are shown in Figure 4 along with TEM data from a companion sample. The results are generally in quite good agreement considering the assumptions and approximations involved in deriving both the TEM and SANS curves.

What appears to be the most significant observation is that all of the information presented in Table II and Figures 3 and 4 [radius of gyration, R_g; swelling, $\Delta V/V$; specific surface, S/V; and size distribution $N(R)$] can be obtained from an analysis of a single scattering curve which can be obtained in a few minutes. At the major small-angle scattering facilities, such as the National Center for Small-Angle Scattering Research in Oak Ridge, the necessary computer programs for such analysis have been developed and are readily available.

SAXS Studies of Voids in Metals Bombarded with Heavy Ions

The use of heavy-ion bombardment to simulate neutron irradiation of metals is of critical importance in reducing the time required to reach a given level of damage as compared to neutron irradiation, and special accelerator facilities have been developed for sophisticated multibeam (heavy ions plus hydrogen or helium) experiments (38,39). The characterization of voids and other defect structures in samples produced in such

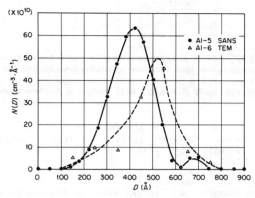

Figure 4. Size distribution for voids in Al-5 as compared with the TEM data of Al-6 (from ref. 21).

facilities poses special problems because the damage layer is very thin (generally less than 1 μm) and very near the specimen surface. Further complications arise because the void sizes are nonuniformly distributed within this damage layer. Considerable skill has been exhibited in developing thinning techniques to examine voids and defects in such samples by TEM.

Because the irradiated volume is so small, small-angle neutron scattering experiments are not possible with heavy-ion bombarded samples. However, a preliminary investigation on a nickel sample damaged to 9 dpa indicated that void sizes could be measured by small-angle x-ray scattering (37). In this experiment, a damaged nickel disc 3 mm in diameter was jet polished from the back via standard TEM preparation techniques to a thickness of 10 μm. The thinned region was about 1.5 mm in diameter and by careful control of the polishing conditions the dimple could be produced with a flat bottom. With such a sample, background scattering from defects other than voids comes from the entire 10 μm thickness while the desired signal, scattering from voids, comes only from the 1 μm layer. If we recall that the optimum thickness for the maximum scattered signal in a transmission experiment is $t_{opt} = \mu^{-1}$ where μ is the linear absorption coefficient, then for nickel studied with CuK_α radiation, $t_{opt} = 27$ μm. Thus, in this experiment, the scattered intensity was roughly 30 times weaker than optimum and the signal-to-noise ratio was 10 times worse than normal. Nevertheless, a quite satisfactory SAXS pattern was obtained on the 10-m SAXS facility in under 1 h. The void sizes determined by SAXS were in good agreement with those obtained by TEM on similar samples. However, a further complication arises if one wishes to obtain the localized swelling ΔV/V. The scattering experiment measures the total scattering from the independently scattering voids, but if the voids are nonuniformly distributed within the sample, as is the case here, there is no way to extract this information from the data. Thus, the local swelling cannot be determined by SAXS alone, but must have accurate theoretical (or experimental) damage distribution profiles.

Following this initial success, Epperson (36) has embarked on a detailed study of the role of oxygen in void formation in nickel-ion bombarded niobium. Here, niobium single crystals were doped with varying concentrations of oxygen and were then irradiated in the ORNL facility (38,39) at various temperatures and to several damage levels. Preliminary SAXS results from the 10-m facility are very encouraging.

Anisotropic Scattering Effects

It has been assumed in all of the preceding discussion that the voids, although not necessarily spherical in shape, were randomly distributed throughout the sample, both in orientation and spatially. Such is not necessarily the case, and two important deviations from these assumptions have yielded additional information from SAS measurements. The first to be discussed will be the effect of randomly distributed, oriented, facetted voids in a single crystal, and the second will be spatially ordered voids (the void lattice).

It has been well known from the first TEM observations that voids in metals are not spherical. Rather, they are facetted with the facet planes usually being low index crystallographic planes. Hendricks et al. (21) recognized that such facetting should produce an anisotropic scattering at higher k values in the small-angle scattering pattern. In the case of voids in neutron-irradiated aluminum, which are essentially octahedra bounded by {111} planes, but which are also truncated on {100} planes, they were able to calculate the structure factor. It showed a strong spherically symmetric central region surrounded by rods of diffuse intensity in <111> and <100> directions as depicted in Figure 5. Such patterns have been observed by Epperson et al. (3) with film techniques and by Spooner and Child (40) with a two-dimensional PSD. Analogous patterns for voids with other shapes have been observed in GaAs (28) and NiAl (32).

The calculations showed that at the higher $\underline{k}$ values the structure factor followed different asymptotic laws as shown in Table III. Thus, we see that although for randomly oriented voids the asymptotic law must average out to follow Porod's law [Eq. (6)], the asymptotic law for oriented voids may be very different in certain specific directions.

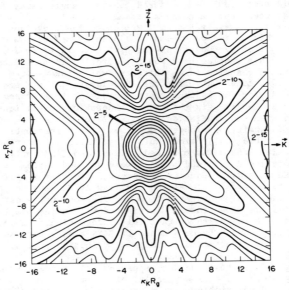

Figure 5. Calculated SAS pattern for truncated octahedral voids in Al. [001] is vertical and [110] is horizontal (from ref. 21).

k∥ to	Asymptotic Law
⟨hkl⟩	$\lvert k \rvert^{-6}$
⟨hk0⟩ ⟩ ⟨h00⟩ ⟩ ⟨hh0⟩ ⟩	$\lvert k \rvert^{-4}$
⟨hhh⟩	$\lvert k \rvert^{-2}$

$$h \neq k \neq l \neq 0$$

To test these ideas the scattering curves were measured at high angles in specific crystallographic directions as shown in Figure 6. Although the data fit the theoretical predictions in ⟨111⟩ directions with remarkable accuracy, it is seen that there is a considerable discrepancy between the anticipated result of k^{-4} in the ⟨100⟩ direction. This was attributed to the presence of a very small volume fraction of cubic voids which can be shown to have a k^{-2} dependence along ⟨100⟩. From these data it was estimated that the volume fraction of cubic voids was 0.18% of that of the octahedral voids. Since the volume fraction (swelling) of the latter was 0.8%, the volume fraction of cubic voids is 0.0014%! Such a small volume fraction of voids can be discerned because the cubic voids scatter most intensely in exactly the directions where the octahedral voids scatter weakly. These effects could certainly not be observed in polycrystalline samples where the scattering is smeared out into the Porod asymptotic average.

In their analysis of the structure factor which led to Figure 5, Hendricks et al. (21) found that at fixed lengths of the scattering vector (which corresponds to a circle drawn on Figure 5), the ratio of the intensity scattered in the ⟨111⟩ and ⟨100⟩ directions depended strongly on the degree

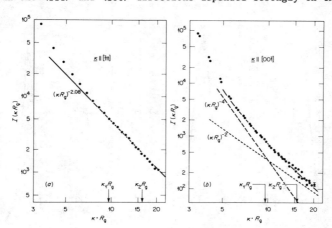

Figure 6. Scattered intensity along (a) [$\bar{1}11$] and (b) [$00\bar{1}$] for sample Al-5 as a function of HR_g. In (b) the void line is a fit of $C_2(kR_g)^{-2} + C_4(kR_g)^{-4}$. (From ref. 22).

of truncation of the void. Further, it was postulated (22) that if the void shape was an equilibrium one, then the truncation parameter was directly proportional to the ratio of the specific surface energies of the {100} and {111} surfaces. This led to an experiment in which the diffracted intensity was recorded simultaneously at two fixed scattering angles while the single crystal sample was rotated as depicted in Figure 7. The resulting data are shown in Figure 8.

From these data, it was shown that the average truncation parameter for this sample was 0.19 ± 0.02, which led to the specific surface energy ratio

$$\frac{\gamma_{100}}{\gamma_{111}} = 1.40 \pm 0.04$$

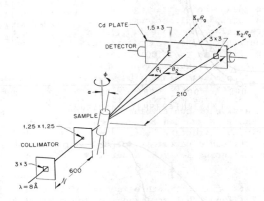

Figure 7. Schematic drawing of the experimental arrangement in the SANS rotation experiment. All dimensions are in cm (from ref. 22).

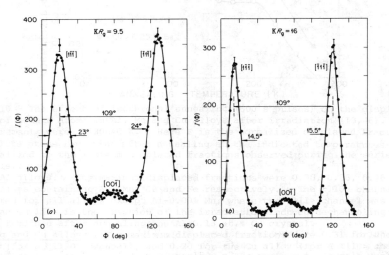

Figure 8. Intensity distribution I(ϕ) for specimen Al-5 measured in the (110) plane at (a) kR_g = 9.5 and (b) kR_g = 16.0 (from ref. 22).

392

This result was considerably higher than that expected for clean aluminum surfaces, and it was postulated that transmutation-produced Si contaminated the surfaces, thus preferentially changing the ratio. This hypothesis was later confirmed when Farrell et al. (41) observed a sheath of precipitated silicon on the surfaces of voids created in samples irradiated to significantly higher fluence than those examined by Hendricks et al.

Effects of Radioactivity

Metals and alloys which are irradiated to high fluences in nuclear reactors show varying degrees of radioactivity and varying half-lives for its decay. In the case of high-purity aluminum, the activity results mostly from ^{22}Na and decays rapidly. Thus, there is little difficulty in handling samples, even the very large ones used in the SANS experiments (20—22). Similarly, there were no significant problems associated with handling small molybdenum (23,24) or nickel samples (27). However, the high level of activity encountered by Schwann et al. (35) in their SANS investigation of stainless steel, and by Spooner in his SANS study of Nb(Zr) (30) required special techniques. In the case of niobium, the small disks used in SAXS experiments (3 mm ϕ and 100 μm thick) produced over 200 cps background in the area detector used in the 10-m SAXS facility at a distance of 5 m. This was ten times greater than the integrated scattering intensity from the voids in the sample. Even worse, the emitted radiation was MoK_α created by K-capture in niobium, and was indistinguishable from the incident MoK_α radiation used in the scattering experiment.

These results have shown that, although induced radioactivity may cause some problems and may preclude the use of either SAXS or SANS, usually one of the techniques can be used for the investigation of voids and void properties.

Summary

Small-angle and x-ray and neutron scattering, often in conjunction with transmission electron microscopy, have proven to be effective tools for obtaining the radius of gyration, size distribution, volume fraction, and specific surfaces of voids in a variety of neutron irradiated and heavy-ion damaged metals and alloys. In comparing the techniques, an important difference to be kept in mind is that TEM is the superior method for obtaining information on single and small numbers of defects, while SAXS and SANS are preferred when bulk averages over very large numbers of defects ($\sim 10^{14}$) are desired. The information on truncation parameters as obtained in the SANS rotation experiment on aluminum could be obtained by other means only with difficulty. In the author's experience, however, one should not emphasize one technique as compared to the other. Rather, SAXS, SANS, and TEM (along with other techniques such as positron annihilation and bulk density) should be viewed as complementary techniques, each providing a different and valuable kind of information in the study of voids. When a description of a void structure can be presented which is consistent with observations from several techniques, one has considerably more confidence in its validity.

References

1. A. Guinier and G. Fournet, Small-Angle Scattering of X-Rays, John Wiley and Sons, New York, 1956.
2. J. Schelten and R. W. Hendricks, J. Appl. Crystallogr. 11(5) (1978) 297-324.

3. J. E. Epperson, R. W. Hendricks, and K. Farrell, Phil. Mag. 30 (1974) 803-817.
4. J. Schelten, Kerntechnik 14 (1972) 86-87.
5. K. Ibel, J. Appl. Crystallogr. 9 (1976) 296-309.
6. R. M. Brugger, J. S. King, S. A. Werner, and W. B. Yelon, private communication, 1976.
7. C. P. Galotto, P. Pizzi, H. Walther, V. Angelastro, N. Cerullo, and G. Cherubini, Nucl. Instr. Methods 134 (1976) 369-378.
8. F. Frisius and M. Naraghi, Atomkernenergie 29 (1977) 139-144.
9. B. P. Schoenborn, J. Alberi, A. M. Saxena, and J. Fischer, J. Appl. Crystallogr. 11 (1978) 455-460.
10. C. Hofmeyer, R. M. Mayer, and D. L. Tillwick, J. Appl. Crystallogr. 12 (1979) 192-200.
11. W. C. Koehler and R. W. Hendricks, J. Appl. Phys. 50 (1979) 1951.
12. H. R. Child and S. Spooner, J. Appl. Crystallogr. 13 (1980) 259-264.
13. B. Farnoux, Les Spectrometres pour la diffusion aux petits angles Laboratoire Leon Brillouin, Report LLB/79/250/BF/MLC, 18 Dec. 1979; 91190 Gif-sur-Yvette, France.
14. C. J. Glinka, J. M. Roe, and J. G. LaRock, NBS Reactor: Summary of Activities July 1979 to June 1980, NBS Technical Note 1142, pp. 66-69 Washington, DC 20234. (May 1981).
15. J. Schelten and R. W. Hendricks, J. Appl. Crystallogr. 8 (1975) 421-429.
16. R. W. Hendricks, J. Appl. Crystallogr. 11 (1978) 15-30.
17. V. Gerold and G. Kostorz, J. Appl. Crystallogr. 11 (1978) 376-404.
18. V. W. Lindberg, J. D. McGervey, and W. Triftshauser, Phil. Mag. 36 (1977) 117-128.
19. J. D. McGervey, V. W. Lindberg, and R. W. Hendricks, J. Nucl. Mater. 69&70 (1978) 809-812.
20. H. A. Mook, J. Appl. Phys. 45 (1974) 43.
21. R. W. Hendricks, J. Schelten, and W. Schmatz, Phil. Mag. 30 (1974) 819-837.
22. R. W. Hendricks, J. Schelten, and G. Lippmann, Phil. Mag. 36 (1977) 907-921.
23. S. Liu, J. Moteff, J. S. Lin, and R. W. Hendricks, J. Appl. Cryst. 11 (1978) 597-602.
24. J. S. Lin, R. W. Hendricks, J. Bentley, and F. W. Wiffen (unpublished research).
25. C. Hofmeyer, K. Isbeck, and R. M. Mayer, J. Appl. Cryst. 8 (1975) 193.
26. C. Hofmeyer, R. M. Mayer, and E. T. Morris, Ann. Cont. S. Afr. Inst. Phys. (1976).
27. J. S. Lin and R. W. Hendricks (unpublished research).
28. S. Gupta, E.W.J. Mitchell, and R. J. Stewart, Phil. Mag. A 37 (1978) 227-243.
29. R. J. Lauf (private communication).
30. S. Spooner, this conference.
31. J. B. Bates, R. W. Hendricks, and L. B. Shaffer, J. Chem. Phys. 61 (1974) 4163-4176.
32. J. E. Epperson, K. W. Gerstenberg, D. Berner, G. Kostorz, and C. Ortiz, Phil. Mag. A 38 (1978) 529.
33. D. G. Martin and J. Caisley, Some Studies of the Effect of Irradiation on the Neutron SAS from Graphite, AERE-R-8515, Atomic Energy Research Establishment, Harwell, Great Britain.
34. P. Krautwasser (private communication).
35. D. Schwahn, D. Pachur, and J. Schelten, Neutron Scattering on Neutron Irradiated Steel, Report Jül-1543, October 1978, Kernforschungsanlage, Jülich GmbH, Julich, West Germany.
36. J. E. Epperson (private communication).
37. R. W. Hendricks (unpublished research).
38. M. B. Lewis, N. H. Packan, G. F. Wells, and R. A. Buhl, Nucl. Instrum. Methods 167 (1980) 233.

39. N. H. Packan and R. A. Buhl, A Multispecimen Dual-Beam Irradiation Damage Chamber, Oak Ridge National Laboratory Report ORNL/TM-7276, Oak Ridge, Tennessee (June 1980).
40. S. Spooner and H. R. Child (private communication).
41. K. Farrell, J. Bentley, and D. N. Braski, Scripta Met. 11 (1977) 243.

Fig. 13 : Isochronal annealing (600-s pulse anneals) of a water-quenched Al-0.03 at.% in crystal after irradiation with 1 MeV He ions at 35 K to a fluence of 1.0×10^{16} cm^{-2}. In the lower part of the figure, the effect of annealing on χ_{Al}, χ_{In}, and $f_{dA}^{<110>}$ is shown. In the upper part, the fractional recovery R_D of the irradiation-induced dechanneling increment is shown (77).

PRELIMINARY SMALL ANGLE NEUTRON SCATTERING INVESTIGATIONS OF

NEUTRON IRRADIATION PRODUCED VOIDS

S. Spooner (1), W. E. Reitz (1) and H. R. Child (2)
(1) Metallurgy Program, Georgia Institute of Technology
Atlanta, Georgia 30332
(2) Solid State Division, Oak Ridge National Laboratory
Oak Ridge, Tennessee 37830

The preliminary structure analysis of neutron irradiation-produced voids has been undertaken with small angle neutron scattering (SANS) at the Oak Ridge Research Reactor (ORR) at Oak Ridge National Laboratory. Two investigations are discussed with emphasis on experimental technique. Scattering from a set of niobium-zirconium alloys doped with boron to produce helium during irradiation was measured to estimate the average void radius and void fraction in samples irradiated under various conditions. The scattering measured from an irradiated 304 stainless steel sample exhibited intervoid interference effects. Magnetic fields were applied to the steel sample to try to find evidence of scattering from ferromagnetic precipitates. Such scattering was not observed at a significant level.

*Research sponsored by the Division of Materials Sciences, U. S. Department of Energy under contract W-7405-eng-26 with the Union Carbide Corporation.

Introduction

This paper reports on two preliminary void structure investigations which were performed with the small angle neutron scattering (SANS) machine at the Oak Ridge Research Reactor (ORR). Experiments on irradiated niobium samples illustrate a small angle scattering technique for samples where a correction for a significant background coming from radioactivity must be made. Experiments on irradiated stainless steel are discussed to show the method for differentiating the scattering from magnetic precipitates from the scattering due to voids. SANS experiments are nondestructive which is advantageous in radiation damage investigations since expensive hot cell sample preparations are minimized. The present work, done on a prototype SANS machine, represents a very preliminary survey of scattering effects in the materials. The possibilities for void investigations on the newer SANS facilities (especially the SANS machine at the HFIR which is part of the National Center for Small Angle Scattering Research) are very much greater and should be explored in the near future.

Experimental

SANS Apparatus

The ORR SANS machine uses point collimation with a neutron wavelength of 4.82 A, which is longer than the Bragg diffraction limit for most metals and alloys. Thus, no double-Bragg scattering occurs in the experiments reported here. A reactor beam is first passed through a liquid nitrogen cooled beryllium filter which transmits neutrons of wavelength greater than 4 A. The filtered beam is then diffracted twice from the (002) planes of three pairs of pyrolytic graphite crystals to produce the monochromatic beam traveling parallel to, but displaced from, the original reactor beam. The mnochromated neutrons travel through a 4.5 meter evacuated flight path to the specimen mounting area. The apertures placed at the ends of the flight path define the incident beam collimation. The scattered (and transmitted) beam travels through an evacuated beam path to a 20 cm x 20 cm area sensitive proportional counter. A cadmium beam stop is placed at the end of the flight path in front of the area detector. The scattered beam resolution is determined by the scattering sample area and the detector resolution (0.5 cm diameter on the detector area). The incident beam flux on the specimen depends on the monochromator collimation aperture and for a 2 cm diameter aperture is 1400 n/cm^2/sec. Typical sample areas vary from 0.5 cm diameter to 2 cm diameter. A recent paper describes the system in more detail (1).

For measurements on the radioactive samples lead shielding was installed around the sample to reduce the general gamma background to an acceptable level. Although the proportional counter discriminates against gamma rays, the gamma signal can overlap the neutron produced signal to a certain degree depending on gamma ray energy. The gamma background from the radioactivity contribution to background scattering is measured with a cadmium sheet blocking the incident beam in order to eliminate the thermal neutron scattering. The difference between counts from a radioactive sample and an empty sample holder is due to the effect of radioactivity from the sample.

Neutron scattering events are recorded as in any proportional counter but the location of the event on the area detector is determined by a rise-time encoding method described by Borkowski and Kopp (2). The location is encoded then converted to a digital computer address, so that the location and number of events can be accumulated within a computer as a histogram of neutron scattering events (counts). In the ORR SANS machine a Hewlett-Packard 21 MX digital computer is used to control the data collection. A low-efficiency fission detector is used to monitor the incident beam intensity. A predetermined total count from the monitor detector is used for timing the data collection. The data (in the form of a 32 x 32 array of neutron counts) can be displayed on a Tektronix scope during collection and at the end of collection the data can be punched out on paper tape or transferred to a segment of computer memory reserved for the purpose. The paper tape can be transferred to a central data storage system shared by other small angle scattering systems (3). After requisite data sets are collected, data can be manipulated by multiplication and addition for data correction. The 32 x 32 array is reduced to a scattering intensity versus magnitude of scattering vector by radial averaging. Although these procedures require the interruption of data collection, the interruption is less than 15 minutes and represents a small fraction of neutron beam time used for data collection. In this investigation scattering data sets for a niobium experiment took 24 hours which included a 12 hour radioactivity background run. The stainless steel experiment required only a few hours of data collection.

Scattering Specimen and Specimen Mounting

There is an optimum area and thickness for samples which are determined by angular resolution and scattering intensity requirements. The scattering angle variable is given by $\kappa = 4\pi\sin\theta/\lambda$ and the scattering resolution is $\Delta\kappa$. Generally $\Delta\kappa$ equals the minimum measurable κ in the scattering pattern (4). The minimum κ, in turn, is determined by the size of the transmitted beam on the detector. This is dictated by the sample area and the lengths of the incident and scattered beam paths and the area detector resolution. Cutting samples down and lengthening beam paths will reduce the minimum κ, but scattering intensities will be reduced in proportion to solid angle factors. The sample thickness can be increased to improve the scattering intensity up to the optimum limit given by the reciprocal of the attenuation factor, where attenuation occurs by both scattering and absorption. For long neutron wavelengths absorption generally dominates the attenuation process.

Because of the radioactivity of reactor irradiated samples, one prefers to work with specimens as supplied. Thus, sample areas and thicknesses are fixed at dimensions typically smaller than those needed to optimize the experiment. In the case of experiments on niobium alloys, the shoulder section of microtensile specimens having a diameter and length of 4 mm (volume = 0.05 cm^3) were used. In the case of the stainless steel specimen, a plate 1 mm thick was supplied with sufficient area so that a 1 cm diameter beam could be used (volume = 0.08 cm^3).

The niobium samples were taken from a set of niobium and niobium-1% zirconium microtensile specimens alloyed with boron. The boron is consumed during irradiation through the (n, α) nuclear reaction which thereby provides a source of interstitial helium gas to affect the void nucleation process. The samples were irradiated in the HFIR reactor at ORNL under a variety of fluences and temperatures. The levels of radioactivity varied, but three years after removal from the ractor some samples were measured at 3 R at contact and sample handling was done under health physics supervision. The principal gamma rays were 0.713 and 0.873 MeV. MoK$_\alpha$ radiation was also generated by the flourescence of molybdenum (which occurs by transmutation of niobium) induced by the beta decay electrons occuring in the specimen. The highest energy gamma ray generates a pulse in the neutron detector which overlaps the pulse generated in the neutron counting process. The gamma rays provided a significant radioactivity background. Niobium specimens were mounted in a lead shield having conical entrance and exit ports and a top loading access hole shown in Figure 1. A shoulder section from each sample set was loaded, with gage section down, into a quartz tube with a necked interior diameter which fixed the sample position. The quartz holder was loaded from the top. Specimen positioning was checked by making a neutron photograph of the incident beam passing by an opaque (boron nitride) dummy specimen of the same dimensions as the niobium samples.

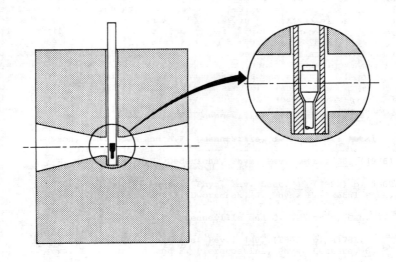

Fig. 1. Lead Shielding used to surround the niobium microtensile specimens. Sample is held in a quartz device loaded from the top of the lead cylindrical shield.

The irradiated stainless steel specimen was in the form of a plate which was taken from a 304 stainless steel control rod thimble in the EBR-II reactor at reactor core mid-plane. The radioactivity remaining five years after removal from the reactor arose principally from Co^{60} and was less than 2 R at contact. The gamma rays (1.17 and 1.33 MeV) were removed by energy discrimination in the detector so that no significant radioactivity background was detected. In the study of the effect of magnetic field, the sample was mounted into an aluminum holder with quartz windows. The holder was placed between the pole pieces of an electromagnet. Lead blocks added to the pole pieces formed the gamma ray shield for the steel specimen. The direction of applied field was perpendicular to the incident beam.

Data Collection

Corrected scattering data when calibrated in absolute units give the most complete structural analysis obtainable from scattering experiments. The total scattering, in counts per time, P^*, from a specimen having a radioactivity background can be expressed,

$$P^*_t = P_s + Tr \cdot (P_{mt} - P_{Cd}) \tag{1}$$

where P_s is the true scattering rate, P^*_{Cd} is the background from the radioactive specimen with the incident beam blocked with cadmium (Cd). Tr is the ratio of neutrons transmitted through the specimen to neutron incident on the sample P_{mt} is the background from the empty sample holder and P_{Cd} is the background measured with cadmium blocking the beam. The term, Tr. $(P_{mt} - P_{Cd})$ is the scattering from air and the sample holder, which is attenuated by passage of the incident beam or the scattered beam through the specimen. Thus, four measurements are needed for the determination of completely corrected scattering rate: P^* , P_{Cd} , P_{mt} , and P_{Cd}. The latter two measurements are determined only once for a fixed scattering arrangement. P^* and P_{Cd}^* have to be measured for each radioactive specimen. The duration of data collection is determined by the accumulation of counts on the incident beam monitor detector rather than the accumulation of time. Thus, the duration of measurement of P^* amd P_{Cd}^* must be the same to measure accurately the radioactivity background since the background counts depend on real time.

The scattering intensity can be expressed in terms of the characteristics of the scattering apparatus and absolute cross section $d\Sigma/d\Omega$ of the sample as follows,

$$P_s = I_o \cdot V_s \cdot Tr \cdot \frac{d\Sigma}{d\Omega} \cdot \Delta\Omega , \qquad (2)$$

Where I_o is the incident neutron intensity in counts per monitor detector per area, V_s is the sample volume, is the solid angle element given by

$$\Delta\Omega = A_d/R^2 , \qquad (3)$$

where A_d is the area of the detector element, R is the detector to sample distance, and $d\Sigma/d\Omega$ is the absolute macroscopic cross section in cm^{-1} given by

$$\frac{d\Sigma}{d\Omega} = N_n \frac{d\sigma}{d\Omega} , \qquad (4)$$

where N_n is the nuclear density and $d\sigma/d\Omega$ is the differential scattering cross section per steradian per nucleus. Thus, the absolute cross section is obtained from measurement of I_o , V_s , Tr and $\Delta\Omega$. Alternatively , a scatterer with a known macroscopic cross section can be measured to establish the instrumental constant. Measurement of the incident beam intensity is a more precise calibration procedure in our experimental arrangement because the counting rate from known scattering standards , such as vanadium or water, is too low to be measured with adequate precision in a short time.

Data Reduction

Scattering from voids in polycrystalline samples is independent of the azimuthal scattering angle. The data were corrected according to Eq. (1) and statistical counting errors were calculated for each of the points in the 32 by 32 array. Then, the data were sorted into a list of intensities in order of increasing scattering angle. The list was subdivided into increments of κ , then the κ values and the intensities were each averaged to give give an $I(\kappa)$ vs κ graph. Alternatively, intensities for points on a circle of fixed were interpolated from the 32 by 32 array. In this method, each intensity was determined by the interpolation of three points. The latter procedure was used to explore intensity variations with azimuthal angle as will be discussed below in connection with magnetic field effects.

Data points near the beam center were set to zero to eliminate meaningless neutron count. Limits on the horizontal and vertical channel numbers in the array were assigned to exclude invalid data at the edges. A direct examination of data points around the beam center is done to determine data compromised by beam stop attenuation. A true incident beam center can be found by examination of intensities calculated for fixed κ as a function of azimuthal angle. Center coordinates are chosen so that intensities at fixed scattering angle exhibit minimum variation with azimuthal angle. The center can be determined to within a tenth of a detector channel width by this method. This consideration is important when the electronics drift causing a horizontal shift in the apparent beam center.

Magnetic and Nuclear Scattering Cross Sections

The magnetic neutron scattering interaction is expressed in terms of a magnetic scattering amplitude per atom (5)

$$P = \frac{e^2\gamma}{2mc^2}\, \mu \ , \tag{5}$$

where $(e^2\gamma/2mc^2)\mu$, $= 0.269 \times 10^{-12}$ cm is the component of the atomic magnetic moment which is given by the projection of the total moment onto a plane perpendicular to the scattering vector, $\vec{\kappa}$, where $\vec{\kappa} = \vec{k}_o - \vec{k}'$ with $\vec{k}_o$ the incident neutron wave detector and k' the scattered neutron wave vector. When a magnetic field aligns the magnetization parallel to the field, the effective magnetic amplitude is

$$P = \frac{e^2\gamma}{2mc^2}\, \mu \sin \alpha \ , \tag{6}$$

where is the angle between $\vec{\kappa}$ and the direction of moment alignment. In small angle scattering experiments, the magnetic field is applied in a direction perpendicular to κ and in the plane of the scattering vector. The experimental arrangement is schematized in Figure 2. In the case where magnetic small angle scattering and nuclear small angle scattering are independent of one another, the macroscopic cross section is a sum of the two effects. If we assume a simple case of scattering from a sample containing voids and magnetic particles, both of which have the same average nuclear scattering amplitude nuclear scattering amplitude in a nonmagnetic matrix, the cross section is given by

$$\frac{d\Sigma}{d\Omega} = \frac{d\Sigma}{d\Omega}\, \text{void} + \frac{d\Sigma}{d\Omega}\, \text{magnetic} \ , \tag{7}$$

where

$$\frac{d\Sigma}{d\Omega}\, \text{void} = (N_n b)^2 \cdot N_v \cdot V_{void}^2 \cdot |F_v(\kappa)|^2 \ . \tag{8}$$

Here N_n is the nuclear density of the metal, b is the nuclear scattering length and $F_v(\kappa)$ is the scattering form factor for the void given by

$$F_v(\kappa) = \frac{1}{V} \int_V \exp(i\vec{\kappa}\cdot r)\, dr \ , \tag{9}$$

where integration is carried out over the limits of the void volume, V. The magnetic cross section for a collection of independently scattering ferromagnetic particles is given by analogy to Eq. 8,

$$\frac{d\Sigma}{d\Omega} = (N_n \cdot \frac{\gamma e^2}{2mc^2})^2\, \mu^2 \cdot N_p \cdot V_{mag}^2 \cdot F_m(\kappa)^2 \cdot \sin \alpha \ , \tag{10}$$

where is the magnetic moment per atom, N_p is the density of magnetic particles of volume v_{mag} and F_m is the particle form factor evaluated as in Eq. 9. It is assumed that there is a discontinuous change in magnetization

density at the magnetic particle boundary. The small angle scattering
observed from a system containing both nuclear and magnetic scatterers under

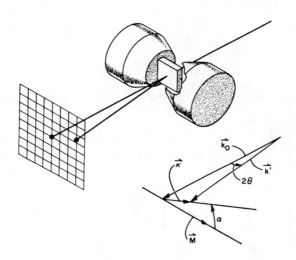

Fig. 2. Vector diagram for magnetic small angle scattering. Incident and
scattered wave vectors are $\vec{k}_o$ and $\vec{k}$ respectively. $\vec{H}$ is the applied
field direction and the magnetization is assumed to be saturated and
parallel to $\vec{H}$. The angle between the scattering vector ($\vec{k}-\vec{k}_o$) and
H is α.

a magnetic field which is perpendicular to $\vec{k}_o$ is written using Eq. 2.

$$P_s(\kappa) = P_n(\kappa) + \sin^2 \alpha \, P_m(\kappa) \quad , \tag{11}$$

In the absence of a magnetic field, the $\sin^2\alpha$ factor would normally be
directionally averaged to a value of 2/3.

Niobium

The corrected scattering intensities are plotted in Figure 3 and Figure
4. Scattering intensity which is measurably larger that background is found
in the range of between 0.005 A^{-1} and 0.025 A^{-1}. The data tend toward a
κ^{-4} dependence over most of the measurement range. Log (I) versus κ^2 plots
are curved and estimates of the radius of gyration from such plots are
larger than the reciprocal of the minimum κ value. This means that the
range of κ in the experiment does not go to a low enough value for a
meaningful determination of the radius of gyration.

A less satisfactory determination of void radius, in the face of
incomplete information, can be made from the larger angle data. When the

403

κ^{-4} dependence can be quantitatively established, a Porod radius is calculated from

$$R^* = 3 Q_o /\pi P ,\qquad (12)$$

where the integrated intensity Q_o is

$$Q_o = 1/2\pi^2 \int \kappa^2 \frac{d\Sigma}{d\Omega} (\kappa) \; d\kappa ,\qquad (13)$$

and

$$P = \lim_{\kappa\to\infty} \kappa^4 \cdot \frac{d\Sigma}{d\Omega}(\kappa) .\qquad (14)$$

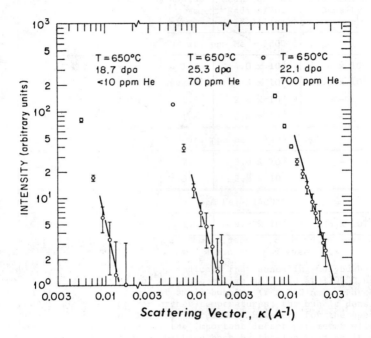

Fig. 3. Scattering intensity from niobium irradiated at 650 C with varying helium contents.

Quantitative measurements of R^* require that the κ^{-4} dependence apply to the large angle data and that the measurements in the range of the experiment accurately define Q_o. In these preliminary experiments, the foregoing requirements are not completely satisfied, hence the results are only qualitative. However, in the evalation of Q_o, the integration of the scattering data in the Guinier range contributes only a small fraction to the total. The void fraction can be estimated from Q_o alone using

$$\Delta V/V = \frac{Q_o}{2\pi^2 (N_n b)^2} ,\qquad (15)$$

Note that the Porod radius is obtained from a ratio of scattering quantities so that a calibration for absolute scattering cross sections is not required. Void fraction determination does require such a calibration, however.

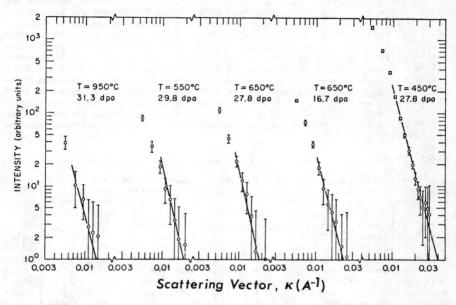

Fig. 4. Scattering intensity from niobium alloy with fixed helium content irradiated for various temperatures. Irradiation fluence is indicated in displacements per atom (dpa).

The results of the niobium void scattering are summarized in Table I. Error limits on the calculated quantities were estimated on the basis of statistical variance in measured counts in both specimen and background experiments and accumalated errors range from 30 to 100%.

Table I. Niobium Void Parameters Calculated from SANS

Specimen	Helium Content (ppm)	Irradiation Temperature ($^{\circ}$C)	Displacements per Atom	Porod Radius (A)	Void Fraction (%)
5A-2	134	650	16.7	181	1.4
5A-4	134	450	27.8	208	12.0
5A-5	134	550	29.8	210	0.6
5A-6	134	950	31.3	79	0.5
5A-7	134	650	27.8	346	0.7
7A-3	700	650	22.1	144	4.0
6A-3	70	650	25.3	137	1.3
4A-1	<10	650	18.7	367	0.4

The analysis is qualitative and rests primarily on the assumption of κ^{-4} dependence in the larger angle data. In order to evaluate the legitimacy of this analysis, modeling of the scattering curves was done with a calculation of scattering from voids whose sizes were distributed with a log-normal distribution (7). The mean size in the distribution is μ and the width of the distribution is σ. The Porod radius can be expresses as

$$\ln R^* = \ln(\ ) + 2.5 \ln^2 \ . \tag{16}$$

Values of σ were assumed in the range from 1.1 to 1.5 and with estimated values of R^*, μ was obtained. Scattering from a distribution of spheres characterized by σ and μ was calculated with Eq. (8) for comparison with measured intensity. It was found that the calculated scattering indeed exhibited κ^{-4} dependence in the range of measured κ and that smaller angle data were approximated best with a large (i.e. a broad distribution). It appears that these preliminary experiments can give useful estimates of void structure characteristics.

Stainless Steel

The scattering from voids in stainless steel was measured with a short and a long scattering path configuration and the two data sets were overlapped in order to cover a wide κ-range. These data are shown in Figure 5. An interference peak is found in the large angle range. The origin of the interference can come from : (1) mutual exclusion of voids over a certain range, (2) an interference associated with a change in alloy composition around each void (only Cr variation could produce a solute depletion effect because of the relative neutron scattering amplitude values of Fe, Cr and Ni) or (3) the formation of a coating of precipitate at the void-matrix interface. A scattering model for the case of spatial exclusion between voids was used to fit the stainless steel void scattering. In this model a Gaussian scattering form factor for voids was assumed to be modulated by an interference function appropriate to a collection of scatterers whose centers can approach no closer that a fixed exclusion distance. A second Gaussian function of wider scattering distribution than the first was added to account for scattering at larger angles. The curve shown in Figure 5 is a least squares fit to the data. A radius of gyration was obtained from the first Gaussian function according to the Guinier approximation for scattering from spheres (6). This radius was 107 A and the exclusion distance obtained from the modulation function was 340 A.

The larger angle data were examined to determine the Porod radius in the manner described previously for niobium voids. The scattering curve at large angles was fitted to the function $P\kappa^{-4} + C$ where P is the Porod constant. The integrated intensity, Q_o, together with P gives a Porod radius of 88 A. The difference between radius of gyration and Porod radius is not unusual since each radius is obtained from different moments of the void size distribution.

The void fraction estimated from Q_o was found to be 3 to 5 times what could be reasonably expected in this material under the conditions of irradiation. Sources for this discrepancy include the scattering from objects other than voids and multiple scattering which redistributes the high intensity scattering at very small angles to larger angles. The scattering density, $N_n \Delta b$, is especially large in stainless steel owing to the large nuclear scattering amplitudes of Fe and Ni. A rough calculation of multiple scattering effects was undertaken using the procedure of Schelten and Schmatz (8). It is clear in our rough calculations that although the interference peak is essentially unchanged, there is a significant redistribution of scattering to higher angles. This would add an important contribution to the computation of Q_o and thereby increase the apparent void fraction to unreasonable values.

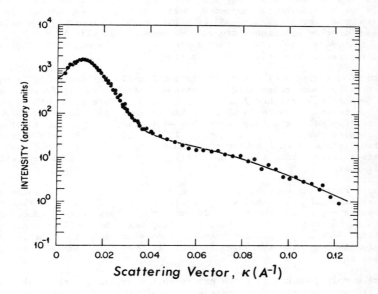

Fig. 5. Scattering intensity from irradiated 304 stainless steel. Note the apparent interference maximum. The data are fit with a model based on a mutual exclusion of voids at high density.

Finally, scattering from the stainless steel specimen under a magnetic field of 15 KOe was examined to estimate the fraction of scattering sources from ferromagetic precipitates or phases (9). There appeared to be little or no magnetic effects on the scattering in the direction parallel and perpendicular to the applied field direction. This result indicates that no magnetic scattering over the range of scattering angles of this experiment experiment magnetic scattering over the range of scattering angles of this experiment can be no greater that a few percent of the void scattering. At the very least this suggests that there are no ferromagnetic precipitates having the size of voids. Non-magnetic precipitates cannot be distinquished from voids in this experiment. Large ferromagnetic precipitates (greater than 1000 A) would contribute scattering at angles smaller than those reached in this experiment.

Summary

Despite low incident neutron intensity, small sample volume (by neutron scattering standards) and a significant background from sample radioactivity, a useful characteriztion of irradiation effects on niobium was obtained from these preliminary SANS experiments. It is shown that low temperature and high helium content promote a large void fraction. The void radius is in the range of 100 A to 200 A and the void fraction approaches 10%. Estimates of the Porod radius and void fraction were extracted from the scattering data. Use of the Harkness, Hren and Gould model (7) for log normal size distribution helped to establish some credence in the approximate analysis attempted. On the basis of these experiments and the now known performance of the 30 meter SANS machine at the HFIR (part of the National Center for Small Angle Scattering at ORNL) it is clear that SANS analysis can be done on the 30 meter SANS machine. The higher incident intensity and lower background of that machine removes the principal impediment in this prliminary study. Furthermore, it will be far easier to access the very low scattering angles which is necessary to undertake a radius of gyration analysis of this samples.

The SANS analysis of scattering from a heavily irradiated 304 stainless steel reveals some remarkable void structure features. Voids of 84 A mean radius seem to be distributed so that the voids exclude each other over a center-to-center distance of 340 A. The estimate of void fraction appears to be severely compromised by multiple scattering and thinner stainless steel samples would help to bring the problem of the very strong void scattering into control. No magnetic scattering could be found in these experiments from which it can be concluded that there are no ferromagnetic precipitates.

These studies have demonstrated the advantages of the SANS technique in the investigation of irradiated material. First, thermal neutrons have a high penetrating power, therefore, it was not necessary to reduce sample dimensions for these experiments. Thus, expensive hot-cell procedures were minimized. Second, the neutron's interaction with magnetic moments permit an evaluation of the presense of small ferromagnetic phases in irradiated stainless steel.

The sampling volume in these experiments provides a large statistical base for precise structure parameter analysis. In the present ORR SANS machine arrangement, the statistical precision of accumulated counts was so limited that the expectation of high precision in the structure analysis could not be fulfilled. In an optimized SANS machine (such as the one at the HFIR) precise quantitative analysis can be achieved.

Although a statistically precise analysis can be obtained from scattering experiments, the scattering curves cannot yield a unique and complete structure interpretation without the aid of direct image information obtained by transmission electron microscopy (TEM). Both TEM and small angle scattering are required in most cases to obtain an unambiguous interpretation of structure. In the instance of irradiated materials, the inconveniences of radioactive TEM sample preparation can be significantly reduced.

Acknowledgements

The authors thank F. W. Wiffen for supplying the niobium samples and his interest in these investigations. We also thank J. O. Steigler for supplying the stainless steel samples.

References

1. H. R. Child and S. Spooner, "New Small Angle Neutron Scattering (SANS) Instrument at ORNL Using a Position Sensitive Area Detector," Journ. Appl. Cryst. , 12 (1980) pp. 7-13.

2. C. J. Borkowski and M. K. Kopp, "New Type of Position Sensitive Radiation Detector of Ionizing Radiation Using Risetime Measurement," Rev. Sci. Instrm. , 39 (1968) pp. 1515-1522.

4. W. Schmatz, T. Springer, J. Schelten and K. Ibel, "Neutron Small Angle Scattering: Experimental Techniques and Applications," Journ. Appl. Cryst. , 7 (1974) pp. 96-113.

5. G. E. Bacon, " Neutron Diffraction, " 2nd ed. , Chapter VI, pp. 145-182, Oxford Press, London 1962.

6. A. Guinier and G. Fournet, " Small Angle Scattering of X-rays , " translation by C. B. Walker, Chapter 2, pp. 5-78; John Wiley and Sons, Inc., New York, N. Y. 1955.

7. S. D. Harkness, R. W. Gould and J. J. Hren, "A Critical Evaluation of X-ray Small Angle Scattering Parameters by Transmission Electron Microscopy; GP Zones in Al Alloys," Phil. Mag. , 19 (1969) pp. 115-128.

8. J. Schelten and W. Schmatz, "Multiple-Scattering Treatment for Small-Angle Scattering Problems," Journ. Appl. Cryst. , 13 (1980) pp.385-390.

9. J. T. Stanley and L. E. Hendrick, "Ferrite Formation in Neutron-Irradiated Austenite Stainless Steel," J. Nucl. Mat. , 80 (1979) pp. 69-78.

Atomic and Nuclear Interactions

THE CHARACTERIZATION OF DEFECTS IN METALS

BY POSITRON ANNIHILATION SPECTROSCOPY[†]

R. W. Siegel

Materials Science Division
Argonne National Laboratory
Argonne, Illinois 60439

The application of positron annihilation spectroscopy (PAS) to the characterization and study of defects in metals has grown rapidly and increasingly useful in recent years. Owing to the ability of the positron to annihilate from a variety of defect-trapped states in metals, PAS can yield defect-specific information which, by itself or in conjunction with more traditional experimental techniques, has already made a significant impact upon our knowledge regarding lattice defect properties in metals. This has been especially true for vacancy defects, as a result of the positron's affinity for lower-than-average electron-density regions in the metal. The physical basis for the positron annihilation techniques is presented in this paper; and the experimental techniques, lifetime, Doppler broadening, and angular correlation, are briefly described and compared with respect to the information that can be obtained from each of them. A number of examples of the application of PAS to the characterization of atomic defects and their agglomerates are presented. The particular examples, chosen from the areas of equilibrium vacancy formation and atomic-defect recovery, were selected with a view toward elucidating the particular advantages of PAS over more traditional defect-characterization techniques. Nevertheless, the limitations of PAS are also pointed out. It is evident that PAS has become a valuable and unique probe of the defect structure of metals and, as such, has now taken its place among the characterization tools of the materials scientist.

†This work was supported by the U.S. Department of Energy.

413

Introduction

The application of positron annihilation spectroscopy (PAS) to the study of lattice defects in metals has grown rapidly since the first realizations (1-3) that positrons are sensitive to these defects. A positron can be trapped in rather highly localized states in a variety of lattice defects, from which the positron subsequently annihilates with unique, defect-specific characteristics. This ability has enabled the positron to be used as a sensitive and localized probe of these defect sites. The unique aspects of PAS arise from the fact that the positron-electron pair annihilation process, which proceeds by the emission of γ-rays, can yield detailed information regarding both the electron density and the electron momenta in the region from which the positron annihilates.

The physical basis of positron annihilation in condensed matter and the many applications of PAS to the study of solids have been extensively reviewed elsewhere (4-8). In addition, several more specific reviews of the application of PAS to the study of defects in materials (9-17) have been published. It is clear that PAS has already made significant contributions to the characterization of defects in metals, and that it will continue to do so. These contributions have primarily been in two areas: (i) the determination of atomic defect properties, and (ii) the monitoring and characterization of microstructure development, such as that occuring during post-irradiation annealing. In addition to its value as an independent analytical tool, PAS has demonstrated its utility as a complement to the more traditional tools for defect characterization in metals.

The present paper is intended as a brief introduction to positron annihilation, and its recent literature, and to the applications of PAS to the characterization of defects in metals. Selected examples from the literature are used to elucidate the particular advantages of PAS for defect characterization vis-à-vis the more traditional tools of the materials scientist. In addition, the limitations of PAS are discussed and some of the future potential of PAS for the study of defects will be considered. The present review is based, in part, on previous reviews (8,17) by this author.

Positron Annihilation

The physical basis for PAS and the type of information that positron annihilation in condensed matter can yield may be described with the help of the schematic representation shown in Figure 1. When an energetic positron (e^+) is injected into a solid from a radioactive source such as ^{22}Na, for example, it becomes thermalized within a few picoseconds by a succession of ionizing collisions, plasmon and electron-hole excitations, and phonon interactions. This results in a positron that is in a periodically extended Bloch-like free state (see Figure 2a) in the lattice, in which its density is highest in the interstitial regions owing to the repulsion of the positron from the positively charged ion cores of the atoms. If lattice defects such as vacancies, vacancy clusters, or dislocations are present in the material in sufficient concentration, the positron can subsequently be trapped in a bound state in such a defect. In this defect-trapped state the positron may be highly localized (see Figure 2b, for example), the degree of localization depending upon the binding energy of the positron to the defect and the extent of the defect itself.

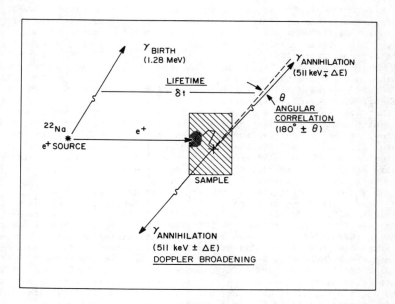

Fig. 1. Schematic representation of positron annihilation. The basis for
the three PAS experimental techniques, lifetime, Doppler
broadening, and angular correlation, is indicated. From (8).

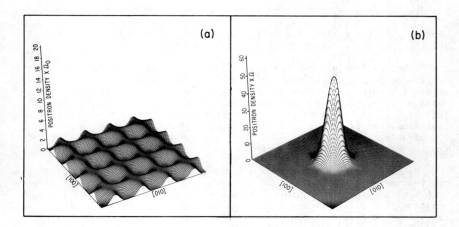

Fig. 2. The positron density in tungsten shown on a network of 25 atom
sites in the (001) plane (a) containing no vacancies and (b) with
a vacancy at the center. The calculated Bloch-state positron
density (a) is normalized with respect to Ω_o, the volume of the
primitive unit cell, and is thus in relative units. The
calculated vacancy-trapped positron density (b) is localized
within the supercell volume, $\Omega = 27\ \Omega_o$, in which it is
normalized, and is thus in absolute units. From (18).

Positron implantation ranges in metals depend upon the particular combination of metal and source, but are generally of the order of 100 to 200 μm for normally-used radioisotope positron sources (e.g., ^{22}Na, ^{64}Cu, ^{58}Co). These sources yield positrons with a spectrum of energies, which range up to end-point energies of less than about 1 MeV, depending upon the isotope. Such implantation ranges guarantee that the information obtained using PAS with these sources comes predominantly from the bulk of the metal, with essentially negligible contributions from the surface. Recently, high-yield slow-positron-beam sources have become available (19-24) that are capable, by means of variable accelerating voltages, of implanting positrons at varying depths below a surface. These positron sources, now being used for surface investigations, will allow in the future for depth dependent, near-surface studies of metals, as well as for studies of interfaces. The present paper will only be concerned with PAS results using the standard radioisotope positron sources.

At a time δt distributed from zero to some hundreds of picoseconds after positron injection into the metal, the positron will annihilate with an electron (its antiparticle) yielding, almost always, two γ-rays. One- and three-γ annihilations are also possible, but are rare. Although the positron lifetimes and source intensities are such that no more than one positron usually exists in the sample at a given instant, the PAS measurement techniques (lifetime, Doppler broadening, angular correlation) integrate over a large number (often ~ 10^6) of annihilation events. The distribution of the δt values for a number of these events, measured in a PAS lifetime experiment, yields information regarding the total electron density in the region of positron-electron annihilation. This is because the rate of positron annihilation, λ, equal to the reciprocal of the positron lifetime, is given by the overlap integral of the electron and positron densities as

$$\lambda = \pi r_0^2 c \int \rho^-(\vec{r})\, \rho^+(\vec{r})\, d\vec{r} \qquad (10)$$

where r_0 is the classical electron radius, c is the velocity of light, and $\rho^+(\vec{r})$ is the positron density in the crystal.

The lifetimes of free (untrapped) positrons are characteristic of the particular metal, and in most metals are of the order of 100 to 250 ps, with alkali metals yielding somewhat higher values. The presence in a metal or alloy of defects, at which the positron is trapped leads to increases in the observed positron lifetimes relative to that of the free positron. This is a result of the trapped positron subsequently annihilating in a region of lower-than-average electron density near the defect (e.g., vacancy, dislocation, void).

Because energy and momentum are conserved in the annihilation process, the two γ-rays resulting from the usual electron-positron pair annihilation each have an energy equal to the rest-mass energy of an electron. However, for detailed quantitative interpretation, direct numerical integration of Eqs. (3,4,5) is necessary. In these calculations an energy increment ΔE; in this paper, two rays propagate in opposite directions plus or minus an angular deviation θ, as shown in Figure 1,2. The deviations ΔE and θ arise from the net momentum (in the laboratory reference frame) of the annihilating positron-electron pair. However, since the positrons have only thermal energies (≈kT), the measured values of ΔE and θ correspond to essentially only the momenta of the annihilating valence or core electrons. The higher-momentum core electrons contribute proportionately more than the valence electrons to the largest values of ΔE and θ. Therefore, since there are fewer core electrons at a defect site at which the positron is trapped and from which it subsequently annihilates, angular-correlation curves [N(θ) vs. θ] and Doppler-broadened spectra [P(ΔE) vs. ΔE] are both more sharply peaked for a metal containing defects that trap positrons. These measurements correspond to the geometry sketched in Fig. (la). A useful variation of the scattering geometry is shown in Fig. (lb), in which the scattering from dislocation loops is studied using a wide-open (that is, no collimator defining the diffracted beam) type (see below) is rather similar. This can be easily seen by comparing the expressions for N(θ) and P(ΔE) in terms of the independent-particle model (IPM) probability per unit time, $R(\vec{p})$, that positron-electron annihilation yields 2γ-emission with total momentum $\vec{p}$:

$$R(\vec{p}) = (r_o^2 c/8\pi^2) \sum_k n_k \mid \iiint e^{-i\vec{p}\cdot\vec{r}} \psi_+(\vec{r}) \psi_k(\vec{r}) d^3r \mid^2 , \qquad (2)$$

where $\psi_+(\vec{r})$ and $\psi_k(\vec{r})$ are the positron and electron wavefunctions, respectively, n_k is the Fermi function, and k represents both the electron wavevector $\vec{k}$ and the band index. The expressions for N(θ) and P(ΔE) are then

$$N(\theta_z) = \iint R(\vec{p}) dp_x dp_y \qquad (3)$$

and

$$P(\Delta E_x) = \iint R(\vec{p}) dp_y dp_z , \qquad (4)$$

where the direction of the emitted γ-rays has been taken as ± x in a Cartesian coordinate system, and the momenta, angles and energies are related by $p_z = mc\theta$ and $2\Delta E_i = cp_i$, where m is the electron mass. Although the available information is similar, angular-correlation experiments may be carried out with greater resolution than Doppler-broadening experiments and can, therefore, yield considerably more detailed information regarding the region from which the positron annihilates, but at the expense of increased data-collection time.

Positron lifetime measurements yield information that is complementary to that from Doppler-broadening and angular-correlation experiments, but it is of an even more-integral nature than that from the momentum measurements regarding the region from which the positron annihilates. This can be seen from the expression for the positron lifetime, in terms of $R(\vec{p})$ as

$$\tau^{-1} \equiv \lambda = \iiint R(\vec{p}) d^3p , \qquad (5)$$

which is equivalent to Equation (1) within the IPM. However, in the case of a sample containing defects, since the average electron density at a defect site can be rather defect specific, positron lifetime measurements have offered a unique method for investigating, for example, vacancy-clustering processes at the smallest cluster sizes in quenched or irradiated materials. This is further facilitated by the relative ease of theoretically calculating positron lifetimes in various defect-trapped states, as compared with analogous positron-annihilation parameters from Doppler-broadening and angular-correlation experiments.

The momentum measurements (angular correlation or Doppler-broadening) can also yield detailed defect-specific information, as indicated by Equations (3) and (4), and a few attempts have been made to apply this information to defect studies. Some notable examples of these are presented below. For the most part, however, detailed electron-momentum information has been used only for the study of the electronic structure of essentially defect-free materials. This application of PAS has been reviewed (25,26) recently. Considerable information regarding the detailed electronic structure of materials is beginning to come from the advent (27) of two-dimensional angular-correlation studies on metals and alloys. The annihilation spectrum obtained from such experiments,

$$N(\theta_z, \theta_y) = \int R(\vec{p}) \, dp_x , \qquad (6)$$

yields the least-integral, and hence most detailed, information presently available regarding the region from which the positron annihilates. Such measurements are already being successfully used for the study of the electronic structure of the defect-free lattice. They should, in the near future, be able to yield unique information regarding the detailed electronic structure and symmetries of lattice defects as well, if applied to relatively simple defect systems as available, for example, in equilibrium vacancy ensembles at high temperatures.

It should be pointed out here that the IPM, upon which Equations (2-6) are based, ignores the effects of positron-electron correlation in the solid, assuming that these particles act independently of one another, which is of course not correct. The positron in the solid, in reality, significantly enhances the electron density in its surrounding region. This enhancement results in considerably shorter positron lifetimes than would be given by Equation (5), as well as more subtle modifications to the momentum (θ or ΔE) distributions given by Equations (3), (4), and (6). Fortunately, the measured positron-annihilation rates and the gross momentum-distribution characteristics can be reasonably accounted for if a constant, multiplicative factor (the enhancement factor) is used to take the many-body effects into account. However, different enhancement factors must be used for valence and core electrons consistent with their degree of tight-binding. For accurate treatment of the more subtle momentum-distribution characteristics, the positron-electron correlation must be taken into account directly in the theoretical formulation.

PAS Experimental Techniques

Lifetime

A typical positron lifetime apparatus is shown schematically in Figure 3. The most commonly used sample-source configuration for lifetime experiments, shown in Figure 3, consists of a positron source, normally ~ 10 μCi of ^{22}Na, sandwiched between two identical samples to be investigated. The samples must be sufficiently thick (usually greater than 250 μm) to stop all of the injected positrons, in this case with an end-point energy of 0.5 MeV. However, the samples need only be of the order of 1 cm in lateral extent to facilitate specimen handling and source deposition. The source, on the other hand, should be as massless as possible to minimize the source contribution to the measured lifetime spectrum (28). Typically, the ^{22}Na source is deposited in the form of a chloride dried from aqueous solution; however, vacuum-evaporated (29) or ion-implanted (30) positron sources are preferable from the standpoint of minimal source contribution.

The lifetime spectrometer shown in Figure 3 is of the fast-slow coincidence type (31,32). The spectrometer measures the time delay, δt, between the detection of the 1.23 MeV birth γ-ray, emitted essentially simultaneously with the positron emitted from the decaying ^{22}Na nucleus, and the subsequent detection of one of the 511 keV γ-rays resulting from the positron-electron annihilation process (see Figure 1). The spectrometer then converts this delay time to an output pulse of

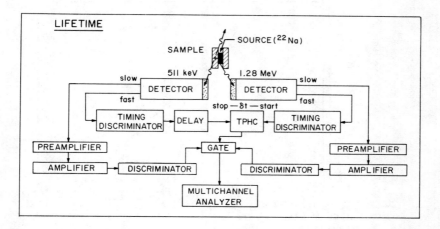

Fig. 3.   Schematic diagram of a PAS lifetime apparatus. TPHC = time-to-
pulse-height converter. From (8).

proportionate height (amplitude). The detectors used are fast plastic
scintillators coupled to appropriate photomultiplier tubes. The fast
channels are used to establish δt as precisely as feasible, while the slow
channels utilizing energy selection are used to drive a linear gate to the
multichannel analyzer. This allows, in principle, only correlated
1.28 MeV-511 keV events to be stored in the collected lifetime spectrum.
The multichannel analyzer can be replaced by an on-line computer to
facilitate data acquisition and analysis.

Typical experimental lifetime spectra, each the sum of a large
number (6 x 10^5) of annihilation events, are shown in Figure 4 for well-
annealed α-iron and electron-irradiated α-iron that has undergone
subsequent annealing to 180 K and 230 K (33). The expected exponential
decay [see below, Equation (8)] in these spectra is observed, but is
convoluted with the instrumental resolution function having, in this case,
a time resolution of 290 ps full width at half maximum (FWHM). Such a
resolution is rather typical for lifetime spectrometers in use today. The
positron decay rate is seen to decrease with the introduction, via
irradiation and subsequent isochronal annealing, of positron trapping sites
at vacancies and vacancy clusters (33). The effects of the instrumental
resolution function, caused by inherent random and systematic timing errors
in the measurement system, are most clearly seen in the smearing of the
distribution of δt values about δt = 0, especially for δt < 0.

Complete deconvolution analyses (34,35) of such spectra can be carried
out in order to obtain the positron lifetime, or lifetimes and relative
intensities in cases in which distinguishable positron states occur.
However, this deconvolution generally requires very high-quality data, and
can be a difficult process when the positron lifetimes in the different
states are similar in magnitude. This can often be the case when a complex
defect ensemble is present in the metal. A more straightforward, but
intrinsically less informative, analysis method is the characterization of
a lifetime spectrum by the time shift or displacement of its centroid

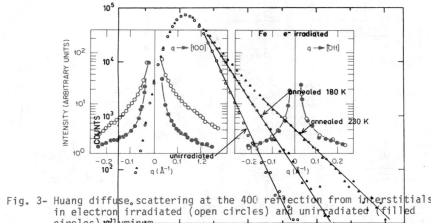

Fig. 3- Huang diffuse scattering at the 400 reflection from interstitials in electron irradiated (open circles) and unirradiated (filled circles) aluminum.

An application of the results of Table I is shown in Fig. (3). Interstitial scattering from interstitials, after radiation background was subtracted, in electron-irradiated high-purity aluminum at various stages, was found to be high along the [011] direction at this reflection. Of the cases listed in Table I, cubic and tetragonal symmetry would be allowed, and other measurements indicated that all configurations other than the <100>-split were inactive; that the octahedral and simultaneously collected reference configuration. The same configuration was found for interstitials in copper (5) and Fig. (4) indicates the large angle scattering results.

Fig. 4- Large angle scattering from interstitials in electron irradiated copper with simulated scattering for <100>-split and octahedral interstitials.

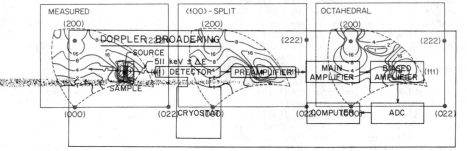

Fig. 5. Schematic diagram of a PAS Doppler-broadening system. From (8).

digital converter (ADC), into an on-line computer for data storage and analysis. Although not shown, a pulse pile-up rejection circuit is best used in parallel with this to avoid problems from signal overlap. The measurements often utilize a sample-source configuration similar to that used in the diffuse scattering experiments with similar source intensities. However, since in this case it is unnecessary to detect a birth γ-ray from the source isotope, a greater variety of positron-emitting isotopes can be used as sources; for example ^{68}Ge or ^{58}Co.

Clustering of Point Defects

The best energy-sensitive detectors available today have an energy resolution of only about 1 keV at 511 keV. 3-dimensional resolution is lower by an order of magnitude than that available with angular correlation measurements. However, for many applications of PAS in which detailed electron momentum information is not required, this limitation can be more than compensated by the considerable (about 100 times) increase in statistical efficiency available in Doppler-broadening systems. The increased efficiency here results from the absence of the requirement for measuring coincident events and the short sample-detector distance possible with this technique. As such, Doppler-broadening measurements have become most popular for defect studies in materials in which the changing state (either concentration or type) of the defect population is being monitored, since a distribution of a particular spectrum can be collected in less than one hour, albeit containing less detailed electron-momentum information, but quite often sufficient for the purpose of the experiment.

A typical example of Doppler-broadened spectra, from annealed and electron-irradiated copper (37), is shown in Figure 6, in which two

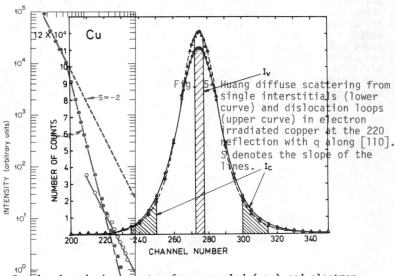

Fig. 5 Huang diffuse scattering from single interstitials (lower curve) and dislocation loops (upper curve) in electron irradiated copper at the 220 reflection with q along [110]. S denotes the slope of the lines.

Fig. 6. Doppler-broadening spectra from annealed (−o−) and electron irradiated (−▲−) copper. Two lineshape parameters I_V and I_C are indicated. The Doppler-broadened line is centered at 511 keV and the width of each channel is ≈0.01 keV. From (37).

lineshape parameters I_v and I_c are defined that emphasize information regarding positron annihilations with valence and core electrons, respectively. The peak parameter I_v is defined here for a region approximately 0.5 keV wide about the peak located at 511 keV; this would be analogous to a region approximately 2 mrad wide about an angular-correlation peak (cf. Figure 8). The shape parameter I_c is the sum of the two segments on either side of the peak corresponding to the regions from about 10 to 20 mrad. Both I_v and I_c are normalized with respect to the total area under the spectrum. A variety of lineshape parameters similar to these have been used to monitor relative changes in materials by means of PAS Doppler-broadening measurements. The utility of such parameters for monitoring defect behavior can be easily seen from the change in I_v upon electron irradiation of the copper indicated in Figure 6. While the absolute values of such parameters themselves contain little physical information, appropriately normalized values of their ratios (37,38) can yield defect-specific information for the characterization of microstructures, as shown in the cases of quenched and irradiated copper discussed below.

Angular Correlation

A schematic diagram of a typical angular-correlation apparatus is shown in Figure 7. In this case the positron source, usually ^{64}Cu, ^{22}Na or ^{58}Co, is normally situated apart from the sample and has a considerably greater activity (from about 0.01 to 1 Ci) than for either lifetime or Doppler-broadening measurements. This is necessitated by the large distances ($\sim$ 3 m) between sample and detectors required for good angular resolution. The source-intensity problem can, however, be partially alleviated by the use of magnetic focusing of the positrons onto the sample. The coincident 511 keV annihilation γ-rays are detected by NaI(Tl) scintillator crystals coupled to appropriate photomultiplier tubes, and the coincidence circuitry is set via the discriminators to pass only 511 keV-511 keV events. One of the detectors is scanned through varying angles θ in order to collect an angular-correlation curve. A number of scans are normally made in order to minimize the effects from source decay and drift in the electronics. High angular resolution (about 0.2 mrad) can be achieved in such a system by means of narrow collimator openings combined with large sample-detector separations, but the resulting coincidence-count rates are rather low as a result. These low count rates are partially compensated, however, by the long ($\sim$ 0.3 m) detector

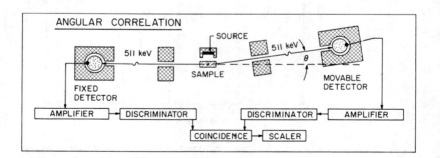

Fig. 7.   Schematic diagram of a PAS angular-correlation system. The cross-hatched forms are lead collimators or shields. From (8).

crystals (e.g., extended into the page in Figure 7), which subtend angles of about 10^{-1} rad at the sample, and the correspondingly long collimating slits, most often used in the so-called long-slit configuration for angular-correlation measurements. In this case, the electron momentum information is integrated over two momentum components [see Equation (3)], since the energy resolution of the detectors is broader than the Doppler-shifts ΔE and the angle subtended at the sample by the detector (along its long axis) is greater than the full angular distribution of the 2γ-annihilation radiation.

Typical angular correlation curves obtained using a long-slit detector geometry are shown in Figure 8 for single-crystal aluminum with and without defects (39). The dramatic increases in the normalized peak count (at $\theta = 0$), with the concomitant changes in the shape of the spectrum, owing to the presence of either vacancies or voids is clear evidence for the utility of such measurements for the characterization of lattice defects. Such angular correlation curves are normally rather well represented by the sum of a parabolic contribution from positron annihilations with valence electrons and an essentially Gaussian contribution from annihilations with core electrons; although in this case of the aluminum containing voids, the strong peaking near $\theta \approx 0$ does not fit this simplified picture well. Considerably greater detail than that shown in Figure 8 regarding the electronic structure sampled by the annihilating positrons is now being made available by the use of two-dimensional detector arrays (40) in place of the one-dimensional, long-slit detector shown in Figure 7. The applicability of this type of information to the characterization of defects will be discussed in the next section.

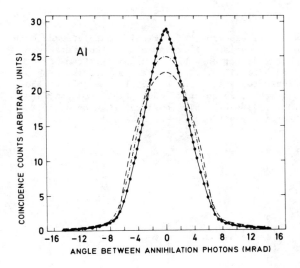

Fig. 8.   Angular correlation curves from single-crystal aluminum: well-annealed, at 100 K (---); containing equilibrium vacancies at 873 K (-•-); and containing voids ($d_v \approx$ 125-495 Å, $N_v \simeq 4 \times 10^{14}$ cm^{-3}) at 100 K produced by neutron irradiation (-•-). From (39).

Fig. 8- Size distri-

The positron is not equally sensitive to all lattice defects. Indeed, this is one of the positive attributes of PAS for the study of defects, especially when compared with the lack of defect specificity of such measurements as residual electrical resistivity. The positron is very sensitive in most metals to local regions of significantly lower-than-average electron density, such as vacancies, small vacancy clusters, and voids, at which the positron can form rather deeply bound states. It is also sensitive to dislocations, grain boundaries, and interfaces, which in general represent somewhat smaller negative perturbations to the average electron density in a metal than these vacancy-type defects, but which may themselves contain defects that can trap positrons more deeply. As example of this would be jogs on dislocation lines, the positron response to which has been recently considered (41). The positron, on the other hand, is essentially insensitive to regions of higher-than-average electron density, as found at an interstitial or small interstitial cluster.† Thus, for example, PAS has been able to differentiate between the earliest stages of vacancy and interstitial clustering in irradiated metals in a manner not directly possible by electrical resistometry or transmission electronical microscopy (TEM). The positron does not appear to become trapped at vacancy-like defects in some metals, such as the alkali metals [see (10,15) for reviews], for which the positron-defect binding is theoretically expected to be rather weak. However, this lack of observed trapping may simply be a result of positron detrapping from relatively weakly-bound states in the defects in these metals under the experimental conditions so far investigated, as suggested elsewhere (43). Low-temperature measurements on alkali metals containing defects could clarify these observations.

Trapping Model

The results of PAS experiments on metals containing defects that trap positrons are normally analyzed in terms of the two-state trapping model (44-46), although the defect ensemble examined is often more complex than this model would seem to allow. This model assumes that the positron exists in one of only two states in the metal, the bulk or Bloch state and the defect-trapped state, in relative concentrations $n_b(t)$ and $n_t(t)$, respectively, with $n_t(0) = 0$. This model can thus be described by the set of coupled differential equations

$$\frac{dn_b(t)}{dt} = - n_b(t)\, \lambda_b - n_b(t)\, \kappa_t \qquad (7a)$$

and

$$\frac{dn_t(t)}{dt} = + n_b(t)\, \kappa_t - n_t(t)\, \lambda_t \;, \qquad (7b)$$

†The positron is expected to be repelled from such regions; however, the interaction of a positron with the strain fields associated with such defect clusters has not yet been carefully considered. Nevertheless, this interaction is probably subtle and would most likely lead to annihilation characteristics between those in the lattice and a dislocation or vacancy at most.

reasonable, but the correspondences not very good for small sizes, which are predominantly composed of vacancy loops. ... in the x-ray transition. A direct comparison of the vacancy loop distribution determined by x-rays with vacancy loops imaged ... low energy (30 keV) copper ion irradiated copper (21) at 293 K, is made in Fig. (8). These microscope results are scaled arbitrarily, so only the size of the loops and the shape of the distribution are of significance. Although the microscope results indicate slightly larger sizes than does the diffuse scattering, both methods show rather narrow distributions. Considering the differences in irradiation conditions, the close similarity in the results seems encouraging. However, this is the first direct comparison of the size distributions of the individual dislocation loop types determined by x-rays with those from TEM, so firm conclusions about the above results may be somewhat premature.

Loops and precipitates

Huang diffuse scattering can be used to differentiate between planar loops and three dimensional precipitate clusters in a manner ... applied in point defect symmetry determinations (see Table I). An example of this is shown in Fig. (10) where the Huang diffuse scattering from dislocation loops in neutron irradiated copper and cobalt precipitates in Cu(1%)Co are shown (22). Characteristic differences can be seen between the scattering in the two cases. The scattering is rather isotropic from the dislocation loops while the precipitate scattering perpendicular to the [100] direction tends to vanish. In addition, the asymmetry in the scattering, along the [100] direction, has the opposite sense for the cobalt precipitates, compared to that for the loops. From Table I it can be seen that these measurements would rule out cubic or tetragonal defects for the neutron irradiation case and rule out trigonal or orthorhombic defect symmetry in the precipitate case. Measurements at the ... have completed the characterization as cubic or spherical symmetry for the precipitates and {111} loops in the case of the neutron irradiated copper. Both of these results ...

The mean positron lifetime, τ, is given by

$$\tau = \int_0^\infty N(t)\, dt = \tau_b \left(\frac{1 + \tau_t \kappa_t}{1 + \tau_b \kappa_t}\right) \tag{10a}$$

$$= (1 - A_t)\,\tau_b + A_t\,\tau_t , \tag{10b}$$

where $A_t = \int_0^\infty \lambda_t\, n_t(t)\, dt = \dfrac{\kappa_t}{\lambda_b + \kappa_t}$ (11)

Fig. 10- Huang scattering from dislocation loops (upper) in neutron irradiated copper and precipitates (lower) in 570°C aged Cu(1%)Co.

is the probability that the positron annihilates from the defect-trapped state. The characteristic shape parameters (denoted generically by F) normally defined for the analyses of angular-correlaton or Doppler-broadening spectra are linear in the trapping probability A_t and, thus, an expression analogous to Equation (10b) can be written for them also:

$$F = (1 - A_t)\,F_b + A_t\,F_t , \tag{12}$$

---

†Use is made here of the ergodic hypothesis, which draws the equivalence between a sequence of individual events (e.g., positron-electron pair annihilations) and the time average or sum of these events (e.g., an experimentally collected lifetime or momentum spectrum).

where A_t is given by Equation (11) and F_b and F_t are the spectrum shape parameters characteristic of positron annihilations from the bulk and defect-trapped states, respectively. Therefore, the total positron trapping rate κ_t is related to either F or τ by

$$\kappa_t = \lambda_b (F - F_b)/(F_t - F) = \lambda_b(\tau - \tau_b)/(\tau_t - \tau). \qquad (13)$$

The trapping rate κ_t within the trapping model [Equations (9c) and (13)] is proportional to the concentration of defect traps. For a single defect type, it is simply $\kappa_t = \mu_t c_t$, where μ_t is the (specific) positron trapping rate per unit defect concentration and c_t is the defect-trap concentration. For an ensemble of various defects (j) that trap positrons, $\kappa_t = \Sigma_j \mu_j c_j$, where μ_j and c_j are the specific positron trapping rate at the jth-type defect and the concentration of this defect, respectively. In such a multidefect-trap situation, the values of τ_t and F_t obtained from two-state trapping model analyses of PAS experiments are essentially weighted averages of the τ_j's or F_j's for the individual defect types present in the system. Furthermore, in this case κ_t is proportional to the total concentration of defects that act as positron traps, where the proportionality constant is an appropriately weighted average of the specific trapping rates for the defect ensemble.

It should be pointed out here, however, that the extraction of unique information regarding the various defects comprising such a multidefect-trap ensemble can be a difficult task. If only two dominant defect types are present, with distinctly different trapped-positron annihilation characteristics, a suitable deconvolution of the PAS data can usually be made. For more complex multidefect situations, only ensemble averages are likely to be available from the PAS spectra in any unique fashion. Nevertheless, owing to the positron's selective trapping at vacancy-like defects, the utility of PAS in studying, for example, the often complex post-irradiation defect annealing behavior in metals has been quite evident, and represents a significant advance over, and complement to, the more traditional resistivity measurements for such investigations.

Although positrons are extremely sensitive to the presence of small concentrations of vacancy-like defects, being able to detect as few as $\sim 10^{-7}$ atomic fraction of vacancies in most metals, their trapping reaches a saturation at concentrations of the order of 10^{-4}. This saturation behavior can be seen in the forms of Equations (10) and (12). Therefore, changes in defect concentration above this saturation limit, which depends specifically for each defect type on its value of μ_j, cannot be monitored by PAS. However, even beyond the saturation of positron trapping in the defect ensemble, the changing nature of the dominant defect traps for positrons can be followed through the changes in the observed positron annihilation characteristics τ_t or F_t.

Thus, within the limitations briefly cited here, the concentration of defects to which the positron is sensitive (such as vacancies, vacancy clusters, or voids) can be monitored directly by experimental PAS determinations of κ_t, or somewhat less directly through I_2 [see Equations (8) and (9)]. In addition, the nature of the defect being monitored by the positron can be deduced by means of its characteristic positron-annihilation parameters $\tau_t \equiv \tau_2$ or F_t. This combination of available information regarding both the concentration and nature of the defects to which the positron is sensitive forms the basis for the application of PAS to the study and characterization of defects in metals.

Vacancy Formation

A major application of PAS has been in the study of the physical properties of vacancies in metals relating to both their formation and migration. The measurement of vacancy formation enthalpies, H_v^F, by PAS has recently been reviewed and compared with other available techniques (42). A number of questions regarding the behavior of the positron in its bulk state, in the temperature region below which observable concentrations of vacancies form, have been considered [see (14) and (47) for recent reviews] in relation to their effect on the use of PAS for vacancy-formation studies. It would appear at present, however, that none of these pose serious problems in the application of PAS to H_v^F measurements. Indeed, a number of these spurious low-temperature effects now seem (48-51) to have been caused by positron trapping at extrinsic lattice defects, such as dislocations, which may be removed by sufficiently careful annealing treatments. The remaining temperature dependences of the positron annihilation parameters in this temperature region appear to be theoretically explainable (52,53) in terms of a combination of lattice expansion and phonon-coupling effects.

A typical set of experimental PAS Doppler-broadening data (54) for the measurement of H_v^F in copper is shown in Figure 9, in which a lineshape parameter F(T), defined by a given normalized region around $\Delta E = 0$ in each Doppler-broadened spectrum, is plotted as a function of temperature T. At temperatures below about 500°C, the positron is insensitive to the rather low equilibrium concentration of vacancies ($c_v < 10^{-7}$) and, hence, F(T) is simply $F_b(T)$, the lineshape parameter corresponding to the positron annihilating in its bulk state. The parameter $F_b(T)$ varies linearly with temperature in this region, as indicated in Figure 9, which is theoretically expected (52,53). At the highest temperatures, F(T) approaches a saturation, indicated by the upper dashed line, which is

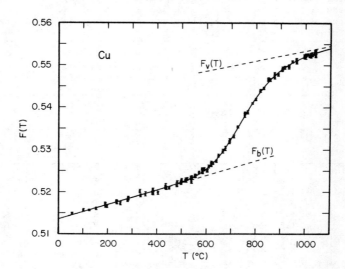

Fig. 9.   Doppler-broadening lineshape parameter F(T) for copper as a function of temperature T, and the two-state trapping model fit to the data.   After (54).

representative of the lineshape parameter $F_v(T)$ and corresponds to the position annihilating from its vacancy-trapped state. The lineshape parameter $F_v(T)$ is also linearly temperature dependent, although more weakly so than $F_g(T)$ since premarily lattice expansion is expected to affect its temperature dependence, with atomic vibrations playing a smaller role in this localized state. The temperature dependence of $F(T)$ in the transition region between these two limiting regions is controlled by the trapping rate

$$K_{1v}(T) = \mu_{1v} c_{1v} \exp(S_{1v}/k) \exp(-H_{1v}^F/kT)$$

where the subscript $1v$ refers to the monovacancy, S_{1v} is the entropy of vacancy formation, and the other quantities have their usual meanings. A two-state trapping model fit to these data has yielded a value of $H_{1v}^F = 1.31 \pm 0.05$ eV for copper.

Results similar to those shown in Figure 9 are available from PAS lifetime and angular-correlation measurements. Each of these techniques has certain advantages for vacancy-formation studies. Positron lifetime measurements can yield the most direct measurement of H_{1v}^F, as recently discussed in detail for aluminum, but detailed angular-correlation measurements contain intrinsically more information regarding the nature of the vacancy defects present in the system. Nevertheless, the facility with which Doppler-broadening measurements can be carried out with high statistics has made this the most widely used technique. In order to make a rough overall evaluation of the diffuse scattering technique compared to the electron microscope technique for studying these defects, a limited set of results from both techniques are in several metals.

Table I is presented in Table I. These are compared with the results from quenching experiments, to which they are most directly comparable owing to the temperature range from which the data are obtained. It should be emphasized that the PAS measurements are made on the vacancy ensemble under conditions of thermodynamic equilibrium, whereas the quenching measurements are performed on a nonequilibrium vacancy ensemble rapidly frozen-in from

Table I. Monovacancy formation enthalpies, H_{1v}^F, from PAS compared with those from quenching.

Metal	H_{1v}^F(eV) PAS	Ref.	H_{1v}^F(eV) Quenching	Ref.
Al	0.66 ± 0.02	56	0.66	60
Ag	1.16 ± 0.02	57	1.10	61
Au	0.97 ± 0.01	57	0.94	62
Cu	1.31 ± 0.05	54	1.30	63
Ni	1.8 ± 0.1	58	1.6	64
V	2.1 ± 0.2	59		
Nb	2.6 ± 0.3	59		
Mo	3.0 ± 0.2	59	3.2	65
Ta	2.8 ± 0.3	59		
W		59		

From this table it would appear that initial survey work, investigations of larger defects, and studies where concentration levels are likely to be low, would clearly favor the use of electron microscopy. For studies of small defects, investigations where subtle changes in defect sizes or concentrations are of interest, and when measurements are required after sequential sample treatments, such as isochronal annealing or ageing, the diffuse scattering technique offers significant advantages. An overall examination of Table II indicates that in many ways, diffuse scattering and electron microscopy are complementary tools. Although the study of point defects and very small clusters by diffuse scattering requires a significant

Table II. Positive, Negative, and Neutral Aspects of Diffuse Scattering and Electron Microscopy.

X-Ray Diffuse Scattering			Electron Microscopy
Non-destructive technique	Single crystals required	Polycrystalline or single crystal	Destructive technique
Good statistics inherent	higher concentrations required	Lower concentrations possible	Good statistics tedious
Sensitive to point defects	>500 Å sizes difficult	Larger sizes possible (sample limited)	>10 Å sizes required
Bulk property measurement	Results require analysis	Direct observation (usually)	Thin samples (non-bulk)

... samples such as obtained from structural components. Rather, the power of diffuse scattering is realized by devising experiments that will provide information regarding vacancy formation from PAS. In addition to this information regarding vacancy formation from PAS, the values for H_{1v}^F, when subtracted from the values of the activation enthalpy for diffusion via a monovacancy mechanism Q_1 (= $H_{1v}^F + H_{1v}^M$) from tracer self-diffusion measurements, also yield values of the monovacancy migration enthalpy H_{1v}^M. Such combined measurements often provide the most reliable source of information regarding monovacancy mobility in metals. Nevertheless, when complications arise in the interpretation of either the PAS measurement of H_{1v}^F or the tracer self-diffusion measurement of Q_1 which can occur, for example, when a metal undergoes high-temperature phase transitions, a more direct PAS measurement of monovacancy mobility can be preferable, as has recently been shown for iron (33).

It should be pointed out here that, while the trapping rate κ in equilibrium vacancy formation experiments is largely dominated by monovacancies owing to the low-temperature range in which these measurements are most sensitive (54), some attempts have been made [for example, in copper (67,68), gold (67), and nickel (68)] to extract information regarding divacancies from PAS data of the type shown in Figure 9 by means of multiparameter fitting schemes. Invariably, the simplifying assumptions required in such parameter optimizations of F(T) data are sufficient to raise doubts (42) regarding the uniqueness of the resulting defect parameters. Indeed, a recent (54) detailed monovacancy-divacancy, two-state trapping model analysis of the data shown in Figure 9 has led to the conclusion that no unique divacancy information could be deduced from such data in copper. It was further concluded in this case that any information regarding divacancies was masked by uncertainties regarding the annihilation characteristics of the positron in its monovacancy- or divacancy-trapped states, $H_{1v}(T)$ and $H_{2v}(T)$, respectively, which are averaged by the fitted parameter $F_v(T)$. Fortunately, this dilemma may be resolved in the near future with the combined availability of (i) realistic calculations of the positron annihilation characteristics in monovacancy and divacancy trapped states in metals based upon a band-theoretical approach (18,55,69) and (ii) measurements of the temperature dependence of positron annihilation in these metals containing equilibrium vacancy defects using the two-dimensional angular correlation (40) technique. This combination of theoretical and experimental information should provide the first opportunity for atomic-defect spectroscopy in metals.

In addition to the investigations of vacancy formation in pure metals, a number of PAS measurements of vacancy formation in both dilute and concentrated alloys have also been carried out, which have yielded effective vacancy formation enthalpies in these alloys [see (14) for a recent review]. Such investigations can yield information regarding the interaction energies between a vacancy and a solute atom in the systems investigated, since this interaction is responsible for an enhanced vacancy concentration at a given temperature in the alloy. In the most complete study of this type (70), a systematic decrease in the effective vacancy formation enthalpy with increasing electron-to-atom ratio was demonstrated for a number of concentrated copper alloys. However, a fuller understanding of such results, which tend to be rather analysis-model dependent (16), must await a better appreciation of how the positron density is distributed in such alloy systems and, as such, how it samples the defects present.

Atomic-Defect Recovery

Atomic-defect recovery in metals after energetic-particle irradiation has also been studied rather extensively using PAS. These investigations have been performed both to study the mobility of the atomic defects produced by the irradiation and to investigate the nature of the post-irradiation vacancy-precipitation process and the resulting defect structures. In addition to these, a number of PAS experiments have been carried out to study the defect recovery in metals after rapid quenching from high temperatures and after deformation. A wide variety of defect recovery experiments have been performed using PAS, some of them correlated with investigations using more traditional techniques, and many examples can be found in the recent literature. It is most useful here to look at only a few selected examples from this class of experiments, in order to point out the types of defect information available from PAS, and to elucidate the particular advantages of PAS over more traditional techniques for such defect studies. For additional examples, see (7,71-84) and the following papers in these Proceedings.

Results from a Doppler-broadening study (37,38) of copper, irradiated at 10 K with 3 MeV electrons to an initial Frenkel-pair concentration of $\sim 3 \times 10^{-4}$, are presented in Figure 10, and in Figure 11a. This study is particularly interesting in that it utilized for the first time the defect specificity of the electron-momentum information available from PAS in investigating the developing microstructure during post-irradiation annealing. A comparison was made between the isochronal-annealing behavior

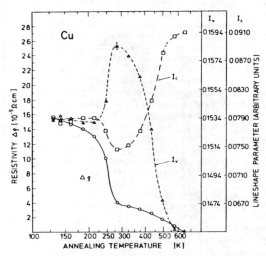

Fig. 10. Doppler-broadening study of electron-irradiated copper. Isochronal annealing of the lineshape parameters I_v and I_c defined in Figure 6 and the electrical-resistivity increment $\Delta\rho$. From (38).

of I_v and I_c, two Doppler-broadening shape parameters representative of predominantly valence- and core-electron annihilations (Figure 6), and that of the residual electrical resistivity of the sample. The resistivity measurements can only indicate the loss of electron-scattering centers (vacancies or interstitials or both) during stage-III annealing at ~ 250 K, and do not in themselves indicate which defect species has become mobile in this temperature range to effect this loss. The PAS-parameter measurements, on the other hand, which are primarily sensitive to vacancy-like defects, rather clearly indicate the loss of vacancies via precipitation into void-like clusters. This can be seen in the rapid increase in the peak-count parameter I_v, with concomitant decrease in I_c, at ~ 250 K that signals the change in the dominant positron traps from monovacancies to the high-order vacancy clusters. The decrease (increase) in I_v (I_c) at higher temperatures then indicates the subsequent dissolution of these clusters.

Based upon these data from electron-irradiated copper, a ratio of measured lineshape-parameter differences was introduced (38), which was designed to be defect-type specific but defect-concentration independent, as long as a given type of defect was the dominant positron trap. This ratio, $R = |(I_v-I_{vo})/(I_c-I_{co})|$, where the terms with o subscripts refer to a reference state, is basically a characteristic shape parameter of the Doppler-broadened (or similarly, angular correlation) annihilation spectrum, which is rather sensitive to the particular defect environment from which the positrons annihilate. The utility of the R-parameter for defect studies can be seen from Figure 11, in which the R-parameter values as a function of annealing temperature for the electron-irradiated copper data from (38) are compared with a similar plot for copper quenched from 1288 K (85). It is evident that both the irradiated copper and the quenched copper, which of course contains only vacancies that precipitate upon ageing around room temperature (63), undergo similar defect annealing as seen by PAS.

The R-parameter value at the lower temperatures in Figure 11 was shown (38) to be equivalent, within experimental uncertainty, to that for vacancies in thermal equilibrium in copper, as well as to that from dislocation loops. This latter equivalence is interesting in that it mimics the equivalence of the positron lifetimes from these two trap states usually found in metals. This equivalence may be a further indication of the positron's affinity for jog-sites along dislocation lines (41), since their electronic structure is expected to be closer to that of a vacancy than that of the straight dislocation line. Although the resulting absolute values of the R-parameter are somewhat different for the electron-irradiated and quenched copper after vacancy clustering, they appear to be quite consistent with the observed (37) decrease of R-parameter value, in the high-temperature limit, with decreasing initial defect concentration. While the R-parameter is clearly useful for defect studies, it should be possible in the future to define parameters of the PAS momentum distributions that yield even more defect-specific information than that shown in Figure 11, based upon an increased knowledge of the positron-annihilation spectra for various defects. Such knowledge will result both from increasingly realistic theoretical calculations of these spectra for given defects and from further calibration experiments on samples containing well-characterized defects. The PAS Doppler-broadening and angular-correlation techniques will then become even more powerful tools than they are at present for the study of lattice defects.

Positron lifetime measurements on metals containing previously undefined defect ensembles currently have the distinct advantage that the

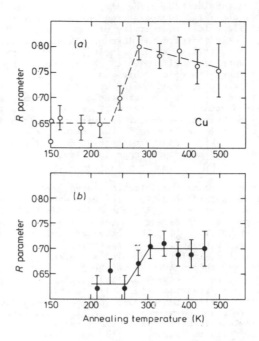

Fig. 11. Ratio (R-parameter) of Doppler-broadening lineshape-parameter differences for (a) electron-irradiated copper as a function of annealing temperature, compared with that for (b) quenched copper. From (85).

positron lifetime in a particular defect-trapped state is a better-defined physical quantity than the various momentum-distribution shape parameters in use, and it can also be theoretically estimated with reasonable certainty. Thus, the positron lifetimes in vacancies and voids are reasonably well established quantities that can be used in evaluating the results of lifetime experiments on metals containing nonequilibrium ensembles of defects. An example of this type of application of PAS is a set of investigations of void formation during isochronal annealing following irradiaton of molybdenum by 1.5×10^{18} cm^{-2} fast neutrons (86) or 2×10^{18} cm^{-2} 10 MeV electrons (87), the results of which are shown in Figure 12. The data for the electron-irradiated molybdenum were subsequently quantitatively analyzed (89) in terms of the vacancy-clustering process.

Figure 12 shows the post-irradiation annealing behavior of the defect-trapped positron lifetime τ_2 in molybdenum from (86,87). It can be seen that while τ_2 in the electron-irradiated case has an initial value representative of the vacancy in molybdenum, the initial value of τ_2 in the neutron-irradiated case is already considerably higher; this indicates the void-like nature of the depleted zones present after neutron irradiation. Upon ageing at elevated temperatures, vacancy precipitation into voids is indicated by the continuous increase of τ_2 up to about 450 ps; this void formation was confirmed by transmission electron microscopy (87), which showed the presence of ~ 1 to 5×10^{13} voids cm^{-3}, with diameters of approximately 30 Å. The rise of τ_2 above 450 ps at higher temperatures was

432

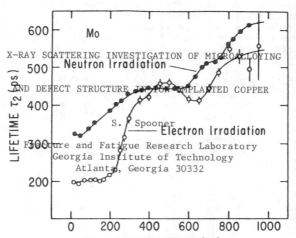

X-RAY SCATTERING INVESTIGATION OF MICROALLOYING
AND DEFECT STRUCTURE IN ION-IMPLANTED COPPER

S. Spooner

Fracture and Fatigue Research Laboratory
Georgia Institute of Technology
Atlanta, Georgia 30332

The double-crystal method for x-ray scattering analysis of radiation described by B. C. Larson (1) has been applied to the investigation of ... molybdenum ... is based on (86) and electron (87) irradiated molybdenum. After (88)] ... and the presence of dislocation loops which originate from implantation damage. The subsequently shown (88) to have been ... to impurity contamination of the voids ... latter effect, to which PAS is ... sensitive ... and for which PAS is a potentially useful investigative tool ... should prove to be a useful one for further study of the effect of impurities upon void formation in metals and the limitations of the technique are discussed.

In an attempt to make PAS lifetime data, such as those from (86) and (87), more quantitatively useful in investigating the void-formation process, theoretical estimates using the jellium model have been made (89) of the positron lifetime in its void-trapped state in aluminum and molybdenum as a function of the number (1...50) of vacancies in the clusters. These calculations are somewhat oversimplified in the case of the smallest clusters (e.g., divacancies and trivacancies), since the theoretical model employed necessitated the assumption that the clusters were spherical. They nevertheless reveal clearly the additional potential of PAS as a detailed quantitative tool for the analysis of vacancy-clustering processes in metals. This potential will be more fully realized in the future with the availability of more sophisticated theoretical models for positron annihilation in small vacancy clusters based on the atomic nature of a lattice; models that will be capable of predicting both positron lifetimes and PAS momentum spectra from positrons trapped in these clusters.

The PAS investigations considered here point up a particular advantage of PAS in the study and characterization of defects in irradiated or quenched metals; namely, the sensitivity of the positron and its annihilation characteristics to both individual vacancies and their clusters, up to sizes large enough to be considered well-defined voids, and to be observed by complementary techniques such as transmission electron microscopy. This broad range of sensitivity is unique to PAS in comparison with the more traditional techniques (e.g., electrical resistometry, field ion microscopy, TEM, etc.) for the monitoring and characterization of defects in metals. ...

N00014-78-C-0270.

have been used to study the early stages of the clustering of quenched-in vacancies in gold (90). Such measurements are less sensitive than PAS, and are likely to be less widely applicable than PAS in a variety of metals. However, as a tool for the study of interstitials and their clusters (91), they do provide an excellent complement to PAS studies of irradiated metals.

With the present knowledge of positron lifetimes in the various defect-trapped states in metals, it is quite possible to utilize PAS to address and shed light on controversial problems in the characterization of defects. An example of such an application (33) is presented in Figure 13, in which the isochronal annealing behavior of the defect-trapped positron lifetime τ_2 and the intensity I_2 of this component are shown for high-purity α-iron electron irradiated to two different doses (initial Frenkel-pair concentrations of about 0.7×10^{-4} and 7×10^{-4}). The annealing behavior of I_2 in the low-dose sample, along with the absence of any significant change in τ_2 near 140 K, indicates that vacancy-interstitial recombination has occurred at about 140 K. In addition, the rapid increase of τ_2 in both samples at about 220 K, with the concomitant decrease in I_2,

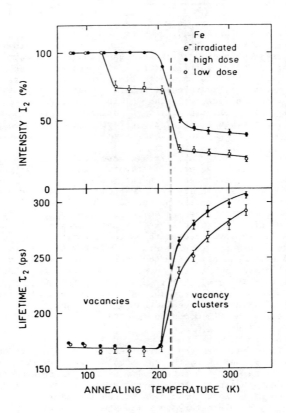

Fig. 13. The lifetime τ_2 of trapped positrons, and the intensity I_2 of this component, in electron-irradiated α-Fe as a function of annealing temperature. From (33).

indicates rather clearly that small vacancy clusters are being formed in
α-iron at these temperatures by the precipitation of supersaturated
vacancies, which yield lifetimes of approximately 175 ps (89). These
vacancy clusters, for which τ_2 lies in the approximate range of 250 to 300
ps, are apparently rather small, containing only about 2 to 5 vacancies,
since well-defined three-dimensional voids, which might be of sufficient
size to be also observed by TEM, would yield a greater lifetime of about
450 ps (89,92) for positrons trapped therein.

Exceptions (93,94) to the interpretation of these data in terms of
vacancy precipitation near 220 K have been taken, suggesting alternatively
that the observed increase in τ_2 in this temperature range is due to the
formation of interstitial agglomerates that can trap positrons. However,
no theoretical or experimental evidence exists for interstitial-type
positron traps that could yield lifetimes up to 300 ps in iron, and it
appears at present that the PAS results from (33), along with other recent
PAS (92,95,96), magnetic after-effect (97) and resistivity-recovery (98)
experiments that have confirmed these results, yield the most direct
information regarding monovacancy mobility in α-iron presently available.

Conclusions

It is evident that positron annihilation spectroscopy has become a
valuable technique for the study and characterization of defects in
metals. The ability of the positron to take up a variety of specific
defect-trapped states in metals, especially those of a vacancy-like nature,
from which the positron's subsequent annihilation yields information
regarding its electronic environment, has made the positron a unique probe
of lattice defects. It is worthwhile at this point to summarize a few of
the strengths, and weaknesses, of PAS as a method for the characterization
of defect microstructures vis-à-vis some of the other available tools of
the materials scientist that are discussed in these Proceedings.

PAS is highly sensitive to the presence of vacancy defects in most
metals under both high-temperature equilibrium and nonequilibrium
conditions. The positron annihilation characteristics are also defect
specific, and be may used as a truely spectroscopic probe of metal defects
under certain conditions. Resistometry, by comparison, is also very
sensitive to defects, and can follow changes in their concentrations beyond
the saturation limit of PAS. However, it is primarily useful as a low-
temperature tool for defect detection; and, as a further limitation, it is
essentially integral in nature, not being able to distinguish among the
scattering centers in a multidefect ensemble. Field ion microscopy, on the
other hand, is highly defect specific, especially with respect to vacancy
defects, but produces data with limited statistics, and is only applicable
under conditions of rather high defect concentrations and low
temperatures.

PAS can selectively follow the vacancy-clustering process in either
quenched or irradiated metals continuously from an ensemble of dispersed
monovacancies to one of voids or dislocation loops, which could be observed
by TEM. Only diffuse-scattering techniques with x-rays, or neutrons,
presently share this capability with PAS. These techniques are, however,
rather complementary to PAS in terms of both their range of sensitivity,
which is shifted to higher defect concentrations by at least an order of
magnitude, and their ability to follow interstitial clustering as well.
For defect clusters that are large enough, and at a sufficient density, to
be characterized by TEM in terms of their nature and distribution, neither

PAS nor the diffuse-scattering techniques can or need compete with TEM and
its associated analytical probes.

In conclusion, PAS has already made a considerable impact on the study
and characterization of defects in metals, owing to the unique
characteristics of the positron as a probe of the defect solid state. This
impact has resulted from PAS investigations that have been carried out both
alone and in conjunction with other complementary techniques. It is now
clear that PAS has taken its place among the more traditional tools of the
materials scientist for the characterization of microstructures, and that
PAS will continue to expand its areas of usefulness in the future as its
potential for the study of materials is more fully realized.

Acknowledgments

The author wishes to thank S. Mantl and L. C. Smedskjaer for reading
the manuscript.

References

1. I. Ya. Dekhtyar, D. A. Levina, and V. S. Mikhalenkov, "Electron-
   Positron Annihilation in Plastically Deformed Metals," Soviet Physics-
   Doklady, 9 (1964) pp. 492-94.

2. S. Berko and J. C. Erskine, "Angular Distribution of Annihilation
   Radiation from Plastically Deformed Aluminum," Phys. Rev. Lett., 19
   (1967) pp. 307-9.

3. I. K. MacKenzie, T. L. Khoo, A. B. McDonald, and B. T. A. McKee,
   "Temperature Dependence of Positron Mean Lives in Metals," Phys. Rev.
   Lett., 19 (1967) pp. 946-48.

4. R. N. West, "Positron Studies of Condensed Matter," Adv. in Phys., 22
   (1973) pp. 263-383.

5. I. Ya. Dekhtyar, "The Use of Positrons for the Study of Solids," Phys.
   Reports (Phys. Lett. C), 9 (1974) pp. 243-353.

6. P. Hautojärvi, ed., Positrons in Solids, 255 pp.; Springer-Verlag,
   Heidelberg, 1979.

7. R. R. Hasiguti and K. Fujiwara, eds., Positron Annihilation (Proc.
   Fifth Intl. Conf. on Positron Annihilation, Lake Yamanaka, Japan,
   1979); The Japan Institute of Metals, Sendai, 1979.

8. R. W. Siegel, "Positron Annihilation Spectroscopy," pp. 393-425 in
   Ann. Rev. of Materials Science, 10 int., Annual Reviews, Palo Alto, 1980.

9. A. Seeger, "The Study of Defects in Crystals by Positron Annihilation,"
   Appl. Phys., 4 (1974) pp. 183-199.

10. A. Seeger, "The Investigation of Point Defects in Equilibrium
    Concentrations with Particular Reference to Positron Annihilation
    Techniques," J. Phys. F: Metal Phys., 3 (1973) pp. 248-94.

11. M. Doyama and R. R. Hasiguti, "Studies of Lattice Defects by Means of
    Positron Annihilation," Crystal Lattice Defects, 4 (1973) pp. 139-63.

a subsidiary peak of 1.4 percent reflectivity is seen at a Bragg angle
displaced to a lower angle than the substrate Bragg angle corresponding to
the expanded lattice parameter. The small peak width is approximately 2
minutes of arc. The reflectivity is the order of the ratio of implanted
layer thickness to the x-ray penetration thickness, $1/\mu$, where μ is the
linear absorption parameter.

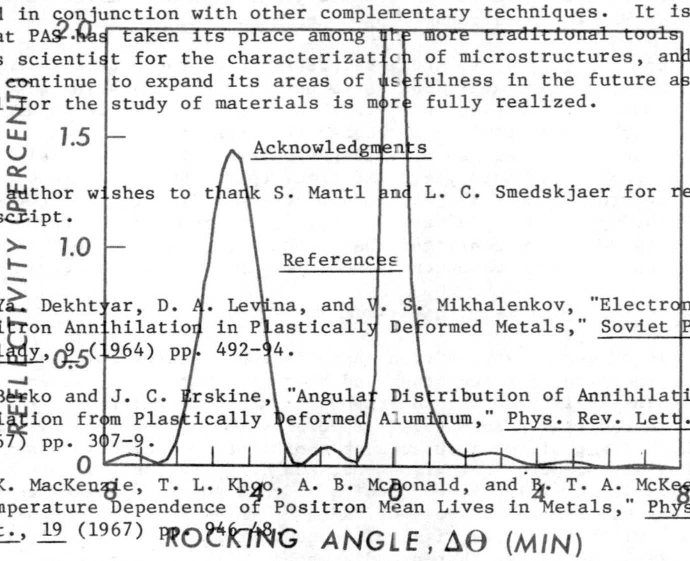

Fig. 1 Calculated reflectivity from a surface implanted to 2 atomic
percent of aluminum in copper to a depth of approximately 1000 A The
subsidiary peak appears at an angle appropriate for the lattice
parameter of this composition.

Consider now the calculation of the scattering from defect clusters in
a crystal of uniform lattice parameter. In this case kinematic diffraction
theory is used to calculate the scattering intensity from an isolated defect
cluster. The scattering signal from a collection of defects is the sum
of the intensities. This implies that no scattering interference occurs
between amplitudes coming from each defect cluster. In (1)
summarized Rev. of Materials Science, intensity from defect clusters.
The experimental geometry used in experiments is shown in Figure 2 where
the scattered x-rays are received by a large detector. Each of the
scattering vectors is associated with a scattering space vector, q, going
from Equilibrium of the Equilibrium concentration of point defects, scattering
sphere Crystal Lattice Defects, intensity is averaged over the
scattering space vectors, q. q_0 is the shortest vector between the Bragg
position and the diffraction. The measured
intensity as a function of rocking angle of the crystal in the same geometry
used for measurement of dynamical diffraction effects described above.
The diffuse scattering from dislocation loops measured close to the
Bragg peak is attributed to long range strain fields around the loop and is
called Huang scattering. Scattering measured farther away from the Bragg

12. A. Seeger, "The Study of Defects in Crystals by Positron Annihilation," Appl. Phys., 4 (1974) pp. 183-99.

13. W. Triftshäuser, "Positron Studies of Metals," Festkörperprobleme (Adv. in Solid State Phys.), XV (1975) pp. 381-410.

14. R. N. West, "Positron Studies of Lattice Defects in Metals," pp. 89-144 in Ref. 6, 1979.

15. R. M. Nieminen and M. J. Manninen, "Positrons in Imperfect Solids: Theory," pp. 145-95 in Ref. 6, 1979.

16. M. Doyama, "Studies of Lattice Defects and Phase Transitions by Positron Annihilation," pp. 13-30 in Ref. 7, 1979.

17. R. W. Siegel, "Positron Annihilation-A Localized Probe of Lattice Defects in Metals," Scripta Met., 14 (1980) pp. 15-22.

18. R. P. Gupta and R. W. Siegel, "Positron Trapping and Annihilation at Vacancies in BCC Refractory Metals," J. Phys. F: Metal Phys., 10 (1980) pp. L7-13.

19. S. Pendyala, P. W. Zitzewitz, J. W. McGowan, and P.H.R. Orth, "Low-Energy Positrons From Metallic Moderators in a Back Scattering Mode," Phys. Lett., 43A (1973) pp. 298-300.

20. P. Coleman, T. C. Griffith, and G. R. Heyland, "A Time of Flight Method of Investigating the Emission of Low Energy Positrons from Metal Surfaces," Proc. Roy. Soc. Lond. A, 331 (1973) pp. 561-69.

21. K. F. Canter, A. P. Mills, Jr., and S. Berko, "Efficient Positronium Formation by Slow Positrons Incident on Solid Targets," Phys. Rev. Lett., 33 (1974) pp. 7-10.

22. K. G. Lynn and H. Lutz, "Positron Interaction with Solid Surfaces," pp. 219-44 in Radiation Effects on Solid Surfaces (Adv. in Chem. Series 158), M. Kaminsky, ed.; American Chemical Society, Washington, D. C., 1976.

23. A. P. Mills, Jr., P. M. Platzman, and B. L. Brown, "Slow-Positron Emission from Metal Surfaces," Phys. Rev. Lett., 41 (1978) pp. 1076-79.

24. K. G. Lynn, "Slow Positron Studies on Metals," Scripta Met., 14 (1980) pp. 9-14.

25. S. Berko, "Fermi Surface Studies in Disordered Alloys: Positron Annihilation Experiments," pp. 239-91 in Electrons in Disordered Metals and at Metallic Surfaces, P. Phariseau, B. L. Györffy, and L. Scheire, eds.; Plenum, New York, 1979.

26. P. E. Mijnarends, "Electron Momentum Densities in Metals and Alloys," pp. 24-88 in Ref. 6, 1979.

27. S. Berko, M. Haghgooie, and J. J. Mader, "Momentum Density Measurements with a new Multicounter Two-Dimensional Angular Correlation of Annihilation Radiation Apparatus," Phys. Lett., 63A (1977) pp. 335-38.

28. H. Weisberg and S. Berko, "Positron Lifetimes in Metals," _Phys. Rev._, 154 (1967) pp. 249-57.

29. D. Herlach and K. Maier, "Integrated Source-Specimen System for High-Temperature Positron Annihilation Experiments," _Appl. Phys._, 11 (1976) pp. 197-99.

30. M. J. Fluss and L. C. Smedskjaer, "A New Positron Source for Positron Annihilation Lifetime Experiments," _Appl. Phys._, 18 (1979) pp. 305-6.

31. D. A. Gedcke and W. J. McDonald, "Design of the Constant Fraction of Pulse Height Trigger for Optimum Time Resolution," _Nucl. Instr. Methods_, 58 (1968) pp. 253-60.

32. R. Myllylä, "A Modern Positron Lifetime Spectrometer," _Nucl. Inst. Methods_, 148 (1978) pp. 267-71.

33. P. Hautojärvi, T. Judin, A. Vehanen, J. Yli-Kauppila, J. Johansson, J. Verdone, and P. Moser, "Annealing of Vacancies in Electron-Irradiated α-Iron," _Solid State Commun._, 29 (1979) pp. 855-58.

34. P. Kirkegaard and M. Eldrup, "POSITRONFIT: A Versitile Program for Analysing Positron Lifetime Spectra," _Comput. Phys. Commun._, 3 (1972) pp. 240-50.

35. P. Kirkegaard, "POSITRONFIT EXTENDED: A New Version of a Program for Analysing Positron Lifetime Spectra,"_Comput. Phys. Commun._, 7 (1974) pp. 401-5.

36. V.H.C. Crisp, I. K. MacKenzie, and R. N. West, "An Improved Positron Lifetime Spectrometer," _J. Phys. E: Sci. Instr._, 6 (1973) pp. 1191-93.

37. S. Mantl and W. Triftshäuser, "Defect Annealing Studies on Metals by Positron Annihilation and Electrical Resistivity Measurements," _Phys. Rev. B_, 17 (1978) pp. 1645-52.

38. S. Mantl and W. Triftshäuser, "Direct Evidence for Vacancy Clustering in Electron-Irradiated Copper by Positron Annihilation," _Phys. Rev. Lett._, 34 (1975) pp. 1554-57.

39. W. Triftshäuser, J. D. McGervey, and R. W. Hendricks, "Positron-Annihilation Studies of Voids in Neutron-Irradiated Aluminum Single Crystals," _Phys. Rev. B_, 9 (1974) 3321-24.

40. S. Berko, "Two-Dimensional Angular Correlation of Annihilation Radiation Experiments," pp. 65-87 in Ref. 7, 1979.

41. L. C. Smedskjaer, M. Manninen, and M. J. Fluss, "An Alternative Interpretation of Positron Annihilation in Dislocations," _J. Phys. F: Metal Phys._, 10 (1980) pp. 2237-49; see also these Proceedings.

42. R. W. Siegel, "Vacancy Concentrations in Metals," (Proc. Intl. Conf. on the Properties of Atomic Defects in Metals, Argonne 1976) _J. Nucl. Mater._, 69 & 70 (1978) pp. 117-46.

43. S. W. Tam and R. W. Siegel, "On the Effect of Vacancy Migration upon the Annihilation of a Trapped Positron in Metals," _J. Phys. F: Metal Phys._, 7 (1977) pp. 877-84.

44. W. Brandt, "Positron Annihilation in Molecular Substances and Ionic Crystals," pp. 155-82 in Positron Annihilation, A. T. Stewart and L. O. Roellig, eds.; Academic, New York, 1967.

45. B. Bergersen and M. J. Stott, "The Effect of Vacancy Formation on the Temperature Dependence of the Positron Lifetime", Solid State Commun., 7 (1969) pp. 1203-05.

46. D. C. Conners and R. N. West, "Positron Annihilation and Defects in Metals," Phys. Lett., 30A (1969) pp. 24-25.

47. M. J. Fluss, R. P. Gupta, L. C. Smedskjaer, and R. W. Siegel, "Temperature-Dependent Behavior of Positron Annihilation in Metals," pp. 243-70 in Positronium and Muonium Chemistry (Adv. in Chem. Series 175), H. J. Ache, ed.; American Chemical Society, Washington, D.C., 1979.

48. L. C. Smedskjaer, M. J. Fluss, M. K. Chason, D. G. Legnini, and R. W. Siegel, "Positron Annihilation in Gold Between 27 K and 592 K," J. Phys. F: Metal Phys., 9 (1979) pp. 1815-20.

49. L. C. Smedskjaer, M. J. Fluss, R. W. Siegel, M. K. Chason, and D. G. Legnini, "Low-temperature Effects in Metals as Observed by Positron Annihilation," pp. 197-200 in Ref. 7, 1979.

50. L. C. Smedskjaer, D. G. Legnini, and R. W. Siegel, "On Low-temperature Positron Trapping in Cadmium," J. Phys. F: Metal Phys., 10 (1980) pp. L1-6.

51. L. C. Smedskjaer, M. J. Fluss, R. W. Siegel, M. K. Chason, and D. G. Legnini, "Observations of the Prevacancy Temperature Dependence of Positron Annihilation in Copper," J. Phys. F: Metal Phys., 10 (1980) pp. 559-69.

52. S. W. Tam, S. K. Sinha, and R. W. Siegel, "Theory of the Temperature Dependence of Positron Bulk Lifetimes-Implications for Vacancy Formation Enthalpy Measurements via Positron Experiments," (Proc. Intl. Conf. on the Properties of Atomic Defects in Metals, Argonne 1976) J. Nucl. Mater., 69 & 70 (1978) pp. 596-99; Erratum, J. Nucl. Mater., (1981) in press.

53. M. J. Stott and R. N. West, "The Positron Density Distribution in Metals: Temperature Effects," J. Phys. F: Metal Phys., 8 (1978) pp. 635-50.

54. M. J. Fluss, L. C. Smedskjaer, R. W. Siegel, D. G. Legnini, and M. K. Chason, "Positron Annihilation Measurement of the Vacancy Formation Enthalpy in Copper," J. Phys. F: Metal Phys., 10 (1980) pp. 1763-74.

55. R. P. Gupta and R. W. Siegel, "Electron and Positron Densities and the Temperature Dependence of the Positron Lifetime in a Vacancy in Aluminum," Phys. Rev. Lett., 39 (1977) pp. 1212-15.

56. M. J. Fluss, L. C. Smedskjaer, M. K. Chason, D. G. Legnini, and R. W. Siegel, "Measurements of the Vacancy Formation Enthalpy in Aluminum Using Positron Annihilation Spectroscopy," Phys. Rev. B, 17 (1978) pp. 3444-55.

57. W. Triftshäuser and J. D. McGervey, "Monovacancy Formation Energy in Copper, Silver and Gold by Positron Annihilation," Appl. Phys., 6 (1975) pp. 177-80.

58. L. C. Smedskjaer, M. J. Fluss, D. G. Legnini, M. K. Chason, and R. W. Siegel, "The Vacancy Formation Enthalpy in Ni Determined by Positron Annihilation," J. Phys. F: Metal Phys., 11 (1981) in press.

59. K. Maier, M. Peo, B. Saile, H. E. Schaefer, and A. Seeger, "High-temperature Positron Annihilation and Vacancy Formation in Refractory Metals," Phil. Mag. A, 40 (1979) pp. 701-28.

60. A. S. Berger, S. T. Ockers, M. K. Chason, and R. W. Siegel, "A Study of Vacancy-Iron Interactions in Quenched Aluminum," (Proc. Intl. Conf. on the Properties of Atomic Defects in Metals, Argonne, 1976) J. Nucl. and Mater., 69 & 70 (1978) pp. 734-37.

61. L. J. Cuddy and E. S. Machlin, "Quenching-in and Annealing of Point Defects in Silver," Phil. Mag., 7 (1962) pp.745-61; M. Doyama and J. S. Koehler, "Quenching and Annealing of Lattice Vacancies in Pure Silver," Phys. Rev., 127 (1962) pp. 21-31.

62. R. P. Sahu, K. C. Jain, and R. W. Siegel, "Vacancy Properties in Gold," (Proc. Intl. Conf. on the Properties of Atomic Defects in Metals, Argonne 1976) J. Nucl. Mater., 69 & 70 (1978) pp. 264-76.

63. A. S. Berger, S. T. Ockers, and R. W. Siegel, "Measurements of the Monovacancy Formation Enthalpy in Copper," J. Phys. F: Metal Phys., 9 (1979) pp. 1023-33.

64. W. Wycisk and M. Feller-Kniepmeier, "Quenching Experiments on High-Purity Nickel," phys. stat. sol. (a), 37 (1976) pp. 183-91.

65. M. Suezawa and H. Kimura, "Quenched-in Vacancies in Molybdenum," Phil Mag. 28 (1973) pp. 901-14; I. A. Schwirtlich and H. Schultz, "Quenching and Recovery Experiments on Molybdenum," Phil. Mag. A, 42 (1980) pp. 601-11.

66. K.-D. Rasch, R. W. Siegel, and H. Schultz, "Quenching and Recovery Investigations of Vacancies in Tungsten," Phil. Mag. A, 41 (1980) pp. 91-117 for the sample

67. G. Dlubek, O. Brümmer, and K. Meyendorf, "A Study of Mono and Divacancies in Cu and Au by Positron Annihilation," Appl. Phys., 13 (1977) pp. 67-70.

68. S. Nanao, K. Kuribayashi, S. Tanigawa, and M. Doyama, "Studies of Defects at Thermal Equilibrium and Melting in Cu and Ni by Positron Annihilation," J. Phys. F: Metal Phys., 7 (1977) pp. 1403-19.

69. R. P. Gupta and R. W. Siegel, "Annihilation of a Positron in a Vacancy in Aluminum," Rev. Rev. B, 22 (1980) pp. 4572-89.

70. H. Fukushima and M. Doyama, "The Formation Energies of a Vacancy in Pure Cu, Cu-Si, Cu-Ga and Cu-Mn Solid Solutions by Positron Annihilation," J. Phys. F: Metal Phys., 6 (1976) pp. 677-85; H. Fukushima and M. Doyama, "Positron Annihilation Study of the Vacancies in Copper Concentrated Alloys," pp. 219-22 in Ref. 7, 1979.

71. M. Weller, J. Diehl, and W. Triftshäuser, "Investigations of Neutron Irradiated Iron by Positron Annihilation and Correlated Internal Friction Measurements," Solid State Commun., 17 (1975) pp. 1223-26.

72. P. Hautojärvi, J. Johansson, A. Vehanen, J. Yli-Kauppila, and P. Moser, Vacancy-Carbon Interaction in Iron, Phys. Rev. Lett., 44 (1980) pp. 1326-29.

73. C. L. Snead Jr., A. N. Goland, and F. W. Wiffen, "Tracing the Evolution of Bubbles in Helium-Injected Aluminum by Means of Positron Annihilation," J. Nucl. Mater., 64 (1977) pp. 195-205.

74. R. Grynszpan, K. G. Lynn, C. L. Snead, Jr., A. N. Goland, and F. W. Wiffen, "Positron-Annihilation Investigation of High-Temperature Neutron Irradiated Molybdenum," Phys. Lett., 62A (1977) pp. 459-62.

75. S. Shiga, M. Hasegawa, S. Koike, K. Abe, and S. Morozumi, "Effect of Alloying Elements on Neutron Irradiation Damage in Molybdenum Studied by Positron Annihilation," pp. 497-500 in Ref. 7, 1979.

76. S. Mantl, B. D. Sharma, and R. Poerschke, "Defect Recovery Studies on TiNi13Cr17Fe, 70Cu30Zn and NiCu Alloys by Positron Annihilation," pp. 245-48 in Ref. 7, 1979.

77. K. Hinode, S. Tanigawa, M. Doyama, K. Shiraishi, and N. Shiotani, "Positron Lifetime Study of Gas Bubbles, pp. 733-36 in Ref. 7, 1979.

78. W. R. Wampler and W. B. Gauster, "Positron Annihilation Studies of Quenched Aluminum," J. Phys. F: Metal Phys., 8 (1978) pp. L1-5.

79. Y. Shirai, N. Nakata, N. Narita, K. Furukawa, and J. Takamura, "A Positron Lifetime Study of Vacancy-Solute Complexes in Dilute Au-Sn Alloys," pp. 209-12 in Ref. 7, 1979.

80. S. Saimoto, B.T.A. McKee, and A. T. Stewart, "Sensitivity of Positrons to Defects in the As-Deformed State of Copper," phys. stat. sol. (a), 21 (1974) pp. 623-26.

81. K. G. Lynn, R. Ure, and J. G. Byrne, "The Effect of Plastic Deformation on Positron Annihilation in Copper of Varying Grain Size," Acta Met., 22 (1974) pp. 1075-77.

82. K. Hinode, S. Tanigawa, and M. Doyama, "An Isochronal Annealing Study of Deformed Copper by Means of Positron Lifetime," J. Phys. Soc. Japan, 42 (1977) pp. 1591-93.

83. C. K. Hu, S. Berko, G. R. Gruzalski, and D. Turnbull, "Evidence for Microporosity and Dislocations in Pb(Cd) Alloys from Positron Annihilation Studies," Solid State Commun. 31 (1979) pp. 65-68.

84. G. Dlubek, O. Brümmer, N. Meyendorf, P. Hautojärvi, A. Vehanen, and J. Yli-Kauppila, "Impurity-Induced Vacancy Clustering in Cold-worked Nickel," J. Phys. F: Metal Phys., 9 (1979) pp. 1961-73.

85. B. Lengeler, S. Mantl, and W. Triftshäuser, "Interaction of Hydrogen and Vacancies in Copper Investigated by Positron Annihilation," J. Phys. F: Metal Phys., 8 (1978) pp. 1691-98.

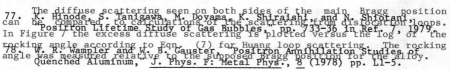

ROCKING ANGLE, Δθ (MIN)

86. K. Petersen, N. Thrane, and R.M.J. Cotterill, "A Positron Annihilation Study of the Annealing of, and Void Formation in, Neutron-Irradiated Molybdenum," Phil. Mag., 29 (1974) pp. 9–23.

87. M. Eldrup, O. E. Mogensen, and J. H. Evans, "A Positron Annihilation Study of the Annealing of Electron Irradiated Molybdenum," J. Phys. F: Metal Phys., 6 (1976) pp. 499–521.

88. N. Thrane and J. H. Evans, "The Effect of Impurities on the Lifetime of Positrons in Voids in Molybdenum," Proc. Fourth Intl. Conf. on Positron Annihilation, Helsingør, Denmark 1976, part 2, E19, pp. 100–5, unpublished.

89. P. Hautojärvi, J. Heiniö, M. Manninen, and R. Nieminen, "The Effect of Microvoid Size on Positron Annihilation Characteristics and Residual Resistivity in Metals," Phil. Mag., 35 (1977) pp. 973–81.

90. P. Ehrhart, H. D. Carstanjen, A. M. Fattah, and J. B. Roberto, "Diffuse-scattering Study of Vacancies in Quenched Gold," Phil. Mag. A, 40 (1979) pp. 843–58.

91. P. Ehrhart, "The Configuration of Atomic Defects as Determined from Scattering Studies," (Proc. Intl. Conf. on the Properties of Atomic Defects in Metals, Argonne 1976) J. Nucl. Mater., 69 & 70 (1978) pp. 200–14.

92. E. Kuramoto, K. Kitajima, M. Hasegawa, and S. Koike, "Studies of Vacancies in α-Iron by Positron Annihilation Technique," pp. 505–8 in Ref. 7, 1979.

93. H. E. Schaefer, P. Valenta, and K. Maier, "Investigation of Vacancies in High-Purity α-Iron by Means of Positron Annihilation," pp. 509–12 in Ref. 7, 1979.

94. W. Frank, A. Seeger, and M. Weller, "Interpretation of Positron-Annihilation Experiments on Electron-Irradiated α-Iron in Terms of Self-Interstitial Migration in Stage III," Rad. Effects, 55 (1981) pp. 111–18.

95. S. Tanigawa, K. Hinode, N. Owada, M. Doyama, and S. Okuda, "The Study of Recovery of Irradiated Iron by Positron Annihilation," pp. 501–4 in Ref. 7, 1979.

96. A. Vehanen, P. Hautojärvi, J. Johansscn, J. Yli-Kauppila, and P. Moser, "Vacancies and Carbon Impurities in α-Iron, Part I: Electron Irradiations," Phys. Rev. B, submitted (1981).

97. J. Verdone, P. Moser, W. Chambron, J. Johansson, P. Hautojärvi, and A. Vehanen, "Magnetic After-Effect in Electron Irradiated Iron," J. Magnetism and Magnetic Mater., 19 (1980) pp. 296–98.

98. S. Takaki, J. Fuss, H. Kugler, U. Dedek, and H. Schultz, "The Resistivity Recovery of High Purity and Carbon Doped Iron Following Low Temperature Electron Irradiation," Rad. Effects, submitted (1981).

POSITRON-ANNIHILATION STUDY OF VOIDS AND OTHER NEUTRON-PRODUCED

MICROSTRUCTURAL FEATURES IN Mo AND Mo-0.5 at.% Ti*

C. L. Snead, Jr.,[1] K. G. Lynn,[1] Y. Jean,[1] F. W. Wiffen,[2] and P. Schultz[3]

[1]Brookhaven National Laboratory, Upton, NY   11973
[2]Oak Ridge National Laboratory, Oak Ridge, TN 37830
[3]University of Guelph, Guelph, Ont., Canada N1G 2W1

Specimens of Mo and Mo-0.5 at.% Ti which have been irradiated with neutrons ($\sim 10^{22}$ n/cm^2, E > 0.1 MeV) at temperatures between 425 and 1500°C have been studied using both lifetime and Doppler-broadening measurements. Both the shape parameter and the intensity of the lifetime component from positrons trapped at voids define swelling as a function of temperature in a way that is independent of the neutron fluence. The lifetime of the void component (470 ps) agrees with previous determinations and is reasonably constant for all cases studied. The relative swelling as a function of irradiation temperature and the swelling peak ($\sim$750°C) are well defined, but no information on the magnitude of the void volume is obtainable. In the determination of the shape and peak of the derived swelling curve, the positron analysis is more definitive than similar determinations using transmission electron microscopy. The microscopy is necessary, however, to determine the magnitude of swelling. Concurrent trapping of positrons at voids and other radiation-induced microstructure is observed for all temperatures of irradiation between 425 and 1500°C. In all cases, saturated trapping ($\sim$100% of· positrons annihilate from traps) is achieved. For measurements made between −263 and 100°C an increase in the percentage of positrons trapped at voids with increasing temperature is observed. This is attributed to thermally activated positron detrapping from a shallow species of trap associated with the sample microstructure. Annealing the specimens to 955°C and repeating the measurements results in an increase of the detrapping phenomenon. One deep (nondetrapping) microstructural trap is identified as dislocation loops with diameters less than 70 Å. The shallow traps are tentatively identified as dislocation tangles and the detrapping energy is determined to be $E_a = 0.1 \pm 0.05$ eV. The temperature dependences observed suggest that caution is needed in the interpretation of positron-annihilation results where either the void or the other microstructural elements are being altered in experiments involving annealing or sequential irradiations.

*Work performed under the auspices of the U.S. Department of Energy.

Introduction

The properties of positrons in solids, including their lifetime and annihilation energetics, are determined by the local electron concentrations. As a result, the positron properties are very sensitive to vacancy-type defects, but relatively insensitive to interstitial-type defects or impurity atoms. The general features of positron spectroscopy as a probe of the defect microstructure are reviewed by Siegel (1).

Positron-annihilation techniques have been successfully applied to the study of voids generated by neutron irradiation of metals. The formation of voids in molybdenum (2,3) during annealing following 60°C reactor neutron irradiation and room-temperature electron irradiation (4,5) has been extensively studied. Voids in aluminum neutron irradiated at 50°C (6,7) have also been studied in detail. Positron-annihilation measurements on neutron-irradiated specimens also showed that gaseous impurities significantly affected both the positron lifetime and the momentum distribution in the void (8). The effect of nongaseous impurities segregating at voids has also been demonstrated in aluminum (9). Other than impurity effects, however, it was assumed in most of the work above that voids were the only defects that act as trapping for positrons (6).

Fewer positron results on voids produced by high-temperature irradiations have been reported (10,11). High-temperature neutron irradiations produce voids and a temperature-dependent dislocation distribution (12,13). This high dislocation concentration makes the study of the voids more complex since it also produces positron trapping sites, thereby making the simplifying assumption above invalid. It is the purpose here to show how this dislocation contribution comes into the analysis, how the competition between void and dislocation traps behaves as a function of irradiation temperature, and finally how this competition varies as a function of measuring temperature.

The best method of determining the void volume in irradiated metals is through measurement of the average size and number density of the cavities present by transmission electron microscopy (TEM). The weak points in such analyses include the determination of the "average" size from a size distribution, often with shape irregularities, and establishing the "number," since small voids may not be observed owing to resolution restrictions. Positrons, on the other hand, are sensitive to all sizes of vacancy-defect structures and the probability of trapping of a positron in a void increases with the size of the void (14-17). Monitoring a trapping-sensitive parameter of the positron annihilation then provides a measure of the probability of trapping of positrons at voids. This probability is directly related to the total volume of voids. At present, however, absolute values of swelling are not obtainable without either calibration by other techniques, or by appeal to theoretical calculations. The strength of the positron measurements at present lies in the determination of _relative_ concentrations of vacancy-type defects and the sensitivity to small voids.

Experimental

Specimens of commercially prepared, low-carbon, arc-cast Mo were annealed at 1200°C at 0.1 mPa to produce a fully recrystallized microstructure. These samples were irradiated in EBR-II to high fluences at several temperatures. The samples were in either a helium or argon atmosphere during irradiation. Tabs of irradiated metal were cut and electropolished. Then $\sim$10 μCi of ^{22}NaCl was deposited on one face of a pair of specimens that make up the positron annihilation sample "sandwich." Table I shows the irradiation and defect parameters determined by TEM (12).

Robert W. Hendricks

Technology for Energy Corporation
Knoxville, Tennessee 37922

Table I. Microstructural Parameters for Neutron Irradiated Mo and Mo-0.5 at.% Ti.

Irrad. Temp. (°C)	Fluence (10^{22} n/cm^2 E > 0.1 MeV)	dpa	Void Conc. (10^{22} m^{-3})	Mean Void Diameter (nm)	Void Swelling (%)	Dislocation Conc. (10^9)
			Mo			
425	2.5	11	6.3	2.4	.05	*
585	2.5	11	8.2	5.7	.57	NA
600	0.43	~2	24.0	3.7	.65	26.
790	2.5	11	4.5	5.7	.45	NA
900	0.54	~2	1.6	8.0	.43	.15
1000	4.4	19	0.5	13.1	.59	NA
1100	0.61	~2	0.013	30.0	.18	.05
1300	0.61	~2	5×10^{-4}	100.0	.27	.05
1500	0.54	~2	$<10^{-5}$	>200.0	.05	.04
			Mo-0.5 at.% Ti			
425	2.5	11	11.0	2.1	.05	*
585	2.5	11	16.0	5	NA	
790	2.5	11	6.5	6.1	NA	
1100	2.5	11	1.7	10.2	NA	

*The dominant microstructural feature is a high concentration of small loops.

Small-angle x-ray and neutron scattering are powerful analytical tools for investigating long-range fluctuations in electron (x-rays) or magnetic moment (neutrons) densities in materials. In recent years they have yielded valuable information about voids, void-size distributions, and swelling in aluminum, aluminum alloys, copper, molybdenum, nickel, nickel-aluminum, niobium and niobium alloys, stainless steels, graphite and silicon carbide. In the case of aluminum information concerning the shape of the voids and the ratio of specific surface energies was obtained. The technique of small-angle scattering and its application to the study of voids is reviewed in this paper. Emphasis is placed on the conditions which limit the applicability of the technique, on the interpretation of the data, and on a comparison of the results obtained with comparison techniques such as transmission electron microscopy and bulk density.

Simultaneous measurements of positron lifetime and Doppler broadening were performed at room temperature for most of the results presented. For the low-temperature results (<20°C), similar simultaneous measurements were also made, but the source strength was ~50 µCi. For each set of measurements sample-detector geometry was kept as constant as possible. The fast-fast lifetime system had a prompt full width at half maximum (FWHM) of ~240 ps using ^{60}Co with energy windows set for [22]Na gammas at 0.511 and 1.28-MeV.

Results

Figure 1 shows a plot of the Doppler-broadening shape parameter S (peak to total counts minus that of the unirradiated annealed specimens) vs. temperature for the entire set of irradiated molybdenum specimens. Positron data that are plotted in Figure 1 were determined over a span of two years, using two different Doppler-broadening systems. The disparity in the plotted points for identical irradiation temperatures is then a good indication of the accuracy (reproducibility) of the data. We also plot for comparison (triangles) the void volume as determined from TEM. An increasing value of the shape parameter indicates an increasing of the Doppler-peak, indicative of increased positron trapping, in a shifting of trapping sites in favor of deeper traps. Two things to note from this plot are the occurrence of a peak at ~750°C and the indication of trapping at all temperatures. The peak at 750°C coincides with the swelling peak established by Smidt and Reed for neutron-irradiated molybdenum as established by TEM and immersion density (18). Their work showed no swelling, however, above

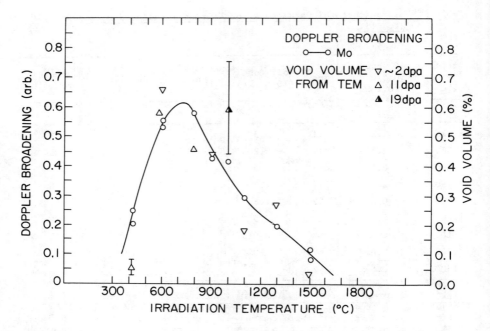

Fig. 1. The Doppler-broadening shape parameter S is plotted as a function
of irradiation temperature for neutron-irradiated Mo (open circles).
The solid line is drawn through the data as a guide only. Scatter
in the data taken for the same irradiation temperatures is an indication of the uncertainty in the points. The void volume as determined from TEM (triangles) is also plotted as a function of
irradiation temperature for comparison. The estimated error in
these values is ±30 – 50% of the value itself.

800°C. Similar relative S-parameter data (not shown) for the Mo-0.5 Ti
irradiated specimens also fall on the solid line that has been drawn only as
a guide for the eye.

Since Doppler-broadening shape-parameter analysis yields essentially
only one independent parameter, defect-specific information is difficult to
infer where more than one type of trapping site is present. (The R parameter
to be discussed later is a qualitative exception.) Lifetime measurements, in
principle, can yield much more information. Each type of defect present
could have as its signature a distinct positron lifetime, and analysis could
separate out the intensity associated with these lifetime components. The
fundamentally difficult analytical problem here is one of reliably resolving
all of the lifetime components that are present in the spectrum. Studies
(19) (Eldrup et al. and Fluss et al.) have been performed to investigate
various types of systematic errors and their effect on the extracted positron
lifetimes and intensities. Extremely good statistics are needed before
different lifetimes simultaneously present in the data can be accurately
extracted.

Figure 2(a) and (b) shows the result of an unconstrained two-lifetime
fit to the data with the source component (^{22}NaCl, 2.45%) subtracted. The
values of the shorter-lifetime component τ_1 are plotted in Figure 2(b).

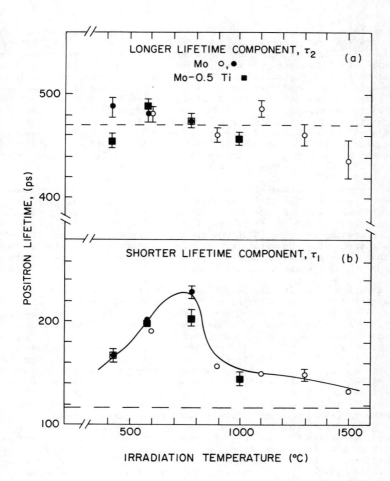

Fig. 2. A two-lifetime fit to the data (background and source effects subtracted) is shown as a function of neutron irradiation temperature
for Mo. In 2(a) the lifetime of positrons in voids, τ_2, is plotted.
The dashed line at 470 ps represents other such lifetime determinations taken from the literature. The error bars are statistical in nature. Figure 2(b) shows the behavior of the lower lifetime component τ_1 vs. irradiation temperature. The lifetime τ_1 is
probably associated with a spectrum of microstructural positron
trapping sites, other than voids. The sum of the intensities of
the τ_1 and τ_2 components is ~100%, consistent with saturation
(almost all positrons trapped). Data of four irradiation temperatures for Mo-0.5 at.% Ti are also plotted to show the similarity
of the results obtained to that for Mo. The dashed line at 116 ps
denotes the lifetime of positrons in "pure" Mo.

The lower dashed line in this figure depicts the value of the lifetime for
unalloyed, well-annealed molybdenum ($\tau_f = 116$ ps). Solid circles represent
the data for the "high-fluence" set (see Table I). The χ^2/ν values range
from 1.0 to 1.4. Data are not plotted for the 950 - 1050°C specimens as
convergence of the program for this sample was not achieved. The dashed

line on this figure for τ_2 (470 ps) is the value representative of large voids (radius > 20 Å) previously found by Grynszpan et al. (11), Petersen et al. (12), Eldrup et al. (6), and Hinode et al. (32). Our reconfirmation of this value is encouraging when one considers the wide ranges of fluences and means of defect production encompassed in these experiments. Indeed, the constant nature of the void lifetime over an incredibly wide range of void sizes and possibly different faceted states is a key factor to the successful determination of swelling from lifetime data. Note that the intensity for the void lifetime component for the 1500°C sample has a value of only 4% (to be discussed later). Based upon the annealing results of Thrane et al. (8) we conclude also that the void lifetimes in this work are not being altered by gases diffusing into the voids from the sample surface. Such effects were shown by Thrane to be substantial.

In general, the computer fits to the data for this two-component analysis were best at high and low irradiation temperatures, and worse for the intermediate region of 600-1000°C. Three-component analyses produced inconsistent results with larger values of χ^2/ν. The conclusions reached were that saturation trapping ($\sim$100% of the positrons were trapped) was achieved for all irradiation temperatures, τ_2 is the value for the lifetime in voids, and that the values of τ_1 near 165 ps represent annihilations from the various types of dislocation traps present.

In order to verify the saturation hypothesis the data were analyzed in terms of the two-state trapping model, by including two traps (3). The fixed parameters in the analysis were the source lifetime and intensity, the lifetime in the perfect lattice, and the right side of the resolution function (21). The parameters adjusted were κ_1 and κ_2, the trapping rates associated with the traps yielding lifetimes τ_1 and τ_2, respectively. The trapping rate κ_i is related to the concentration of defects C_i through the relation $\kappa_i = \mu_i C_i$ where μ_i is the specific trapping rate per unit concentration of defect i. The results of this analysis are plotted in Figure 3(a). The errors associated with the points are approximately equivalent to the absolute value of the trapping rate. The solid and open points denote differences in fluences as in Figure 2, with the circles representing κ_2 (voids) and the squares κ_1 (dislocations). The lines shown in the figure are included as guides and indicate the changing nature of the trap dominance as the irradiation temperature changes. Dislocations prevail at high and low temperatures, with voids dominant in the 600 - 1000°C region. This is the same region where τ_1 deviated above the nominal 165-ps value in the two-component fit (Figure 2b). From the high values for κ it is clear that $\sim$100% trapping is realized for all points. Note that the lifetime component τ_ℓ for annihilations from the lattice (the so-called Bloch state) given by $\tau_\ell = \tau_f/(1 + \kappa\tau_f)$ is vanishingly small and will be unresolved. In a defect-free metal (no positron trapping sites) τ_ℓ is identical to τ_f.

The changing nature of positron traps can also be seen in the behavior of the R parameter from the Doppler-broadening analysis. Using the momentum profiles, Triftshäuser (22) developed a defect-specific parameter, R, which is applicable for both Doppler-broadening and angular-correlation measurements. This parameter can be used to check whether the dominant trapping species is changing from one type to another for different data sets. In the R-parameter calculation one assumes that each type of defect trap has a different and unique momentum spectrum which implies that the conduction and core annihilation rates are unique for each defect type. This calculation uses the small central region of the annihilation photopeak I_v and the wings of the spectrum I_c. Thus, I_v is a measure of annihilation of positrons with conduction and core electrons in the central region and I_c is a measure of annihilation with predominantly core electrons. One can measure a characteristic value of I_v and I_c in the free and trapped state

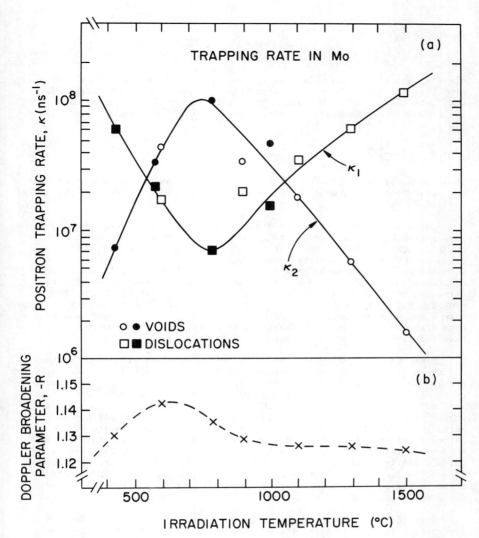

Fig. 3. Figure 3(a) shows the results of fitting the trapping model to the data, but allowing for two distinct trap lifetimes. In this analysis the lifetime for the pure Mo was fixed. Parameters that are adjacent are the trap lifetimes τ_1 and τ_2, their trapping rates κ_1 and κ_2, and the lifetime of positrons that annihilate without being trapped. Note that the peak in the void trapping rate κ_2 at $\sim$750°C coincides with a minimum in the rate κ_1 associated with positron trapping at other microstructural defects. The statistical uncertainty in the values of the points are equal to the values themselves. Figure 3(b) shows the value of the Doppler-broadening R parameter (described in text). The variation of the R parameter with irradiation temperature is interpreted as a change in dominance from one positron trap species to another.

where

$$I_v = I_v^f P_f + I_v^t P_t \tag{1}$$

where I_v^f and I_v^t are the values of I_v, and P_f and P_t are the annihilation probabilities in the free and trapped state, respectively. Following Mantl and Triftshäuser (23) the ratio of the differences of the two regions in the free and trapped state yielding a <u>concentration-independent</u> R parameter (in principal, at least) is

$$R = \frac{I_v^t - F_v^f}{I_c^t - I_c^f}. \tag{2}$$

This parameter is constant in cases where one predominant defect is present and saturation trapping can be produced to find I_v^t and I_c^t for that particular type of defect.

The R-parameter analysis as depicted at the bottom of Figure 3 shows the same peak behavior seen in Figure 1 and for τ_1 in Figure 2. The same value for R is achieved at low and high irradiation temperatures with a higher value at intermediate temperatures. Note that the peak is shifted to slightly lower temperature with respect to the maxima found in the other figures. The interpretation of this behavior is that at low and high temperatures one type of defect (dislocations) dominates the positron trapping process while for the 500 - 1000°C region there is a competition between traps with a second trap (voids) taking over the dominant role. This interpretation also is consistent with the behavior of the κ's of the two traps, resolved from the lifetime measurements.

Since the trapping probability for voids increases with their number and their size, the number of positrons that become trapped at voids reflects the volume of the voids in the specimen. From Figure 3 the behavior of κ_2 associated with void trapping indicates a peak structure centered at $\sim 750°$C. To define a swelling curve for these specimens, we propose that such a plot of positron trapping at voids appears to be a reasonable approach, although the error bars associated with the κ determinations are quite significant. A more accurate determination of void trapping is available from the unconstrained two-lifetime analysis. From the two-component fit we extract the intensity I_2 of the τ_2 component (Figure 2a) as proportional to the void volume. A plot of I_2 as a function of irradiation temperature is shown in Figure 4. The data for the Mo-0.5 Ti specimens are also included for comparison. The lines shown on the figure are guides for the eye only.

With these assumptions we see that void swelling is present for all irradiation temperatures for the molybdenum specimens studied, albeit only at low levels at the lowest and highest temperatures ($I_2 = 9$ and 4%, respectively). For full quantification of the swelling one requires the peak swelling temperature, dependence of the swelling on the irradiation temperature, and magnitudes of the swelling. Our results show that positron trapping at voids is a sensitive measure of the first two properties, but without an independent means of calibration, the magnitude of the swelling is not extractable from the technique.

The microstructure of all of the samples examined by positron annihilation spectroscopy had been characterized by TEM. The void and dislocation parameters from that characterization (12,24) are included in Table I, and the total void volume for the molybdenum samples is plotted as a function of irradiation temperature in Figure 1 and Figure 5. There are two features of the TEM observation that reinforce the relevance of the positron data. First, the general shape of the void volume vs. irradiation temperature is

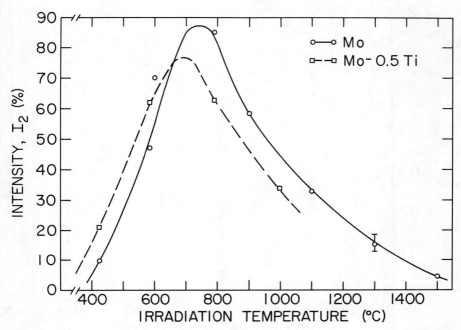

Fig. 4. The intensity of the void component I_2 (associated with τ_2 is determined from a two-component fit to the data) is plotted as a function of the neutron irradiation temperature for Mo and Mo-0.5 at.% Ti. As these data represent the relative change of the void volume in the specimens, the solid curve (drawn by eye) is suggested to represent the "swelling curve" for neutron-irradiated Mo and Mo-0.5 at.% Ti. The peak swelling in Mo is seen to be at $\sim750°C$.

similar to either the Doppler broadening (Figure 1) or I_2, the intensity of annihilation in the void traps (Figure 5). Secondly, the TEM data show little or no dependence of void volume on fluence, an unexpected result that is consistent with the current findings.

All of the results discussed above have been based upon positron measurements made at room temperature. By making all measurements at one temperature, one usually can claim that temperature-dependent positron effects are eliminated from the analysis. In what follows we show that at least for these irradiated specimens, a significant temperature dependence does exist for the annihilation characteristics, and that this assumption cannot be made. The sample that evidenced the largest amount of swelling, the 600°C irradiated set, was selected for positron analysis as a function of measurement temperature. Both Doppler-broadening and lifetime measurements were made as a function of temperature on the as-irradiated specimens from -265 to 225°C. The results of the lifetime measurements are shown in Figure 6(a).

Following these measurements the samples were annealed in vacuo at 955°C for 2 h. The purpose of the anneal was to reduce the dislocation concentration while leaving the void distribution relatively intact. The annealing temperature was chosen so as to be above Stage VI recovery as defined by Cornelis et al. (25) for molybdenum. Brimhall et al. (26) have

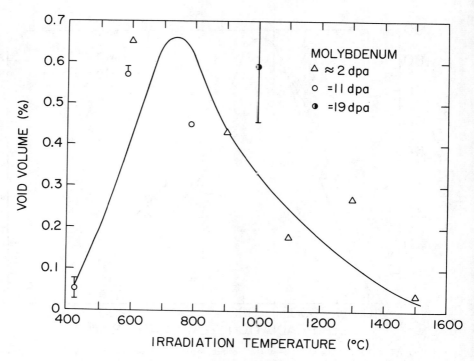

Fig. 5. The void volume as determined by TEM is plotted as a function of
neutron irradiation temperature for Mo. For comparison the solid
line is the result of swelling as determined from the intensity of
the void trap component I_2 (from Figure 4).

shown this anneal to be sufficient for the removal of dislocation loops
whose diameter is less than 70 Å after neutron irradiation at $\sim 10^{20}$ n/cm^2,
E > 0.1 MeV. Following this anneal, positron measurements were again performed, this time at temperatures from -263 to 625°C and the results are
shown in Figure 6(b).

Discussion of Results

We interpret the data shown in Figures 1-4 in the following terms. The
maximum in the S parameter vs. irradiation temperature around 750°C (Figure
1) is indicative of enhanced positron trapping, either from a larger percentage of positrons trapped for specimens irradiated in that temperature range,
or from a shifting of the predominant trapping sites from one type of trap
whose ratio of I_v/I_v^f is less than that of another with a larger ratio. From
the life-time data of Figure 2, the component τ_2 of ~ 470 ps is attributed to
annihilations from voids. This is most likely due to annihilation from surface traps associated with the voids (14). The constancy of this positron
lifetime (τ_2) over the entire irradiation temperature range indicates that
(a) the majority of the voids doing the trapping are > 20 Å in diameter (17),
and (b) no appreciable amount of gas is contained in the voids (8).

Also from Figure 2 we conclude that since τ_1 is everywhere greater than
τ_f, a second trapping component is present with a mean lifetime of ~ 165 ps.
From the behavior of τ_1 with temperature, there are two likely possibilities
involving these traps. First, the τ_1 computer-fitted component could be an

452

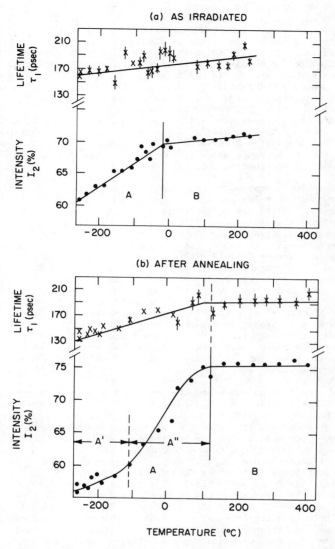

Fig. 6. The dependence of positron annihilation characteristics in Mo with voids present is shown in a plot of (a) τ_1 and I_2 as a function of measuring temperature from −263 to 250°C. The specimen is Mo irradiated at 600°C to ∿2 dpa (see Table I). Figure 6(b) shows a repeat of the 6(a) measurements but after annealing the specimens to 955°C and extending the measurement temperature to 400°C. The behavior in Region A is interpreted as indicative of positron release from shallow traps (dislocations) and subsequent retrapping at voids. The 955°C anneal removed a large portion of the deep dislocation trap, thereby enhancing the shallow-trap contribution. All data were obtained from a two-lifetime fit with the lifetime of the void component fixed at $\tau_2 = 480$ ps.

average of two separate components, τ_d, the component of the trapping defect and τ_ℓ, the average lifetime for the annihilation of positrons that do not get trapped (annihilate from the Bloch state). From the trapping model as we saw before $\tau_\ell = \tau_f(1+\kappa\tau_f)^{-1}$ and τ_ℓ decreases with increasing κ. In all of our data analysis τ_ℓ has not been extracted from the data except for the well-annealed molybdenum specimen where κ was assumed to be equal to 0. In the present studies where κ is sufficiently large (the saturation condition) τ_ℓ approaches zero and is therefore neglected. For relatively high values of κ, τ_ℓ is effectively zero since it cannot be extracted from the resolution function. The second possibility is that if saturation is approached then τ_1 reflects a composite of different lifetimes associated with traps other than voids. The changing τ_1 then would reflect a changing dominance of the different lifetimes involved (i.e., trap depths). The higher values of the χ^2/ν over the range where τ_1 is significantly higher than 165 ps could support either possibility.

The evidence presented in Figure 3, however, strongly favors the second possibility. At both low and high temperatures where the first model would predict minimum values for defect trapping probabilities, the κ's associated with the temperature extremes are, themselves, consistent with the saturation trapping condition. The value $\kappa \simeq 10^{17}$ s^{-1} is extremely high trapping rates when compared with those, for instance, of Hall et al. (21) who determined trapping rate of 3.5×10^{14} s^{-1} at single vacancies in aluminum. This was obtaned for a vacancy concentration of 5×10^{-5} (at 450°C) at which point the lifetime vs. temperature curve is approaching saturation. McKee et al. (27) have estimated $\kappa = 2.9 \times 10^{15}$ s^{-1} for dislocations in Cu. Cotterill (28) has pointed out that jogs on dislocation lines might act as deeper traps than the edge dislocation line segment itself. Our high values of κ when compared with those for vacancies or dislocations (albeit in a different metal) are consistent with 100% positron trapping. We noted before that the error bars associated with the κ values are large owing to correlation among the parameters and assign values for the uncertainty equal to the values of κ themselves.

Based upon this evidence and the extensive knowledge of the microstructure after irradiation (12,24) we attribute the traps associated with τ_1 to dislocations and dislocation loops. The κ behavior of Figure 3 is then seen as a competition between dislocation and void traps, with the dislocation-type traps dominating the trapping at high and low irradiation temperatures, the voids dominating the trapping at $\sim$750°C, and contributions from both at intermediate temperatures. The behavior of the R parameter, as discussed previously, reinforces this interpretation. The TEM data in Table I also qualitatively support this interpretation at the higher temperatures. In increasing the irradiation temperature from 1100 to 1500°C, the void volume is reduced by a factor of $\sim$1000, while the dislocation concentration changes little. (Alternatively, one could ascribe the τ_1 defects to small voids instead of dislocations. This, however, would require a bimodal distribution of voids, with smaller ones sharply peaked about a value of $\sim$10 Å, and with no appreciable effect of the dislocation microstructure. There is no plausible reason to assume such a void dislocation.)

To better define the void swelling we use the intensity of the void component I_2 as determined from a simple two-lifetime fit (Figure 4) rather than the inherently more complicated trapping-model of κ for voids. Here, the region where voids dominate the trapping is delineated with much less uncertainty. The peak in I_2 associated with the maximum in swelling is defined at $\sim$750°C for both Mo and Mo-0.5 at.% Ti, with swelling falling off on both sides, but more slowly on the high-temperature side. The limitation here is that the absolute magnitude of the swelling is not obtainable from these data. However, the general shape of the swelling vs. irradiation temperature curve defined by positron annihilation and by TEM is the same. The surprising lack of fluence sensitivity in the positron annihilation data was first interpreted as a weakness in the technique, based on the assumption that there was a strong fluence dependence of microstructure in all metals. The TEM results, however, confirmed the weak fluence dependence in the swelling (void volume) of molybdenum. The weakness in the TEM determination of total void volume includes problems or limitations imposed by: (a) the limited area of sample viewed, (b) the limited number of voids measured, (c) the approximation made in handling the exact geometric shape of the voids, and (d) inaccuracies in treating long-range differences between areas of the same sample, including differences between grain interiors and grain-boundary-adjacent regions. The result is that the swelling curve defined by the positron annihilation measurements is valid, and may even be a more accurate definition of the curve shape, but not the magnitude, than the data determined by TEM.

Temperature-Dependent Effects

The dependence of the positron lifetime characteristics on measuring temperature as shown in Figure 6 is quite striking. The solid lines are drawn only as guides for the eye. Because these data were taken in a cryostat, the count rates for the lifetime system were limited by geometry and the statistical accuracy is less than for the room-temperature measurements. The fitted timing resolution for these measurements was 400 ± 20 ps FWHM whereas a ^{60}Co prompt fit gave 330 ps. The void lifetime here was $\tau_2 = 480$ ps and remained constant over the entire temperature range investigated (note our choice of 470 ps in Figure 2 could equally be 480 ps estimated by eye). In the two-component fit here τ_2 was fixed. In spite of the broader resolution function, poorer statistics, and a completely different lifetime apparatus, the I_2 value at room temperature here (71%) is identical to that obtained previously (Figure 2).

The interesting feature of Figure 6(a) is the apparent break in the curve at $\sim -25°C$ and the steep decrease of I_2 with decreasing temperature from that temperature down to $-263°C$. This decrease appears linear   Since this specimen (irradiated at 600°C) has the highest concentration of both voids and dislocations (see Table I) the saturated trapping condition obtained throughout the entire irradiation set for much lower concentrations of both defects is certainly maintained through these temperature-dependent measurements. This means that the increase in I_2 with increasing temperature starting at $-263°C$ represents more positrons trapped at voids, and fewer at dislocation-type traps, with increasing temperature.

We propose that there are two major types of dislocation and/or impurity traps; shallow and deep. As the temperature is increased from $-263°C$, the smaller positron binding energy to the shallow traps allows the positron to be thermally detrapped. Because of the high concentration of remaining traps, the positron either finds a deep dislocation trap or a void. This release and subsequent competition among traps during subsequent (very-short-time migration) can explain the increase in I_2 from 60 to 71% upon going from -263 to 25°C.

The results of annealing the specimens to 955°C and repeating the measurements as shown in Figure 6(b) can give us definitive information as to the identity of one of the dislocation traps; namely, the deep trap. After the specimens were annealed, τ_2 increased to 564 ps, most likely indicating the infusion of impurities at the void surface (7). After the anneal, the range of the change in I_2 with temperature is significantly increased from 55 to 75% rather than 60 to 70%. This translates to a higher number of positrons being trapped in the shallow traps at low temperatures

after the anneal than were trapped there prior to the anneal. The annealing
results of Brimhall et al. (24) which show that the major result of the
anneal to 955°C is to remove the majority of the dislocation loops present
with radius r < 70 Å is compelling evidence for these loops to be the _deep_
dislocation traps. With the small loops removed, competition for trapping
of positrons is between coarser dislocation distributions and voids. This
enhances the percentage of positrons trapped at the shallow traps, thereby
increasing the number of detrapped with increasing temperature.

The behavior of τ_1 with temperature between -263 and 130°C is better
resolved after the anneal then before. An increase from ∿140 to ∿205 ps is
observed after annealing whereas before annealing the maximum increase is
∿170 to ∿205 ps. This can be interpreted as a shift in the contribution of
dislocation trapping from shallow (smaller τ) traps to deep (larger τ)
traps. One could equally interpret the τ_1 behavior as reflecting the change
in I_2 through the trapping model where τ_1 here is an average of τ (deep trap)
and τ_ℓ. The former interpretation is more acceptable if trapping is still
saturated at low temperatures following the anneal. The latter explanation
holds if saturation is not realized at low temperature.

For thermally activated detrapping (the region A' + A" following the
anneal) the activation energy for the process can be estimated. An Arrhenius
plot of $\ell n(\Delta I_2)$ vs. $1/kT$ yields a value of $E_a = 0.1 \pm 0.05$ eV. The scatter in
the data account for the unusually large uncertainty. Detrapping at shallow
traps (grain boundaries) has been suggested (29) for low-temperature trapping
phenomena. More recently Smedskjaer et al. (30) have attributed low-
temperature positron trapping (and detrapping) to dislocation. It is
interesting that the binding energy they calculate for a positron to a dislocation line is ≤ 0.1 eV.

This type of temperature dependence of the annihilation parameters
associated with positron-void trapping has also been observed in neutron-
irradiated aluminum (31). Nieminen et al. (31) explained their results
using a mechanism involving diffusion-limited trapping of the positron at
voids. In this interpretation shallow traps are not necessary to explain
the observations as a temperature-dependent trapped-state population evolves
from the theory. There are several reasons why we do not think that this
model can explain our results. (1) The void radii in our specimens are too
small to fit the criterion of the Nieminen model; (2) the presence of traps
other than voids that produce saturation trapping even when the void
trapping probability is too low to do so certainly must be included in the
analysis; and (3) the Nieminen model cannot reproduce the curvature of the
low-temperature I_2 data presented in Figure 6.

Conclusions

(a) Positron-annihilation, Doppler-broadening, and (especially) lifetime measurements of molybdenum irradiated with neutrons at temperatures
where swelling can result from void formation define the _shape_ of the curve
depicting void swelling vs. irradiation temperature more consistently than
does TEM. This appears to be true because at the defect levels investigated
here, the fate of all positrons was to be trapped at irradiation-produced
defects. As the void volume changed, the number of positrons that are
trapped at voids changed accordingly. Thus, a consistent measure of the
relative void volume could be determined over a wide range of irradiation
temperature and fluence. This relative-void-volume determination has the
additional advantage of being apparently fluence independent.

(b) A single peak in the swelling curve for molybdenum is found at
∿750°C. The swelling peak for Mo-0.5 at.% Ti appears to be located slightly

lower than 750°C, but the precision of this determination is poor since only four irradiation temperatures are available.

(c) The <u>magnitude</u> of the swelling (void volume) is <u>not</u> determinable from this analysis--only relative swelling behavior can be inferred. Whether magnitude information is obtainable from positron studies performed on lower-fluence irradiated specimens where all the positrons are not trapped is, at this point, conjectural.

(d) Evaluation, using positron techniques, is a bulk-sampling measurement which is sensitive to vacancy-type defects. It is a nondestructive technique, except for surface preparation, and samples, after the positron measurements, are available for other measurements.

(e) For neutron-irradiated specimens the positron-annihilation results are sensitive not only to the presence of the dislocation microstructure that was generated by the irradiation, but, if temperature-dependent measurements are performed, also to the <u>details</u> of the microstructure.

(f) Thermally assisted detrapping of positrons from shallow traps (probably dislocation tangles produced during irradiation, but other traps such as impurity clusters have not been ruled out) in the temperature range -263 to 100°C has been observed. Part of the traps that make up the non-detrapping microstructural traps are identified as small dislocation loops (r < 70 Å).

(g) Because of the detrapping phenomenon, which produces an intensity of positrons trapped at voids that is dependent upon the temperature at which the measurement is performed, <u>and</u> the annealing history of the irradiated specimen as well, care must be exercised in such experiments on annealing studies so as not to attribute a changing I_2 to the wrong experimental parameter.

(h) The model proposed by Nieminen et al. does not fit the data for high-fluence neutron-irradiated molybdenum as well as the one described herein where microstructural trapping of positrons is shown to be very important.

(i) The energy for detrapping of positrons from shallow microstructural traps as determined over the temperature range -263 to 100°C is $E_a = 0.1 \pm 0.05$ eV in agreement with the model proposed by Smedskjaer et al. (30).

Acknowledgments

We would like to thank A. N. Goland for his interest and suggestions throughout the course of this work. We appreciate the critical reading of the manuscript by I. K. MacKenzie and J. O. Steigler.

References

1. R. W. Siegel, "The Characterization of Defects in Metals by Positron Annihilation Spectroscopy," this proceedings.

2. K. Petersen, M. Knudsen, and R. M. J. Cotterill, "Changes in the Positron Annihilation Characteristics in Molybdenum Induced by Neutron Irradiation," <u>Phil. Mag.</u> <u>32</u>(3) (1975) pp. 417-426, and references therein.

3. K. Hinode, S. Tanigawa, H. Kumakura, M. Doyama, and K. Shiraishi, "Positron Lifetime Study of Neutron-irradiated Molybdenum," <u>J. Phys. Soc. of Japan</u> <u>45</u> (6) (1978) pp. 1858-1866.

4. M. Eldrup, O. E. Mogensen, and J. H. Evans, "A Positron Annihilation study of the Annealing of Electron Irradiated Molybdenum," J. Phys. F: Metal Physics 6 (4) (1976) pp. 499-521.

5. M. Eldrup, B. T. A. McKee, and A. T. Stewart, "A Positron Trapping Study of Molybdenum Irradiated with 2, 10, and 50 MeV electrons," J. Phys. F: Metal Physics 9 (4) (1979) pp. 637-644.

6. W. Triftshäuser, J. D. McGervey, and R. W. Hendricks, "Positron-Annihilation Studies of Voids in Neutron-irradiated Aluminum Single Crystals," Phys. Rev. B 9 (8) (1974) pp. 3321-3324.

7. K. Petersen, N. Thrane, G. Trumpy, and R. W. Hendricks, "Positron Annihilation of Voids in a Neutron Irradiated Aluminum Single Crystal," Appl. Phys. 10 (1976) pp. 1-6.

8. N. Thrane and J. H. Evans, "The Effect of Impurities on the Lifetime of Positrons in Voids in Molybdenum," Appl. Phys. 12 (1977) pp. 183-185.

9. J. E. Epperson, R. W. Hendricks, and K. Farrell, "Studies of Voids in Neutron-irradiated Al Single Crystals. I. Small-angle X-ray Scattering and Transmission Electron Microscopy," Phil. Mag. 30 (1974) pp. 803-810.

10. L. J. Cheng, P. Sen, I. K. MacKenzie, and H. E. Kissinger, "Positron Lifetime Study of Molybdenum Neutron-irradiated at Various Temperatures," Sol. St. Comm. 20 (1976) pp. 953-955.

11. R. Grynszpan, K. G. Lynn, C. L. Snead, Jr., A. N. Goland, and F. W. Wiffen, "Positron-annihilation Investigation of High-temperature Neutron-irradiated Molybdenum," Phys. Lett. 62A (6) (1977) pp. 459-462.

12. J. Bentley and F. W. Wiffen, "Neutron-irradiated Effects in Molybdenum and Molybdenum Alloys," Proc. of the Second ANS Topical Meeting on the Technology of Controlled Nuclear Fusion, Sept. 21-23, 1976, Richland, Washington, pp. 209-218.

13. D. M. Maher, B. L. Eyre, and A. F. Bartlett, "Neutron-irradiation Damage in Molybdenum, Part IV. A Quantitative Correlation between Irradiated and Irradiated-annealed Structures," Phil. Mag. 24 (190) (1971) pp. 745-765.

14. C. H. Hodges and M. J. Stott, "Positrons in Metals with Voids, Vacancies, and Surfaces," Sol. St. Comm. 12 (1973) pp. 1153-1156.

15. P. Jena, A. K. Gupta, and K. S. Singwi, "Positron Annihilation in Small Metal Voids," Sol. St. Comm. 21 (1977) pp. 293-296.

16. N. Thrane, K. Petersen, and J. H. Evans, "The Relationship between Void Size and Positron Lifetime in Neutron Irradiated Molybdenum," Appl. Phys. 12 (1977) pp. 187-189.

17. P. Hautojärvi, J. Heiniö, M. Manninen, and R. Nieminen, "The Effect of Microvoid Size on Positron Annihilation Characteristics and Residual Resistivity in Metals," Phil. Mag. 35 (4) (1977) pp. 973-981.

18. F. A. Smidt, Jr. and J. R. Reed, Controlled Thermonuclear Reactor Materials Program Annual Progress Report 1975 (NRL Memorandum Report) 3293 (1976) pp. 28-39.

19. (a) M. Eldrup, Y. M. Huang, and B. T. McKee, Appl. Phys. 15 (1978) pp. 65-71.
(b) M. J. Fluss, L. C. Smedskjaer, M. K. Chason, D. G. Legnini, and R. W. Siegel, Phys. Rev. B 17 (1978) pp. 3444-3455.

20. K. Hinode, S. Tanigawa, M. Doyama, and K. Shiraishi, "A Study of the Annealing Behavior of High-temperature Neutron-irradiated Molybdenum by Means of Positron Lifetime," J. Nucl. Mat. 66 (1977) pp. 212-214.

21. T. Hall, A. N. Goland, and C. L. Snead, Jr., "Applications of Positron-Lifetime Measurements to the Study of Defects in Metals," Phys. Rev. B 10 (8) (1974) pp. 3062-3074.

22. W. Triftshäuser, "Positron Trapping in Solid and Liquid Metals," Phys. Rev. B 12 (11) (1975) pp. 4634-4639.

23. S. Mantl and W. Triftshäuser, "Defect Annealing Studies on Metals by Positron Annihilation and Electrical Resistivity," Phys. Rev. B 17 (1978) pp. 1645-1651.

24. J. Bentley and F. W. Wiffen, unpublished results presented at the AIME Annual Meeting in Denver, CO, Feb. 26 - Mar. 2, 1978.

25. J. Cornelis, P. DeMeester, L. Stals, and J. Nihoul, "The Influence of Dislocations on Neutron-irradiated Damage in Molybdenum," Phys. Stat. Sol. (a) 18 (1973) pp. 515-522.

26. J. L. Brimhall, B. Mastel, and T. K. Bierlein, "Thermal Stability of Radiation Produced Defects in Molybdenum," Acta Met. 16 (1968) pp. 781-788.

27. B. T. A. McKee, S. Saimoto, A. T. Stewart, and M. J. Stott, "Positron Trapping at Dislocations in Copper," Can. J. Phys. 52 (1974) pp. 759-765.

28. Private communication, R. M. J. Cotterill, 1974.

29. I. K. MacKenzie, "Comments on 'Evidence for Temperature Dependence of Positron Trapping Rate in Plastically Deformed Copper'" by P. Rice-Evans, T. Hlaing, and I. Chaglar, Phys. Rev. B 16 (1977) pp. 4705-4706.

30. L. C. Smedskjaer, M. Manninen, and M. J. Fluss, "Positron Annihilation as a Technique for Dislocation Line Studies," this proceedings, and J. Phys. F: Metal Physics (to be published).

31. R. M. Nieminen, J. Laakkonen, P. Hautojärvi, and A. Vehanen, "Temperature Dependence of Positron Trapping at Voids in Metals," Phys. Rev. B 19 (1979) pp. 1397-1402.

POSITRON LIFETIME AND DOPPLER BROADENING TECHNIQUES

APPLIED TO IRRADIATION DAMAGED SILVER

Richard H. Howell

Lawrence Livermore Laboratory
Livermore, California 94550

Positron lifetime and Doppler broadening measurements have been used to study defect production resulting from room temperature irradiation of pure silver by D-T fusion neutrons or energetic protons. Use of the positron annihilation analysis has established that the surviving defects from both irradiations have the same dose dependence and that defect concentration can be quantitatively measured and compared to damage models. The relative merit of the lifetime and Doppler broadening measurements in arriving at these conclusions will be discussed in this report along with some practical aspects of the measurements. In the proton damaged samples the trapping rate approaches saturation, a circumstance which could be misinterpreted if a less extensive data set containing only Doppler broadening data were available. Some remarks about the analysis of positron data and general conclusions about the defect structure will be given.

Introduction

Positron measurements of the Doppler broadening, angular correlation or lifetime of positrons in pure metals and other solids have become increasingly popular for studying defects in those materials (1). The following discussion is intended to examine the advantages and limitations of one positron technique as compared to another. Since the initial observation that positrons may trap at some defect sites in metallic systems, the use of positron analysis to probe defect characteristics has become more sophisticated. It is now possible to identify some types of defect structure and to observe changes in the structure in several systems. In well characterized systems quantative statements can be made about the relative and, in some cases, the absolute concentration of defects in the sample using positron annihilation techniques.

The positron annihilation measurements can be separated into electron momentum sensitive or electron density sensitive experiments. Electron-positron momentum spectra are measured by Doppler broadening and the angular correlation measurement. The local electron-positron density is measured in a positron lifetime experiment. Each of these approaches has unique characteristics and requirements which determine the ease with which quantative measurements can be made or defect identification performed. A detailed comparison of the Doppler broadening measurement and the lifetime measurement is given below. The comparison is made to illustrate the strengths of each technique in following qualitative changes in otherwise uncharacterized samples, in identifying defect types and changes in defect type, and in supplying unambiguous data on the concentration of defects in the samples. To aid in this discussion, data from a recently completed experiment with irradiated silver will be used.

Basic Considerations

The basic physics of the electron-positron momentum and lifetime measurements has been discussed elsewhere (2). Some practical aspects of the techniques that affect the choice of the experiment are reviewed here. The aspects include convenience in developing the apparatus, time required to obtain a useful spectrum, difficulty in reducing the spectrum to commonly used parameters, and the constraints placed or sample geometry during measurement.

Experimental Techniques

The measurements of electron-positron momentum, the angular correlation measurement and the Doppler broadening measurement, share many characteristics in the context of this discussion. In present studies of defects in materials, the data obtained are similar. Since the Doppler broadening measurement can be performed with a system comprised of finished commerical components and the angular correlation apparatus must be built, the following discussion will concentrate on the Doppler broadening method. The Doppler broadening measurement is performed with an electronically stablized, high resolution lithium-drifted germanium detector. Since the measurement is of the energy of a single annihilation gamma ray there are few constraints on system geometry. The positrons may be derived from any positron emitter and may be transported over some distance to the sample, which may be a single piece of material. The detector must view the sample but large detector-sample distances can be compensated with little penalty by increasing the source strength. The time needed to obtain a useful spectrum is limited by the acceptable counting rate in the Ge(Li) detector and can be less than one hour.

The shape of the energy spectrum detected in the Doppler broadening experiment reflects the momentum distribution of the electrons near the positron. Data from separate samples are compared either by comparing the total shape of the broadening peak or by the reduction of the spectrum to one of a set of commonly used parameters: S, the ratio of peak intensity to total intensity, W, the ratio of intensity in the wings to total intensity, and R the ratio of the change in S to the change in W between two samples (3). The S and W parameters reflect low and high momentum components of the electron distribution respectively. A change in defect concentration results in changes in S which are proportional to changes in W. A change in defect type may not affect S and W proportionately. The R parameter can typify a defect type and can be compared among several samples in a set to investigate changes in defect structure. This comparison requires greater statistical accuracy as the R parameter is obtained from the difference of similar quantities. Thus longer data collection times, on the order of several hours, are required to obtain useful values of the R parameter.

The positron lifetime experiment is performed with a fast coincidence system. The detector materials and electronics components can be obtained commercially but the detectors and in some systems the fast discriminators must be assembled . The measurement is of the time interval between the arrival of a gamma ray which signals the production of a positron and the arrival of an annihilation gamma ray. There are only a few radioactive sources that produce a gamma ray during positron decay and can be used in the lifetime measurement. The most commonly used is ^{22}Na. Since the measurement requires a coincidence between gamma ray detection in two detectors the sample-detector geometry is restricted. The detectors must be close to the sample for high counting rates and unfavorable geometric conditions can not be compensated by increased source strength without a penalty of higher backgrounds. The source sample geometry is also restricted by the requirement that the positron emitter be inside the sample. This leads to various source-sample sandwiches and the necessity of preparing sample material for both sides of the source. The data rates are limited by acceptable background levels and the electronic stability of the fast electronics. The time required for obtaining a useful spectrum is often several hours.

The shape of the lifetime distribution obtained in the lifetime measurement is determined by the system resolution and the annihilation rates for the positrons. The reduction of these data to characteristic parameters is done by computer fitting of the spectra to models of the shape containing those parameters plus parameters describing the system. Often several lifetime components can be obtained from a single spectrum. These can then identify defect types in the trapping model discussed below.

Trapping Model

The basis for the analysis of defects by positron annihilation is the trapping model (4). In this model the positron is assumed to annihilate either in a region of the material which can be considered defect-free or in the defect trap to which it is bound. Therefore a measurement of positron annihilation characteristics in a material that contains defects has components characteristic of both the trapped and untrapped positrons. In a general sense the electron density near a trapped positron is lower and the proportion of low momentum electrons is higher. Thus there exists the potential for detecting defect concentration and type in both momentum and lifetime measurements.

The simplest application of the trapping model to the results of a positron measurement is made when a parameter of the measurement is thought to vary in a linear fashion with the concentration of traps. In this case the trapping rate for the positrons, K, can be written.

$$\frac{K(C)}{\lambda_f} = \frac{F(0) - F(C)}{F(C) - F(Sat)} \tag{1}$$

where λ_f is the annihilation rate of the positrons in defect free material and F(x) is a parameter extracted from positron data from samples which are defect free, (x = 0), at defect concentration C, (x = C), and at sufficiently high defect concentration that all of the positrons are trapped at the time of annihilation, a condition that we call positron trapping saturation, (x = Sat). Examples of F defined above include the average lifetime, and S and W parameters from Doppler broadening and the peak counting rate in the angular correlation experiment.

The trapping rate of the positrons depends on the defect concentration and on the defect type with different defects having different rates for the same concentration. In quantative analysis of positron data it is necessary either to reduce the data from each spectrum to values for the individual trapping rates or to fit the unreduced data set with a function based on a model of the performance of the trapping rate for which the trapping rate variation is known. The second approach has been used sucessfully in studing the equilibrium temperature dependence of single vacancies in a number of metals (5). In cases for which the dependence of the trapping rate on experimental parameters is not well known, the results of equation 1 must be applied.

The lifetime spectra can often be reduced to parameters containing more detail than the single lifetime. Several lifetime components are often identified in a lifetime spectrum. The existence of more than one lifetime component does not require that the trapping model be applied; however in defect studies where the trapping model is applicable, it is possible to relate the parameters of the trapping model to the results of a multilifetime analysis through the solution to a set of rate equations derived from the trapping model. The solution to the trapping model rate equations for n(t), the number of positrons at time t, is given by (5)

$$n(t) = n_0 \left[1 - \sum_i \frac{K_i}{\lambda_f - \lambda_i + e} \right] \exp - (\lambda_f + e)t$$

$$+ \sum_i \frac{n_0 K_i}{\lambda_f - \lambda_i + e} \exp - \lambda_i t \tag{2}$$

$$\text{where } e = \sum_i K_i$$

where λ_f and λ_i are the annihilation rates for the free and trapped positrons respectively and the K_i are the trapping rates into each of the i traps. There are two methods of applying this result to the analysis of a positron lifetime spectrum. A two or more lifetime model can be fit to the spectrum and the results can be related algebracially to equation 2. A second method is to fit the form of equation 2 directly to the data (6). The second method is preferable in those situations where the trapping model has been established since physically reasonable constraints on the free lifetime and trap lifetime can be applied in the fit. In practice this equation must be folded with the resolution of the lifetime coincidence apparatus and then fit to the data. Differences in the description of the resolution function and fitting to the data represent a source of small systematic differences among researchers.

Since the trapping model can be directly applied to each spectrum it is possible, in principle ,to obtain values for all of the annihilation rates and trapping rates from the analysis of each spectrum. This is true even if there are several traps with separate lifetimes. However,in practice sometimes it is difficult to obtain consistent values for several trap lifetimes when there are multiple traps. It is sometimes necessary to constrain the value of the free lifetime to that measured in a defect free sample or to constrain the value of one or more of the trap lifetimes to an average value. In the best cases it is possible to determine all of the parameters from one spectrum without applying any constraints. The value of a trap lifetime is directly related to the physical condition of the defect and its local surrondings. Typical values for the positron lifetimes of some well defined defects have been identified experimentally in several metals and calculated theoretically (6). Trap lifetimes for the defects so identified include vacancies ($\sim$200 psec), voids ($\sim$500 psec) and dislocations ($\sim$200 psec) (5). Thus the positron lifetime is a sensitive parameter,with a meaningful value,which can be compared with theory and among samples of the same material containing defects of varying concentration.

The corresponding method for applying the trapping model so directly to an individual Doppler broadening or angular correlation spectrum is not well developed and does not relate as directly to the positron trapping and annihilation rates. Because of this difference characterization of the traps in a sample can sometimes be done on a single sample measurement in the lifetime experiment, however a set of samples is allways necessary if trap characterization is to be obtained from the R parameter in the Doppler broadening experiment. The Doppler broadening measurement is simpler and faster and so can follow qualitative changes in a sample set with less effort on the part of the experimenter. However, to obtain quantative results describing the changes in a sample material requires that the full range from very low defect concentration to saturation with all of the positrons trapped be represented in the sample set. If there are multiple defects or if positron trapping-saturation cannot be obtained, then quantative data cannot be simply obtained in the Doppler broadening measurement. These limitations do not necessarily apply to the lifetime measurement which can provide a quantative value of the trapping rate from the fit of each spectrum.

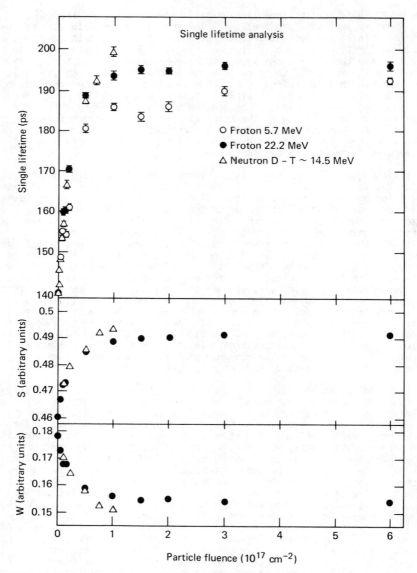

Fig. 1. - Positron parameter for silver irradiated by three sources for samples of different doses. The lifetime values are the result of a single lifetime fit to the lifetime spectrum. The definitions of the S and W parameters are given in the text. The data at the highest doses appear to have reached saturation; however, the saturation values are different for each of the sources suggesting that the effect is not positron trapping saturation.

Silver Experiment

The comparison between Doppler broadening and lifetime measurements can be illustrated by some of the data from a recent study of pure silver irradiated by D-T fusion neutrons or protons in the energy range between 5 and 22.25 MeV or deformed by cold rolling. The samples were analyzed by both positron lifetime and Doppler broadening measurements as a function of radiation dose or deformation. The data were then reduced so that defect type and concentration could be compared among the samples irradiated with different energies or sources. A full description of the results of this experiment is found in (7) and a similar experiment on copper is described in (8,9).

The damage produced by irradiation of silver in this experiment is expected to grow in concentration with dose and to be comprised of extended defects,some of which can be observed in transmission electron microscopy(10). The positron sensitivity to these extended defects includes vacancy loops and dislocation loops but not interstitial clusters or other defects which have a large positive charge. If individual vacancies survived at room temperatures the positron measurements would be sensitive to them as well.

Qualitative Analysis

The data obtained as a function of dose for irradiated silver are shown in Fig. 1. Both the S and W parameters from Doppler broadening measurements and the value obtained by fitting a single lifetime to the lifetime spectra are given. The data collection times were comparable for the two measurements. The characteristics of this data seem similar to many other data sets obtained as the concentration of defects change from a very low value to one sufficiently high that all the positrons annihilate in the trapped state. If the Doppler broadening data were unsupported by the lifetime data the only option available to obtain a quantative relationship of the trapping rate with radiation dose would be to use equation 1 with the unirradiated and highest dose S or W values for F(0) and F(Sat) respectively. In many well characterized systems this approach is sufficient.

Defect Characterization

Since the trap lifetime is one of the values which serve to characterize a defect and the R parameter, while not specific, can indicate changes in defect character, we expect to see changes in the R parameter when the lifetime of the trap is seen to change. In the analysis of the silver data both the trap lifetime and the R parameter values were constent for all of the samples measured regardless of the method of introducing damage into the samples. This may be taken as an indication that the traps detected in these measurements share similar features. The second feature observed in this data is that the trap lifetime values (200 +/- 5 psec) determined independently for each spectrum are uniformly larger than the value of the saturated lifetime found for the same data set. This indicates that the saturation effect observed is not due to an overabundance of traps, trapping all the positrons before annihilation. The saturation effect is due rather to some limit on the number of traps which can be introduced in the sample by irradiation. This interpretation is supported by the comparison of the saturation value of the single lifetime analysis of data from highly deformed samples (202 +/- 4 psec) with the generally lower values found in irradiated samples,all of which had similar trap lifetime values at high defect density.

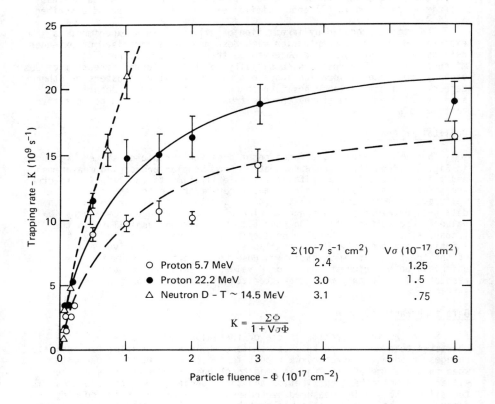

Fig. 2. - Trapping rate values for samples of different doses irradiated by three sources. The curves through each data set represent the prediction of a model based on defect recombination with a dose dependent recombination volume (13). The results for the trapping rate are quantative and can be compared among the irradiation sources.

The value of the trap lifetime is typical of that observed for single vacancies(11). Since the defects are known to be clusters or vacancy loops it is puzzling that they have a single vacancy lifetime. Some explaination of this may be found in (12) where it is suggested that vacancies which are the final positron trap are pinned to the structure of the extended defect.

Quantative Analysis

Since these samples are not prepared by irradiation so that positron trapping saturation is obtained, Doppler broadening data can not be used to obtain a quantative relationship between trapping rate and dose. This limitation does not apply to the lifetime analysis which produces a trapping rate value for each spectrum. The results of the trapping model analysis for these same data are shown in Fig. 2. The saturation in trapping rate can be described by a model (13) which has a defect recombination volume the size of which depends on the defect concentration. From this model and data describing the radiation characteristics, both the recombination volume and the trap production rate can be obtained. Similar data for each radiation source are shown in Fig. 2. and the relative trap production rates can be compared with damage energy values calculated from recoil production data. For the trapping rates and damage energy production to be consistent requires a threshold which excludes low energy recoils from the damage production process.

A set of data obtained by trapping model analysis of lifetime spectra obtained from silver samples that were irradiated by an equal fluence of protons ($2x10^{16}$ p/cm^2) of different energies is presented in Fig. 3. The trapping rates are quantative and the relative values can be used to chose between defect calculations based on different assumptions. A set of damage energy calculations with different thresholds set on the lower energy of the silver recoil energy distribution are compared with the data. The calculations are normalized to the high proton energy data and the differences in shape are used to chose the best threshold value. The best threshold values are found to be in the range of 50 to 100 keV in the recoil spectrum. This is also the threshold for the onset of multiple cascade production.

The volume obtained from the analysis of the data shown in Fig. 2 must be calculated with some value for the recoil production cross section Consistent results between the various irradiation sources is obtained only when a threshold similar to that found in the preceeding paragraph is used. In that case the values of the volume($8-16 \times 10^{-17}$ cm^3) are consistent with published values and are roughly the same magnitude as the volume of the multiple cascade region of the silver recoil.

Absolute values of defect concentration can be derived from the trapping rate data only if the trapping rate per unit defect is known for the trap in this system. Trapping rate per unit defect values are available from deformation studies; however, the defects in the two experiments may not be exactly the same and so the values derived from the deformation studies (14) must be applied with some caution. Using the values obtained for deformed samples the efficiency of survival of the primary defects was calculated to be about 0.1 . This value for defect survival is consistent with other work(10,15) on silver at room temperature.

In this experiment the application of the lifetime measurement produced a more detailed interpretation of the defect production characteristics.This

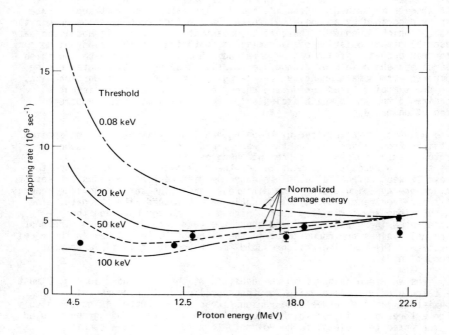

Fig. 3. - Trapping rate values for samples irradiated by protons of different energies to the same dose (2×10^{16} p/cm^2). The curves are predictions of damage energy calculations with various threshold set in the recoil distribution normalized to the trapping rate at high proton energy. The trapping rate values are quantative and can be used to select one of the threshold values from the defect production calculation.

was possible because the analysis of the lifetime data can produce values for the trapping rate, trap lifetime and bulk lifetime for each individual sample. The independent determination of the trap lifetime prevented the values for the parameters S, W and single lifetime determined at saturation from being misinterpreted.

Additional Considerations

There are still limitations in defect identification and quantification. In order to calculate the concentration of a particular defect from the trapping rate obtained in positron measurements requires that the trapping rate for a unit quantity of that defect be known. These values are only available for a few defect types in a few materials. An added complication is that our understanding of the trapping mechanism for extended defects such as dislocations and cascade clusters is still developing(12). Until the basic trapping mechanism is better understood, the comparison of data taken in such unrelated experiments as irradiation damage and deformation experiments in the same sample material will be suspect. The identification of defect types is similarly limited by this lack of understanding. While different trap lifetimes are indicative of different characteristics in a trap, the same lifetime does not indicate that the trap is necessarily the same. In particular it has been noted that vacancies and dislocations often have very similar lifetimes even though they are quite different in structure.

Conclusion

The comparison between the Doppler broadening measurement and the lifetime measurement has been made and an example experiment discussed. It can be seen that, although the Doppler broadening experiment is more easily constructed and faster in producing data, some care must be exercised in the analysis of the data for trapping rates or defect identification in circumstances for which the basic system is not well characterized. The positron lifetime experiment, while less convenient to perform, results in less ambiguous results for defect identification and trapping rate values. In cases where positron trapping saturation is not reached or where there are multiple traps, the Doppler broadening data alone would be difficult or impossible to quantify. However the limitations in geometry and counting rate may preclude the use of the lifetime measurement on a sample set in some experiments. Also there exist situations that have only one defect trap and do reach trapping saturation. For these cases the Doppler technique can result in quantative data but the identification of defect changes may be less sensitive than that obtained from a lifetime study.

"Work performed under the auspices of the U.S. Department of Energy by the Lawrence Livermore Laboratory under contract number W-7405-ENG-48"

471

References

1. Please see the article by R. Siegel in this report for a general description of positron trapping in defects.

2. P. Hautojarvi and A. Vehamen, "Introduction to Positron Annihilation" pp. 1-22 in Positrons in Solids, ed. P. Hautojarvi; Springer-Verlag, New York, 1979.

3. S. Mantl and W. Triftshauser, "Defect Annealing Studies on Metals by Positron Annihilation and Electrical Resistivity Measurements", Phys. Rev. B, 17, 1978, pp. 1645-1652.

4. W. Brant, "Positron Annihilation in Molecular Substances and Ionic Crystals", pp. 155-182 in Positron Annihilation, Proc. intern. Conf. Positron Annihilation, ed. A.T. Stweart, L.O. Roellig; Academic Press, New York, 1967.

5. R. West, "Positron Studies of Lattice Defects in Metals" pp. 89-139 in Positrons in Solids, ed. P. Hautojarvi; Springer-Verlag, New York, 1979.

6. R.M. Nieminer and M.J. Manninen, "Positrons in Imperfect Solids: Theory" pp. 145-192 in Positrons in Solids, ed P. Hautojarvi, Springer-Verlag, New York, 1979, and references therein.

7. R.H. Howell, "A Positron Annihilation Study of Silver Irradiated by Energetic Protons or Neutrons", Phys. Rev. B in press.

8. R.H. Howell, "A Positron Lifetime Study of Irradiation Effects in Copper Irradiated with Energetic Protons" Phys. Rev. B, 18, 1978, pp. 3015-3025.

9. R.H. Howell," A Positron Lifetime Study of Copper Irradiated by D-T Fusion Neutrons", Phys. Rev. B,22,1980, p1722.

10. R.L. Lyles and K.L. Merkle," 14 MeV Neutron Damage in Silver and Gold",Vol 1.,p191,Radiation Effects and Tritium Technology for Fusion Reactors,NTIS,Springfield,VA,1976.

11. P. Hautojarvi,J. Johansson,A. Vehanen,J. Yli-Kauppila,P. Gerard and C. Minier,"Annealing of Vacancies in Electron Irradiated Silver",unpublished.

12. L. Smedskjaer and M. Fluss,in this volume.

13. A. Sosin and W. Bauer, "Atomic Displacement Mechanism in Metals and Semiconductors", pp. 153-321 in Studies in Radiation Efects in Solids, Vol. 3, ed. G.J. Dienes; Gordon and Breach, New York 1969.

14. J. Baram and M. Rosen, "Annihilation of Positrons in F.C.C. Cold-Worked Polycrystals", Phys. Stat. Sol., 16, 1973, pp. 263-272.

15. K.L. Merkle and R.S. Averback,"Energetic Displacement Cascades at Elevated Temperatures",p161,Fundemental Aspects of Radiation Damage in Metals;NTIS,Springfield,VA,1976.

POSITRON ANNIHILATION AS A TECHNIQUE FOR

DISLOCATION LINE STUDIES*

Lars C. Smedskjaer, Matti Manninen†, and Michael J. Fluss

Materials Science Division
Argonne National Laboratory
Argonne, IL 60439

Positron annihilation spectroscopy (PAS) is an established technique for the study of point-defect behavior. Following the initial observation of positron annihilation in voids [Cotterill et al., Nature, 239 (1972) pp 99-101], PAS has also become useful for the study of defect clusters. Additionally, PAS may be a useful technique for the study of dislocations. Recently observed anomalies in the temperature dependence of PAS signals at low temperatures in annealed metals can tentatively be attributed to a positron-dislocation interaction, in which the detailed properties of the dislocation-line structure and its associated defects (e.g., jogs) alter the PAS signal and its temperature dependence. This new interpretation of the positron-dislocation interaction assumes that the positron-dislocation line binding energy is small ($\lesssim$ 0.1 eV). It will be shown that such an assumption may lead to a strong temperature dependence of the positron trapping rate at low temperatures. Since such a temperature dependence is sensitive to the number of jogs per unit length of the dislocation line, PAS may be a new technique for studying dislocation-line structures such as jogs, thereby augmenting current microscopy methods.

---

*Work supported by the U.S. Department of Energy.
†Physics Department, Michigan Technological University, Houghton, Michigan 49931; Present address: NORDITA, Blegdamsvej 17, 2100 Copenhagen Ø, Denmark

Introduction

In recent years positron annihilation has become established as a technique for the study of defects in metals (1-4). Considerable contributions using positron annihilation have been made in the understanding of point defect behavior (e.g., vacancies) as well as in the formation and behavior of more extended defects (e.g., voids). Positron annihilation has also been applied in annealing studies performed either after cold working or irradiation where the latter type of experiment may lead to information about vacancy migration and clustering.

Considerably less is known with respect to positron annihilation in the presence of dislocations, since only a few experiments have been performed under circumstances where the dislocation structure was well characterized. An example is the experiment of Cotterill et al. (5) who studied the effects of dislocation loops in aluminum and another is the investigations in copper by Saimoto et al. (6) and McKee et al. (7).

Dislocations are expected to be variable in their properties (e.g. width of the dislocation core, jog concentration). It is not clear whether the positron can distinguish in some way among dislocations with different properties, since no experiments to elucidate this particular question have been performed. One may actually summarize most of the present knowledge with respect to positron-dislocation interations by stating that when dislocations are present in a sample one observes a long positron lifetime, similar to that in a vacancy, and that the positron trapping rate is proportional to the dislocation density. For dislocation loops in aluminum, Cotterill et al. (5) found a specific trapping rate of ~ 0.06 cm^2/s, while McKee et al. (7) found a rate of 1.5 cm^2/s for dislocations produced by tensile straining of a copper single crystal. Johnson et al. (8) measured the specific trapping rate for dislocations formed by bending of a copper single crystal and arrived at a specific trapping rate of 8 cm^2/s, which is somewhat higher than that of McKee et al. (7). Considering these findings, more insight is needed into the details of the positron-dislocation interaction in anticipation of future applications.

In the present work the positron-dislocation interaction will be discussed from a theoretical viewpoint [see also (9)]. Our considerations have however been precipitated by the observation of "low-temperature effects" first reported by Lichtenberger et al. (10) for several annealed, high-purity metal samples. These effects manifest themselves as an abnormal temperature dependence of the positron parameters (lifetime or lineshape) for temperatures below room temperature. Recently, Smedskjaer et al. (11-13) have shown that this temperature effect is due to positron trapping in extrinsic defects. A more complete account for these effects has recently been given by Smedskjaer (23). Here we will demonstrate how these defects could be dislocations and which dislocation properties might be deduced from the nature of the temperature dependence of the positron signal.

The Dislocation as a Positron Trap

Previous considerations about the positron-dislocation interaction have assumed that the dislocation acts as a cylindrical potential well to the positron [e.g. Arponen et al. (14)]. By making the further assumption that the long lifetime observed in the presence of dislocations was due to annihilation while the positron was capturerd in this well, one deduced the positron-dislocation binding energy. Arponen et al. (14) found, in this way, a binding energy of ~ 3 eV for dislocations in aluminum.

Further calculations by Bergersen and McMullen (15) showed that these two assumptions resulted in a specific trapping rate of ~ 1 cm^2/s due to electron-hole pair generation. The results with respect to the binding energy and the trapping rate rely, however, on the assumption that the long lifetime observed is the lifetime of a positron in the cylindrical well. There is, at present, no experimental evidence which uniquely supports this assumption and it may therefore be justified to consider other possible origins for the long lifetime.

A dislocation should not be described as a simple line-like defect alone, but rather should be considered as containing (or surrounded by) point-like defects along its length (e.g., jogs). Therefore, it is possible that the long positron lifetime is due to these point-like defects rather than the line defect itself. In the present work we shall assume this is the case. The question about the positron-dislocation binding energy therefore is to be reconsidered from this new point of view.

Martin and Paetsch (16) calculated the positron-dislocation binding energy for different types of core configurations in aluminum. In contrast to Arponen et al. (14), they found that the binding energy was always small ($\lesssim 0.1$ eV) depending on the core configuration. In their calculations however, Martin and Paetsch (16) did not include the enhancement of the electron density at the positron's position. This enhancement, in the case of positron-vacancy binding gives a significant contribution to the binding energy. Therefore, one can not conclude from the calculations of Martin and Paetsch that the binding energy is small. However, if it is indeed assumed that the binding energy is small, then one may neglect the enhancement effects and use the calculations of Martin and Paetsch (16) to determine how small the binding energy could be.

Our considerations of the dislocation as a positron trap will be based upon the following two assumptions:

1)   The positron-dislocation binding energy is low ($E_b \lesssim 0.1$ eV) and the positron annihilation signal (lifetime or lineshape) from annihilations in the dislocations is not easily distinguished from that of annihilations taking place in the defect-free material.

2)   The long positron lifetimes observed for samples containing dislocations are due to annihilations in defects associated with the dislocations; additionally, the positron-defect binding energy is high (a few eV).

These assumptions are a reasonable alternative to current interpretations of positron trapping in dislocations. Since, to the authors' knowledge, there are no contradictory experimental results to date, it is useful to investigate the consequences of these assumptions within the context of the trapping model for positrons in defects.

The Trapping Model

Figure 1 shows a graphical representation of the trapping model corresponding to the previous two assumptions. The possible positron states are represented by horizontal lines. Three positron states are shown: 1) the bulk state in which the positron moves almost freely in the defect free part of the lattice, 2) the dislocation line bound state in which the positron is weakly bound to the dislocation line and 3) the strong defect bound state in which the positron is strongly bound to a point defect associated with the dislocation line. In the dislocation

line bound state, the positron is found within one or a few atomic
distances from the line but is otherwise free to move along the line. In
the strong defect bound state, however, the positron is found mainly
within the atomic volume containing the defect. The vertical arrows
connecting the states represent the transition rates, while the vertical
arrows ending with the 2γ symbol represent the decay rates, and the F_x
symbols beneath indicate the lineshape which would be observed as a result
of each of the decays. In the figure, η' is the positron trapping rate
from the defect free material (bulk) directly into the point-like defect
(e.g., jog), κ is the trapping rate from the defect free material into the
dislocation line bound state, δ is the corresponding detrapping rate, η is
the trapping rate from the dislocation line bound state into the point-
like defect, λ_b is the decay rate of the positron in the defect-free
material and also the approximated decay rate for a positron in the
dislocation line bound state, while λ_t is the decay rate of a positron
already trapped in a point-like defect. Note that Fig. 1 takes into
account the detrapping rate δ for the positron-dislocation line trapped
state, which is important, since the binding energy is low. In contrast,
detrapping from the positron-defect trapped state is not important, since
this binding energy is high. The similarity between the annihilation
signal from the bulk and that from the dislocation line is accounted for
in Fig. 1 by assigning the same decay rates and lineshapes to these two
processes. This is, of course, an approximation which may not always be
justified, although this should have little or no consequence in the
current investigation.

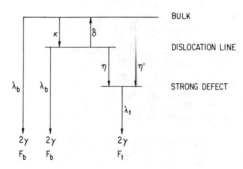

Fig. 1.   Decay scheme for a dislocation as a weak linetrap accompanied by
           stronger binding point-like defects. Note that detrapping from
           the weak linetrap is taken into account.

        Since the following considerations are based on quantum-mechanics, it
is helpful to describe the less correct classical representation of the
present model. In this representation the model describes how a positron
diffusing through an otherwise defect-free lattice can be captured
(trapped) by a dislocation and its associated defects. The positron,
being initially in the defect-free lattice, may be captured either
directly by the point defects associated with the dislocation (rate η') or
by the dislocation line itself (rate κ), or if none of the above, it may
annihilate with a decay rate λ_b. If the positron is directly captured by
one of the point defects associated with the dislocation nothing more will
happen except that the positron annihilates with a decay-rate λ_t. If, on

the other hand, the positron is captured by the dislocation line (rate κ), it can diffuse rapidly along the line until it either encounters one of the point defects (e.g., jog) where it will become trapped (rate η), or it collides with a phonon, which can result in a detrapping (rate δ) from the dislocation line, or, if none of the above, it annihilates with the decay rate $\sim \lambda_b$.

By solving the rate equations corresponding to Fig. 1, one obtains the probability, A_t, that a positron will be trapped by, and annihilated from, the point-defect-like state:

$$A_t = \frac{\eta'(\lambda_b + \delta + \eta) + \eta\kappa}{(\lambda_b + \kappa + \eta')(\lambda_b + \eta) + \delta(\lambda_b + \eta')} \qquad (1)$$

The objective now is to determine the magnitude and temperature dependence of A_t. Since the binding energy of the defect trapped state is high, one may take η and η' as being temperature independent provided that the defect concentration along the dislocation line does not depend on temperature for the temperature region considered. Since the decay rates λ_b and λ_t are only weakly temperature dependent as well, the major part of the temperature dependence of A_t is due to the strong temperature dependences of κ and especially δ.

In the following sections, κ, η, ν' and δ will be calculated. The temperature dependence of A_t will then be illustrated and the experimental consequences will be discussed.

The Positron Trapping Rate at Ideal Dislocations

The ideal dislocation line is assumed to contain no defects and to be infinitely long, leading to a two-dimensional positron-dislocation potential. For low binding energies the dominant trapping mechanism is the emission of an acoustic phonon, as shown by Bergersen and McMullen (15), who calculated the dislocation trapping rate for positron binding energies $E_b > 0.5$ eV and traps of a few Å in diameter. A more complete account of the dependence of the trapping rate on the binding energy, trap diameter and temperature will be given below. The two dimensional potential is a reasonable approximation if the dislocation line is sufficiently long to contain a large number of bound states. This criterion is fulfilled if the length ℓ of the dislocation line satisfies $\ell \gg 2\pi h(2mE_b)^{-1/2}$, where $-E_b$ is the energy of the lowest positron state in the trap and m is the mass of the positron.

In the core region, the positron potential is very difficult to estimate. Rather than studying any speicific form of the trapping potential, we consider a class of initial and final positron states. The initial states are approximated by plane wave $\psi_i = L^{-3/2} \exp(-i\vec{k}\cdot\vec{r})$, while the final states are approximated by radially symmetric wavefunctions

$$\psi_f = \sigma^{-1}(\pi L)^{-1/2} \exp[-(x^2 + y^2)/2\sigma^2]\exp(-iK_z z).$$

In the above expressions L^3 is the normalization volume, σ is the radial spread of the final state and K_z is the wavevector component along the dislocation line. Thus,

$$|K_z| < (2mE_b/\hbar^2)^{1/2}$$

thereby ensuring that ψ_f represents a bound state.

The transition rate is obtained from

$$\kappa = \frac{2\pi}{\hbar} \sum_i P_i |M_{if}|^2 \delta(\epsilon_i - \epsilon_f) \tag{2}$$

where P_i is the probability of finding the positron in the initial state i, M_{if} is the matrix element between initial and final states, and the summation is extended over all possible initial and final states fulfilling both energy and momentum conservation. The matrix element M_{if} is expressed as

$$M_{if} = \int d\bar{r} \, e^{-i\bar{q}\cdot\bar{r}} \, \psi_i^*(\bar{r}) \, \psi_f(\bar{r}) \left[\frac{E_d^2 \hbar q}{2\rho c_s L^3} \right]^{1/2} (n_{\bar{q}} + 1)^{1/2} \tag{3}$$

where E_d is the deformation-potential constant [Bergersen et al. (17)], ρ is the density of the metal, c_s is the velocity of sound, $\bar{q}$ is the wavevector of the emitted phonon ($|\bar{q}| < q_D$ where q_D is the Debye wavevector), and $n_{\bar{q}}$ is the phonon occupation number. Thus, the following calculations contain the inherent approximations of the Debye model. The specific trapping rate $\nu = \kappa/D$, where D is the dislocation density, has been calculated from Eq. 2 and Eq. 3 for aluminum at T = 0 K using the value for E_d obtained by Bergersen et al. (17). The result is shown in Fig. 2 as a function of σ for different choices of E_b = 1.0, 0.1, 0.03 and 0.01 eV.

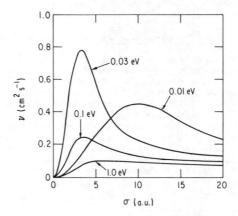

Fig. 2.   Specific positron trapping rate ν into an ideal dislocation as function of the radial spread σ for different values of the positron-dislocation line binding energy E_b at T = 0 K (see text).

For high binding energies (E_b) and small radial spreads (σ) the results are in agreement with those of Bergersen and McMullen (15). Figure 2 also shows that the trapping rate depends strongly on both E_b and σ, and that surprisingly high trapping rates may occur for low values of E_b. It is emphasized that the phonon-induced trapping rates for low binding energies are similar to the electron-hole pair-induced trapping rates obtained for high binding energies by Bergersen and McMullen (15).

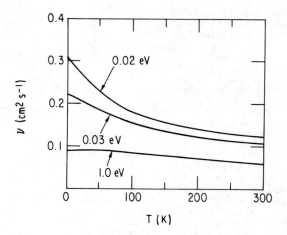

Fig. 3. Specific positron trapping rate ν into an ideal dislocation as function of temperature T for E_b = 0.02 eV, E_b = 0.03 eV and E_b =1 eV, while σ = 10 a.u. in all cases

In Fig. 3 the calculated specific trapping rate ν is shown as a function of temperature for three different binding energies. The figure demonstrates that at low binding energies the trapping rate decreases rather rapidly as the temperature increases, whereas at higher binding energies the trapping rate is almost constant.

Owing to its strong dependence on the radial spread σ and binding energy E_b, the specific trapping rate ν would be expected to be sensitive to whether a Gaussian or a more realistic wavefunction is used. In addition, ν depends quite strongly on the choice of the parameters E_d, c_s, and the Debye temperature θ_D. Further, the Debye approximation by itself may be critical for the estimation of ν, especially when $E_b \sim k_B \theta_D$. Thus, the present calculation can only serve as an estimate of the magnitude of ν.

Trapping into Strong Traps on the Dislocation Line

The purpose of the present section is to estimate the trapping rates η and η' assuming that the specific trapping rate of the delocalized (free) positron at the strong trap is similar to that at a vacancy. The trapping rate of a positron at a vacancy depends only weakly on the positron wavelength, if the wavelength is larger than the typical dimensions of the vacancy. The trapping rate is therefore essentially proportional to the probability amplitude of the free positron at the position of the trap. This principle may be applied to estimate the trapping rate for the case where the positron is localized on the dislocation line, since the radial extent of the dislocation line-trapped positron is expected to be larger than the radial extent of a positron in a strong trap. In this fashion one obtains

$$\eta \overset{\sim}{=} \mu_v \frac{\Omega}{\pi \sigma^2} \, a$$

$$\eta' = \mu_v \, \Omega \, D \, a$$

(4)

479

where μ_v is the specific trapping rate of the free positron at a vacancy ($\sim 10^{15} s^{-1}$ atom^{-1}), Ω is the atomic volume, a is the number of strong defects per unit dislocation line length and D is the dislocation density. From Eq. 4 it is seen that for wide ranges of a ($10^{-1} > a > 10^{-4} \unicode{0x212B}^{-1}$) and D ($10^{10}$ cm$^{-2} > D > 0$) the rates η and η' fulfill $\eta \gg \lambda_b$ and $\eta' \ll \lambda_b$, where typical values for λ_b are $\lesssim 10^{10} s^{-1}$. Thus, for a $>10^{-4} \unicode{0x212B}^{-1}$, strong traps on or along the dislocation line have a very high trapping rate for positrons already localized at the dislocation. However, this also means that the decay scheme shown in Fig. 1 is inadequate for high densities of strong defects along the dislocation line, since η may become so large that the dislocation line bound state ψ_f can no longer be considered as a realistic eigenstate. Within the context of the present picture of a dislocation as a line with associated defects, it is anticipated that a dislocation line possessing a high defect density should be described as a continuum of strongly bound states thereby resulting in a description of the positron states in a dislocation similar to that used by Bergesen and McMullen (15). In the present work, however, only lines with low defect densities will be considered, and we shall thus approximate the dislocation as if it were a weakly binding line trap accompanied by a few strongly binding trapping centers. This approximation will apply better to situations where long and straight dislocations are present (e.g., in annealed samples) than to conditions where the dislocation network is tangled (e.g., in deformed samples).

Detrapping of the Positron

By defining the function

$$S(T) = \frac{\kappa}{\kappa + \delta} \qquad (5)$$

It is seen that S(T) would be the probability of finding the positron in the dislocation line bound state if λ_b and η were both zero. Thus S(T) can be calculated from thermodynamics if the phonon system is in thermal equilibrium at the time of the trapping of the positron. In the following, the actual continuous spectrum of bound states will be approximated with an M-fold degenerate bound state of energy $-E_b$ and S(T) is then obtained as

$$S(T) \simeq [1 + \frac{m}{4\hbar^2} \pi^{-1/2} D^{-1} E_b^{-1/2} (k_B T)^{3/2} \exp\{-E_B/(k_B T)\}]^{-1} \qquad (6)$$

The Temperature Dependence of the Trapping Probability $A_t(T)$

It was previously shown that $\eta \gg \lambda_b$ and $\eta' \ll \lambda_b$ for a wide range of densities a and D. We therefore make the approximations that λ_b can be neglected when compared to η and that $\eta' \simeq 0$. With these approximations, and the detrapping described by the function S(T), the probability A_t that the positron will annihilate in a strong trap can be written as

$$A_t \simeq \frac{\eta \kappa S(T)}{\eta(\lambda_b + \kappa)S(T) + \kappa \lambda_b} \qquad (7)$$

In Fig. 4 A_t is shown as a function of the temperature neglecting the temperature dependences of η, κ, and λ_b, which are weak compared to that of S(T). In the figure, A_t is shown for $\kappa = 0.05 \lambda_b$, $E_b = 0.03$ eV, $D = 10^8$ cm^{-2} and for three different values of η. It is seen that the transition-temperature region for $A_t(T)$ depends on η, such that increasing values of η result in higher transition temperatures. It should also be noted that increasing values of η result in a weaker temperature dependence of $A_t(T)$. Thus, in conclusion, $A_t(T)$ depends on both E_b and η, such that high values of E_b and η result in a less pronounced temperature

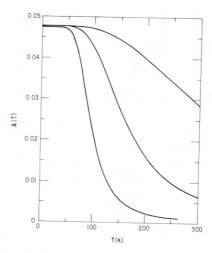

Fig. 4. The probability $A_t(T)$ (see text) as function of temperature T
for different values of $\eta = 10\,\lambda_b$, $100\,\lambda_b$, $1000\,\lambda_b$ (left to
right) while $E_b = 0.03$ eV, $D = 10^8$ cm^{-2} and $\kappa = 0.05\,\lambda_b$. The
temperature dependences of κ, λ_b and η have been neglected.

dependence at higher temperatures, while small values of E_b and η give
rise to a more dramatic temperature dependence at lower temperatures.

Discussion

The most important feature of the present model is the phenomenon of
detrapping of positrons. The effects of detrapping may be observed below
room temperature for the present choices of binding energies and
dislocation densities.

Figure 5 illustrates how a mean lifetime, $\bar{\tau}$, vs temperature, T, for
aluminum can be affected by the presence of a dislocation line state with
a low binding energy. For this illustration the positron bulk lifetime τ_b
has been taken to be proportional to the thermal expansion of the lattice
such that $\tau_b = 168$ ps and $d\tau_b/dT = 5$ ps/200 K for $T = 300$ K in agreement
with experimental results for aluminum [Fluss et al. (18)]. The positron
lifetime in the strong traps was taken to be 244 ps throughout the
temperature range $0 < T < 500$ K and it was further assumed that a possible
temperature dependence of the trapping rate η could be disregarded.
Clearly, this assumption is only valid at low temperatures, since at
higher temperatures migration and subsequent annihilation of the strong
traps (e.g., jogs) can be anticipated. Selection of the dislocation line
trapping rate, κ, was guided by experimental results [Smedskjaer et al.
(11,12)] rather than by theory. For the current example,
$\kappa = \kappa_o\, e^{-\gamma T}$, where $\kappa_o = 5 \times 10^{-2}\,\lambda_b(T = 0$ K) and $\gamma = 2.3 \times 10^{-3}$K^{-1}, thus
resulting in a temperature dependence similar to that shown in Fig. 3 for
$E_b \sim 0.03$ eV and $D = 10^8$ cm^{-2}. The present choice of κ_o and D corresponds
to $\nu \sim 3$ cm^2/s, which compares reasonably well with the present theory
considering its inherent uncertainties. Positron trapping in dislocations
for supposedly well annealed aluminum has previously been suggested [Hall
et al. (19)]. For a lifetime experiment aimed at deducing the vacancy
formation enthalpy it was found that the data were consistent with the
presence of a rather high dislocation density ($\sim 10^9$ cm^{-2}).

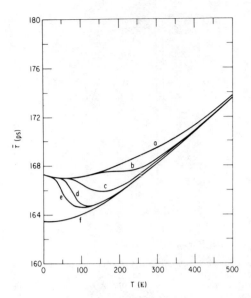

Fig. 5.   The calculated mean-lifetime $\bar{\tau}$ for aluminum for selected values
of E_b and η as function of temperature T.   a) $E_b = 0.1$ eV,
$\eta = 10^3 \lambda_b$, b) $E_b = 0.1$ eV $\eta = 10^2 \lambda_b$, c) $E_b = 0.03$ eV $\eta = 10^3$
λ_b, d) $E_b = 0.03$ eV $\eta = 10^2 \lambda_b$, e) $E_b = 0.01$ eV $\eta = 10^3 \lambda_b$, (f)
Bulk lifetime $\lambda_b(T)$ in aluminum as obtained from thermal
expansion (see text). $\kappa = \kappa_o \exp(-\gamma T)$ and $D = 10^8$ cm^{-2} for all
curves shown.

The mean lifetime vs temperature curves shown in Fig. 5 demonstrate
the effects of varying E_b and η, while keeping all other parameters
constant.   The curves shown can be divided into three main temperature
regions: a low-temperature region where the temperature dependence is
essentially due to $\kappa(T)$, an intermediate region where detrapping dominates
and a high temperature region where the detrapping is almost complete,
thereby resulting in a temperature dependence equal to that of the bulk
lifetime.   The three temperature regions are easily recognizable in curve
"a" ($E_b = 0.1$ eV, $\eta/\lambda_b = 10^3$) and for this particular case they should be
taken as T $\lesssim$ 130 K, 130 $\lesssim$ T $\lesssim$ 400 K and T $\gtrsim$ 400 K.   The three temperature
regions, however, may not always be clearly revealed, as seen from curve
"b" ($E_b = 0.1$ eV, $\eta/\lambda_b = 10^2$), which is essentially flat for T $\lesssim$ 250 K.
Figure 5 also shows that small values of E_b and η both tend to result in
large negative temperature dependences of $\bar{\tau}$ in the respective
intermediate-temperature regions, while larger values of E_b and η both
result in an apparently weaker temperature dependence.

It should be noted that the present model predicts strong temperature
dependences at low temperature only if η is sufficiently small.   For high
values of η, which may be found in samples with high dislocation
densities, the model would suggest a weak temperature dependence of the
positron annihilation parameters.

Finally, attention should be drawn to the possible relationship
between the present model and the low-temperature effects recently
reported for some annealed metal samples [e.g., Rice-Evans et al. (20,21)
Herlach et al. (22), Smedskjaer et al. (12)].   These effects manifest

themselves as a "flattening" in the temperature dependence of the positron parameters (e.g., lifetime, lineshape) for temperatures below 200-300 K. A more detailed account for these effects is found in the work by Smedskjaer (23). If in fact these effects are due to dislocation line trapping, a variability among the observations for different samples would be expected, since the details of the dislocation structure will depend upon the details of the history of the sample. The sample dependence of these low-temperature effects in gold has been demonstrated by Smedskjaer et al. (13) by comparison of results from different laboratories.

Ideally, one could support the idea that the positron is trapped first in a shallow trap and then subsequently in a deeper trap by observing both a temperature dependence of the trapped state population and the annihilation characteristics indicative of a deep trapped state. These two related behaviors have in fact been observed in cadmium [Smedskjaer et al. (12)]. With respect to gold, temperature dependent prevacancy trapping has been found. Unfortunately nothing is known about the annihilation characteristics of the trapped state in the case of gold. For annealed copper, prevacancy trapping has been observed as well by Smedskjaer et al. (11), where the annihilation characteristics of the trapped state were found to be similar to those for a vacancy. However, in this case, the temperature dependence of the trapped population was not measured at sufficiently low temperatures to establish it independent of the presence of equilibrium vacancies. Thus, at present, the observations of both the temperature dependence of the population and of the annihilation characteristics has only been made for cadmium, where the results are in qualitative agreement with the present model. With respect to gold and copper, the experimental results do not exclude the possibility of an interpretation similar to that made for cadmium.

The consistancy between the present model and current experimental data does not rule out alternative explanations for the so called low temperature anomalies [e.g. Bergersen and McMullen(15)]. A need for a firm explanation for present observations is incumbent on those involved in the study of defects and electronic structure by positron annihilation. Indeed, the experiments needed to verify or refute the present model are in fact the beginning of a new application of positron annihilation. Experiments where long dislocations can be both produced and characterized could provide the basis for a continued investigation of the proposed positron-dislocation model.

Conclusion

In concluding, one may speculate on the use of positron annihilation as a future tool for complementary studies of the microstructure of dislocations.

The low temperature dependence of the positron signal in the presence of dislocations depends on the structure of the dislocation core and on the point defect density found along the dislocation line. For dislocations with low jog densities, this temperature dependence is mainly due to the competition between the strongly temperature sensitive detrapping process and the trapping process into jogs. Thus at low temperatures, where it may be assumed that the jog density is temperature independent, one may deduce the jog density along the dislocation line from the temperature dependence of the positron annihilation signal. Therefore positron annihilation may serve as a tool complementing high resolution microscopy studies of jogs on dislocations.

One might further, as does Smedskjaer (23), consider whether positrons can be used to determine the jog formation enthalpy. In cases where most of the jogs present along a dislocation line are thermally activated and the jog formation enthalpy is significantly below that of a vacancy (e.g., by a factor of two), the present model predicts a shoulder like temperature dependence of the positron signal at temperatures just below those where positron trapping in thermally generated vacancies is observed. It is perhaps fortuitous that such shoulder-like effects sometimes are observed for supposedly well annealed samples (e.g. Segers et al. (24), Maier et al. (25) Jean et al. (26)), but if they are indeed due to thermally activated jogs, then the positron data should provide an estimate of the jog formation enthalpy.

References

1. "Positrons in Solids", Topics in Current Physics (Springer-Verlag, Berlin, Heidelberg, New York, P. Hautojärvi, ed.) 12 (1979).

2. "Positron Annihilation" Proceedings of the Fifth International Conference on Positron Annihilation, Japan Inst. of Metals, R. R. Hasiguti and K. Fujiwara ed. (1979).

3. M. J. Fluss, R. P. Gupta, L. C. Smedskjaer and R. W. Siegel, "Temperature-Dependent Behavior of Positron Annihilation in Metals" in Positronium and Muonium Chemistry, Advances in Chemistry Series American Chemical Society, Hans J. Ache ed. 175 (1979) pp. 243-270.

4. R. W. Siegel in these Proceedings.

5. R.M.J. Cotterill, K. Petersen, G. Trumpy and J. Träff, J. Phys. F: Metal Phys., 2 (1972) pp. 459-467.

6. S. Saimoto, B.T.A. McKee and A. T. Stewart, Phys. Stat. Sol (a), 21 (1974) pp. 623-626.

7. B.T.A. McKee, S. Saimoto, A. T. Stewart and M. J. Stott, Can. J. Phys., 52 (1974) pp. 759-765.

8. M. L. Johnson, S. F. Saterlie and J. G. Byrne, Met. Trans., 9A (1978) pp. 841-845.

9. L. C. Smedskjaer, M. Manninen and M. J. Fluss, J. Phys. F: Metal Phys., 10 (1980) pp. 2237-2249.

10. C. Lichtenberger, C. W. Schulte and I. K. MacKenzie, Appl. Phys., 6 (1975) pp. 305-307.

11. L. C. Smedskjaer, M. J. Fluss, R. W. Siegel, M. K. Chason and D. G. Legnini, J. Phys. F: Metal Physics, 10 (1980) pp.559-569.

12. L. C. Smedskjaer, D. G. Legnini and R. W. Siegel, J. Phys. F: Metal Physics, 10 (1980) pp. L1-L6.

13. L. C. Smedskjaer, M. J. Fluss, M. K. Chason, D. G. Legnini and R. W. Siegel, J. Phys. F: Metal Physics, 9 (1979) pp. 1815-1820.

14. J. Arponen, P. Hautojärvi, R. Nieminen and E. Pajanne, J. Phys. F: Metal Physics, 3 (1973) pp. 2092-2108.

15. B. Bergersen and T. McMullen, Solid State Comm., 24 (1977) pp. 421-424.

16. J. W. Martin and R. Paetsch, J. Phys. F: Metal Physics, 2 (1972) pp. 997-1008.

17. B. Bergersen, E. Pajanne, P. Kubica, M. J. Stott and C. H. Hodges, Solid St. Comm., 15 (1974) pp. 1377-1380.

18. M. J. Fluss, L. C. Smedskjaer, M. K. Chason, D. G. Legnini and R. W. Siegel, Phys. Rev. B, 17 (1978) pp. 3444-3455.

19. T. M. Hall, A. N. Goland and C. L. Snead Jr., Phys. Rev. B, 10 (1974) pp. 3062-3074.

20. P. Rice-Evans, I. Chaglar and F. El Khangi, Phys. Rev. Lett., 40 (1978) pp. 716-719.

21. P. Rice-Evans, I. Chaglar and F. El Khangi, Phil. Mag. A, 38 (1978) pp. 543-558.

22. D. Herlach, H. Stoll, W. Trost, H. Metz, T. E. Jackmann, K. Maier, H. E. Schaefer and A. Seeger, Appl. Phys., 12 (1977) pp. 59-67.

23. L. C. Smedskjaer "Positron Prevacancy Effects in Pure Annealed Metals", in Proc. Int. School of Physics "Enrico Fermi" Bologna, Italy, July 14-24, 1981, ed. W. Brandt and A. Dupasquier, to be published.

24. D. Segers, L. Dorikens-Vanpraet and M. Dorikens, Appl. Phys., 13 (1977) pp. 51-54.

25. K. Maier, H. Metz, D. Herlach, H. E. Schaefer and A. Seeger, Phys. Rev. Lett., 39 (1977) pp. 484-487.

26. Y. C. Jean, K. G. Lynn and A. N. Goland, Phys. Rev. B, 23 (1981) pp. 5719-5724.

NUCLEAR MICROANALYSIS AS A PROBE OF IMPURITY-DEFECT INTERACTIONS*

M. B. Lewis and K. Farrell

Oak Ridge National Laboratory, Oak Ridge, Tennessee 37830 USA

It is shown that nuclear microanalysis offers unique opportunities for probing impurity migration and impurity-defect interactions in irradiated materials. The principles and practice of the technique are described and the special advantages and limitations are discussed. Procedures are outlined for extracting (1) impurity diffusion coefficients, (2) impurity-defect binding energies, and (3) trap generation coefficient of heavy ions used to create displacement damage. Examples involving the impurities deuterium and helium in austenitic stainless steels and nickel are described. Preliminary values are given for: the bulk diffusion coefficient of deuterium in austenite at 25°C (1.4×10^{-12} cm^2/s); the binding energies of deuterium with point defects in austenite and of helium-3 in nickel (0.33 and ~2.1 eV, respectively); and a room-temperature trap generation coefficient for deuterium in nickel-ion-bombarded austenite of ~15 per incident nickel ion.

---

*Research sponsored by the Division of Materials Sciences, U.S. Department of Energy, under contract No. W-7405-eng-26 with the Union Carbide Corporation.

Introduction

There are at least four ways the use of ion accelerators can broaden the range of techniques for studying displacive radiation damage and can provide information not readily available from more conventional techniques. First it is a means for generating atomic displacements. Second, it is a doping method for implanting controlled quantities of specified foreign atoms to investigate impurity effects. Third, it is a microanalysis technique for detecting impurities within the implanted or ion damaged region, and describing their depth distribution. Fourth, it provides a way of using ion channeling with appropriately oriented crystals to detect the lattice location of impurities and small shifts in the displacement of atoms from their normal lattice sites. This paper addresses the third topic, nuclear microanalysis (NMA), and discusses its advantages and potential use in radiation damage studies.

Special Advantages, Potential Applications, and Limitations of NMA

NMA is a relatively new analytical technique described in detail in several recent books (1-4). Briefly, NMA utilizes nuclear reactions and scattering between a probing ion beam and impurity atoms in a target to confirm the identity, the quantity, and the spatial distribution of implanted or residual foreign elements at or below the surface of the target. It is a nondestructive technique that allows the specimen to be used for further experiments once the impurity profile is characterized. It can detect almost any element, and is especially useful for the lighter, low Z elements such as carbon and the gases hydrogen, helium, oxygen and nitrogen which are difficult to follow nondestructively on a microscopic scale by other analytical techniques and which are intimately involved in the nucleation of damage microstructure (5).

NMA should not be considered merely as a complementary adjunct to ion implantation. It has other, special advantages. NMA can measure changes in the original position and shape of the impurity profile with respect to time, temperature, displacive irradiation, and other impurities. It thereby yields information on kinetics of migration of the impurity and on the effects of other impurities and point defects. Thus one can extract basic properties such as diffusion rates, trapping coefficients, and impurity-defect binding energies that are so essential to a true understanding of the development of radiation damage microstructure. Moreover, because the depth resolution of NMA is higher (5 to 100 nm) than most diffusion measurement techniques, migration data can be measured at much shorter annealing times or at lower temperatures than are needed for the more conventional techniques. This means that real diffusion rates can be determined at relatively low temperatures in the range for reactor operation, typically below $0.5\ T_m$, rather than having to use data extrapolated from measurements made at $T > 0.5\ T_m$. This is particularly important for complex alloys which may be in a single phase condition at high temperatures where conventional diffusion data are gathered, but may become multiphase with an entirely different matrix composition during irradiation. In such cases, use of extrapolated high-temperature diffusion data could be highly misleading.

NMA is most powerful for examining light impurities and impurities that have a higher atomic mass than their host, but it is not limited to such impurities. It can also probe implanted isotopes of impurities or even of the host material which have unique reaction ion energies when probed with suitable light ion beams. Thus the movement of major alloying elements or the host atoms can be studied by adding a relatively small quantity of the appropriate isotope.

NMA is least effective for examining high mass impurities in high mass substrates and is also limited to depths of 1 to 10 µm. The range of atomic mass and depth can be increased using higher energy accelerators, but the subsurface, nondestructive depth resolution is limited by factors (straggling and multiple scattering) which are nearly independent of bombarding energy or instrumentation. Depths greater than 10 µm can be probed by combining NMA with some surface removal technique. NMA does not have high lateral resolution because it is difficult to focus ion beams to small diameters. This follows from the fact that an ion (gas) source image is generally much larger than an electron source image. Although NMA is non-destructive in the sense that the physical shape of the specimen is retained, the implantation process and the probing ion beam do create point defects in the target. This makes the technique nondestructive in a limited sense. For radiation damage studies this displacement damage can be used to advantage.

Principles

When a probing beam of ions enters a target containing a concentration, ρ, of foreign atoms at depth, x, as depicted in Figure 1, the ions interact with the substrate atoms and the foreign atoms and are scattered. The atomic interaction may be elastic, in which case the mass of the ion remains unchanged, or it may be a nuclear reaction with the creation of new elements of masses different from those of the incident ion and target. A fraction of these reaction products will be backscattered out of the target and can be detected and identified with a suitable mass discriminator. From a knowledge of the nuclear reactions, the concentration of foreign atoms can be deduced.

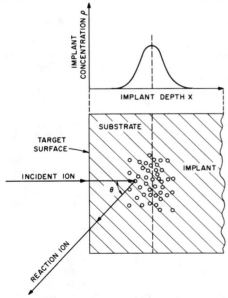

The energy of the reaction ion depends on the path in the target traversed by the incident ion before the nuclear collision, on the nuclear reaction itself, and on the length of the backscattering path of the reaction ion after the collision. Where the incident and collision energies are known, the measured energy of the reaction ion contains information on the depth at which the reaction took place. Thus, the energy spectrum, or yield Y(E), of the backscattered ions is related to a specific density-depth, $\rho(x)$, profile in the substrate.

Equipment and Practice

The equipment and instrumentation consist of a particle accelerator to provide the probing beam, and a target chamber and associated measuring devices and data handling machines, as shown in Figure 2.

Figure 1. Schematic illustration of ion beam and target geometry showing a target previously implanted with impurity.

The beam is collimated to about 2 mm^2 by an aperture outside the evacuated chamber. A second insulated collimator of slightly larger size is positioned at the chamber and is negatively charged to repel stray

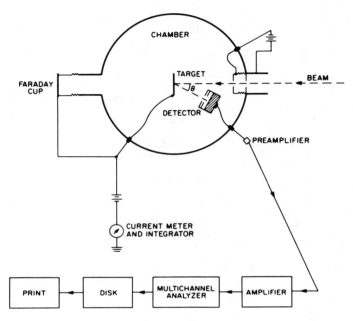

Figure 2. Schematic layout of experimental configuration and data collection.

electrons from entering the chamber along with the positively charged beam. The collimated ion beam subsequently strikes the target and is completely absorbed. A positive bias is applied to the target so that secondary electrons created by the ion beam stopping in the target cannot escape the target surface. This ensures that the integrated current measured between the target and ground is true beam integrated current from which one can deduce the total number of incident ions striking the target during the measurement. The energy and number of reaction ions are measured by a solid state detector with typical energy resolution 10-20 keV. A collimator in front of the detector determines the solid angle between the target and the detector. In some cases, a Mylar window is placed in front of the detector to prevent low energy elastically scattered beam ions from striking the detector. This minimizes pulse pileup spectrum distortion but limits the overall spectral resolution to about 20 keV. Alternatively, one can reject unwanted ions with a magnet or reduce such distortion by an electronic pileup rejector, but the efficiency of the electronic instrument is low for pulses less than about 200 keV.

The pulses created by the reaction ions are amplified and routed to a multichannel analyzer. The pulse height or area of each pulse is proportional to the energy of the reaction ion. The multichannel analyzer converts each pulse to its digital equivalent and stores the results in a histogram of energy "channels." At the end of the experiment, the resulting pulse height spectrum along with the integrated charge of the incident beam is used to deduce the absolute concentration-depth profile(s).

A photograph of the inside of one of our scattering chambers is shown in Figure 3. The beam enters from the lower left of the figure (B) and strikes the target (T). The target is heated from behind by a filament heater (TH) surrounded by a water-cooled shroud to prevent the detector (D) from overheating. The gear mechanism allows the experimenter to move the target about two orthogonal axes and to move the detector about the target itself. The angular precision is typically 0.1 deg. If the target is a high-grade crystal, the orientation of the target relative to the beam can be made such that ion channeling along the crystal axis can also be measured.

Pulse Handling and Data Extraction

The pulse height data output is analyzed and stored with a Tennecomp TP-50 computer system. These data, stored as counts (yield) per channel (energy) must be unfolded to determine the density profiles. This is accomplished by the method of convolution integrals described elsewhere (6). Briefly, one first estimates or calculates an approximate profile, ρ. A computer code then estimates energy losses, straggling, multiple scattering, and other relevant kinematic information. The convolution integral,

$$Y(E)/\Delta E = C\int_S P_i \rho \sigma_R P_r dS \qquad (1)$$

can then be calculated. In Eq. (1) the subscripts i and r refer, respectively, to the incident and reaction ion involved in the nuclear reaction; $Y(E)/\Delta E$ is the yield (counts/channel) of a reaction product, C is the normalizing constant including beam flux on the target, the P's are Gaussian probability functions whose widths are determined by the overall resolution, σ_R is the nuclear cross section, and S represents the space of energy and volume over which the reaction takes place. The stopping powers are taken from ref. (7), and Bragg's rule (1) is used to estimate stopping powers in the target. For non-Gaussian distributions error function curve fitting is employed using appropriate equations for the expected distribution. Data are plotted either as pulse counts versus channel numbers, or as the deduced concentration-depth profile. An example for deuterium-electrocharged 310 stainless steel analyzed 11 min after charging is shown in Figure 4. The lower part of the figure displays the measured reaction spectrum from the D(d,p) reaction,

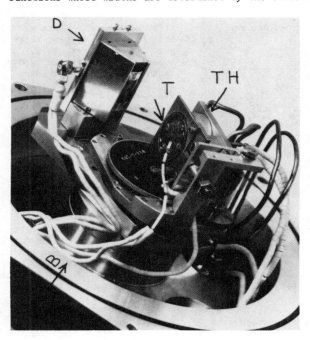

Figure 3. Photograph of the inside of the scattering chamber used in this work. Indicated are beam direction (B), target, (T), target heater (TH), and detector (D).

where p is the reaction ion. The upper part shows the deduced concentration profile for the best-fit curve to the reaction spectrum. Note that the
higher channel numbers at the right-hand side of the spectrum reflect the D
concentration at the surface of the target. The corresponding concentration
profile reveals considerable loss of D from the immediate surface layers,
representing escape of D during the time interval between charging and
analyzing. We will return for further discussion of this figure later.

Resolution and Sensitivity

The method so far described is a quantitative analytical technique which
can be used to probe impurity elements to depths of a few microns with resolutions of 5-100 nm and sensitivities of 10-100 appm. Resolution, range, and
sensitivity can be improved by combining the technique with sputtering,
electropolishing or some other surface removal technique and by measuring
Auger and x-ray yields. But then the nondestructive nature of NMA is lost.

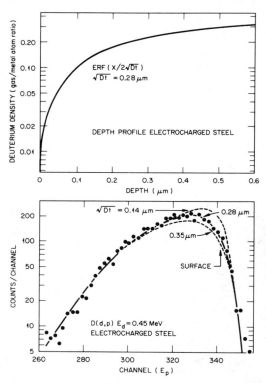

Fig. 4. Data and calculated fits for a
D(d,p)T reaction spectrum from a 0.45-MeV
deuterium beam incident on a stainless steel
sample which has been electrocharged with
deuterium. The corresponding concentration
profile is shown above the data. The right
edge of the reaction spectrum corresponds to
the surface of the specimen, i.e., depth
= 0 in the upper figure.

Selected Applications

Since NMA is most attrac-
tive for the study of light
impurities in host materials,
it is particularly well
suited for examination of the
role of gases in evolution of
radiation damage microstructure. Transmission electron
microscopy studies have shown
that hydrogen, oxygen, nitrogen, and the inert gases,
especially helium, are important in the nucleation and
stability of point defect
clusters (5). We are, therefore, using NMA to investigate the interaction of some
of these gases with point
defects.

Deuterium Migration and Trapping

As an isotope of hydro-
gen, deuterium (D) is especially suitable for NMA, both
as an impurity and as a probing beam. Its use as an impurity atom allows a clear separation from interference by atmo-
spheric hydrogen, and the nuclear
reactions D(d,p) or D(^{3}He,α)* are
specific and well characterized
at low bombarding energies.

----
*Typical nuclear reaction
notation A(a,b)B refers to
target nucleus, A, bombarded
by ion, a, from which ion, b,
is released, and B (if given)
is the residual nucleus.

Deuterium can be introduced into targets from a gas phase, by ion implantation or by cathodic electrolysis. Introduction from the gas phase avoids co-generation of point defects, but the gas levels are low and more difficult to probe. Ion implantation injects suitable hydrogen levels but atomic displacements and trapping occur concurrently and it is not easy to saturate the traps to measure bulk diffusion. Cathodic electrolysis is not defect free; it introduces damage in the form of slip dislocations. But the concentrations of deuterium can be high enough to saturate the dislocations and permit bulk diffusion. Therefore, a comparison of migration of deuterium introduced by ion implantation and with that from cathodic electrolysis permits a relative evaluation of interaction of deuterium with radiation damage defects and with dislocations. If the radiation damage is done at low temperature where a large fraction of point defects are retained, the comparison is one of trapping by point and line defects.

Experiments were done on disks of annealed commercial type 310 and 316 austenitic stainless steels (SS) and pure nickel measuring 5 mm dia and 0.3 mm thick. These were prepared by mechanical polishing through 0.1 μm diamond abrasive to give an optically flat surface. The type 316 SS and nickel specimens were also electropolished. The average grain diameter was about 35 μm for the SS and 50 μm for the nickel specimens. Chemical compositions are given in Table I.

Table I. Chemical Compositions of Targets (Weight Percent)

Material	Fe	Ni	Cr	Mo	Mn	Si	P	C	O	N
316 SS	Bal.	12	17	2.4	1.6	0.7	0.035	0.06	--	0.05
310 SS	Bal.	21.1	23.4	0.2	2.1	1.0	0.018	0.06	0.021	0.05
Nickel	0.0009	Bal.	--	--	--	--	--	0.0002	0.0003	--

For ion implantation of deuterium the incident beam energies were 85 and 140 keV, and the probing beam was 450 keV $_2$D, all at 45° to the target. Typical beam currents were 0.1 μA over a 3 mm^2 spot. The atomic displacements associated with these beams in nickel were calculated using the projected range distribution values from ref. 7 and the EDEP code (8), and are shown as functions of depth in Figure 5. Beam heating was found to be about 4°C during implantation and about 10°C during probing. In some cases, specimens were predamaged with a 2 MeV Ni^{++} bombardment in an ultrahigh vacuum (~2 × 10^{-6} Pa) chamber with the beam normal to the target; beam levels were either 300 or 1500 μC/cm^2, corresponding to peak damage levels of 1 and 5 dpa at a depth of about 0.4 μm, as shown in Figure 5.

Electrocharged Deuterium

For electrocharging experiments, each sample was mounted in an electrolytic cell with a 1N D$_2$SO$_4$ solution poisoned with typically 2.5 mg/ℓ of arsenic. Platinum foil was used as the anode from which the electric field was directed normal to the specimens surface. Charging took place at a cathodic current density of 0.1 A/cm^2 for various times up to 100 min. After charging, the specimen was quickly transferred to a scattering chamber and the chamber was immediately evacuated. The transfer and evacuation time was typically 11 min.

During cathodic electrolysis of aqueous solutions the hydrogen (deuterium) fugacity at the cathode can be extremely high, resulting in high concentrations of the gas in the cathode. It is expected that the immediate subsurface layers will reach the equilibrium concentration almost instantly,

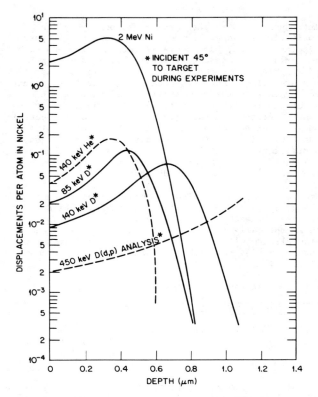

DEPTH (μm)

Figure 5. Calculated damage-depth profiles
for the various ion beams used in these studies
during typical experiments.

and this concentration will advance into the cathode as a "front" with increasing charging time. The role of poisons in the electrolyte is not known explicitly, but is suspected to involve depression of formation of molecular gas, thus allowing greater concentrations of the atomic species in the cathode.

As soon as the cathodic current is cut off, deuterium begins to escape through the specimen surfaces. By the time the specimen is ready for NMA, usually 11 min after charging, considerable loss of deuterium has occurred from the surface layers. This can be seen readily in Figure 4. The data in this example are best fitted by an error function solution of Fick's diffusion laws, from which we can extract a deuterium diffusion coefficient, in this particular case $D = 1.42 \times 10^{-12}$ cm^2/s at room temperature (9). This coefficient is in good agreement with extrapolation of higher temperature data measured in various austenitic steels by more conventional techniques. This indicates that migration of the escaping deuterium is unhindered by the dislocations formed during electrocharging, and the diffusion coefficient represents free bulk diffusion behavior. Further experiments now being made at longer degassing periods are showing that as the deuterium concentration in the surface layers, damaged during electrocharging, is reduced its mobility is retarded. The implication is that when there is insufficient deuterium to saturate the dislocations, which temporarily trap the deuterium, migration is reduced and determined by the rate of release from the dislocations.

Trapping by point defects is illustrated in Figure 6. The specimen was 316 SS which had been previously bombarded at room temperature with a 2 MeV nickel beam to a fluence of 300 μC/cm^2. After electrocharging for 10 min the specimen was allowed to partially degas for 1.5 h at room temperature and its deuterium profile was measured and compared with that for a specimen that did not receive ion damage. Deuterium retention near the surface is enhanced at least tenfold by the nickel ion bombardment. The curve showing the difference between the two conditions corresponds in depth with that for

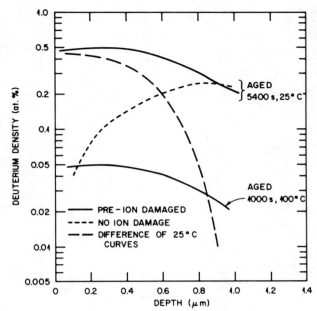

Figure 6. Density (or concentration) versus depth profile for deuterium-charged 316 SS samples, one of which was irradiated (predamaged) by 2 MeV ions.

nickel ion damage in Figure 5. So, despite the presence of an intense dislocation structure from the electrocharge, ion bombardment causes stronger trapping. Whether this is due to a higher density of traps or to a stronger trapping energy for the ion damage is not yet clear, but experiments and calculations are in hand to resolve this question. Aging at 100°C results in release of most of the trapped deuterium, indicating that trapping by ion damage is strongly dependent on temperature.

Ion-Implanted Deuterium

The trapping of D by the ion damage in the 316 steel in the electrocharging and release experiment above suggests that ion-implanted deuterium may become trapped in defects created by the ion implantation itself. In order to examine this question further, samples of nickel and 316 SS were each implanted with 85 or 140 keV D^+ ions to a fluence of 6×10^{-3} C/cm^2. When the implantation was complete, the accelerator energy was raised to 450 keV for analysis. In this way a profile could be made with only about 500 s delay after implantation. These implantations and measurements were carried out at ambient temperature, ~25°C, in a scattering chamber pumped by an oil diffusion pump with a liquid nitrogen trap near the target surface. The typical pressure was 10^{-4} Pa. This was not sufficient to prevent contamination of the target surface by carbon impurities, which were monitored by observation of counts in the ^{12}C(d,p) reaction. The profiles from these experiments are presented in Figures 7 and 8, together with data from specimens that were predamaged with nickel ions. Ignoring the latter for the moment, several features of the supposedly undamaged specimens, marked "no predamage," are noteworthy.

First, the deuterium does not disperse easily. Despite the lengthy aging periods and the high bulk diffusion rates for deuterium, the implanted deuterium remains centered around the expected ion damage profile for the implantation (cf., Figure 5). Diffusion would be expected to broaden the profile according to $\sigma^2 = \sigma_0^2 + 2Dt$, where σ is the standard deviation of the implanted profile, D is the diffusion coefficient, and t is time. Taking a room temperature D of 5×10^{-10} cm^2/s for deuterium in nickel (10), the expected diffusion length $\sqrt{2Dt}$ in 500 s would be about 7 µm. It is clear from Figure 7 that the widths (standard deviation) of the profiles for nickel are only ~0.3 µm. Even for the 316 steel, with a lower bulk diffusion coefficient of 1.4×10^{-12} cm^2/s as measured for our 310 SS, the diffusion length would be 0.38 µm for the 500 s age and 1.7 µm at 10,000 s. Obviously, the migration of the residual D is restricted by the implantation damage.

495

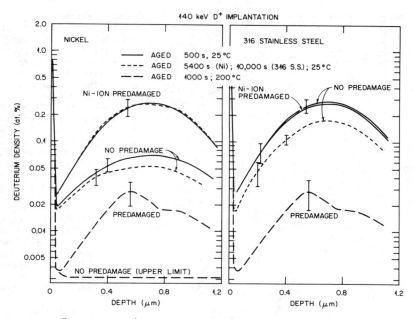

Figure 7. Density (or concentration) versus depth profiles for deuterium-implanted 316 SS and nickel samples with and without predamage.

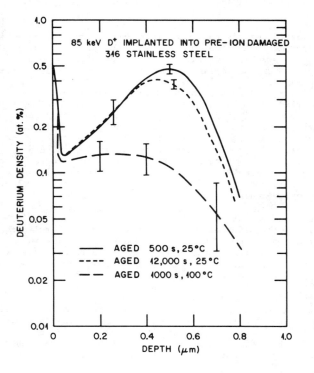

Second, the nickel and the steel, although receiving essentially similar deuterium profiles, do not show equal retention of deuterium. Whereas only about 30% of the integrated implanted D remained in the undamaged nickel specimen at the time of the first measurement, almost 90% was retained in the steel. Based upon theoretical calculations, this difference is believed to be attributable to the different bulk deuterium diffusion coefficients.

Third, there is marked accumulation of

Figure 8. Density (or concentration) versus depth profiles for deuterium-implanted and aged 316 SS samples predamaged.

deuterium just below the incident surface of the implanted targets. The present nuclear technique was able to determine only that this region was located within the first 0.05 μm of the surface. However, one of the samples was probed in a secondary mass spectrometry experiment (11) and the region was found to be ~0.01 μm wide. Based upon this information and the present D(d,p) scattering data, the magnitude of the surface peak was determined as shown in Figures 7 and 8. We observed no change in the surface peak even after a 200°C anneal, implying that binding of D in the surface layer is much greater than the trapping farther beneath the surface. This would be consistent with chemical bonding, as in the formation of a deuteride compound, possibly by reaction with the contamination film from the vacuum. Such chemical bonding might very well have been assisted by the high degree of ionization damage produced in the near-surface depths during deuterium implantation. The formation of this surface concentration of deuterium raises the question of whether it affects the shapes of the measured profiles deeper in the specimen. It is certainly not responsible for the profiles. A surface barrier to deuterium egress would not prevent decay of the profiles by inward migration. That such dispersion does not occur despite favorable bulk diffusion conditions is strong testimony for considerable trapping by the implantation damage. The earlier experiments showing trapping in the ion-damaged electrocharged steel, where there was no detectable surface contamination or deuterium buildup, support this contention.

This conclusion is further strengthened by examination of the profiles in Figure 7 for similar specimens that were predamaged with 1.5×10^{-3} C/cm^2 2 MeV Ni^{2+} ions. These curves indicate a higher degree of internal trapping. Moreover, the predamaged nickel and steel specimens trap equally almost all of the implanted deuterium, and the degree of release in both materials is significantly reduced. The trapping is now so strong as to overcome the advantage of the higher bulk diffusion coefficient in the nickel. Only when the targets are heated well above room temperature do the release rates greatly increase.

Despite the presence of the surface layer of deuterium and the possibility that it may retard deuterium escape through the surface, it does not seriously interfere with the recognition and extraction of meaningful internal trapping data. This is shown more clearly by another experiment in which 85 keV D ions were implanted into nickel ion-damaged 316 SS and allowed to age at room temperature and 100°C (Figure 8). Due to better statistical accuracy, we were able in this case to observe a distinct asymmetry in the annealed profiles compared to the first room-temperature profile. In particular, the mean depth of the residual gas profile appeared to move from about 0.5 μm to about 0.3 μm after the 100°C anneal while much of the total D escaped the probed region (0-1 μm). The defect density from predamage and D implantation combined is not expected to be greater in the 0.3 μm region compared to the 0.5 μm region. Although possible surface layer retardation may also contribute to the shift, theoretical calculations were able to qualitatively reproduce this shift by a release-followed-by-retrapping mechanism.

Ion-Implanted Helium

Preliminary experiments have been made on nickel to examine helium profiles under conditions of:
(a) implantation at room temperature,
(b) implantation at an elevated temperature near that at which heavy-ion damage is usually performed, and
(c) simultaneous helium implantation and heavy-ion displacement damage at the temperature in (b).

For these experiments the isotope ^{3}He is useful since it can be probed with the ^{3}He(d,α) reaction.

In order to determine the approximate helium density distribution expected from the implantation of monoenergetic ions, we utilized the code, EDEP (8), and the ion implantation tables of Brice (12). The calculations for range X_0 and standard deviation are given in Table II. Implantation was done at 300 keV to an approximate peak helium concentration of 0.5 at. %. The elevated-temperature bombardments were carried out at 700°C for a period of 1 h. There was no subsequent aging at 700°C (0.56 T_m) since the implantation period itself was considered an adequate diffusion period for our purposes. The profiles were examined with the ^{3}He(d,α) reaction at E = 400 keV and φ = 155 (lab). The results are shown in Figure 9.

Table II. Comparison of Range Parameters from Theory and Measurement[*]

		EDEP	BRICE[†]	Measured	
200 keV		X_0 = 0.48, σ = 0.081	(0.48, 0.104)		
300 keV	^{3}He in Ni	0.65, 0.087	(0.65, 0.114)	0.65	0.10
400 keV		0.79, 0.090	(0.79, 0.120)	±0.03	±0.02
200 keV		X_0 = 0.54, = 0.095	X_0 = 0.53, = 0.122		
300 keV	^{4}He in Ni	= 0.72, = 0.102	= 0.72, = 0.134		
400 keV		= 0.88, = 0.106	= 0.88, = 0.141		

[*]Units are in microns.
[†]Values in parentheses extrapolated from ^{4}He using ^{3}He/^{4}He ratios in the EDEP column.

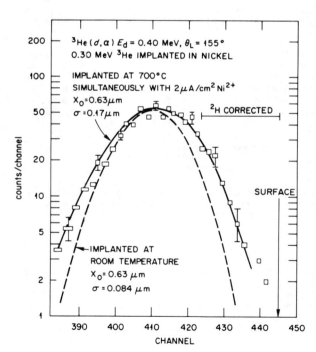

The broken line in Figure 9 represents condition (a), implantation at room temperature. The data points and the continuous line through them are for condition (c), simultaneous helium implantation and heavy-ion bombardment (4 MeV Ni^{+2} to a peak damage level of 20 dpa) at 700°C. The curve for condition (b), helium implantation only at 700°C, fell between the other two curves and is omitted for clarity. In curve fitting, the shapes of the profiles were assumed to be Gaussian distributions centered at the depths, X_0,

Figure 9. Distribution of ^{3}He in nickel after implantation alone at room temperature or simultaneously with 4 MeV Ni ion damage at 700°C.

498

with standard deviations, σ, as shown. In the calculations, allowance was made for the fact that the helium was implanted 15° off the target normal. Our final estimate and uncertainty for the ^{3}He implant at 300 keV is shown in Table II. It is obvious from Table II that the measured profile for the room-temperature implantation is in very good agreement with the expected value. Implantation at elevated temperature, with or without simultaneous displacement damage, gave only slightly broader profiles and little loss (~15%) of integrated helium. The conclusion is that there is very little transport at 0.56 T_m, no more than 80 nm, even in the presence of considerable heavy-ion displacement damage. This conclusion agrees with reports of restricted helium migration in nickel by Phillips et al. (13), Whitmell and Nelson (14) at elevated temperatures, and with the limited helium release measured by Thomas et al. (15) at lower temperatures at which interstitial diffusion was assumed. The implication is that the helium is strongly bound, probably in clusters with vacancies, which is expected to be the lowest energy configuration for helium in metals (16).

Extraction of Impurity-Defect Binding Energies

In order to quantitatively analyze the transport of the D and ^{3}He with time and temperature, we have applied the basic approach of diffusion with trapping suggested by McNabb and Foster (17). Their differential equation involves a diffusive and a sink (trap) term:

$$\frac{\partial c}{\partial t} = D\nabla^2 C - N \frac{\partial n}{\partial t} + G , \qquad (2)$$

where C is the concentration of dissolved atoms, D is the coefficient for uninhibited diffusion, N is the density of traps, n is the fractional occupation of the traps, G is a generator (ion beam flux), and t is time. The flow of atoms into and out of the traps is then

$$\frac{\partial n}{\partial t} = kC (1 - n) - pn , \qquad (3)$$

where k is a model parameter and p is the release rate of atoms from the traps. We have applied this approach in a manner similar to that of Myers and Picraux (18) by numerically solving these coupled differential equations in which the latter can be parameterized as

$$\frac{\partial n}{\partial t} = 4\pi RD [C(N-n) - nz \exp (-Q_T/kT)], \qquad (4)$$

where R is the trapping radius and Q_T is the trapping energy or energy below that of normal equilibrium sites of density z.

If we write the trap density $N(x) = N_0 f(x)$ such that $f(x)dx = 1$, N_0 becomes the total number of traps. If F is the predamage ion fluence, then $g = N_0/F$ is the number of traps generated per incident ion and is a dimensionless coefficient. This number can be determined only in the electro-charging experiments because only in this case are the traps saturated without additional displacement damage. We assume that the functional form of $f(x)$ is given by the displacement energy deposition code EDEP (8).

The binding energy Q_T is to be determined by comparing the solution of Eqs. (2) and (4) to the measured profiles of trapped D and ^{3}He. All solutions to the differential equations were characterized by rapid ($\tau < 1$ s) migration of impurity (D or ^{3}He) from solution (assumed at t = 0) to traps. This was followed by slow ($\tau \gg 1$ s) release from and frequent recapture into the traps. The solutions were not very responsive to the value of R, but were sensitive to the value of the trapped impurity binding energy Q_T.

The example solution in Figure 10 corresponds to data like that in Figure 8 where the initial, implanted deuterium, C, is allowed to migrate into a predamaged region of defect density, N. After 1000 s at 100°C the solution or profile of dissolved (C) plus trapped (Nn) gas is to be compared to the experimentally measured deuterium profile. Notice that the asymmetry and shift of the mean depth of the originally implanted profile qualitatively agree with the data in Figure 8. While the profile in the 0–0.2 μm region is sensitive to the boundary condition imposed at the surface (i.e., the degree of surface blocking, $\partial C/\partial x = 0$), the region >0.4 μm is not. For this reason we believe that a deuterium-defect binding energy can be deduced independent of surface contamination.

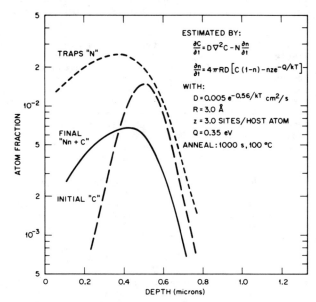

In support of this we find close agreement between the value of Q_T deduced from the electrocharge data (0.36 eV) and the ion implantation data (0.32 eV). These analyses yield a trap generation coefficient for 2 MeV Ni ions in 316 SS of ~15/ion.

Figure 10. Estimated effects of profiles of traps, N, on redistribution (final N_n + C) at 100°C of initial concentration profiles, C, of deuterium in irradiated stainless steel.

In applying this analyses to our helium data a primary uncertainty lies in the fact that a helium diffusion constant is not available for 700°C. We have assumed a value extrapolated from the 500°C interstitial diffusion coefficient in ref. 15. Our preliminary estimates are summarized in Table III. Details of these analyses will be given in subsequent publications.

Table III. Information Deduced from Analyses of Deuterium and Helium Concentration Versus Depth Profiles

Impurity	Target	Deduced Quantity
D	310 SS	$D_D(300\ K) = (1.4 \pm 0.2) \times 10^{-12}\ cm^2/s$
D	316, 310 SS	$Q_T = 0.33 \pm 0.03$ eV
D	316 SS	g = 15 $\pm$ 5 traps/ion
^{3}He	Nickel	Q_T = 2.1 eV[*]

D_D = Diffusion coefficient for deuterium. Q_T = impurity-defect binding energy.   g = the trap generation coefficient of nickel ion damage for deuterium, stated as the number of deuterium traps per incident 2 MeV Ni ion in 316 SS at room temperature.
[*]Based on D_{He} extrapolated from ref. 15.

The values for Q_T in Table III were found to be in good agreement with those determined by gas release measurements for deuterium in 316 SS (19) and helium in nickel (15,20).

Conclusions

It is shown that nuclear microanalysis is a particularly favorable method of measuring impurity migration and impurity-defect interactions in irradiated solids. Examples involving the impurities deuterium and helium briefly indicate how one can measure: (1) diffusion coefficients, (2) impurity-defect binding energies, and (3) trap generation coefficients of heavy ions used to create displacement damage.

The methods and procedures outlined here are general so that they can be applied to other light impurities such as carbon, nitrogen, or oxygen and to isotopes of such elements. Preliminary values are given for the bulk diffusion coefficient of deuterium in austenite at room temperature (1.4×10^{-12} cm^2/s), for the impurity-defect binding energy of deuterium in austenite (0.33 eV) and ^3He in nickel (2.1 eV), and the generation coefficient of nickel ion damage for deuterium traps at room temperature (~15 traps/incident nickel ion).

References

1. W. K. Chu, J. W. Mayer and M. A. Nicolet, Backscattering Spectroscopy, Academic Press, New York, 1978.
2. J. P. Thomas and A. Cachard, eds., Material Characterization Using Ion Beams, Plenum Press, New York, 1978.
3. J. F. Ziegler, ed., New Uses of Ion Accelerators, Plenum Press, New York, 1975.
4. P. D. Townsend, J. C. Kelly, and N.E.W. Hartley, Ion Implantation, Sputtering, and Their Applications, Academic Press, New York, 1976.
5. K. Farrell, "Experimental Effects of Helium on Cavity Formation During Irradiation — A Review," Rad. Effects, 53 (1980) 175.
6. M. B. Lewis, Analysis of Nuclear Backscattering and Reaction Data by the Method of Convolution Integrals, ORNL/TM-6697, Oak Ridge National Laboratory (1979); "A Deconvolution Technique for Depth Profiling with Nuclear Microanalysis," Nucl. Inst. Meth. (in press).
7. H. H. Anderson and J. F. Ziegler, Hydrogen Stopping Powers and Ranges, Pergamon Press, New York, 1977.
8. I. Manning and G. P. Meuller, Comp. Phys. Comm. 7 (1974) 85.
9. M. B. Lewis and K. Farrell, "An Ion Beam Technique for Measurement of Deuterium Diffusion Coefficients," Appl. Phys. Letters 36 (1980) 819.
10. J. Volkl and G. Alefeld, in Diffusion in Solids: Recent Developments, ed. by A. S. Nowick and J. J. Burton, Academic Press, New York, 1975, p. 232.
11. Private communication, S. S. Cristy, 1979.
12. D. K. Brice, Ion Implantation Range and Energy Deposition, Vol. I, IFI, Plenum Data Company, New York, 1975.
13. V. Philipps, K. Sonnenberg, and J. M. Williams, "Diffusion of Helium in Nickel at High Temperatures," pp. 173—186 in Proceedings of Consultant Symposium on Rare Gases in Metals and Ionic Solids, Harwell, AERE-R-9733, March 1980.
14. D. S. Whitmell and R. S. Nelson, Rad. Effects 14 (1972) 249.
15. G. J. Thomas, A. W. Swansiger, and M. I. Baskes, J. Appl. Phys. 50 (1979) 6942.
16. J. E. Inglesfield and J. B. Pendry, Phil. Mag. 8(12) (1965) 997.
17. A. McNabb and P. K. Foster, Trans. AIME 227 (1963) 618.
18. S. M. Myers and S. T. Picraux, J. Appl. Phys. 50 (1979) 5710.
19. K. L. Wilson and M. I. Baskes, J. Nucl. Mater. 76&77 (1978) 291.
20. D. Edwards, Jr., and E. V. Kornelsen, Surface Sci. 44 (1974) 1.

Subject Index

Author Index